T0202494

# Undergraduate Lecture Notes in Physics

Undergraduate Lecture Notes in Physics (ULNP) publishes authoritative texts covering topics throughout pure and applied physics. Each title in the series is suitable as a basis for undergraduate instruction, typically containing practice problems, worked examples, chapter summaries, and suggestions for further reading.

ULNP titles must provide at least one of the following:

- An exceptionally clear and concise treatment of a standard undergraduate subject.
- A solid undergraduate-level introduction to a graduate, advanced, or non-standard subject.
- A novel perspective or an unusual approach to teaching a subject.

ULNP especially encourages new, original, and idiosyncratic approaches to physics teaching at the undergraduate level.

The purpose of ULNP is to provide intriguing, absorbing books that will continue to be the reader's preferred reference throughout their academic career.



More information about this series at http://www.springer.com/series/8917

Matthew J. Benacquista · Joseph D. Romano

# Classical Mechanics

Matthew J. Benacquista
Department of Physics and Astronomy
University of Texas Rio Grande Valley
Brownsville, TX
USA

Joseph D. Romano
Department of Physics and Astronomy
University of Texas Rio Grande Valley
Brownsville, TX
USA

ISSN 2192-4791 ISSN 2192-4805 (electronic)
Undergraduate Lecture Notes in Physics
ISBN 978-3-319-68779-7 ISBN 978-3-319-68780-3 (eBook)
https://doi.org/10.1007/978-3-319-68780-3

Library of Congress Control Number: 2017955247



Printed on acid-free paper

This Springer imprint is published by Springer Nature
The registered company is Springer International Publishing AG
The registered company address is: Gewerbestrasse 11, 6330 Cham, Switzerland

# Preface

This book represents our attempt to provide an introduction to the subject of classical mechanics at a level *intermediate* between that presented in standard undergraduate-level textbooks (e.g., *Classical Dynamics of Particles and Systems* by Marion and Thorton) and advanced graduate-level textbooks (e.g., *Classical Mechanics* by Goldstein, Safko, and Poole or *Mechanics* by Landau and Lifshitz). We develop the background and tools of classical mechanics for use in fields of modern physics, such as quantum mechanics, astrophysics, particle physics, and relativity. Students who have had basic undergraduate classical mechanics or who have a good understanding of the mathematical methods of physics should benefit from this book.

We envision the target audience to be advanced undergraduates and first-year graduate students. As such, we anticipate that the reader will have had a mathematical methods course at the level of, e.g., *Mathematical Methods in the Physical Sciences* by Boas, but we provide a thorough refresher of relevant material (e.g., vector calculus, differential forms, calculus of variations, linear algebra, and special functions) in the appendices. The inclusion of these appendices should allow instructors to tailor their course to the specific mathematical preparation of their students, especially for advanced undergraduate students.

We have found that the major challenge of teaching classical mechanics is in introducing the power of the mathematical tools without getting lost in the details of the mathematical formalism. We interleave physical applications with the introduction of mathematical principles so that the students develop a strong physical intuition about the use of these powerful tools. Lagrangian and Hamiltonian methods are introduced early on, so they can be used to solve problems related to central force motion, rigid body motion, small oscillations, etc. We have also included optional chapters on continuous systems and special relativity, extending the standard formalism to classical fields and relativistic systems.

Exercises are given throughout each chapter to reinforce material as it is presented in the text. Additional (somewhat longer) problems are provided at the end of each chapter, which bring together multiple concepts introduced in the chapter. The exercises will let students assess their own understanding of individual concepts introduced in the chapter. The longer problems will assess the students' ability to synthesize their skills. Although we do not provide worked solutions to the exercises and problems, we often give hints to guide students toward a solution if they get "lost" along the way. We do not view these hints as *crutches* for solving a problem, but rather as *suggestions* for attacking a problem in a relatively efficient manner. (We realize, of course, that there are usually many different ways of solving a problem.) Problem-solving is a skill that one develops over time with plenty of practice, and we recommend that the student works through as many exercises and problems as possible while reading the book.

The amount of material included in this text is appropriate for a one-semester course, allowing some freedom in the choice of topics that are covered. Although the material is developed more-or-less linearly, with some of the later chapters depending on previous ones, the order in which the chapters are covered need not follow the order that we have chosen. The first three chapters form the basis of the book and should be covered first. But Chaps. 4 and 5 (on central forces and scattering), Chaps. 6 and 7 (on rigid body motion), and Chaps. 8 and 9 (on small oscillations and waves) are three separate applications of Chaps. 1–3 and can be covered in *any* order, e.g., Chaps. 6 and 7 before Chaps. 4 and 5, etc. In addition, Chap. 10 (on Lagrangian and Hamiltonian formulations of continuous systems and fields) and Chap. 11 (on special relativity) are both optional chapters in the sense that no other chapter depends on the material discussed in those chapters. We think that Chap. 10 is best taught after Chaps. 8 and 9, which transition from discrete to continuous systems, while Chap. 11 can actually be taught at *any* time during the semester, after Chaps. 1–3.

A typical sequence of chapters for a one-semester course for advanced undergraduates, which includes an in-depth review of the relevant mathematical methods presented in the appendices, is: Appendix A, Chap. 1, Appendix B, Appendix C, Chap. 2, Chap. 3, Chap. 4, Chap. 5, Appendix D, Chap. 6, Chap. 7, and Chap. 8, with Appendix E referred to as needed. A one-semester course for beginning graduate students is Chaps. 1–10, with Chap. 11 optional, and with the appendices referred to only as needed by individual students. We believe that this introduction to classical mechanics will benefit students in whatever branch of physics they decide to pursue.

Red Lodge, MT, USA — Matthew J. Benacquista
Brownsville, TX, USA — Joseph D. Romano
August 2017

# Acknowledgements

First and foremost, we acknowledge the influence of the many *excellent* textbooks that we have used over the years, both when learning and then teaching classical mechanics: Goldstein et al. (2002), Fetter-Walecka (1980), Lanczos (1949), Landau-Lifshitz (1976), and Marion-Thornton (1995) for the classical mechanics material, and Boas (2006), Mathews-Walker (1970), Schey (1996), and appendices from Griffiths (1999), Griffiths (2005) for the associated mathematical methods. These texts have definitely shaped the presentation in our book. In short, we have taken what we found to be best of all these texts and packaged it together in a way that will hopefully be useful both to instructors and students who use our book. We do not aspire to *replace* any of these classic texts, but rather to *add* to the existing literature in a way that may resonate with some of our readers.

Secondly, we acknowledge our former teachers and mentors from whom we first learned classical mechanics: N. David Mermin, A.P. Balachandran, and Karel Kuchăr (for J.D.R.); David Griffiths, Nicholas Wheeler, and Richard Robiscoe (for M.J.B.). Their enthusiasm and passion for teaching is something we try to imitate when we are in the classroom.

Last, but certainly not least, we acknowledge all of our former students and colleagues who were kind enough to read through early drafts of the book: Andres Cuellar, Mike Disney, Sam Finn, Jeff Hazboun, Richard Price, Joel Solis, and Charles Torre. Special thanks go to students in PHYS5421 and PHY5310 (Graduate Classical Mechanics) at the University of Texas at Brownsville, and the University of Texas Rio Grande Valley, and to Karel Kuchăr, who shared his unpublished lecture notes and problem book for the graduate classical mechanics classes he taught at the University of Utah. And we cannot thank enough Jolien Creighton, who *field tested* a draft of this book in his graduate classical mechanics class in Spring 2017. He found numerous errors and inconsistencies, and made many suggestions, which have led to the addition of new material and (hopefully!) improvements in the overall presentation of book. Of course, we take full responsibility for all other errors that remain. Finally, J.D.R. thanks the Artemis Group at the Observatoire de la Côte d'Azur in Nice, France, and the Albert Einstein Institute in Hannover, Germany, for their hospitality during the final months of editing the book.

## Contents

# Math Conventions Used

| | |
|---|---|
| $\equiv$ | Definition |
| $\mathbf{F}_{IJ}$ | Force that particle $I$ exerts on particle $J$ |
| $\mathbf{r}_{IJ}$ | Position vector of particle $I$ relative to particle $J$, so $\mathbf{r}_{IJ} \equiv \mathbf{r}_I - \mathbf{r}_J$ |
| $\nabla_I$ | Gradient with respect to the coordinates of position vector $\mathbf{r}_I$ |
| $\nabla_{IJ}$ | Gradient with respect to the coordinates of $\mathbf{r}_{IJ}$ |
| $\sum_{I,J}$ | Summation over both $I$ and $J$, so $\sum_{I,J} \equiv \sum_I \sum_J$ |
| $\mathbb{R}$ | Set of real numbers |
| $\mathbb{C}$ | Set of complex numbers |
| $\mathrm{Re}\,(z)$ | Real part of a complex number $z$ |
| $\mathrm{Im}\,(z)$ | Imaginary part of a complex number $z$ |
| $z^*$ | Complex conjugate of a complex number $z$ |
| $\hat{\mathbf{u}}$ | Unit vectors are denoted with a hat ˆ |
| $f(x)$ | Ordinary functions are denoted with round brackets ( ) |
| $\mathrm{d}f/\mathrm{d}x$ | Ordinary derivative of $f(x)$ with respect to $x$ |
| $I[y]$ | Functionals are denoted with square brackets [ ] |
| $\delta I/\delta y(x)$ | Functional derivative of $I[y]$ with respect to the function $y(x)$ |
| $\mathrm{d}^3x$ | Coordinate volume element, e.g., $\mathrm{d}x\mathrm{d}y\mathrm{d}z$ or $\mathrm{d}r\mathrm{d}\theta\mathrm{d}\phi$, etc. |
| $\mathrm{d}V$ | Invariant volume element, e.g., $\mathrm{d}x\mathrm{d}y\mathrm{d}z$ or $r^2 \sin\theta\,\mathrm{d}r\mathrm{d}\theta\mathrm{d}\phi$, etc. |
| $I, J, \ldots$ | Indices labeling a system of particles, $I = 1, 2, \ldots, N$ |
| $a, b, \ldots$ | Indices labeling generalized coordinates $q^a$, where $a = 1, 2, \ldots, n$ |
| $i, j, \ldots$ | Spatial indices or indices for abstract $n$-dimensional vectors |
| $\alpha, \beta, \ldots$ | Indices labeling normal mode frequencies and eigenvectors in Chap. 8, and the components of spacetime vectors in Chap. 11 |
| $\mathscr{C}$ | Configuration space for a system of particles |
| $Q$ | Subspace of the configuration space $\mathscr{C}$ spanned by generalized coordinates $q^a$, where $a = 1, 2, \ldots, n$ |
| $\Gamma$ | Phase space for a system of particles |

# Chapter 1
# Elementary Newtonian Mechanics

Much of classical mechanics was developed to provide powerful mathematical tools for obtaining the equations of motion for systems of objects subject to external and internal forces. These include *Newton's laws*, the *principle of virtual work*, and *Hamilton's principle*, which we shall discuss, in turn, in the first three chapters. These tools let us choose coordinates that are most suitable for the solution of a given problem; they also allow us to describe motion when observed from non-inertial reference frames, such as the rotating surface of the Earth. A deeper study of these mathematical tools and how they respond to different transformations of the system (e.g., translations or rotations of the coordinates) leads to a better understanding of the nature of Newtonian mechanics, and points the way to the modern physics of quantum mechanics and special relativity.

For the greater part of this book, we will concentrate on Newton's formulation of mechanics, in which the universe exists in a flat, three-dimensional space described by Euclidean geometry. Changes in this Newtonian universe are measured using a standard clock that ticks at a uniform rate over all space. Adapting Newtonian mechanics to the non-Euclidean geometry of special relativity will be discussed at the end of the text in Chap. 11.

In this chapter, we review some of the basic methods familiar from introductory physics for obtaining and solving the equations of motion for *single particles* and then *systems of particles*, with and without constraints on their motion.



M.J. Benacquista and J.D. Romano, *Classical Mechanics*, Undergraduate Lecture Notes in Physics, https://doi.org/10.1007/978-3-319-68780-3_1

## 1.1 Newton's Laws of Motion

From introductory physics, we are familiar with Newton's laws of motion. The first law describes the motion of an object with respect to an **inertial reference frame**:

> **Newton's 1st law:** Unless acted on by an outside force the natural motion of an object is constant velocity.

One way of thinking about this law is that it provides us with a procedure for determining if we are using an inertial reference frame. That is, if we can find a way to turn off (or shield) all external and internal forces from a system, and we find that all particles in the system are moving with constant velocities, then we will know that we are describing the system in an inertial frame of reference.

Once we have determined that we are in an inertial reference frame, Newton's 2nd law tells us how an applied force will alter this natural motion:

> **Newton's 2nd law:** The effect of an applied force $\mathbf{F}$ upon an object of mass $m$ is to induce an acceleration $\mathbf{a}$ such that
>
> $$\mathbf{F} = m\mathbf{a}\,. \tag{1.1}$$

This simple form of Newton's 2nd law assumes that the mass is constant, but we can include the effect of a varying mass by writing Newton's 2nd law in terms of **momentum** $\mathbf{p} \equiv m\mathbf{v}$, so that

$$\mathbf{F} = \dot{\mathbf{p}} \equiv \frac{\mathrm{d}\mathbf{p}}{\mathrm{d}t}\,. \tag{1.2}$$

Note that unless specifically stated otherwise, we will assume throughout this text that the mass of an object is constant, for which $\mathbf{F} = m\mathbf{a}$ and $\mathbf{F} = \dot{\mathbf{p}}$ are equivalent statements of Newton's 2nd law.

When there are multiple objects exchanging forces between themselves within a system, Newton's 3rd law describes how the forces of interaction behave:

> **Newton's 3rd law:** If an object applies a force $\mathbf{F}$ on a second object, then the second object applies an equal and opposite force $-\mathbf{F}$ on the first object.

In its simplest form, the 3rd law insures that the internal forces between particles in a system do not provide an unbalanced force on the system as a whole, which would allow the system to spontaneously accelerate away in the absence of external forces. Note that not all forces obey Newton's 3rd law, but these involve a field which can carry away momentum.[1] There is also a **strong form** of Newton's 3rd

[1] A simple example of such a force is the *electromagnetic* force between two moving point charges; see, e.g., Sect. 8.2.1 of Griffiths (1999).

law, which requires that $\mathbf{F}_{IJ}$, the interparticle forces between particles $I$ and $J$, not only satisfy $\mathbf{F}_{JI} = -\mathbf{F}_{IJ}$, but also point in the direction of the lines connecting pairs of particles—i.e., $\mathbf{F}_{JI} \propto \mathbf{r}_{IJ}$, where $\mathbf{r}_{IJ} \equiv \mathbf{r}_I - \mathbf{r}_J$ is the displacement vector joining particles $I$ and $J$. Such forces are called **central** (or **radial**) forces. The strong form of Newton's 3rd law is needed for conservation of angular momentum, as we will explore in more detail in Sect. 1.4.

***Example 1.1*** Consider a rocket moving in interstellar space, free of all external forces, as shown in Fig. 1.1. We want to determine the velocity $v$ of the rocket as a function of time, assuming that its mass decreases at a constant rate, $\mathrm{d}m/\mathrm{d}t \equiv -\alpha$ (where $\alpha$ is positive so $\mathrm{d}m/\mathrm{d}t$ is explicitly negative), as it expels exhaust gases through the nozzle of the rocket engine.

To do this calculation, we need to use Newton's 2nd law in the form $F = \mathrm{d}p/\mathrm{d}t$, since the mass of the rocket is not constant. (We have dropped the vector symbols in this equation since this is a 1-dimensional problem.) Let's assume that at time $t$ the rocket has mass $m$, and that it is moving vertically upward with velocity $v$. At time $t + \mathrm{d}t$, the rocket will have lost mass $\mathrm{d}m' \equiv -\mathrm{d}m > 0$ (the exhaust gases), and will have changed its velocity to $v + \mathrm{d}v$. We will assume that the exhaust gases $\mathrm{d}m'$ exit the rocket with *constant* velocity $-u$ with respect to the rocket, so that with respect to the fixed inertial frame, the exhaust gases are moving with velocity $v - u$. The change in the total momentum of the system over the time interval $t$ to $t + \mathrm{d}t$ is then

$$\begin{aligned}
\mathrm{d}p &= p(t + \mathrm{d}t) - p(t) \\
&= \left[(m - \mathrm{d}m')(v + \mathrm{d}v) + \mathrm{d}m'(v - u)\right] - mv \\
&= m\,\mathrm{d}v - u\,\mathrm{d}m' \\
&= m\,\mathrm{d}v + u\,\mathrm{d}m\,,
\end{aligned} \tag{1.3}$$

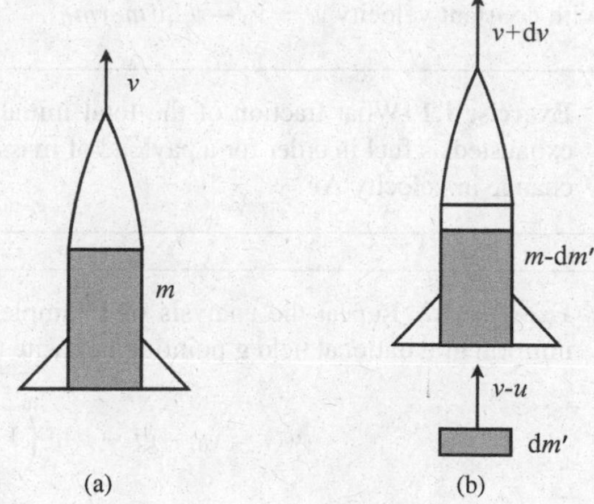


**Fig. 1.1** A rocket moving in interstellar space, free of all external forces. Panel (a): Rocket at time $t$ (mass $m$, velocity $v$). Panel (b): Rocket and exhaust at time $t + \mathrm{d}t$ (mass $m - \mathrm{d}m'$, velocity $v + \mathrm{d}v$; mass $\mathrm{d}m'$, velocity $v - u$)

where we ignored the $-\mathrm{d}m'\,\mathrm{d}v$ term (since it is 2nd-order small) to get the third line, and switched back to $\mathrm{d}m$ to get the last line. But note, however, that there are no external forces acting on the system, so $\mathrm{d}p/\mathrm{d}t = F = 0$, which implies

$$0 = m\,\mathrm{d}v + u\,\mathrm{d}m\,, \tag{1.4}$$

or, equivalently,

$$\mathrm{d}v = -u\,\frac{\mathrm{d}m}{m}\,. \tag{1.5}$$

This is a separable differential equation, which can be immediately integrated, subject to the initial condition that $v = v_0$ when $m = m_0$:

$$v - v_0 = -u\ln(m/m_0)\,. \tag{1.6}$$

To get the time dependence of $v$, we make use of the assumption that the mass-loss rate is constant,

$$\frac{\mathrm{d}m}{\mathrm{d}t} \equiv -\alpha = \text{const}\,, \tag{1.7}$$

which implies

$$m(t) = m_0 - \alpha t\,. \tag{1.8}$$

Making this substitution into (1.6), we have

$$v(t) = v_0 - u\ln\left(1 - \frac{\alpha t}{m_0}\right). \tag{1.9}$$

Note that this equation is valid only up to time $t_\mathrm{f}$, when all of the fuel has been exhausted, and the mass of the rocket is $m_\mathrm{f}$ ($> 0$). After that time, the rocket moves with constant velocity $v_\mathrm{f} = v_0 - u\ln(m_\mathrm{f}/m_0)$. □

**Exercise 1.1** What fraction of the total initial mass $m_0$ of a rocket must be exhausted as fuel in order for a payload of mass $m_\mathrm{f}$ to be accelerated through a change in velocity $\Delta v$?

**Exercise 1.2** Repeat the analysis of Example 1.1 for a rocket moving in a uniform gravitational field $\mathbf{g}$ pointing opposite to $\mathbf{v}$. You should find

$$v(t) = v_0 - gt - u\ln\left(1 - \frac{\alpha t}{m_0}\right). \tag{1.10}$$

## 1.2 Single-Particle Mechanics

In this section, we will discuss the motion of a single object (a *particle*) that is subject to external forces. Our use of the term "particle" implies that the object has no internal structure and no physical extent (i.e., it is effectively a zero-dimensional point). This will allow us to focus simply on its motion without having to consider the influence that the external forces may have on the internal structure or orientation of the object. (We will treat real three-dimensional objects later in Chaps. 6 and 7, in the context of rotational motion.) Note that we can use the particle approximation even for extended objects provided the changes in internal energy or rotational state of the object are negligible. In these cases, we simply use a point within the object as a stand-in for the particle's position.

Let's first consider a particle viewed in an inertial frame of reference. Within this frame, the position of the particle is defined by a time-dependent vector $\mathbf{r}(t)$, and its linear momentum is $\mathbf{p} = m\dot{\mathbf{r}} = m\mathbf{v}$. Since we are in an inertial frame, any variation in $\mathbf{p}$ will be due to an impressed external force, so $\mathbf{F} = \dot{\mathbf{p}}$.

**Exercise 1.3** Let a particle's position be given by $\mathbf{r}(t)$ in an inertial frame $O$, and let the mass be constant, so that $\mathbf{F} = \dot{\mathbf{p}} = m\mathbf{a} = m\ddot{\mathbf{r}}$. Transform to a new reference frame $O'$ that is moving at constant velocity $\mathbf{u}$ with respect to the original one, so $\mathbf{r}'(t) = \mathbf{r}(t) - \mathbf{u}\,t$. Show that $m\ddot{\mathbf{r}}' = m\ddot{\mathbf{r}} = \mathbf{F}$, so that Newton's 2nd law has the same form in this new reference frame. Thus, the new reference frame is also inertial.

In a single particle universe, if mass is conserved, then the mass of the particle must be constant. The impressed force $\mathbf{F} \equiv \mathbf{F}(\mathbf{r}, \dot{\mathbf{r}}, t)$ then governs the acceleration of the particle, and we obtain a 2nd-order differential equation, which must be solved in order to determine the motion $\mathbf{r}(t)$. In the remainder of this section, we will review the fundamentals of single-particle mechanics and recover some of the familiar conservation laws.

***Example 1.2*** Air resistance can be modeled as a velocity-dependent force with $\mathbf{F} = -b\mathbf{v}$, where $b$ is a real, positive proportionality constant. If a particle starts with an initial velocity $\mathbf{v}_0$, how far does it go before coming to rest under the influence of air resistance alone?

We can obtain the equation of motion from $\mathbf{F} = m\mathbf{a}$ and solve for $\mathbf{r}(t)$, but we are more interested in $\mathbf{v}$ as a function of $\mathbf{r}$. Note that the problem is essentially one-dimensional, so let's choose a coordinate system with an $x$-axis that lies along the initial velocity, so we can dispense with the boldface vector notation. Then Newton's 2nd law reads:

$$F = -bv = ma = m\frac{\mathrm{d}v}{\mathrm{d}t} = m\frac{\mathrm{d}x}{\mathrm{d}t}\frac{\mathrm{d}v}{\mathrm{d}x} = mv\frac{\mathrm{d}v}{\mathrm{d}x}\,. \tag{1.11}$$

This leaves us with the simple differential equation

$$-\frac{b}{m} = \frac{\mathrm{d}v}{\mathrm{d}x}, \tag{1.12}$$

which is solved by $v = v_0 - bx/m$. Consequently the distance traveled by the particle is the value of $x$ for which $v = 0$. This is $x = mv_0/b$. □

### 1.2.1 Work

When a particle is subject to an external force, the force does **work** on the particle as it moves along a path $\mathbf{s}(t)$ according to the line integral

$$W_{12} \equiv \int_{\wp_1}^{\wp_2} \mathbf{F} \cdot \mathrm{d}\mathbf{s}, \tag{1.13}$$

where $\wp_1$ and $\wp_2$ are the endpoints of the path, corresponding to the times $t = t_1$ and $t = t_2$. (See Fig. 1.2.) The work can be thought of as the amount of energy deposited into the particle by the agent producing the force. Note that, in general, the work done by a force will be dependent upon the path taken by the particle.

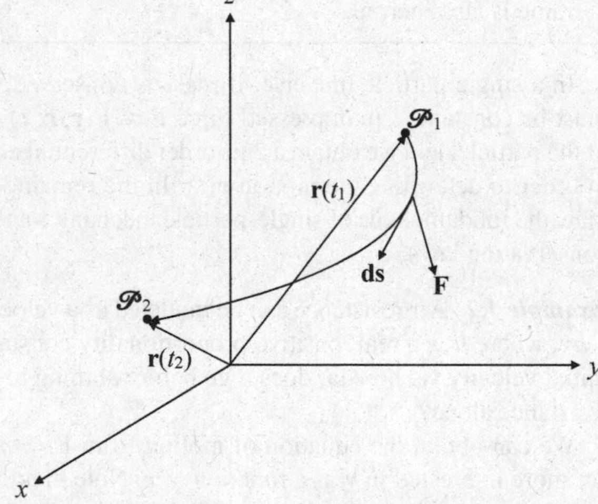


**Fig. 1.2** The work $W_{12}$ is calculated for the particular path that a particle takes in moving from point $\wp_1$ to point $\wp_2$

**Exercise 1.4** A particle of mass $m$ is subject to a force that is dependent upon its velocity, $\mathbf{F} = -b\mathbf{v}$, where $b$ is a real, positive proportionality constant. (a) Calculate the work done by the force as the particle moves with constant velocity along the $x$-axis from $x = -a$ to $x = +a$. (b) Calculate the work done by the force if the particle moves with constant speed along a semicircle of radius $a$ from $x = -a$ to $x = +a$. (c) Along which path does the force do the most work?

**Exercise 1.5** A particle of mass $m$ is subject to a force that is dependent upon its velocity, $\mathbf{F} = -bv^2\hat{\mathbf{v}}$, where $b$ is a real, positive proportionality constant and $\hat{\mathbf{v}}$ is a unit vector in the direction of $\mathbf{v}$. Assuming that this is the *only* force acting on the particle, show that the work done by this force as the particle moves a distance $a$ along a straight line is

$$W = \frac{1}{2}mv_0^2\left(e^{-2ba/m} - 1\right), \tag{1.14}$$

where $v_0$ is the initial velocity. (*Hint*: Treat this as a 1-dimensional problem and use $\mathbf{F} = m\mathbf{a}$ to solve for $v$ as a function of $x$.)

### 1.2.2 *Work-Energy Theorem*

The expression for the **kinetic energy** of a particle,

$$T \equiv \frac{1}{2}mv^2, \tag{1.15}$$

arises naturally if one calculates the work done on the particle by the net force in moving it from one location to another. To see this, assume that the mass $m$ of the particle is constant, so that the net force is given by $\mathbf{F} = m\mathbf{a} = m\mathrm{d}\mathbf{v}/\mathrm{d}t$. Then

$$\int_{\wp_1}^{\wp_2} \mathbf{F}\cdot\mathrm{d}\mathbf{s} = \int_{\wp_1}^{\wp_2} m\frac{\mathrm{d}\mathbf{v}}{\mathrm{d}t}\cdot\mathrm{d}\mathbf{s} = \int_{\wp_1}^{\wp_2} m\mathrm{d}\mathbf{v}\cdot\mathbf{v} = \int_{\wp_1}^{\wp_2} \mathrm{d}\left(\frac{1}{2}mv^2\right). \tag{1.16}$$

But this last integral is trivial to evaluate, so

$$W_{12} \equiv \int_{\wp_1}^{\wp_2} \mathbf{F}\cdot\mathrm{d}\mathbf{s} = \frac{1}{2}mv_2^2 - \frac{1}{2}mv_1^2 \equiv T_2 - T_1, \tag{1.17}$$

where $v_i$ is the velocity of the particle at point $\wp_i$. This is the **work-energy theorem** for a single particle, which relates the work done on a particle by the net force to its change in kinetic energy.

### 1.2.3 Conservative Forces

There is a certain class of forces for which the work done is *independent* of the path and depends only upon the endpoints. These forces are called **conservative forces**. In order for a line integral to be independent of the path, the integrand must be expressible as the gradient of a scalar function. Specifically, if $\mathbf{F} = -\nabla U(\mathbf{r})$ for some function $U$, then $\mathbf{F}$ is conservative and $U(\mathbf{r})$ is the **potential energy** for the force $\mathbf{F}$. For a conservative force, the work done is the difference between the values of the potential energy at the endpoints:

$$W_{12} = \int_{\wp_1}^{\wp_2} \mathbf{F} \cdot d\mathbf{s} = -\int_{\wp_1}^{\wp_2} \nabla U \cdot d\mathbf{s} = -(U_2 - U_1) \,. \tag{1.18}$$

If we combine the above result for a conservative force with (1.17), which holds in general, we see that $U_1 - U_2 = T_2 - T_1$ or, equivalently, $T_1 + U_1 = T_2 + U_2$, so the quantity

$$E \equiv T + U, \tag{1.19}$$

called the **mechanical energy** of the particle, is constant throughout the motion. *Thus, the mechanical energy of a particle is conserved if the external forces are conservative.*

### 1.2.4 Angular Momentum

We can also define angular momentum about a preferred point, even in a single particle universe. If we place the origin of our coordinate system at this preferred point, then the **angular momentum** is defined as

$$\boldsymbol{\ell} \equiv \mathbf{r} \times \mathbf{p} \,. \tag{1.20}$$

The time derivative of $\boldsymbol{\ell}$ is

$$\dot{\boldsymbol{\ell}} = \dot{\mathbf{r}} \times \mathbf{p} + \mathbf{r} \times \dot{\mathbf{p}} = \mathbf{r} \times \mathbf{F}, \tag{1.21}$$

where we have used Newton's 2nd law and the fact that $\dot{\mathbf{r}} \times \mathbf{p} = \dot{\mathbf{r}} \times (m\dot{\mathbf{r}}) = 0$. *Thus, the angular momentum is conserved if the* **torque** $\boldsymbol{\tau} \equiv \mathbf{r} \times \mathbf{F}$ *is zero.*

**Exercise 1.6** In an inertial frame with a Cartesian coordinate system, a particle of mass $m$ starts at rest with an initial position of $\mathbf{r}_0 = x_0\hat{\mathbf{x}} + y_0\hat{\mathbf{y}}$. At $t = 0$ the particle experiences a force $\mathbf{F} = F\hat{\mathbf{x}}$. (a) Using Newton's 2nd law $\mathbf{F} = \dot{\mathbf{p}}$, solve the equation of motion to obtain $\mathbf{r}(t)$ and $\mathbf{p}(t)$. (b) Determine the angular momentum about the origin and show that it satisfies $\boldsymbol{\tau} = \dot{\boldsymbol{\ell}}$. (c) Now choose a new coordinate system that is translated in the $y$ direction by $y_0$, so that $\mathbf{r}_0' = x_0\hat{\mathbf{x}}$. Repeat part (a) and calculate the new torque $\boldsymbol{\tau}$. Is angular momentum conserved in this coordinate system?

## 1.3 Systems of Particles

When we expand our scope to include systems with multiple particles, we must take into account **interparticle forces** and the apparent bulk motion of the entire system. For a system of $N$ particles, the momentum $\mathbf{p}_I$ of the $I$th particle can change due to interactions with other particles as well as to impressed external forces. Thus, Newton's 2nd law reads

$$\frac{\mathrm{d}\mathbf{p}_I}{\mathrm{d}t} = \mathbf{F}_I^{(\mathrm{e})} + \sum_{J \neq I} \mathbf{F}_{JI}\,, \qquad I = 1, 2, \cdots, N\,, \tag{1.22}$$

where the sum is over all other particles in the system ($J$ runs from 1 to $N$, excluding $I$), and $\mathbf{F}_{JI}$ is the force that particle $J$ exerts on particle $I$. (To simplify the notation in what follows, we will define $\mathbf{F}_{II} = 0$ so that such sums can run over *all* indices, including $J = I$.) The **total linear momentum** of the system is then the sum of the individual particle momenta,

$$\mathbf{P} \equiv \sum_I \mathbf{p}_I\,. \tag{1.23}$$

The change in the total linear momentum is then

$$\frac{\mathrm{d}\mathbf{P}}{\mathrm{d}t} = \frac{\mathrm{d}}{\mathrm{d}t}\sum_I \mathbf{p}_I = \sum_I \frac{\mathrm{d}\mathbf{p}_I}{\mathrm{d}t} = \sum_I \mathbf{F}_I^{(\mathrm{e})} + \sum_{I,J} \mathbf{F}_{JI}\,, \tag{1.24}$$

where the double summation $\sum_{I,J} \equiv \sum_I \sum_J$ counts each particle twice (once as $I$ and once as $J$). But from Newton's 3rd law, $\mathbf{F}_{JI} = -\mathbf{F}_{IJ}$, so the interparticle forces sum to zero. Defining the net external force to be $\mathbf{F}^{(e)} \equiv \sum_I \mathbf{F}_I^{(e)}$, we have

$$\frac{\mathrm{d}\mathbf{P}}{\mathrm{d}t} = \mathbf{F}^{(e)}, \tag{1.25}$$

which shows that *the total linear momentum of a system is conserved if the net external force on the system is zero.*

### 1.3.1 Center of Mass

The total momentum of a system of particles acts as if the system were a single particle under the influence of the net external applied force. Thus, it is possible to define a single position for the system. This position is known as the **center of mass**, which is defined by

$$\mathbf{R} \equiv \frac{1}{M} \sum_I m_I \mathbf{r}_I, \tag{1.26}$$

where $M \equiv \sum m_I$ is the total mass of the system.

**Exercise 1.7** Show that the total momentum can be expressed as $\mathbf{P} = M\dot{\mathbf{R}}$. (Note that we assume that the masses of the individual particles are constant.)

### 1.3.2 Angular Momentum

In a similar fashion to the definition of the total (linear) momentum, we can define the **total angular momentum** of a system of particles to be the sum of the individual angular momenta,

$$\mathbf{L} \equiv \sum_I \boldsymbol{\ell}_I. \tag{1.27}$$

If the interparticle forces are *central* (i.e., they are all directed along the line segments joining pairs of particles), then the total angular momentum responds to the action of the net external torque in the same way that a single particle does, i.e.,

$$\frac{d\mathbf{L}}{dt} = \boldsymbol{\tau}^{(e)}, \tag{1.28}$$

where $\boldsymbol{\tau}^{(e)} \equiv \sum_I \boldsymbol{\tau}_I^{(e)}$ is the sum of the external torques on the individual particles. *Thus, we see that the total angular momentum of a system is conserved if the interparticle forces are central and the net external torque on the system is zero.*

**Exercise 1.8** Verify (1.28). (*Hint*: You will need to assume that the interparticle forces are central (i.e., $\mathbf{F}_{JI} \propto \mathbf{r}_{IJ} \equiv \mathbf{r}_I - \mathbf{r}_J$) in order to have only the *external* torques $\boldsymbol{\tau}_I^{(e)} \equiv \mathbf{r}_I \times \mathbf{F}_I^{(e)}$ contribute to the final sum.)

**Exercise 1.9** For a system of particles, we can write the position of particle $I$ as $\mathbf{r}_I = \mathbf{R} + \mathbf{r}'_I$, where $\mathbf{r}'_I$ is the position of the particle relative to $\mathbf{R}$—the location of the center of mass. Show that

$$\mathbf{L} = \mathbf{R} \times \mathbf{P} + \sum_I \mathbf{r}'_I \times \mathbf{p}'_I, \tag{1.29}$$

where $\mathbf{p}'_I \equiv m_I \dot{\mathbf{r}}'_I$.

### *1.3.3 Work*

The time evolution of a single particle is described by the path traced-out in three dimensions by its position vector $\mathbf{r}(t)$. For a system of $N$ particles, each particle traces out a different path $\mathbf{r}_I(t)$, where $I = 1, 2, \cdots, N$, so time evolution of a system corresponds to motion of a point in an abstract $3N$-dimensional space, called the **configuration space** of the system. Thus, the instantaneous positions of all the particles of the system correspond to a *single point* in configuration space. As the system evolves, this point traces out a (1-dimensional) curve in configuration space. The work done on a system of particles as it goes from configuration 1 to configuration 2 is the sum of the work done on each individual particle in the system. Thus,

$$W_{12} = \sum_I \int_1^2 \mathbf{F}_I \cdot d\mathbf{s}_I. \tag{1.30}$$

Defining the total kinetic energy of the system of particles to be

$$T \equiv \sum_I \frac{1}{2} m_I v_I^2, \tag{1.31}$$

we find that the total work done on a system of particles is equal to the change in the total kinetic energy, so that

$$W_{12} = T_2 - T_1 . \tag{1.32}$$

This is the **work-energy theorem** in the context of a system of particles.

### 1.3.4 Conservative Forces

For a single particle, a force is conservative if it is the gradient of a potential. This can be simply expressed as $\mathbf{F} = -\nabla U(\mathbf{r})$, where the independent variable $\mathbf{r}$ is the position of the particle. In multi-particle systems, there are coordinates $\mathbf{r}_I$ for each particle in the system. Thus, net external forces are conservative if and only if the external force on *each* particle is conservative, i.e.,

$$\mathbf{F}_I^{(e)} = -\nabla_I U_I^{(e)}(\mathbf{r}_1, \mathbf{r}_2, \cdots, \mathbf{r}_N) , \tag{1.33}$$

where $\nabla_I$ means the gradient of the potential with respect to the coordinate position of particle $I$. Note that the potential itself carries a subscript $I$ and may depend on the properties (e.g., position, mass, charge, $\cdots$) of each individual particle, but it does not explicitly depend on the velocities $\dot{\mathbf{r}}_1, \dot{\mathbf{r}}_2, \cdots$ or the time $t$.

Let's look at the work done and the conditions that are placed on the forces in order for us to be able to describe a well-defined potential energy for a system of particles. In general, the work done is

$$W_{12} = \sum_I \int_1^2 \mathbf{F}_I \cdot d\mathbf{s}_I = \sum_I \int_1^2 \left( \mathbf{F}_I^{(e)} + \sum_J \mathbf{F}_{JI} \right) \cdot d\mathbf{s}_I , \tag{1.34}$$

where the sum over all particles $J$ describes the work done by the interparticle forces. (Recall that we have defined $\mathbf{F}_{II} = 0$.) Thus, the work done on the system splits into two parts—the work done by external forces and the work done by interparticle forces. If the external forces are conservative, then the work done by them is simply minus the change in the external potential from configuration 1 to configuration 2—i.e., $-\Delta U^{(e)} \equiv U_1^{(e)} - U_2^{(e)}$, where $U^{(e)} \equiv \sum_I U_I^{(e)}$. The interparticle forces that appear in the second term of (1.34) may depend on the position of particle $J$, and may contribute to the work in a path-dependent way. We can make this dependence on the position of particle $J$ explicit by noticing that the double sum over $I$ and $J$ counts each pair of particles twice—once as experiencing a force and once as exerting a force. Thus, we can write the sum as

$$\sum_{I,J} \int_1^2 \mathbf{F}_{JI} \cdot d\mathbf{s}_I = \frac{1}{2} \sum_{I,J} \left[ \int_1^2 \mathbf{F}_{JI} \cdot d\mathbf{s}_I + \int_1^2 \mathbf{F}_{IJ} \cdot d\mathbf{s}_J \right]. \tag{1.35}$$

Because of Newton's third law, $\mathbf{F}_{IJ} = -\mathbf{F}_{JI}$, we then have

$$\sum_{I,J} \int_1^2 \mathbf{F}_{JI} \cdot d\mathbf{s}_I = \frac{1}{2} \sum_{I,J} \int_1^2 \mathbf{F}_{JI} \cdot (d\mathbf{s}_I - d\mathbf{s}_J) = \frac{1}{2} \sum_{I,J} \int_1^2 \mathbf{F}_{JI} \cdot d\mathbf{r}_{IJ}, \tag{1.36}$$

where $d\mathbf{r}_{IJ}$ is the change in the relative separation between particles $I$ and $J$, which we denote by $\mathbf{r}_{IJ} \equiv \mathbf{r}_I - \mathbf{r}_J$. If the interparticle forces can also be described as a gradient of a potential, so that $\mathbf{F}_{JI} = -\nabla_{IJ} U_{IJ}(\mathbf{r}_{IJ})$, then the integral becomes path-independent and the total work done is

$$W_{12} = U_1 - U_2, \tag{1.37}$$

where

$$U \equiv \sum_I U_I^{(e)}(\mathbf{r}_1, \mathbf{r}_2, \cdots, \mathbf{r}_N) + \frac{1}{2} \sum_{I,J} U_{IJ}(\mathbf{r}_{IJ}). \tag{1.38}$$

This total potential is the sum of the external potential energies of each particle as well as the internal potential energies due to interparticle interactions. *Thus, if all the forces (both external and internal) are conservative, then the total mechanical energy $E = T + U$ is conserved for the system.*

***Example 1.3*** Let's consider the effects of an interparticle force that is not directed along the line joining the two particles. Let two particles, with $m_1 = m_2 \equiv m$, lie at rest in the $xy$-plane, separated by an initial distance $2a$. These two particles feel no external force, but are subject to an interparticle force given by $\mathbf{F}_{21} = k\hat{\mathbf{z}} \times \mathbf{r}_{12}$, where $\hat{\mathbf{z}}$ is the usual unit vector in the $z$-direction in cylindrical coordinates and $k$ is a constant (units of N/m). This force will still obey the weak form of Newton's 3rd law, so that $\mathbf{F}_{12} = -\mathbf{F}_{21}$. Let's choose a reference frame in which the center of mass lies at the origin, as shown in Fig. 1.3. Since the net force on the two particles is zero, the total momentum is conserved and the center of mass will remain at the origin. The force on particle 1 is then $\mathbf{F}_{21} = 2kr\hat{\boldsymbol{\phi}}$, where $r \equiv |\mathbf{r}_1| = |\mathbf{r}_2|$. Recalling that $\hat{\boldsymbol{\phi}}$ changes direction as we move from point to point, this problem is easier to solve using Cartesian coordinates. Newton's 2nd law gives the following coupled equations:

$$\begin{aligned} m\ddot{x} &= -2ky, \\ m\ddot{y} &= +2kx. \end{aligned} \tag{1.39}$$

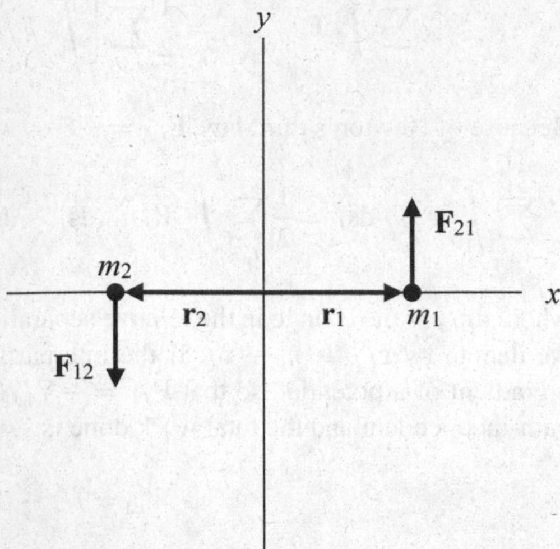


**Fig. 1.3** Initial positions of the particles in Example 1.3. The interparticle forces obey $\mathbf{F}_{21} = k\hat{\mathbf{z}} \times \mathbf{r}_{12}$

We can combine these two equations by defining the complex function $\zeta \equiv x + \mathrm{i}y$, giving the single complex differential equation

$$m\ddot{\zeta} = 2\mathrm{i}k\zeta\,. \tag{1.40}$$

The solution to this equation is simply

$$\zeta(t) = A\mathrm{e}^{t\sqrt{2\mathrm{i}k/m}} + B\mathrm{e}^{-t\sqrt{2\mathrm{i}k/m}}\,. \tag{1.41}$$

The initial conditions for this problem are that $\zeta(0) = a$ and $\dot{\zeta}(0) = 0$. Imposing these conditions requires $A = B = a/2$. Defining $\omega \equiv \sqrt{k/m}$ and noting that $\sqrt{2\mathrm{i}k/m} = \sqrt{k/m}\,(1+\mathrm{i}) = \omega\,(1+\mathrm{i})$, we find

$$\zeta(t) = \frac{a}{2}\left(\mathrm{e}^{\omega t}\mathrm{e}^{\mathrm{i}\omega t} + \mathrm{e}^{-\omega t}\mathrm{e}^{-\mathrm{i}\omega t}\right). \tag{1.42}$$

Taking its real and imaginary parts:

$$\begin{aligned} x(t) &= \mathrm{Re}\,\zeta(t) = \frac{a}{2}\left(\mathrm{e}^{\omega t}\cos\omega t + \mathrm{e}^{-\omega t}\cos\omega t\right) = a\cos\omega t\cosh\omega t\,,\\ y(t) &= \mathrm{Im}\,\zeta(t) = \frac{a}{2}\left(\mathrm{e}^{\omega t}\sin\omega t - \mathrm{e}^{-\omega t}\sin\omega t\right) = a\sin\omega t\sinh\omega t\,. \end{aligned} \tag{1.43}$$

Thus, these particles spiral away from each other, gaining angular momentum and kinetic energy as shown in Fig. 1.4. The total angular momentum of the system is

$$\mathbf{L} = 2\mathbf{r} \times \mathbf{p} = 2m\,(x\dot{y} - y\dot{x})\,\hat{\mathbf{z}} = ma^2\omega\,[\sin(2\omega t) + \sinh(2\omega t)]\,\hat{\mathbf{z}}\,. \tag{1.44}$$

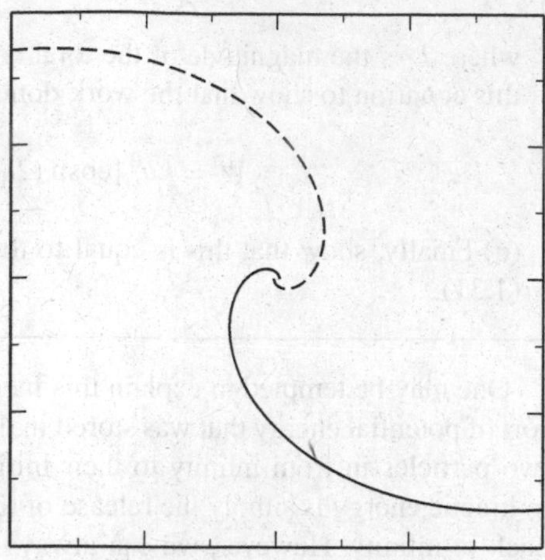

**Fig. 1.4** The trajectories of the particles under the influence of the interparticle force $\mathbf{F}_{JI} = k\hat{\mathbf{z}} \times \mathbf{r}_{IJ}$

This increase is the direct result of the fact that the interparticle forces do not point along the interparticle separation. Although there are no external forces on this system, the net torque on the system is

$$\sum_I \boldsymbol{\tau}_I = 4\mathbf{r} \times \left(k\hat{\mathbf{z}} \times \mathbf{r}\right) = 4kr^2\hat{\mathbf{z}}. \tag{1.45}$$

**Exercise 1.10** Calculate the net torque from (1.45) and show that it is equal to the time derivative of the total angular momentum given in (1.44).

The changing angular momentum in this problem indicates that there is an increase in the kinetic energy of the system. This increase comes from the work done by the interparticle forces. Consider the infinitesimal work $\mathrm{d}W$ done by the forces,

$$\mathrm{d}W = \mathbf{F}_{21} \cdot \mathrm{d}\mathbf{r}_1 + \mathbf{F}_{12} \cdot \mathrm{d}\mathbf{r}_2 . \tag{1.46}$$

Since $\mathbf{r}_1 = -\mathbf{r}_2$, the rate of work done is then

$$\frac{\mathrm{d}W}{\mathrm{d}t} = 4k\left(\hat{\mathbf{z}} \times \mathbf{r}_1\right) \cdot \mathbf{v}_1 . \tag{1.47}$$

**Exercise 1.11** (a) Using the scalar triple product identity (A.9), show that the rate at which work is done can be written as

$$\frac{\mathrm{d}W}{\mathrm{d}t} = 2\omega^2 L , \tag{1.48}$$

where $L$ is the magnitude of the total angular momentum vector. (b) Integrate this equation to show that the work done by the forces as a function of time is

$$W = ka^2 [\cosh(2\omega t) - \cos(2\omega t)] \ . \tag{1.49}$$

(c) Finally, show that this is equal to the total kinetic energy calculated using (1.31).

One may be tempted to explain this increase in kinetic energy by invoking some sort of potential energy that was stored in the system in the process of bringing these two particles in from infinity to their initial positions. Then the apparent increase in kinetic energy is simply the release of this potential energy as the particles spiral back to infinity. However, we can always set up the initial conditions by bringing the particles together from $\pm\infty$ along the $x$-axis. In this way, the interparticle forces are always *perpendicular* to the motion of the particles, and so no work is done. The interparticle forces that are invoked in this example are not conservative, and it is not possible to define a potential energy associated with these forces. The real solution to this apparent conundrum is that we are using nonsensical forces in this example. These forces are the equivalent of frictional forces that point in the direction of motion (as opposed to against the motion).

**Exercise 1.12** We know that a necessary and sufficient condition for a force to be described as the gradient of a potential is that the integral $\oint_C \mathbf{F} \cdot d\mathbf{s}$ vanish. Choose a circle of radius $r_0$ centered on one particle and show that this integral is non-zero, thus proving that $\mathbf{F}$ is non-conservative.

□

## 1.4 Conservation Laws

Newton's laws provide us with 2nd-order differential equations for the motion of a system of particles $\mathbf{r}_I(t)$ by relating the accelerations to the known forces acting on the particles. These are known as the equations of motion for the system. If certain combinations of the positions and velocities of the particles can be shown to be time-independent, then these quantities are conserved. Each conserved quantity can reduce the order of the equations of motion by one, so they are also called **integrals of the motion**. The common conserved quantities are the total linear momentum, total angular momentum, and total mechanical energy of the system. Certain conditions are placed on the forces acting on the system in order for these quantities to be conserved. From our analyses in the previous sections, we have seen the following conservation laws:

I. **Conservation of Linear Momentum:** If the net external force on a system is zero, then the total linear momentum is conserved:

$$\sum_I \mathbf{F}_I^{(e)} = 0 \quad \Rightarrow \quad \mathbf{P} \equiv \sum_I m_I \mathbf{v}_I = \text{const}\,. \tag{1.50}$$

II. **Conservation of Angular Momentum:** If the net external torque on a system is zero and the *strong form* of Newton's 3rd law holds (so that $\mathbf{F}_{JI}$ is directed along the line connecting particles $I$ and $J$), then the total angular momentum is conserved:

$$\sum_I \boldsymbol{\tau}_I^{(e)} = 0\,, \quad \mathbf{F}_{JI} \propto \mathbf{r}_{IJ} \quad \Rightarrow \quad \mathbf{L} \equiv \sum_I \mathbf{r}_I \times \mathbf{p}_I = \text{const}\,. \tag{1.51}$$

III. **Conservation of Mechanical Energy:** If both the external forces and interparticle forces are expressible as gradients of scalar potentials,

$$\mathbf{F}_I^{(e)} = -\boldsymbol{\nabla}_I U_I^{(e)}(\mathbf{r}_1, \mathbf{r}_2, \cdots, \mathbf{r}_N)\,, \quad \mathbf{F}_{JI} = -\boldsymbol{\nabla}_{IJ} U_{IJ}(\mathbf{r}_{IJ})\,, \tag{1.52}$$

then the total mechanical energy of the system $E \equiv T + U$ is conserved:

$$E = \frac{1}{2}\sum_I m_I v_I^2 + \sum_I U_I^{(e)}(\mathbf{r}_1, \mathbf{r}_2, \cdots, \mathbf{r}_N) + \frac{1}{2}\sum_{I,J} U_{IJ}(\mathbf{r}_{IJ}) = \text{const}\,. \tag{1.53}$$

We will return to these conservation laws in Sects. 3.3 and 3.6.2, after we have developed the Lagrangian and Hamiltonian formulations of mechanics.

## 1.5 Non-inertial Reference Frames

So far we have restricted our attention to studying the motion of a particle (or a system of particles) as seen from an *inertial* frame of reference. We saw in Exercise 1.3 that inertial reference frames move at constant velocity with respect to one another. We can formalize this relationship as a coordinate transformation (known as a **Galilean transformation**) between the two frames as

$$\mathbf{r} = \mathbf{r}' + \mathbf{u}t\,, \tag{1.54}$$

where the origin of the primed coordinate system is moving with constant velocity $\mathbf{u}$ within the unprimed coordinate system. Recall that, in an inertial frame, a particle moves with constant velocity (i.e., has zero acceleration) if there are no forces acting on it. When we are not in an inertial frame, there will be spurious (or **fictitious**) accelerations arising from the acceleration of the reference frame. These effects can be seen in simple every-day situations such as sitting in a vehicle that is accelerating or rounding a corner. In these situations, loose objects will appear to accelerate relative to the observer or vehicle.

We perceive these accelerations because we effectively carry around with us an origin $O'$ and a set of orthonormal basis vectors $\hat{\mathbf{e}}_{i'}$ that are fixed with respect to us, as shown in Fig. 1.5. The motion of the origin $O'$ is described by the position vector $\mathbf{R}(t)$ as seen in the frame of an inertial observer $O$, with corresponding orthonormal basis vectors $\hat{\mathbf{e}}_i$. The position of a particle located at $\wp$ is described in the inertial and non-inertial reference frames by the displacement vectors $\mathbf{r}(t)$ and $\mathbf{r}'(t)$, respectively, which are defined with respect to the observers $O$ and $O'$. These two displacement vectors are related by the vector $\mathbf{R}(t)$ joining $O$ and $O'$, so that

$$\mathbf{r} = \mathbf{r}' + \mathbf{R}\,. \tag{1.55}$$

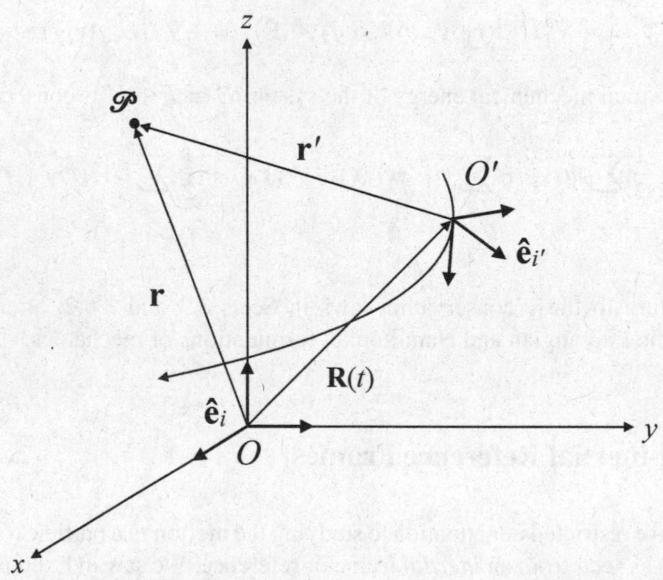


**Fig. 1.5** The motion of a non-inertial observer $O'$ described by $\mathbf{R}(t)$ in the reference frame of inertial observer $O$. $O'$ carries a set of orthonormal basis vectors $\hat{\mathbf{e}}_{i'}$. The position $\wp$ of a particle is described in the inertial and non-inertial reference frames by the displacement vectors $\mathbf{r} = \mathbf{r}(t)$ from $O$ to $\wp$, and $\mathbf{r}' = \mathbf{r}'(t)$ from $O'$ to $\wp$, respectively

Note that both the translational motion of the origin $O'$ and the rotational motion of the basis vectors $\hat{\mathbf{e}}_{i'}$ relative to the fixed (inertial) frame lead to differences in the velocity and acceleration of the particle as seen in these two frames. To determine what these differences are, it is simplest to first *separate* the effects of the translational and rotational motion, and then combine the results at the end to handle the more general case of translational-plus-rotational motion. We do this in the following three subsections.

### 1.5.1 Translational Motion

Let's begin with the simplest scenario, which is to allow $O'$ to move with respect $O$, but to keep the basis vectors of the non-inertial reference frame fixed with respect to the inertial frame—i.e., $\hat{\mathbf{e}}_{i'} = \hat{\mathbf{e}}_i$ for $i = 1, 2, 3$. To relate the accelerations of the particle as measured with respect to both $O$ and $O'$, we simply differentiate (1.55) twice with respect to time—i.e., $\ddot{\mathbf{r}} = \ddot{\mathbf{r}}' + \ddot{\mathbf{R}}$, or, equivalently,

$$\mathbf{a} = \mathbf{a}' + \ddot{\mathbf{R}}\,. \tag{1.56}$$

Thus, Newton's 2nd law, $\mathbf{F} = m\mathbf{a}$, which is valid in the inertial frame $O$, becomes

$$m\mathbf{a}' = \mathbf{F} - m\ddot{\mathbf{R}} \tag{1.57}$$

with respect to $O'$. Note the presence of the *fictitious force* $\mathbf{F}_{\rm accel} \equiv -m\ddot{\mathbf{R}}$, which is non-zero if $O'$ is accelerating with respect to $O$, and which points in the direction *opposite* to the acceleration $\ddot{\mathbf{R}}$.

***Example 1.4*** Consider a reference frame $O'$ that is accelerating with constant linear acceleration—e.g., a car starting up from a stop. Since the basis vectors in the accelerated and inertial frames are identical, the observed acceleration of loose objects in the car is simply $-\ddot{\mathbf{R}}$. We perceive this acceleration to be caused by the fictitious force $\mathbf{F}_{\rm accel} = -m\ddot{\mathbf{R}}$, which points in the direction opposite to the car's acceleration. From the perspective of the passengers in the car, they perceive that they have *zero* acceleration relative to their reference frame, i.e., $\mathbf{a}' = 0$, but this is due to the exact cancellation of two forces. One is the fictitious force $\mathbf{F}_{\rm accel} = -m\ddot{\mathbf{R}}$, which they feel pushing them back in their seats, and the other is the true force $\mathbf{F} = m\mathbf{a}$, which is accelerating them along with the car, but which they perceive as a *normal* force from the seat acting in response to the backward-directed fictitious force. □

### 1.5.2 Rotational Motion

Now let's consider the case where the origins $O$ and $O'$ of the two reference frames occupy the same position in space, but the basis vectors $\hat{\mathbf{e}}_{i'}$ of the non-inertial frame are *rotating* with respect to the basis vectors $\hat{\mathbf{e}}_i$ of the inertial frame. Then

$$\hat{\mathbf{e}}_{i'} = \sum_j R_{i'j}\hat{\mathbf{e}}_j \,, \tag{1.58}$$

where $R_{i'j}$ are the component of a **rotation matrix** $\mathsf{R}$. (Note that $R_{i'j} = \hat{\mathbf{e}}_{i'} \cdot \hat{\mathbf{e}}_j \equiv \cos\theta_{i'j}$, where $\theta_{i'j}$ is the angle between $\hat{\mathbf{e}}_{i'}$ and $\hat{\mathbf{e}}_j$. These are just the **direction cosines** relating the basis vectors of the two frames.) Since rotations preserve the length of vectors, $\mathsf{R}$ is an **orthogonal** matrix, which means that $\mathsf{R}^{-1} = \mathsf{R}^T$ (the transpose of $\mathsf{R}$), or, equivalently,[2]

$$\sum_{i'} R_{i'j}R_{i'k} = \delta_{jk} \,, \qquad \sum_i R_{j'i}R_{k'i} = \delta_{j'k'} \,. \tag{1.59}$$

Using this result, it follows that the components $A_i$ and $A_{i'}$ of a vector $\mathbf{A}$ with respect to the two reference frames are related by

$$A_{i'} = \sum_j R_{i'j}A_j \,, \tag{1.60}$$

which has the same form as the transformation equation (1.58) for the basis vectors.

To calculate the time derivative of $\mathbf{A}$, we will expand $\mathbf{A}$ in the two different reference frames. If we first expand with respect to the inertial frame, we find

$$\frac{\mathrm{d}\mathbf{A}}{\mathrm{d}t} = \frac{\mathrm{d}}{\mathrm{d}t}\left(\sum_i A_i\hat{\mathbf{e}}_i\right) = \sum_i\left(\frac{\mathrm{d}A_i}{\mathrm{d}t}\hat{\mathbf{e}}_i + A_i\frac{\mathrm{d}\hat{\mathbf{e}}_i}{\mathrm{d}t}\right) = \sum_i\frac{\mathrm{d}A_i}{\mathrm{d}t}\hat{\mathbf{e}}_i \,, \tag{1.61}$$

where the last equality follows from the basis vectors $\hat{\mathbf{e}}_i$ being at rest in the inertial frame. If we expand with respect to the rotating frame, we find

$$\frac{\mathrm{d}\mathbf{A}}{\mathrm{d}t} = \frac{\mathrm{d}}{\mathrm{d}t}\left(\sum_{i'} A_{i'}\hat{\mathbf{e}}_{i'}\right) = \sum_{i'}\left(\frac{\mathrm{d}A_{i'}}{\mathrm{d}t}\hat{\mathbf{e}}_{i'} + A_{i'}\frac{\mathrm{d}\hat{\mathbf{e}}_{i'}}{\mathrm{d}t}\right) = \left(\frac{\mathrm{d}\mathbf{A}}{\mathrm{d}t}\right)_{\mathrm{r}} + \sum_{i'} A_{i'}\frac{\mathrm{d}\hat{\mathbf{e}}_{i'}}{\mathrm{d}t} \,, \tag{1.62}$$

[2]These concepts are described in more detail in Chap. 6 and Appendix D.

where

$$\left(\frac{\mathrm{d}\mathbf{A}}{\mathrm{d}t}\right)_{\mathrm{r}} \equiv \sum_{i'} \frac{\mathrm{d}A_{i'}}{\mathrm{d}t}\hat{\mathbf{e}}_{i'} \tag{1.63}$$

is the time derivative of $\mathbf{A}$ as seen in the rotating frame of reference (hence the subscript 'r'). To evaluate the last term in (1.62), we use (1.58) and (1.60) to expand $A_{i'}$ and $\hat{\mathbf{e}}_{i'}$ in terms of $A_i$ and $\hat{\mathbf{e}}_i$. This yields

$$\sum_{i'} A_{i'}\frac{\mathrm{d}\hat{\mathbf{e}}_{i'}}{\mathrm{d}t} = \sum_{i'}\sum_{j} R_{i'j}A_j\frac{\mathrm{d}}{\mathrm{d}t}\left(\sum_k R_{i'k}\hat{\mathbf{e}}_k\right) = \sum_j A_j\sum_k\left(\sum_{i'} R_{i'j}\frac{\mathrm{d}R_{i'k}}{\mathrm{d}t}\right)\hat{\mathbf{e}}_k\,. \tag{1.64}$$

But note that the matrix defined as

$$M_{jk} \equiv \sum_{i'} R_{i'j}\frac{\mathrm{d}R_{i'k}}{\mathrm{d}t} \tag{1.65}$$

is *anti-symmetric* (i.e., $M_{jk} = -M_{kj}$) as a consequence of (1.59). Since an anti-symmetric $3\times 3$ matrix has three independent components, we can define the components $\omega_i$ of a vector $\boldsymbol{\omega}$ in terms of $M_{jk}$ and the (totally anti-symmetric) Levi-Civita symbol $\varepsilon_{ijk}$, defined in (A.7):

$$M_{jk} \equiv \sum_i \omega_i\varepsilon_{ijk} \quad\Leftrightarrow\quad \omega_i = \frac{1}{2}\sum_{j,k}\varepsilon_{ijk}M_{jk}\,. \tag{1.66}$$

Thus,

$$\sum_j A_j\sum_k\left(\sum_{i'} R_{i'j}\frac{\mathrm{d}R_{i'k}}{\mathrm{d}t}\right)\hat{\mathbf{e}}_k = \sum_j A_j\sum_k M_{jk}\hat{\mathbf{e}}_k = \sum_{i,j,k}\omega_i A_j\varepsilon_{ijk}\hat{\mathbf{e}}_k = \boldsymbol{\omega}\times\mathbf{A}\,. \tag{1.67}$$

Putting all these results together,

$$\left(\frac{\mathrm{d}\mathbf{A}}{\mathrm{d}t}\right)_{\mathrm{f}} = \left(\frac{\mathrm{d}\mathbf{A}}{\mathrm{d}t}\right)_{\mathrm{r}} + \boldsymbol{\omega}\times\mathbf{A}\,, \tag{1.68}$$

where we have written $\mathrm{d}\mathbf{A}/\mathrm{d}t = (\mathrm{d}\mathbf{A}/\mathrm{d}t)_{\mathrm{f}}$, which follows from (1.61); the subscript 'f' indicates the fixed (inertial) frame.

**Exercise 1.13** It turns out that $\boldsymbol{\omega}$ defined by (1.66) and (1.65) is the **instantaneous angular velocity vector** of the rotating reference frame relative to the inertial reference frame. Verify that this is indeed the case by calculating $\boldsymbol{\omega}$ for the simple case of a rotation about the $z$-axis with constant angular velocity $\omega$:

$$R_{i'j} = \begin{bmatrix} \cos\omega t & \sin\omega t & 0 \\ -\sin\omega t & \cos\omega t & 0 \\ 0 & 0 & 1 \end{bmatrix}. \tag{1.69}$$

You should find that $\boldsymbol{\omega} = \omega\hat{\mathbf{z}}$.

Equation (1.68) is a general result, so we can apply it to *any* vector $\mathbf{A}$. In particular, if we take $\mathbf{A}$ to be the position vector $\mathbf{r}$ of a particle relative to the shared origin of $O$ and $O'$, then

$$\mathbf{v}_{\rm f} = \mathbf{v}_{\rm r} + \boldsymbol{\omega} \times \mathbf{r}, \tag{1.70}$$

where $\mathbf{v}_{\rm f}$ and $\mathbf{v}_{\rm r}$ are shorthand for $(\mathrm{d}\mathbf{r}/\mathrm{d}t)_{\rm f}$ and $(\mathrm{d}\mathbf{r}/\mathrm{d}t)_{\rm r}$. Similarly, if we apply (1.68) to the angular velocity vector $\boldsymbol{\omega}$, we find

$$\left(\frac{\mathrm{d}\boldsymbol{\omega}}{\mathrm{d}t}\right)_{\rm f} = \left(\frac{\mathrm{d}\boldsymbol{\omega}}{\mathrm{d}t}\right)_{\rm r} \tag{1.71}$$

since $\boldsymbol{\omega} \times \boldsymbol{\omega} = 0$. Thus, we can write $\mathrm{d}\boldsymbol{\omega}/\mathrm{d}t \equiv \dot{\boldsymbol{\omega}}$ without ambiguity. Finally, if we take $\mathbf{A}$ to equal $\mathbf{v}_{\rm f}$ from (1.70), we find

$$\begin{aligned}
\left(\frac{\mathrm{d}\mathbf{v}_{\rm f}}{\mathrm{d}t}\right)_{\rm f} &= \left(\frac{\mathrm{d}\mathbf{v}_{\rm f}}{\mathrm{d}t}\right)_{\rm r} + \boldsymbol{\omega} \times \mathbf{v}_{\rm f} \\
&= \left(\frac{\mathrm{d}}{\mathrm{d}t}(\mathbf{v}_{\rm r} + \boldsymbol{\omega} \times \mathbf{r})\right)_{\rm r} + \boldsymbol{\omega} \times (\mathbf{v}_{\rm r} + \boldsymbol{\omega} \times \mathbf{r}) \\
&= \left(\frac{\mathrm{d}\mathbf{v}_{\rm r}}{\mathrm{d}t}\right)_{\rm r} + \left(\frac{\mathrm{d}\boldsymbol{\omega}}{\mathrm{d}t}\right)_{\rm r} \times \mathbf{r} + \boldsymbol{\omega} \times \left(\frac{\mathrm{d}\mathbf{r}}{\mathrm{d}t}\right)_{\rm r} + \boldsymbol{\omega} \times \mathbf{v}_{\rm r} + \boldsymbol{\omega} \times (\boldsymbol{\omega} \times \mathbf{r}) \\
&= \left(\frac{\mathrm{d}\mathbf{v}_{\rm r}}{\mathrm{d}t}\right)_{\rm r} + \left(\frac{\mathrm{d}\boldsymbol{\omega}}{\mathrm{d}t}\right)_{\rm r} \times \mathbf{r} + 2\boldsymbol{\omega} \times \mathbf{v}_{\rm r} + \boldsymbol{\omega} \times (\boldsymbol{\omega} \times \mathbf{r}),
\end{aligned} \tag{1.72}$$

or, more compactly,

$$\mathbf{a}_{\rm f} = \mathbf{a}_{\rm r} + \dot{\boldsymbol{\omega}} \times \mathbf{r} + 2\boldsymbol{\omega} \times \mathbf{v}_{\rm r} + \boldsymbol{\omega} \times (\boldsymbol{\omega} \times \boldsymbol{r}), \tag{1.73}$$

where $\mathbf{a}_{\rm f}$ and $\mathbf{a}_{\rm r}$ are shorthand for $(\mathrm{d}\mathbf{v}_{\rm f}/\mathrm{d}t)_{\rm f}$ and $(\mathrm{d}\mathbf{v}_{\rm r}/\mathrm{d}t)_{\rm r}$.

Thus, Newton's 2nd law $\mathbf{F} = m\mathbf{a}_{\rm f}$, which is valid in an inertial frame, can be written in a rotating reference frame as

$$m\mathbf{a}_{\rm r} = \mathbf{F} - m\dot{\boldsymbol{\omega}} \times \mathbf{r} - 2m\boldsymbol{\omega} \times \mathbf{v}_{\rm r} - m\boldsymbol{\omega} \times (\boldsymbol{\omega} \times \boldsymbol{r}). \tag{1.74}$$

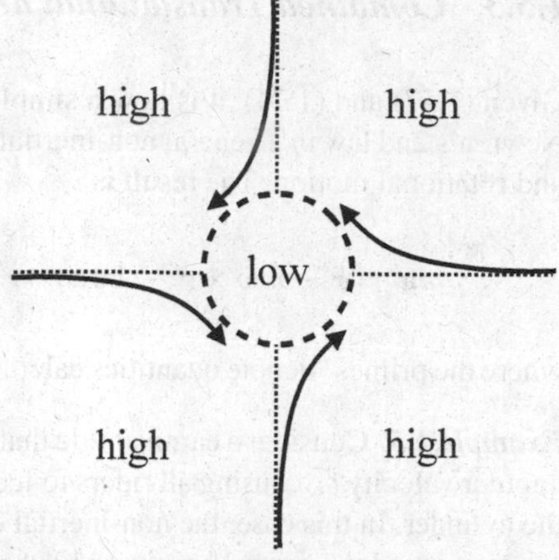


**Fig. 1.6** Cyclonic motion of air currents as seen in the Northern hemisphere, driven by pressure gradients and the Coriolis force associated with Earth's rotational motion

If we interpret these last three terms as additional (fictitious) forces, then Newton's 2nd law in the rotating frame has the more standard looking form, $m\mathbf{a}_r = \mathbf{F}_{\text{eff}}$. The first fictitious force term, $\mathbf{F}_{\text{ang accel}} \equiv -m\dot{\boldsymbol{\omega}} \times \mathbf{r}$, is related to the angular acceleration of the rotating reference frame. For a *uniformly* rotating frame, like a lab attached to the surface of the Earth, $\dot{\boldsymbol{\omega}} = 0$, so this fictitious force vanishes. The last two fictitious force terms are the **Coriolis** and **centrifugal** forces, respectively:

$$\mathbf{F}_{\text{coriolis}} \equiv -2m\boldsymbol{\omega} \times \mathbf{v}_r\,, \qquad \mathbf{F}_{\text{centrifugal}} \equiv -m\boldsymbol{\omega} \times (\boldsymbol{\omega} \times \boldsymbol{r})\,. \tag{1.75}$$

The centrifugal force is directed radially away from the axis of rotation and has magnitude $m\omega^2 r \sin\theta$ where $\theta$ is the angle between $\boldsymbol{\omega}$ and $\mathbf{r}$. The Coriolis force is non-zero only if $\mathbf{v}_r \neq 0$, and is directed perpendicular to both $\mathbf{v}_r$ and $\boldsymbol{\omega}$. As viewed along the direction of $\mathbf{v}_r$, the Coriolis force associated with counter-clockwise rotational motion produces a deflection to the right; for clockwise rotational motion, it produces a deflection to the left. The Coriolis force associated with Earth's rotational motion is responsible for the circulating or **cyclonic** weather patterns associated with hurricanes and cyclones, as illustrated in Fig. 1.6. Basically, a pressure gradient gives rise to air currents that tend to flow from high pressure to low pressure regions. But as the air flows toward the low pressure region, the Coriolis force deflects the air currents away from their straight line paths. Since the projection of $\boldsymbol{\omega}$ perpendicular to the local tangent plane changes sign as one crosses the equator, the direction of the cyclonic motion (either counter-clockwise or clockwise) is different in the Northern and Southern hemispheres.

### 1.5.3 Combined Translational and Rotational Motion

Given (1.57) and (1.74), it is now a simple manner to write down the equivalent of Newton's 2nd law in a general non-inertial reference frame having both translational and rotational motion. The result is

$$m\mathbf{a}' = \mathbf{F} - m\dot{\boldsymbol{\omega}} \times \mathbf{r}' - 2m\boldsymbol{\omega} \times \mathbf{v}' - m\boldsymbol{\omega} \times (\boldsymbol{\omega} \times \mathbf{r}') - m\ddot{\mathbf{R}}\,, \qquad (1.76)$$

where the primes $'$ denote quantities calculated with respect to the non-inertial frame.

***Example 1.5*** Consider a carnival ride that spins a cylinder about its central axis with angular velocity $\omega$, causing all riders to feel that they are pressed against the walls of the cylinder. In this case, the non-inertial observer feels a net acceleration as he/she rotates around the central axis, with his/her basis vectors rotating with respect to the inertial frame at the same rate (See Fig. 1.7). We'd like to know what fictitious forces the rider feels, and if the rider threw a ball in toward the center of the ride, where would it land?

To do this problem, we first note that the basis vectors in $O'$ are related to those in $O$ by the rotation matrix

$$R_{i'j} = \begin{bmatrix} \cos\omega t & \sin\omega t & 0 \\ -\sin\omega t & \cos\omega t & 0 \\ 0 & 0 & 1 \end{bmatrix}. \qquad (1.77)$$

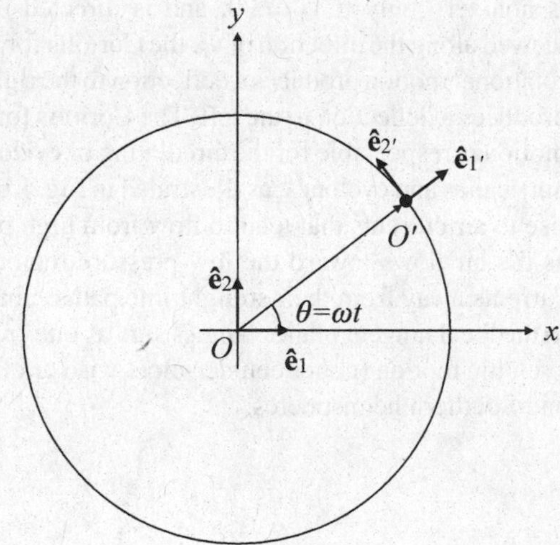


**Fig. 1.7** Basis vectors and reference frames for an inertial observer $O$ at the center of a carnival ride and for a non-inertial observer $O'$ on the ride. The basis vectors of the non-inertial observer, $\hat{\mathbf{e}}_{1'}$, $\hat{\mathbf{e}}_{2'}$, rotate along with the rider at the same rate. Note that $\hat{\mathbf{e}}_3 = \hat{\mathbf{e}}_{3'} = \hat{\mathbf{z}}$ for both observers, which points out of the page

As shown in Exercise 1.13, the associated angular velocity vector $\boldsymbol{\omega}$ is simply $\omega\hat{\mathbf{z}}$, consistent with the motion of the carnival ride. Also, since the position $O'$ of the rider with respect to the inertial frame is

$$\mathbf{R} = R\cos\omega t\,\hat{\mathbf{e}}_1 + R\sin\omega t\,\hat{\mathbf{e}}_2\,, \tag{1.78}$$

then $\ddot{\mathbf{R}} = -\omega^2\mathbf{R}$. Thus, Newton's 2nd law in this non-inertial frame reduces to

$$m\mathbf{a}' = \mathbf{F} - 2m\boldsymbol{\omega}\times\mathbf{v}' + m\omega^2\mathbf{r}' + m\omega^2\mathbf{R}\,, \tag{1.79}$$

where we have also assumed that the position vector $\mathbf{r}'$ of a particle as seen in the non-inertial frame has no $z$-component in order to write the second-to-last term in that form. The last two terms in the above expression can be thought of as centrifugal force terms associated with (i) the rotational motion of the basis vectors $\hat{\mathbf{e}}_{i'}$, and (ii) the rotational motion of the origin $O'$ (i.e., the rider) with respect to $O$. The "origin" centrifugal force is the fictitious force that appears to drive objects out toward the wall from the center of the ride. The "basis" centrifugal force is the fictitious force that appears to drive objects away from the rider. Finally, the Coriolis force $-2m\boldsymbol{\omega}\times\mathbf{v}'$ is the result of the fact that the rider is moving with tangential velocity $\boldsymbol{\omega}\times\mathbf{R}$. Any additional velocity of an observed particle will add to this tangential velocity. Thus, if the rider throws a ball in toward the center, the Coriolis force will cause it to appear to accelerate *in the direction of motion of the rider*. We can also understand this by noting that the ball has an initial velocity of $\mathbf{v}' + \boldsymbol{\omega}\times\mathbf{R}$ with respect to the inertial frame. As the ball moves in toward the center, its tangential velocity is now greater than the comoving tangential velocity, so it will appear to move in the direction of rotation. This is similar to what we saw in Fig. 1.6 for the deflection of air currents due to the Earth's rotational motion. □

### *1.5.4 Foucault's Pendulum*

A simple way to demonstrate the Earth's rotational motion is to show that the plane of a swinging pendulum precesses with time, with a precessional period equal to (1 day)/ $\sin\lambda$, where $\lambda = \pi/2 - \theta$ is the latitude of the pendulum's location.[3] In this subsection, we solve the equations of motion for the swinging pendulum as seen in a rotating reference frame attached to the surface of the Earth, and derive the above expression for the precessional period. Such a demonstration is called **Foucault's pendulum** in honor of the French physicist, Jean Léon Foucault who first exhibited this demonstration in Paris in 1851.

---

[3] $\theta$ is the usual spherical coordinate angle measured from the $z$-axis (the North pole), and is called the *co-latitude*.

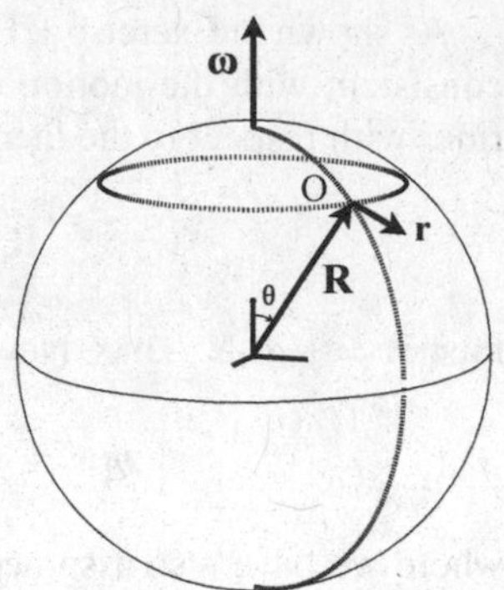


**Fig. 1.8** Definitions of **R** and **r** for a non-inertial reference frame with origin $O$ attached to the surface of the Earth. The size of the pendulum bob displacement **r** relative to the Earth's radius **R** has been exaggerated in this figure for ease of visualization

To simplify the notation in what follows, we will drop the primes on quantities calculated in the non-inertial frame attached to the surface of the Earth, so that (1.76) becomes

$$m\mathbf{a} = \mathbf{F} - m\dot{\boldsymbol{\omega}} \times \mathbf{r} - 2m\boldsymbol{\omega} \times \mathbf{v} - m\boldsymbol{\omega} \times (\boldsymbol{\omega} \times \mathbf{r}) - m\ddot{\mathbf{R}}_\mathrm{f}\,, \tag{1.80}$$

where **R** is the radius vector pointing from the center of the Earth to the origin $O$ of the local reference frame, and **r** is the displacement of the pendulum bob away from equilibrium, as shown in Fig. 1.8. (The subscript "f" on $\ddot{\mathbf{R}}_\mathrm{f}$ is to indicate that this acceleration is calculated with respect to the *fixed* (i.e., inertial) frame.) In addition, $\mathbf{F} = \mathbf{T} + m\mathbf{g}_0$, where **T** is the tension in the string attached to the pendulum bob and $\mathbf{g}_0$ points towards the center of the Earth (in the direction of $-\mathbf{R}$); $\dot{\boldsymbol{\omega}} = 0$, since the angular velocity of the Earth is constant; and $\ddot{\mathbf{R}}_\mathrm{f} = \boldsymbol{\omega} \times (\boldsymbol{\omega} \times \mathbf{R})$, as a consequence of (1.68). Thus,

$$m\mathbf{a} = \mathbf{T} + m\left[\mathbf{g}_0 - \boldsymbol{\omega} \times (\boldsymbol{\omega} \times (\mathbf{r} + \mathbf{R}))\right] - 2m\boldsymbol{\omega} \times \mathbf{v}\,. \tag{1.81}$$

The term in square brackets

$$\mathbf{g} \equiv \mathbf{g}_0 - \boldsymbol{\omega} \times (\boldsymbol{\omega} \times (\mathbf{r} + \mathbf{R}))\,, \tag{1.82}$$

defines the effective local direction of the Earth's gravitational field, which differs from $\mathbf{g}_0$ by the centrifugal acceleration associated with the Earth's rotational motion. Since the displacement $r$ of the pendulum bob away from equilibrium is small compared to the Earth's radius $R$, the centrifugal acceleration is dominated by $-\boldsymbol{\omega} \times (\boldsymbol{\omega} \times \mathbf{R})$. Note that a *plumb line* (i.e., a mass suspended from the end of a string) points in the direction of $\mathbf{g}$ (and not $\mathbf{g}_0$), as shown in Fig. 1.9. Thus, for our analysis, we will define our local coordinate system so that $\hat{\mathbf{z}}$ points along $-\mathbf{g}$. We then choose $\hat{\mathbf{x}}$ perpendicular to $\hat{\mathbf{z}}$, pointing South; and $\hat{\mathbf{y}}$ perpendicular to both $\hat{\mathbf{z}}$ and $\hat{\mathbf{x}}$, pointing East (along the line of constant latitude), as shown in Fig. 1.10.

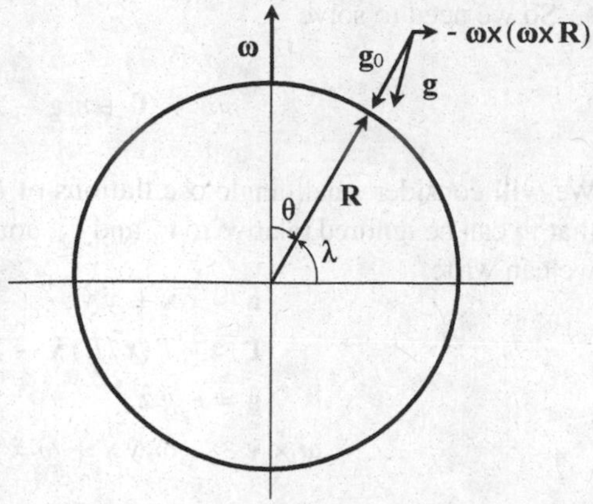

**Fig. 1.9** Change in the direction of the local gravitational field due to Earth's rotational motion. The deviation from $\mathbf{g}_0$ has been exaggerated for ease of visualization

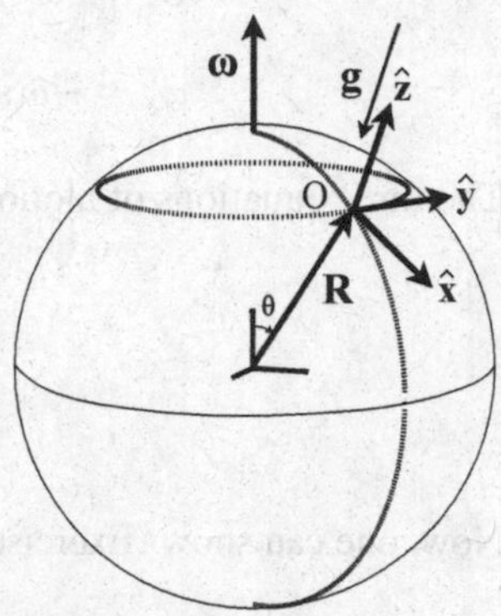

**Fig. 1.10** Definition of a non-inertial reference frame attached to the surface of the Earth. Note that $\hat{\mathbf{z}}$ points along $-\mathbf{g}$, which is opposite the direction of the effective local gravitational field. The direction of $\mathbf{g}$ differs slightly from $\mathbf{g}_0$ (which points in the direction of $-\mathbf{R}$) due to the centrifugal acceleration associated with the Earth's rotational motion, as shown previously in Fig. 1.9

**Exercise 1.14** Show that the angle $\delta$ between $\mathbf{g}$ (the direction of a plumb line at the surface of the Earth at latitude $\lambda = \pi/2 - \theta$) and $\mathbf{g}_0$ (the direction pointing toward the center of the Earth) is given to leading order by

$$\delta \approx \frac{R\omega^2}{g_0} \sin\theta \cos\theta , \tag{1.83}$$

with maximum value $\delta = 0.0017$ rad $\approx 0.1°$ at $\theta = \pi/4$. Show also that the centrifugal acceleration vector $\delta\mathbf{g} \equiv -\boldsymbol{\omega} \times (\boldsymbol{\omega} \times \mathbf{R})$ has maximum magnitude $|\delta\mathbf{g}|/g_0 = 0.003$ at the equator, $\theta = \pi/2$.
*Hint:* First show that the maximum value of $\boldsymbol{\omega} \times (\boldsymbol{\omega} \times \mathbf{R})$ is small compared to $g_0 = 9.8$ m/s$^2$, where $\omega = 2\pi/(1\text{ day})$ and $R = 6400$ km. Then use the law of sines and the small-angle approximation to obtain (1.83).

So we need to solve

$$m\mathbf{a} = \mathbf{T} + m\mathbf{g} - 2m\boldsymbol{\omega} \times \mathbf{v} \,. \tag{1.84}$$

We will consider small-angle oscillations of the pendulum bob in the $xy$-plane, so that $v_z$ can be ignored relative to $v_x$ and $v_y$, and $a_z \approx 0$. Given these approximations, we can write

$$\begin{aligned} \mathbf{a} &\approx \ddot{x}\,\hat{\mathbf{x}} + \ddot{y}\,\hat{\mathbf{y}} \,, \\ \mathbf{T} &\approx -T(x/L)\,\hat{\mathbf{x}} - T(y/L)\,\hat{\mathbf{y}} + T\,\hat{\mathbf{z}} \,, \\ \mathbf{g} &= -g\,\hat{\mathbf{z}} \,, \\ \boldsymbol{\omega} \times \mathbf{v} &\approx -\omega_z \dot{y}\,\hat{\mathbf{x}} + \omega_z \dot{x}\,\hat{\mathbf{y}} + \omega_x \dot{y}\,\hat{\mathbf{z}} \,, \end{aligned} \tag{1.85}$$

where $L$ is the length of the pendulum and

$$\omega_x \approx -\omega \sin\theta \,, \qquad \omega_y = 0 \,, \qquad \omega_z \approx \omega\cos\theta \,. \tag{1.86}$$

The three equations of motion are thus

$$\begin{aligned} m\ddot{x} &\approx -Tx/L + 2m\omega_z \dot{y} \,, \\ m\ddot{y} &\approx -Ty/L - 2m\omega_z \dot{x} \,, \\ 0 &\approx T - mg - 2m\omega_x \dot{y} \,. \end{aligned} \tag{1.87}$$

Now, one can show (Exercise 1.15) that

$$|\omega_x \dot{y}| \ll g \,, \tag{1.88}$$

for a typical pendulum with period $P \ll 1$ day $= 2\pi/\omega$. Thus, we can ignore the last term in the $\ddot{z} \approx 0$ equation in (1.87) and solve it for the tension,

$$T = mg \,, \tag{1.89}$$

giving the expected result. Using this value for $T$, the $\ddot{x}$ and $\ddot{y}$ equations reduce to:

$$\begin{aligned} \ddot{x} &\approx -\Omega^2 x + 2\omega_z \dot{y} \,, \\ \ddot{y} &\approx -\Omega^2 y - 2\omega_z \dot{x} \,, \end{aligned} \tag{1.90}$$

where $\Omega \equiv \sqrt{g/L}$ is the unperturbed frequency of oscillation that we expect for a pendulum of length $L$. Since $\omega \ll \Omega$, the terms proportional to $\omega_z$ in the above equations act as perturbations to the standard simple harmonic oscillator equations $\ddot{x} = -\Omega^2 x$ and $\ddot{y} = -\Omega^2 y$, which describe simple harmonic motion with angular frequency $\Omega$.

**Exercise 1.15** Verify (1.88). *Hint*: If the maximum displacement of the pendulum bob away from equilibrium is $D$, then you can show that $\dot{y}$ is bounded by $D\Omega$, where $\Omega \equiv \sqrt{g/L}$ is the unperturbed (angular) frequency of oscillation. Note that $D \ll L$ to be consistent with the small-angle approximation for the pendulum bob.

To solve the coupled differential equations in (1.90), we perform the same "trick" that we used in Example 1.3 and form the complex combination

$$\zeta \equiv x + \mathrm{i}y\,, \tag{1.91}$$

allowing us to recast the two equations in (1.90) as a *single* complex differential equation

$$\ddot{\zeta} + 2\mathrm{i}\omega_z\dot{\zeta} + \Omega^2\zeta = 0\,. \tag{1.92}$$

This is a 2nd-order ordinary differential equation with constant coefficients, which can be solved in the usual way (See, e.g., Chap. 8 in Boas (2006)). Substituting the trial solution $\zeta(t) = \mathrm{e}^{\lambda t}$, with complex $\lambda$, we obtain a quadratic equation for $\lambda$:

$$\lambda^2 + 2\mathrm{i}\omega_z\lambda + \Omega^2 = 0\,. \tag{1.93}$$

This equation has two complex solutions

$$\lambda_\pm = -\mathrm{i}\left(\omega_z \mp \sqrt{\Omega^2 + \omega_z^2}\right) \approx -\mathrm{i}(\omega_z \mp \Omega)\,, \tag{1.94}$$

where we've used $\omega_z \ll \Omega$ to get the last (approximate) equality. Thus, the general solution to (1.92) is

$$\zeta(t) = A\mathrm{e}^{\lambda_+ t} + B\mathrm{e}^{\lambda_- t}\,, \tag{1.95}$$

where $A$ and $B$ are complex coefficients, to be determined by the initial conditions.

If we assume that the pendulum bob is pulled out a distance $D$ in the $x$-direction and released from rest, then

$$x(0) = D\,,\quad y(0) = 0\,,\quad \dot{x}(0) = 0\,,\quad \dot{y}(0) = 0\,, \tag{1.96}$$

or, equivalently,

$$\zeta(0) = D\,,\qquad \dot{\zeta}(0) = 0\,. \tag{1.97}$$

Imposing these conditions on $\zeta(t)$ determines $A$ and $B$, leading to (Exercise 1.16):

$$\begin{aligned} x(t) &= D\left[\cos\omega_z t\cos\Omega t + \frac{\omega_z}{\Omega}\sin\omega_z t\sin\Omega t\right],\\ y(t) &= D\left[-\sin\omega_z t\cos\Omega t + \frac{\omega_z}{\Omega}\cos\omega_z t\sin\Omega t\right]. \end{aligned} \qquad (1.98)$$

Note that these equations can be written in matrix form

$$\begin{bmatrix} x(t) \\ y(t) \end{bmatrix} = \begin{bmatrix} \cos\omega_z t & \sin\omega_z t \\ -\sin\omega_z t & \cos\omega_z t \end{bmatrix} \begin{bmatrix} \bar{x}(t) \\ \bar{y}(t) \end{bmatrix}, \qquad (1.99)$$

where

$$\bar{x}(t) = D\cos\Omega t\,, \qquad \bar{y}(t) = D\frac{\omega_z}{\Omega}\sin\Omega t\,. \qquad (1.100)$$

The matrix

$$R_z \equiv \begin{bmatrix} \cos\omega_z t & \sin\omega_z t \\ -\sin\omega_z t & \cos\omega_z t \end{bmatrix}, \qquad (1.101)$$

which appears in (1.99), represents a uniform rotation in the $xy$-plane with angular velocity $\omega_z = \omega\cos\theta$. This is just the precessional frequency of the plane of oscillation of the pendulum. The period of the precession is then

$$P_{\text{precession}} = \frac{2\pi}{\omega_z} = \frac{1\text{ day}}{\cos\theta} = \frac{1\text{ day}}{\sin\lambda}, \qquad (1.102)$$

where $\lambda = \pi/2 - \theta$ is the latitude. For example, if we take $\lambda = 49°$, which is the latitude of Paris (where Foucault first did this demonstration), we have a precessional period of 31 hours and 48 minutes. At the equator, the pendulum does *not* precess.

**Exercise 1.16** Verify the solution given in (1.98).

Plots of $(x(t), y(t))$ and $(\bar{x}(t), \bar{y}(t))$ are shown in Figs. 1.11 and 1.12. For these plots, we decreased the angular frequency of oscillation $\Omega$ by a factor of 200 compared to typical values, so as to easily see the precession of the plane of oscillation of the pendulum after only a few oscillations. Typical Foucault pendulum demonstrations have suspensions of order $L = 30$ m (roughly 100 ft). For such an $L$, the angular oscillation frequency $\Omega = \sqrt{g/L} \approx 0.57$ rad/s, which corresponds to an oscillation period of $2\pi/\Omega \approx 11$ s. For these figures, we have oscillation periods of roughly 2200 s $\approx$ 36 min, so only $\sim$ 50 back-and-forth motions of the pendulum bob would be needed for a complete 360° precession at the latitude of Paris.

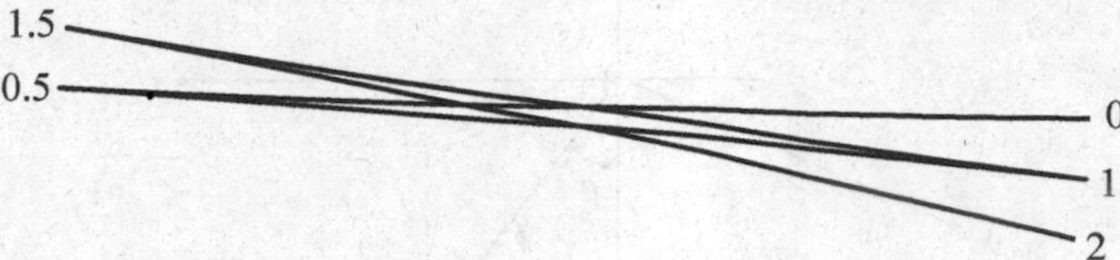


**Fig. 1.11** Motion of the pendulum bob as seen in the non-inertial reference frame. Note that the plane of oscillation of the pendulum precesses. (The numbers correspond to back-and-forth oscillations of the pendulum bob.) As described in the text, the angular frequency of oscillation $\Omega$ has been reduced considerably for this figure so as to easily visualize the precession of the plane of oscillation in just two oscillation periods

**Fig. 1.12** Motion of the pendulum bob as seen in a "corotating frame", which rotates relative to the non-inertial frame with the precession frequency $\omega_z$ of the plane of oscillation of the pendulum. As described in the text, the angular frequency of oscillation $\Omega$ has been reduced considerably for this figure so as to easily visualize the elliptical nature of the motion in this reference frame

## 1.6 Constrained Systems

For some systems, the motion of a particle (or particles) is restricted to a prescribed surface or path. The constraints on the motion are often the result of additional forces acting on the particle (such as normal forces or tension forces) that *adjust their values* in order to maintain the motion on the prescribed surface or path. These forces then become additional unknowns that must be solved for while obtaining the equations of motion for the system.

There are a variety of techniques for dealing with these forces of constraint. We will look at a few examples in order to see how the constraints are imposed on solutions obtained through Newton's laws. In many cases, this involves reducing the number of degrees of freedom in the system by finding an equation relating the coordinates to one another and using it to solve for one or more of the degrees of freedom in terms of the remaining variables. In Chap. 2, we will examine more powerful mathematical tools for handling constrained systems.

***Example 1.6*** A **spherical pendulum** consists of a mass $m$ at the end of a massless rigid rod of length $\ell$. The rod is free to pivot around the other end, so the particle is constrained to move under the influence of gravity on the surface of a sphere of radius $\ell$. Spherical coordinates allow us to easily impose the constraint and reduce the number of degrees of freedom from three to two by requiring that $r = \ell$. In order to allow our solution to be easily compared with the well-known results of the simple pendulum, we orient our coordinates with the $z$-axis is pointing downward, so that the polar angle $\theta$ is measured as a displacement angle from the equilibrium position of the pendulum hanging vertically. The orientation of the coordinates and system are shown in Fig. 1.13

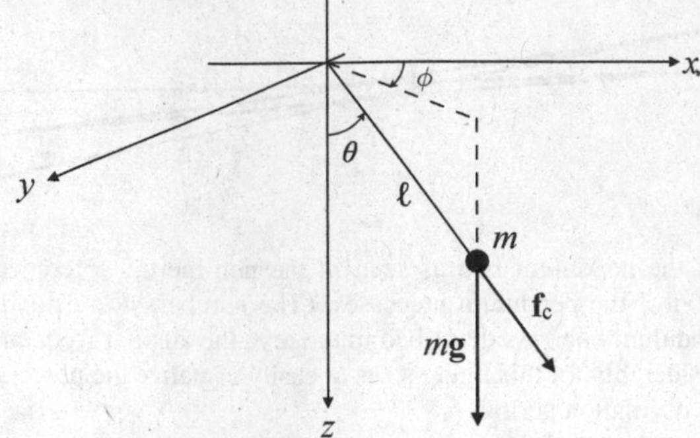


**Fig. 1.13** The spherical pendulum with a mass $m$ on the end of a massless rigid rod of length $\ell$. The force of gravity points down in the positive $z$ direction, and the force of constraint $\mathbf{f}_c$ points along the rod in the direction of $\hat{\mathbf{r}}$. If the constraint force is a tension force, it will point radially inward; if it is a normal force, it will point radially outward. The relevant polar and azimuthal angles are the usual spherical coordinates $\theta$ and $\phi$

In these coordinates, the forces are $\mathbf{f}_c = f_c\hat{\mathbf{r}}$ and $m\mathbf{g} = mg\cos\theta\,\hat{\mathbf{r}} - mg\sin\theta\,\hat{\boldsymbol{\theta}}$, so the net force acting on the particle is

$$\sum \mathbf{F} = (f_c + mg\cos\theta)\,\hat{\mathbf{r}} - mg\sin\theta\,\hat{\boldsymbol{\theta}}\,. \tag{1.103}$$

The position of the particle is simply $\mathbf{r} = \ell\hat{\mathbf{r}}$, but since $\hat{\mathbf{r}}$ points in different directions as the particle moves, we must also include the time dependence of the spherical basis vectors when finding the acceleration. Referring to the definitions in (A.46) and taking the time derivatives explicitly, we find

$$\begin{aligned}\frac{\mathrm{d}\hat{\mathbf{r}}}{\mathrm{d}t} =& \dot{\phi}\sin\theta\left(-\sin\phi\hat{\mathbf{x}} + \cos\phi\hat{\mathbf{y}}\right) \\ &+ \dot{\theta}\left(\cos\phi\cos\theta\hat{\mathbf{x}} + \sin\phi\cos\theta\hat{\mathbf{y}} - \sin\theta\hat{\mathbf{z}}\right)\,,\end{aligned} \tag{1.104}$$

which can be written in terms of the spherical basis vectors as

$$\frac{\mathrm{d}\hat{\mathbf{r}}}{\mathrm{d}t} = \dot{\phi}\sin\theta\hat{\boldsymbol{\phi}} + \dot{\theta}\hat{\boldsymbol{\theta}}\,. \tag{1.105}$$

When we take the second time derivative, we will also need the time derivatives of $\hat{\boldsymbol{\theta}}$ and $\hat{\boldsymbol{\phi}}$, which can be found using the same procedure. Thus, we additionally have

$$\begin{aligned}\frac{\mathrm{d}\hat{\boldsymbol{\theta}}}{\mathrm{d}t} &= -\dot{\theta}\hat{\mathbf{r}} + \dot{\phi}\cos\theta\hat{\boldsymbol{\phi}}\,, \\ \frac{\mathrm{d}\hat{\boldsymbol{\phi}}}{\mathrm{d}t} &= -\dot{\phi}\sin\theta\hat{\mathbf{r}} - \dot{\phi}\cos\theta\hat{\boldsymbol{\theta}}\,.\end{aligned} \tag{1.106}$$

**Exercise 1.17** We can also obtain the above expressions for the time derivatives of $\hat{\mathbf{r}}$, $\hat{\boldsymbol{\theta}}$, $\hat{\boldsymbol{\phi}}$ using directional derivatives, which are discussed in Appendix A.4.2. Note that for any basis vector $\hat{\mathbf{e}}_i$, its time derivative is given by

$$\frac{\mathrm{d}\hat{\mathbf{e}}_i}{\mathrm{d}t} = \frac{\partial \hat{\mathbf{e}}_i}{\partial r}\dot{r} + \frac{\partial \hat{\mathbf{e}}_i}{\partial \theta}\dot{\theta} + \frac{\partial \hat{\mathbf{e}}_i}{\partial \phi}\dot{\phi}\,, \tag{1.107}$$

where

$$\frac{\partial \hat{\mathbf{e}}_i}{\partial r} = \nabla_{\hat{\mathbf{r}}}\hat{\mathbf{e}}_i\,, \qquad \frac{\partial \hat{\mathbf{e}}_i}{\partial \theta} = r\nabla_{\hat{\boldsymbol{\theta}}}\hat{\mathbf{e}}_i\,, \qquad \frac{\partial \hat{\mathbf{e}}_i}{\partial \phi} = r\sin\theta\nabla_{\hat{\boldsymbol{\phi}}}\hat{\mathbf{e}}_i\,. \tag{1.108}$$

Use (A.49) from Example A.2 to recover (1.105) and (1.106).

Taking two time derivatives of $\mathbf{r} = \ell\hat{\mathbf{r}}$, we find that the acceleration is

$$\begin{aligned}\mathbf{a} = &-\ell\left(\dot{\theta}^2 + \dot{\phi}^2\sin^2\theta\right)\hat{\mathbf{r}} + \ell\left(\ddot{\theta} - \dot{\phi}^2\sin\theta\cos\theta\right)\hat{\boldsymbol{\theta}}\\ &+ \ell\left(\ddot{\phi}\sin\theta + 2\dot{\theta}\dot{\phi}\cos\theta\right)\hat{\boldsymbol{\phi}}\,.\end{aligned} \tag{1.109}$$

Newton's 2nd law then gives the three equations of motion

$$f_c + mg\cos\theta = -m\ell\left(\dot{\theta}^2 + \dot{\phi}^2\sin^2\theta\right)\,, \tag{1.110a}$$

$$-mg\sin\theta = m\ell\left(\ddot{\theta} - \dot{\phi}^2\sin\theta\cos\theta\right)\,, \tag{1.110b}$$

$$0 = m\ell\left(\ddot{\phi}\sin\theta + 2\dot{\theta}\dot{\phi}\cos\theta\right)\,. \tag{1.110c}$$

The first equation, (1.110a), gives us the constraint force

$$f_c = -mg\cos\theta - m\ell\left(\dot{\theta}^2 + \dot{\phi}^2\sin^2\theta\right)\,, \tag{1.111}$$

and can easily be seen to be the tension (or normal) force needed to counteract the weight plus the centripetal force needed to keep the particle moving in a circle with speed $v$, where $v^2 = \ell^2\left(\dot{\theta}^2 + \dot{\phi}^2\sin^2\theta\right)$. The second equation, (1.110b), gives us the 2nd-order differential equation governing $\theta$,

$$m\ell\ddot{\theta} = m\ell\dot{\phi}^2\sin\theta\cos\theta - mg\sin\theta\,. \tag{1.112}$$

The third equation, (1.110c), can be shown to be related to the conserved component of the angular momentum. First, note that we can multiply (1.110c) by $\ell\sin\theta$, for which the right-hand side becomes a total time derivative of $m\ell^2\dot{\phi}\sin^2\theta$, which then must be a conserved quantity. However, the net torque on this system is not zero, so we cannot say that the full angular momentum vector is conserved. The angular momentum is $\mathbf{L} = \mathbf{r}\times\mathbf{p}$, which can be expanded as

$$\mathbf{L} = m\ell^2 \left[ -\left(\dot{\theta}\sin\phi + \dot{\phi}\cos\phi\sin\theta\cos\theta\right)\hat{\mathbf{x}} \right.$$
$$\left. + \left(\dot{\theta}\cos\phi - \dot{\phi}\sin\phi\sin\theta\cos\theta\right)\hat{\mathbf{y}} + \dot{\phi}\sin^2\theta\hat{\mathbf{z}} \right] . \tag{1.113}$$

Since the net torque is $\boldsymbol{\tau} = \mathbf{r} \times mg\hat{\mathbf{z}}$, the $\hat{\mathbf{z}}$ component of the torque is zero. This means that $L_z = m\ell^2\dot{\phi}\sin^2\theta$ is a conserved quantity, as demonstrated by (1.110c). We can use this expression for $L_z$ to eliminate $\dot{\phi}$ from (1.112), leaving a 2nd-order differential equation for $\theta$ alone. Unfortunately, (1.112) is non-linear, so it can be solved exactly only for special cases. □

In the spherical pendulum example, the constraints were expressed as an equation relating the three degrees of freedom to the restriction that $r = \ell$. In spherical coordinates, this constraining equation is so simple that it is hard to see it as a non-trivial equation. In Cartesian coordinates, it is more obvious as $\ell^2 = x^2 + y^2 + z^2$. When combined with Newton's laws, the constraint equations give the unknown forces of constraint. In some situations, the constraints are valid only when the forces of constraint lie within a restricted range. Using similar techniques to determine the forces of constraint, we can then determine under which conditions the constraints hold and when the system is no longer constrained.

***Example 1.7*** Consider a skier going down a hemispherical hill. At first the skier is constrained to follow the surface of the hill. As the skier descends and speeds up, she will eventually leave the surface and begin to follow a ballistic trajectory. We want to determine at what angle the skier leaves the slope.

This problem is effectively two-dimensional and cylindrical coordinates are the most appropriate. We will measure the angle $\phi$ off of the (vertical) $y$-axis instead of the $x$-axis, but otherwise these are the standard cylindrical coordinates. The radius of the hill is $R$, so the constraint equation is $R^2 = x^2 + y^2$. The forces and coordinate choices are shown in Fig. 1.14. The constraint force is the normal force that the hill exerts on the skier. When the combined radial force on the skier drops to zero, then there is no centripetal force left to constrain the skier to the surface of the hill. In the cylindrical basis, the forces acting on the body are

$$\begin{aligned} \mathbf{F}_n &= F_n\hat{\boldsymbol{\rho}} , \\ m\mathbf{g} &= -mg\cos\phi\hat{\boldsymbol{\rho}} + mg\sin\phi\hat{\boldsymbol{\phi}} . \end{aligned} \tag{1.114}$$

Remembering that it is easiest to use the Cartesian basis when computing the acceleration, we find

$$\mathbf{a} = -R\dot{\phi}^2\hat{\boldsymbol{\rho}} + R\ddot{\phi}\hat{\boldsymbol{\phi}}. \tag{1.115}$$

Thus, the equations of motion are obtained from Newton's 2nd law, giving

$$F_n - mg\cos\phi = -mR\dot{\phi}^2 \tag{1.116a}$$
$$mg\sin\phi = mR\ddot{\phi} . \tag{1.116b}$$

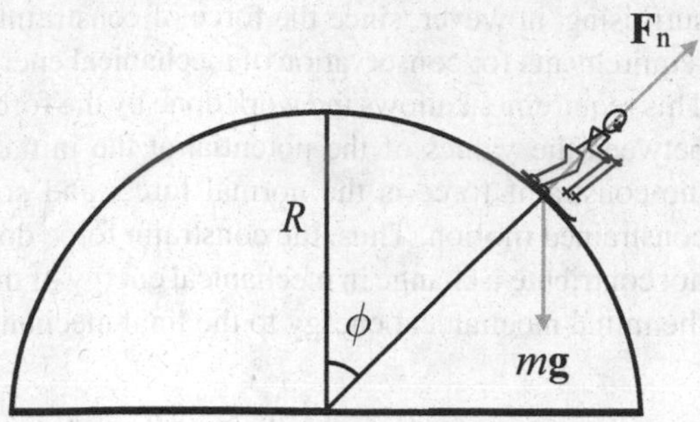


**Fig. 1.14** A skier going down a spherical hill feels a weight force pointing downward ($-\hat{\mathbf{y}}$ direction) and a normal force pointing along the $\hat{\rho}$ direction. The motion is constrained to the surface of the hill

Now, the skier leaves the surface when $F_n = 0$. If we knew $\dot{\phi}$, then we could solve (1.116a) for the value of $\phi$ at which the skier leaves the surface.

Equation (1.116b) provides the differential equation to solve for $\phi$. There is a nice trick for solving this differential equation that involves converting it to a 1st-order differential equation. First, we multiply both sides by $\dot{\phi}$, to find

$$\dot{\phi}\ddot{\phi} = \frac{g}{R}\dot{\phi}\sin\phi \,. \tag{1.117}$$

The left-hand side is the total time derivative of $\frac{1}{2}\dot{\phi}^2$, while the right-hand side is the total time derivative of $-(g/R)\cos\phi$. We can then integrate both sides to find

$$\frac{1}{2}\dot{\phi}^2 = -\frac{g}{R}\cos\phi + C \,, \tag{1.118}$$

where $C$ is an arbitrary constant that is fixed by the initial conditions. The skier starts from rest at $\phi = 0$, so $\dot{\phi} = 0$ when $\cos\phi = 1$. Thus, the arbitrary constant is $C = g/R$. We now have the solution for $\dot{\phi}^2$ that we need for (1.116a),

$$\dot{\phi}^2 = \frac{2g}{R}(1 - \cos\phi) \,. \tag{1.119}$$

Substituting this expression into (1.116a) and setting $F_n = 0$ gives the solution for the angle at which the skier leaves the surface,

$$\cos\phi = \frac{2}{3} \,. \tag{1.120}$$

□

In the previous example, the equation for the angular velocity (1.119) could also be obtained using conservation of mechanical energy, and that is frequently how this problem is solved in introductory physics courses. This should be somewhat surprising, however, since the force of constraint is not conservative and one of the requirements for conservation of mechanical energy is that all forces be conservative. This requirement allows the work done by the forces to be expressed as the difference between the values of the potential at the initial and final states. In this example, the constraint force is the normal force, and so it is always perpendicular to the constrained motion. Thus, the constraint force does no work on the system and does not contribute a change in mechanical energy of the system. Therefore, we can equate the initial mechanical energy to the final mechanical energy,

$$E_{\rm i} = mgR = mgR\cos\phi + \frac{1}{2}mR^2\dot{\phi}^2 = E_{\rm f}\,. \tag{1.121}$$

Solving for $\dot{\phi}^2$, we recover (1.119).

Once we have the equation for $\dot{\phi}^2$, we can also integrate it to find the time-dependent solution $\phi(t)$:

$$\int_0^\phi \frac{{\rm d}\phi}{A\sqrt{1-\cos\phi}} = \int_0^t {\rm d}t\,, \tag{1.122}$$

where $A \equiv \sqrt{2g/R}$. Although it is not trivial, this integral can be done (or looked up in a handbook of integrals), and we get

$$\sqrt{\frac{R}{g}}\left[\ln\left(\tan\left(\frac{\phi}{4}\right)\right) - \ln\left(\tan\left(0\right)\right)\right] = t\,. \tag{1.123}$$

This all looks fine until we try to solve this equation for $\phi(t)$ and notice that $\ln(\tan(0)) \to -\infty$. What does this mean? A direct interpretation is that it will take an infinite amount of time for the skier to reach any angle $\phi$. Upon further reflection, we see that the math is telling us something that we have been overlooking. Namely, the skier starts with an initial velocity of $v = R\dot{\phi} = 0$ at the top of the hill where the forces are in equilibrium, so the acceleration is zero. The skier isn't going anywhere until someone pushes her!

**Exercise 1.18** Redo the problem for the skier in Example 1.7, but allowing for a non-zero initial velocity $v_0$, and determine how the angle at which the skier leaves the slope depends on the initial velocity. What is the maximum value of $v_0$ allowed for the skier to be on the ground at the top of the hill?

You may have noticed that direct application of Newton's laws to constrained systems frequently involves a lot of algebra and the careful solution of multiple equations with multiple unknowns to obtain the equations of motion and the equations for the constraint forces. A great deal of mathematical machinery has been developed to streamline and simplify the analysis of constrained systems with general coordinate systems. We will explore these in the next chapter.

## Suggested References

*Full references are given in the bibliography at the end of the book.*

Fetter and Walecka (1980): Although more advanced, the first two chapters provide a thorough review of mechanics and non-inertial reference frames.
Marion and Thornton (1995): An excellent introductory text on classical mechanics, particularly suited for undergraduates.

## Additional Problems

**Problem 1.1** Extend the calculation of Exercise 1.17 to obtain the acceleration vector in spherical coordinates $(r, \theta, \phi)$ for *unconstrained* motion in three dimensions—that is, allowing the radial coordinate $r$ to also change with time.

**Problem 1.2** Calculate the acceleration vector for unconstrained motion in three dimensions in cylindrical coordinates $(\rho, \phi, z)$.

**Problem 1.3** Consider a simple planar pendulum consisting of a mass $m$ suspended from a (massless) string of length $\ell$ in a uniform gravitational field $\mathbf{g}$. (See Fig. 1.15.) Let $\mathbf{T}$ denote the tension in the string and $v_0$ denote the initial velocity of the pendulum bob—i.e., the tangential velocity at its lowest point $\theta = 0$.

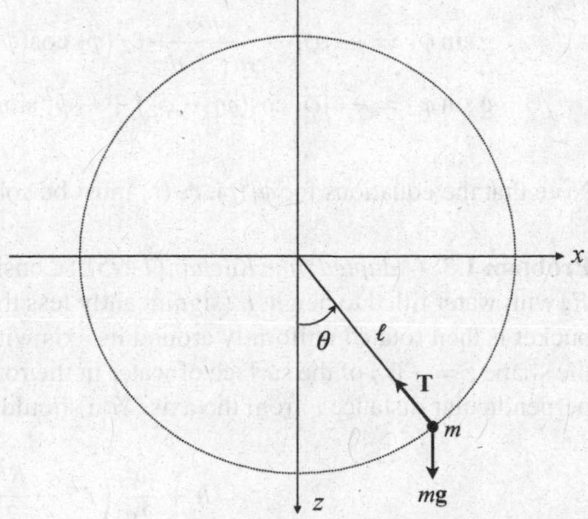


**Fig. 1.15** A pendulum bob of mass $m$ is suspended from a (massless) string of length $\ell$ in a uniform gravitational field $\mathbf{g}$. The gravitational force $m\mathbf{g}$ and the tension $\mathbf{T}$ exerted by the string are shown in the figure

(a) Obtain an equation for the tension $T$ as a function of $\theta$ and $\dot{\theta}$.
(b) Integrate the $\ddot{\theta}$ equation to obtain an equation relating $\theta$ to $\dot{\theta}$. This equation involves an integration constant that can solved for in terms of the initial velocity $v_0$. Interpret the equation in terms of the total energy of the pendulum.
(c) Determine the maximum value of $\theta$ having $\dot{\theta} = 0$ and $T \geq 0$.
(d) Determine the minimum initial velocity $v_0$ needed for the pendulum bob to make a complete loop-the-loop—i.e., to reach the top of the circle ($\theta = \pi$) with $T \geq 0$. What does $\dot{\theta}$ equal at the top of the circle for this minimum-initial-velocity case?

**Problem 1.4** The planar **double pendulum** consists of two point masses ($m_1$ and $m_2$) at the end of two massless rigid rods of lengths $\ell_1$ and $\ell_2$ as shown in Fig. 1.16. If we choose Cartesian coordinates with the $x$-axis pointing down and the $y$-axis pointing to the right, then the constraints can be incorporated into the positions of the particles with

$$\begin{aligned}\mathbf{r}_1 &= \ell_1 \cos\phi_1 \hat{\mathbf{x}} + \ell_1 \sin\phi_1 \hat{\mathbf{y}}\,,\\ \mathbf{r}_2 &= (\ell_1 \cos\phi_1 + \ell_2 \cos\phi_2)\,\hat{\mathbf{x}} + (\ell_1 \sin\phi_1 + \ell_2 \sin\phi_2)\,\hat{\mathbf{y}}\,,\end{aligned} \tag{1.124}$$

which reduces the number of degrees of freedom from four $(x_1, y_1, x_2, y_2)$ to two $(\phi_1, \phi_2)$.

(a) Apply Newton's 2nd law to each mass and show that the magnitudes of the constraint forces $\mathbf{T}_1$ and $\mathbf{T}_2$ obey

$$\begin{aligned}T_1 \sin\phi_1 &= -(m_1+m_2)\ell_1\left(\ddot{\phi}_1\cos\phi_1 - \dot{\phi}_1^2\sin\phi_1\right) - m_2\ell_2\left(\ddot{\phi}_2\cos\phi_2 - \dot{\phi}_2^2\sin\phi_2\right),\\ T_2 \sin\phi_2 &= -m_2\ell_1\left(\ddot{\phi}_1\cos\phi_1 - \dot{\phi}_1^2\sin\phi_1\right) - m_2\ell_2\left(\ddot{\phi}_2\cos\phi_2 - \dot{\phi}_2^2\sin\phi_2\right).\end{aligned} \tag{1.125}$$

(b) Use the result from part (a) to obtain the following equations of motion

$$\begin{aligned}g\sin\phi_1 &= -\ell_1\ddot{\phi}_1 - \frac{m_2}{m_1+m_2}\ell_2\left(\ddot{\phi}_2\cos(\phi_1-\phi_2) + \dot{\phi}_2^2\sin(\phi_1-\phi_2)\right),\\ g\sin\phi_2 &= -\ell_1\ddot{\phi}_1\cos(\phi_1-\phi_2) + \ell_1\dot{\phi}_1^2\sin(\phi_1-\phi_2) - \ell_2\ddot{\phi}_2\,.\end{aligned} \tag{1.126}$$

Note that the equations for $\phi_1(t)$, $\phi_2(t)$ must be solved numerically.

**Problem 1.5** (*Adapted from Kuchǎr (1995).*) Consider a cylindrical bucket of radius $R$, with water filled to height $h$ (significantly less than the height of the bucket). The bucket is then rotated uniformly around its axis with angular velocity $\omega$. Determine the shape $z = f(r)$ of the surface of water in the rotating bucket, as a function of the perpendicular distance $r$ from the axis. You should find

$$z = h + \frac{\omega^2}{2g}\left(r^2 - \frac{R^2}{2}\right). \tag{1.127}$$

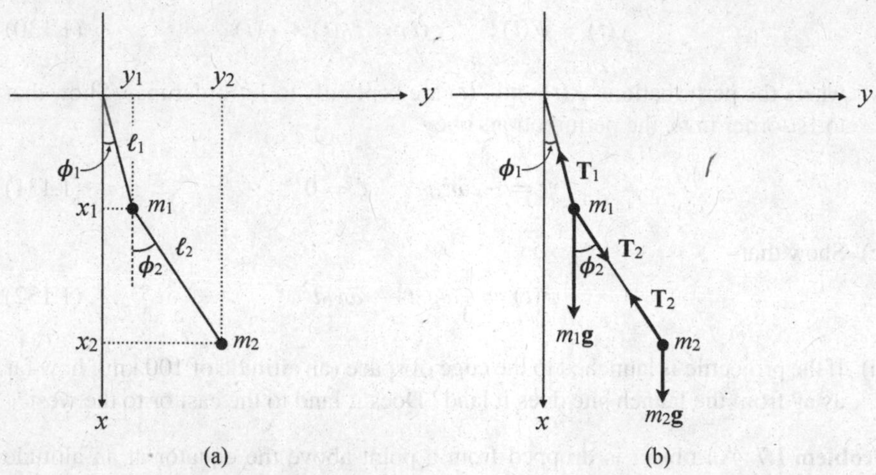


**Fig. 1.16** The double pendulum. Panel (a): The two masses, $m_1$ and $m_2$, are constrained by the two massless rigid rods. The four degrees of freedom can be reduced to two ($\phi_1$ and $\phi_2$) due to these constraints. Panel (b) shows the two constraint forces $\mathbf{T}_1$ and $\mathbf{T}_2$, and the gravitational forces $m_1\mathbf{g}$ and $m_2\mathbf{g}$

*Hint*: Minimize the sum of the gravitational and centrifugal potential energies of a cylindrical shell of water of mass $\mathrm{d}m = \rho 2\pi r z(r)\mathrm{d}r$ in the non-inertial reference frame rotating with the water, subject to the constraint

$$\int_0^R 2\pi r z(r)\mathrm{d}r = \pi R^2 h\,. \tag{1.128}$$

See Appendix C.8 if you need a refresher on variational problems subject to constraints.

**Problem 1.6** A projectile is launched vertically from the equator with an initial speed $v_0$. We want to find out where it will land, assuming that we can approximate the gravitational force as uniform, with $\mathbf{F} = m\mathbf{g}_0$.

(a) Starting with (1.80) and using the coordinates shown in Fig. 1.10 (with $\theta = \pi/2$), show that the equations of motion for the projectile are

$$\begin{aligned} \ddot{x} &= 0\,, \\ \ddot{y} &= \omega^2 y - 2\omega\dot{z}\,, \\ \ddot{z} &= (\omega^2 R - g_0) + \omega^2 z + 2\omega\dot{y}\,. \end{aligned} \tag{1.129}$$

(b) In the absence of the Earth's rotation, $\omega = 0$ and the unperturbed trajectory follows $\ddot{y}_0 = 0$ and $\ddot{z}_0 = -g_0$, so $y_0(t) = 0$ and $z_0(t) = v_0 t - \frac{1}{2}g_0 t^2$. Write the perturbed trajectory as

$$y(t) = \psi(t)\,, \qquad z(t) = z_0(t) + \zeta(t)\,, \tag{1.130}$$

where the perturbations $\psi(t)$ and $\zeta(t)$ are kept only to 1st-order in $\omega$. Show that to 1st-order in $\omega$, the perturbations obey

$$\ddot{\psi} = -2\omega\dot{z}_0\,, \qquad \ddot{\zeta} = 0\,. \tag{1.131}$$

(c) Show that

$$\psi(t) = \frac{1}{3}\omega g_0 t^3 - \omega v_0 t^2\,. \tag{1.132}$$

(d) If the projectile is launched to the edge of space (an altitude of 100 km), how far away from the launch site does it land? Does it land to the east or to the west?

**Problem 1.7** An object is dropped from a point above the equator at an altitude of 100 km. Following the procedure outlined in Problem 1.6, determine where the object lands relative to the point directly below the release point.

**Problem 1.8** Generalize the procedure outlined in Problem 1.6 for arbitrary co-latitude $\theta$. In so doing, define

$$\mathbf{g} \equiv \mathbf{g}_0 - \boldsymbol{\omega} \times (\boldsymbol{\omega} \times \mathbf{R})\,, \tag{1.133}$$

which points in the direction of a plumb line located at the origin of the non-inertial reference frame, and which defines the direction of the local vertical, i.e., $\hat{\mathbf{z}} \propto -\mathbf{g}$. For a particle launched in the $\hat{\mathbf{z}}$ direction to an altitude of $h$, show that:

(a) to 1st-order in $\omega$, the displacement of the projectile in the $y$-direction is

$$\Delta y = -\frac{8}{3}\sqrt{\frac{2h^3}{g}}\,\omega \sin\theta\,, \tag{1.134}$$

with the minus sign indicating that the projectile lands *west* of the launch site.

(b) to 2nd-order in $\omega$, the displacment of the projectile in the $x$-direction is

$$\Delta x = -\frac{8h^2}{g}\omega^2 \sin\theta\cos\theta\,, \tag{1.135}$$

with the minus sign indicating that the projectile lands *north* of the launch site in the Northern hemisphere and south of launch site in the Southern hemisphere.

(c) Calculate the displacements $\Delta x$ and $\Delta y$ for a projectile launched from 26° north latitude and that reaches an altitude of 100 km.

# Chapter 2
# Principle of Virtual Work and Lagrange's Equations

In the previous chapter, we showed several ways that Newton's laws can be used to obtain the equations of motion for systems of particles. We looked at the consequences of generalizing to non-inertial reference frames and at systems that were constrained to move in restricted spaces. Newton's laws are powerful tools, but they require a careful choice of coordinate bases and the manipulation of vectors to obtain the equations of motion. In this chapter, we introduce the principle of virtual work, which is the foundation for all other variational principles of mechanics. We discuss d'Alembert's principle, which is a simple application of the principle of virtual work to the dynamical equations of Newtonian mechanics, and then derive Lagrange's equations of the 1st and 2nd kind. We shall see that the Lagrangian formulation of mechanics allows us to analyze problems that are much harder to solve when approached with a direct application of Newton's laws.

## 2.1 Newtonian Approach to Constrained Systems

The equations of motion that arise from the application of Newton's laws to a system of particles describe trajectories $\mathbf{r}_I(t)$ satisfying

$$\dot{\mathbf{p}}_I = \mathbf{F}_I\,, \qquad I = 1, 2, \cdots, N\,, \tag{2.1}$$

where $\mathbf{F}_I$ represents the resultant force acting on particle $I$ and $\mathbf{p}_I \equiv m\dot{\mathbf{r}}_I$ is the momentum. If there are constraints on the motion of the system, then included in the resultant force $\mathbf{F}_I$ are forces $\mathbf{F}_I^{(c)}$ that constrain the particle positions to remain within restricted surfaces in the configuration space. If we separate out the constraint force contribution to the resultant forces, Newton's 2nd law becomes



M.J. Benacquista and J.D. Romano, *Classical Mechanics*, Undergraduate Lecture Notes in Physics, https://doi.org/10.1007/978-3-319-68780-3_2

$$\dot{\mathbf{p}}_I = \mathbf{F}_I^{(a)} + \mathbf{F}_I^{(c)}, \qquad I = 1, 2, \cdots, N, \tag{2.2}$$

where $\mathbf{F}_I^{(a)}$ is the net *applied* (or *impressed*) force, which is the sum of all the specified external and interparticle forces acting on particle $I$. The constraint forces are *unspecified* forces that react (i.e., adjust themselves) to the motion of the particles so as to keep the system on the constraint surface.

The Newtonian approach to constrained systems consists of working with the vector equations of motion (2.2) for the individual particles, which typically requires solving for the constraint forces $\mathbf{F}^{(c)}$ in order to determine the particle trajectories $\mathbf{r}_I(t)$. Often times, these equations of motion can be simplified by working in terms of coordinates on the configuration space that are well-adapted to the constraints imposed on the system. In what follows, we will denote the full $3N$-dimensional configuration space of the system by $\mathscr{C}$, and arbitrary coordinates on this space by $x^\alpha$, where $\alpha = 1, 2, \cdots, 3N$. (Note that we can always choose Cartesian coordinates $x_I, y_I, z_I$, where $I = 1, 2, \cdots, N$, on the configuration space, but this is not always the best choice.) Constraints on the system can then be described by $M$ $(< 3N)$ kinematical conditions, which either

(i) relate the coordinates on configuration space to one another (so-called **holonomic constraints**), or
(ii) can only be expressed in *infinitesimal* form via relations between the coordinate *differentials* that cannot be integrated to yield relations between the coordinates themselves (so-called **non-holonomic constraints**).[1]

These two types of constraints will be described in more detail in Sects. 2.2.1 and 2.2.2. Holonomic constraints have the advantage that it is possible to work entirely in terms of a *reduced* set of coordinates $q^a$, where $a = 1, 2, \cdots n \equiv (3N - M)$, which represent the *independent* degrees of freedom of the system. The coordinates $q^a$ are called **generalized coordinates** for the system and span a $(3N - M)$-dimensional subspace $Q \subset \mathscr{C}$.

***Example 2.1*** For the spherical pendulum problem discussed in Example 1.6, the configuration space coordinates were the three Cartesian coordinates $x$, $y$, $z$, of the pendulum bob. The constraint equation was

$$\varphi(x, y, z) = \ell^2 - x^2 - y^2 - z^2 = 0, \tag{2.3}$$

which defined the surface of a sphere of radius $\ell$. The two generalized coordinates were $\theta$ and $\phi$, which are coordinates on the surface of that sphere and are related to $x$, $y$, $z$ via

[1]Our definition of non-holonomic constraints follows the convention used by Hertz (2004) and by Lanczos (1949). Other authors, e.g. Fetter and Walecka (1980) and Flannery (2005), use the term *non-integrable* for these constraints, and reserve non-holonomic for the more general class of constraints that may not even be expressible in analytic form.

$$x = \ell \sin\theta \cos\phi\,, \qquad y = \ell \sin\theta \sin\phi\,, \qquad z = \ell \cos\theta\,. \tag{2.4}$$

The force equations included the forces of constraint, as in (2.2), and the equations of motion were found from these equations. The number of equations remained the same at $3N = 3$, allowing us to solve for the $(3N - M) = 2$ equations of motion for the generalized coordinates plus the $M = 1$ constraint force. □

**Exercise 2.1** What are the coordinates on the configuration space $\mathscr{C}$ and constraint surface $Q$ for the planar double pendulum, described in Problem 1.4.

**Exercise 2.2** (*Adapted from Kuchăr (1995).*) A thin bar of length $\ell$ moves in a plane such that its two endpoints are always in contact with a hoop of radius $r$ (with $2r > \ell$). (a) Write down the constraints on the Cartesian coordinates of each endpoint $(x_1, y_1)$ and $(x_2, y_2)$. (b) Write down the corresponding constraints on the Cartesian velocity components. (c) Find a generalized coordinate for the system.

## 2.2 Types of Constraints

As mentioned above, constraints on a system of particles can be described kinematically in terms of either conditions on the coordinates or non-integrable conditions on the coordinate differentials in the configuration space. Here, we look in more detail at the properties of these *holonomic* and *non-holonomic* constraints.

### 2.2.1 Holonomic Constraints

Holonomic constraints are the simplest type of constraints. They are relations between the coordinates $x^\alpha$ on the configuration space $\mathscr{C}$, which, in general, can also depend on time,

$$\varphi^A(x^1, x^2, \cdots, x^{3N}, t) = 0\,, \qquad A = 1, 2, \cdots, M\,, \tag{2.5}$$

with $M < 3N$. The intersection of the sets of coordinate values satisfying $\varphi^A = 0$, for $A = 1, 2, \cdots, M$, defines the **constraint surface** $Q$. It is an $n \equiv (3N - M)$-dimensional subspace of $\mathscr{C}$, which can be parameterized by generalized coordinates $q^a$, where $a = 1, 2, \cdots, n$. The parametric equations

$$\mathbf{r}_I = \mathbf{r}_I(q^1, q^2, \cdots, q^n, t), \qquad I = 1, 2, \cdots, N, \tag{2.6}$$

or, equivalently,

$$x^\alpha = x^\alpha(q^1, q^2, \cdots, q^n, t), \qquad \alpha = 1, 2, \cdots, 3N, \tag{2.7}$$

can be thought of as an **embedding** of the $n$-dimensional subspace $Q$ in the $3N$-dimensional configuration space $\mathscr{C}$.

**Exercise 2.3** Write down the constraint functions and embedding equations for the planar double pendulum described in Problem 1.4.

### 2.2.2 *Non-holonomic Constraints*

Non-holonomic constraints are relations between the coordinate *differentials*

$$\sum_\alpha C^A_\alpha \mathrm{d}x^\alpha = 0, \qquad A = 1, 2, \cdots, M, \tag{2.8}$$

that *cannot be integrated* to yield relations between the coordinates themselves. As we shall discuss in more detail in the following subsection, this means that the constraints cannot be mapped by an invertible transformation to a set of total differentials $\mathrm{d}\varphi^A$ of holonomic constraints $\varphi^A = 0$. Thus, non-holonomic constraints are fundamentally *infinitesimal* in form and cannot be associated with any $(3N - M)$-dimensional subspace $Q$ parametrized by generalized coordinates $q^a$. In addition, as we shall see in Sect. 3.2.2, the equations of motion for systems subject to non-holonomic constraints cannot be obtained by varying an action functional, as is the case for holonomic constraints.

***Example 2.2*** The classical example of a system subject to non-holonomic constraints is a sphere that rolls without slipping or pivoting on a horizontal two-dimensional surface, e.g., Lanczos (1949). Since the sphere is constrained to the surface, the number of degrees of freedom is reduced from six (three for position and three for orientation) to five, as the $z$-component of the center-of-mass of the sphere has a fixed value. The remaining five coordinates are the $x$ and $y$ coordinates of the center-of-mass of the sphere, and three angular coordinates needed to spec-

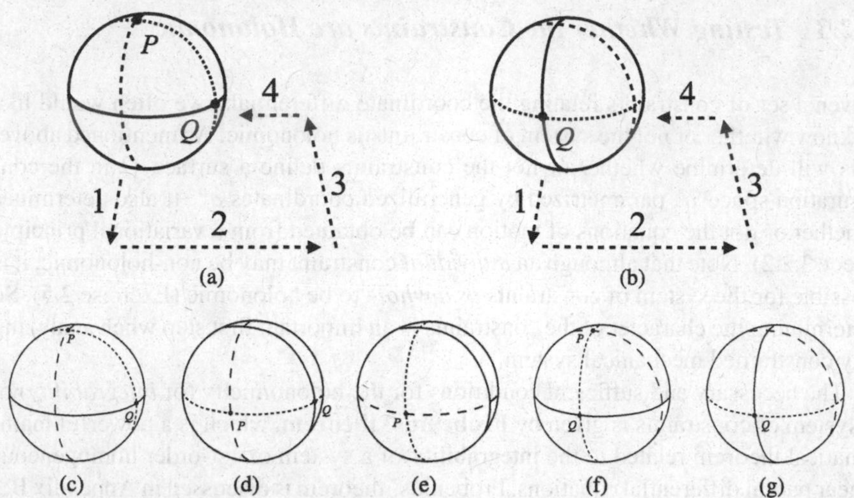


**Fig. 2.1** Demonstration of the non-holonomic nature of the constraints on a sphere that rolls without slipping or pivoting on a horizontal surface. Panels (a) and (b) show the initial and final configurations of the sphere after rolling along the counter-clockwise square path indicated by the dotted arrows 1–4. The sphere rolls through an angle of $\pi/2$ for each side of the square. Panels (c)–(g) show the configuration of the sphere at various points along the path. (c): initial configuration (before rolling); (d): after rolling along side 1; (e): after rolling along side 2; (f): after rolling along side 3; (g): after rolling along side 4 (which is the final configuration). Since the initial and final configurations of the sphere are different, but the $(x, y)$ coordinates of the center-of-mass of the sphere are the same, the rolling-without-slipping-or-pivoting constraints are non-integrable and hence non-holonomic

ify the rotational orientation of the sphere. Pure rolling without slipping or pivoting further reduces the number of degrees of freedom from five to two, as changes in the rotational orientation of the sphere are related to changes in the location of the center-of-mass of the sphere. There are relations between the coordinate differentials for both the angular coordinates and the center of mass, but they cannot be integrated to yield relations between the coordinates themselves (See Problem 6.1).

This non-integrability can also be demonstrated physically by having the sphere roll around a closed curve in the $xy$-plane, as shown in Fig. 2.1. The initial and final $x$ and $y$ coordinates of the sphere will, of course, be the same. But the initial and final rotational orientations of the sphere will, in general, be *different* from one another. Hence the coordinates describing the rotational orientation of the sphere are not uniquely determined by $x$ and $y$, and so cannot be written as functions of the $x$ and $y$ components of the center-of-mass. So the constraints are non-holonomic. □

### 2.2.3 Testing Whether the Constraints are Holonomic

Given a set of constraints relating the coordinate differentials, we often would like to know whether or not the system of constraints is holonomic. As mentioned above, this will determine whether or not the constraints define a surface $Q$ in the configuration space $\mathscr{C}$, parametrized by generalized coordinates $q^a$. It also determines whether or not the equations of motion can be obtained from a variational principle (Sect. 3.2.2). Note that although an *individual* constraint may be non-holonomic, it is possible for the system of constraints *as a whole* to be holonomic (Exercise 2.5). So determining the character of the constraints is an important first step when analyzing any constrained mechanical system.

The necessary and sufficient conditions for the holonomicity (or *integrability*) of a system of constraints is given by **Frobenius' theorem**, which is a powerful mathematical theorem related to the integrability of a system of 1st-order homogeneous linear partial differential equations. Frobenius' theorem is discussed in Appendix B.3 in the context of *differential forms*. Here we cast the key statements of Frobenius' theorem in terms of *anti-symmetric* combinations of partial derivatives. To simplify the notation a bit in this section, we will drop any explicit time dependence of the functions (for reasons that will become more clear in Sect. 2.3) and write the coordinate differentials as $\mathrm{d}x^\alpha$, where for this more general discussion we take $\alpha = 1, 2, \cdots, n$.

So let's start with just a single constraint on the coordinate differentials, which we will write as

$$\sum_{\alpha=1}^{n} C_\alpha \mathrm{d}x^\alpha = 0\,, \tag{2.9}$$

where $C_\alpha \equiv C_\alpha(x^1, x^2, \cdots, x^n)$. As you may recall from a math methods class on differential equations, this equation is *integrable* if and only if there exists a function $\mu \equiv \mu(x^1, x^2, \cdots, x^n)$, called an **integrating factor**, for which $\mu$ times the constraint is a total differential

$$\mathrm{d}\varphi = \mu \sum_\alpha C_\alpha \, \mathrm{d}x^\alpha\,. \tag{2.10}$$

The necessary and sufficient condition for this to hold (at least locally) is

$$\partial_\alpha(\mu C_\beta) - \partial_\beta(\mu C_\alpha) = 0\,, \tag{2.11}$$

since partial derivatives of $\varphi$ commute. (We are using the shorthand notation $\partial_\alpha \equiv \partial/\partial x^\alpha$.) By expanding the derivatives we get

$$\partial_\alpha C_\beta - \partial_\beta C_\alpha = \mu^{-1}\left(C_\alpha \partial_\beta \mu - C_\beta \partial_\alpha \mu\right)\,, \tag{2.12}$$

which can be thought of as a partial differential equation for $\mu$. But rather than try to solve this equation for $\mu$ (which is often hard to do), we can eliminate the $\mu$ terms altogether. First we multiply (2.12) by $C_\gamma$ and then add together the *cyclic permutations* of this equation with respect to the three indices ($\alpha\beta\gamma \to \beta\gamma\alpha \to \gamma\alpha\beta$). The right-hand side then automatically equals zero and the left hand-side can be written as a completely anti-symmetric sum over the three indices using the $n$-dimensional Levi-Civita symbol[2] $\varepsilon_{\alpha\beta\gamma\cdots\delta}$. Thus, we have shown that

$$\sum_{\alpha,\beta,\gamma} \varepsilon_{\alpha\beta\gamma\cdots\delta}(\partial_\alpha C_\beta)C_\gamma = 0 \tag{2.13}$$

are *necessary* conditions for the constraint (2.9) to be integrable.[3] That they are also *sufficient* conditions is the content of **Frobenius' theorem** (which we will not prove here; see e.g., Flanders (1963)). Frobenius' theorem thus tells us that (2.13) are the necessary and sufficient conditions for integrability of the constraint equation (2.9).

**Exercise 2.4** Check if the following constraints in 3-dimensions are individually integrable or not:

$$yz\,\mathrm{d}x + zx\,\mathrm{d}y + xy\,\mathrm{d}z = 0\,, \tag{2.14a}$$
$$-y\,\mathrm{d}x + x\,\mathrm{d}y + \mathrm{d}z = 0\,. \tag{2.14b}$$

Frobenius's theorem can also be extended to a set of constraints

$$\sum_\alpha C^A_\alpha\,\mathrm{d}x^\alpha = 0\,, \qquad A = 1, 2, \cdots, M\,, \tag{2.15}$$

where $M < n$ and $C^A_\alpha \equiv C^A_\alpha(x^1, x^2, \cdots, x^n)$. This set of constraints is *integrable* and thus defines an $(n - M)$-dimensional subspace if and only if there exists an invertible transformation $\mu_{AB} \equiv \mu_{AB}(x^1, x^2, \cdots, x^n)$ that maps the original set of constraints (2.15) to a set of $M$ total differentials,

$$\mathrm{d}\varphi^A = \sum_B \mu_{AB} \sum_\alpha C^B_\alpha\,\mathrm{d}x^\alpha\,, \qquad A = 1, 2, \cdots, M\,. \tag{2.16}$$

[2]The 3-dimensional Levi-Civita symbol is defined in (A.7).

[3]In two dimensions, (2.13) is automatically satisfied, since anti-symmetrizing over three indices in a two-dimensional space identically gives zero. Said another way, there is no non-zero 3-index Levi-Civita symbol $\varepsilon_{\alpha\beta\gamma}$ in two-dimensions. The practical consequence of this is that *any* differential constraint in two-dimensions is integrable (Exercise B.6).

Expanding $\mathrm{d}\varphi^A = \sum_\alpha (\partial_\alpha \varphi^A)\, \mathrm{d}x^\alpha$, and then inverting the equation to solve for the $C^A_\alpha$, we obtain

$$C^A_\alpha = \sum_B \left(\mu^{-1}\right)_{AB} \partial_\alpha \varphi^B . \tag{2.17}$$

Frobenius' theorem then says that the necessary and sufficient conditions for this to be the case is for

$$\sum_{\alpha,\beta,\gamma,\delta,\cdots,\mu} \varepsilon_{\alpha\beta\gamma\delta\cdots\mu\cdots\nu} (\partial_\alpha C^A_\beta) C^1_\gamma C^2_\delta \cdots C^M_\mu = 0 , \quad A = 1, 2, \cdots, M . \tag{2.18}$$

For a single constraint ($M = 1$),, we recover (2.13). As before, (2.18) will tell you whether or not the system of constraints is integrable (and hence holonomic), without having to find the integrating factor $\mu_{AB}$.

**Exercise 2.5** (*Adapted from Kuchǎr (1995).*)

(a) Show that the following constraints on the coordinate differentials

$$\begin{aligned} C^1 &\equiv (x^2 + y^2)\, \mathrm{d}x + xz\, \mathrm{d}z = 0 , \\ C^2 &\equiv (x^2 + y^2)\, \mathrm{d}y + yz\, \mathrm{d}z = 0 , \end{aligned} \tag{2.19}$$

are individually non-holonomic, but that the system of constraints *as a whole* is holonomic.

(b) Since the system as a whole is holonomic, it should be possible to find two functions that directly constrain the coordinates

$$\varphi^1(x, y, z) = 0 , \qquad \varphi^2(x, y, z) = 0 . \tag{2.20}$$

Combine the differential constraints $C^1 = 0$, $C^2 = 0$ to obtain two differential equations that can be integrated to yield (2.20). By doing so you will have found an integrating factor for the original system of constraints. (The functions that you use to combine the constraints are just the components of $\mu_{AB}$ in (2.16).) Note that the functions $\varphi^1$, $\varphi^2$ are not unique, since any function of $\varphi^1$ and $\varphi^2$ times any function of the coordinates will also equal zero. (*Hint*: For this part, it is simplest to first convert to cylindrical coordinates.)

## 2.3 Principle of Virtual Work

An alternative to the Newtonian (individual-particle) approach to mechanics is to treat the system of particles *as a whole*. Different configurations of the system correspond to different points in the $3N$-dimensional configuration space $\mathscr{C}$, and the solution to the equations of motion (subject to suitable boundary conditions) corresponds to a particular *trajectory* in this space. Thus, the dynamics of the entire system is described entirely by a single curve in configuration space.

A key concept in this alternative formulation of classical mechanics is that of a **virtual displacement**. A virtual displacement is defined to be a *fixed-time* displacement of the system that is *consistent* with any constraints on the system. It differs, in general, from an *actual* displacement of the system, which takes place over time. This is illustrated most vividly in the context of a *time-dependent* constraint, such as that for a plane pendulum whose length changes with time $\ell = \ell(t)$, as shown in Fig. 2.2. We will denote a virtual displacement by either $\delta\mathbf{r}_I$ or $\delta x^\alpha$, where $\mathbf{r}_I, I = 1, 2, \cdots, N$ are the position vectors of the $N$ particles, and $x^\alpha, \alpha = 1, 2, \cdots, 3N$ are any set of coordinates on the configuration space $\mathscr{C}$. The explicit relationship between these two representations is

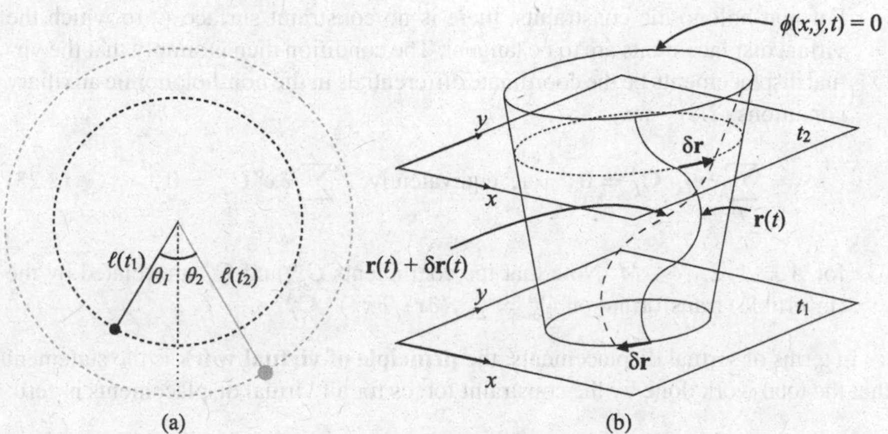


**Fig. 2.2** Panel (a): A pendulum constrained to lie in the $xy$-plane with time-varying length $\ell(t)$. This is a 2-dimensional example and the constraint is $\varphi(x, y, t) = x^2 + y^2 - \ell(t)^2 = 0$, so the surface of constraint is a circle with increasing radius. Panel (b): The 2-dimensional configuration space extended to three dimensions to explicitly show the time dependence of the constraint. The surface of constraint is now a cone with the time axis running along the axis of the cone. Virtual displacements $\delta\mathbf{r}$ lie in a (fixed-time) configuration plane and are tangent to the surface of the cone. The trajectory of the particle is a path $\mathbf{r}(t)$ that runs up the side of the cone

$$\delta \mathbf{r}_I = \sum_\alpha \frac{\partial \mathbf{r}_I}{\partial x^\alpha} \delta x^\alpha , \qquad I = 1, 2, \cdots, N , \tag{2.21}$$

where $\mathbf{r}_I \equiv \mathbf{r}_I(x^1, x^2, \cdots, x^{3N})$.

The statement that a virtual displacement is "consistent with the constraints" means that:

(i) For holonomic constraints, the virtual displacements are *tangent* to the constraint surface $Q$ defined by the intersection of the solutions of the constraint equations $\varphi^A = 0$, for $A = 1, 2, \cdots, M$. Mathematically,

$$\sum_I \delta \mathbf{r}_I \cdot \nabla_I \varphi^A = 0 \quad \text{or, equivalently,} \quad \sum_\alpha \delta x^\alpha \frac{\partial \varphi^A}{\partial x^\alpha} = 0 , \tag{2.22}$$

for $A = 1, 2, \cdots, M$. This result follows from the fact that each constraint equation $\varphi^A = 0$ is a surface of constant $\varphi^A$, and the gradient of $\varphi^A$ in configuration space points in a direction *normal* to these surfaces. Since the virtual displacements are tangent to these surfaces, the dot product of a virtual displacement and the gradient of $\varphi^A$ is zero.

(ii) For non-holonomic constraints, there is no constraint surface $Q$ to which the virtual displacements are to be tangent. The condition then is simply that the virtual displacements be the coordinate differentials in the non-holonomic auxiliary conditions (2.8)—i.e.,

$$\sum_I \delta \mathbf{r}_I \cdot \mathbf{C}_I^A = 0 , \quad \text{or, equivalently,} \quad \sum_\alpha \delta x^\alpha C_\alpha^A = 0 , \tag{2.23}$$

for $A = 1, 2, \cdots, M$. Note that the coefficients $C_\alpha^A$ and $\mathbf{C}_I^A$ are related by the (invertible) transformation $C_\alpha^A = \sum_I (\partial \mathbf{r}_I / \partial x^\alpha) \cdot \mathbf{C}_I^A$.

In terms of virtual displacements, the **principle of virtual work** is the statement that the total work done by the constraint forces for all virtual displacements is zero:

If $\delta \mathbf{r}_I$ or $\delta x^\alpha$ is a virtual displacement, i.e., (2.22) or (2.23), then

$$\sum_I \delta \mathbf{r}_I \cdot \mathbf{F}_I^{(c)} = 0 \quad \text{or, equivalently,} \quad \sum_\alpha \delta x^\alpha F_\alpha^{(c)} = 0 . \tag{2.24}$$

In what follows, we will use the terminology **virtual work** to denote the work done by a force along a virtual displacement. The principle of virtual work is then just the statement that *the total virtual work done by the constraint forces is zero.*

**Exercise 2.6** Use the principle of virtual work to derive the standard equations of **static equilibrium** for a rigid body,

$$\sum_I \mathbf{F}_I^{(a)} = 0\,, \quad \sum_I \mathbf{r}_I \times \mathbf{F}_I^{(a)} = 0\,, \tag{2.25}$$

i.e., the sum of the applied forces and the sum of the applied torques must be zero. *Hint*: Treat the rigid body as a collection of mass points $m_I$, where $I = 1, 2, \cdots, N$, held together by strong, centrally-directed, interparticle forces, which constrain the distances $d_{IJ} \equiv |\mathbf{r}_I - \mathbf{r}_J|$ between pairs of particles to have fixed values. Then the only allowed virtual displacements are *translations* and rigid *rotations*,

$$\begin{aligned} \delta\mathbf{r}_I &= \delta\mathbf{C} \qquad &\text{(translations)}\,, \\ \delta\mathbf{r}_I &= \delta\boldsymbol{\omega} \times \mathbf{r}_I \qquad &\text{(rotations)}\,, \end{aligned} \tag{2.26}$$

where $\delta\mathbf{C}$ and $\delta\boldsymbol{\omega}$ are (infinitesimal) *constant* vectors. By using the above form of $\delta\mathbf{r}_I$, the force equation $\mathbf{F}_I^{(a)} + \mathbf{F}_I^{(c)} = 0$, and the principle of virtual work (2.24), you should end up with (2.25).

### 2.3.1 *A New Principle of Mechanics*

It is important to point out that the principle of virtual work is a *new principle* of mechanics and not a theorem, since (2.24) is *not* derivable from Newton's laws, as discussed in Lanczos (1949). Alternatively, one can think of the principle of virtual work as the *definition* of an *ideal* constraint force. Then what makes the principle of virtual work useful is that the majority of *real* constraint forces that we encounter in mechanics (e.g., macroscopic forces exerted by rigid rods and inextendible strings, normal support forces exerted by smooth surfaces, interparticle central forces that hold the mass points together in a rigid body, etc.) are good approximations to the ideal constraint forces that do satisfy this principle. (One main exception is sliding constraint forces.) In addition, as we shall see later in this chapter and the next, the principle of virtual work is the foundation on which all the other variational principles of mechanics are built, e.g., d'Alembert's principle (Sect. 2.5) and Hamilton's principle of stationary action (Sect. 3.1).

## 2.4 Method of Lagrange Multipliers

You may have noticed that the principle of virtual work (2.24) and the conditions for a virtual displacement, (2.22) or (2.23), suggest that the constraint forces are simply related to the gradient of the functions $\varphi^A$ for holonomic constraints, or to the coefficients $C^A_\alpha$ for non-holonomic constraints. Indeed, for holonomic constraints, it is easy to see that if

$$\mathbf{F}^{(c)}_I = \sum_A \lambda_A \nabla_I \varphi^A \quad \text{or} \quad F^{(c)}_\alpha = \sum_A \lambda_A \frac{\partial \varphi^A}{\partial x^\alpha}, \tag{2.27}$$

then the principle of virtual work (2.24) is satisfied. Similarly, for non-holonomic constraints, if

$$\mathbf{F}^{(c)}_I = \sum_A \lambda_A \mathbf{C}^A_I \quad \text{or} \quad F^{(c)}_\alpha = \sum_A \lambda_A C^A_\alpha, \tag{2.28}$$

then the principle of virtual work is satisfied. This means that (2.27) and (2.28) are *sufficient* conditions for the constraint forces to do no total virtual work (i.e., if (2.27) or (2.28) hold, then (2.24) is automatically satisfied.) What's not so easy to see—but is true nonetheless—is that (2.27) and (2.28) are also *necessary* conditions for constraint forces to do no total virtual work (i.e., if the principle of virtual work (2.24) holds, then (2.27) or (2.28) must also be true). This follows using the **method of Lagrange multipliers**, which we demonstrate in the following proof.

*Proof* Let's work with the coordinate version of (2.24):

$$\sum_{\alpha=1}^{3N} \delta x^\alpha F^{(c)}_\alpha = 0, \tag{2.29}$$

and the virtual displacement condition for non-holonomic constraints,[4]

$$\sum_{\alpha=1}^{3N} \delta x^\alpha C^A_\alpha = 0, \quad A = 1, 2, \cdots, M. \tag{2.30}$$

Now (2.30) implies that the $3N$ components of the virtual displacements $\delta x^\alpha$ are not all independent, but are related by these $M$ equations. Without loss of generality, let us assume that we can solve (2.30) for the last $M$ components of $\delta x^\alpha$ (i.e., $\delta x^\beta$ for $\beta = n+1, n+2, \cdots, n+M$, where $n \equiv (3N - M)$) in terms of the other

[4]The proof for holonomic constraints would be exactly the same; one simply replaces $C^A_\alpha$ with the partial derivatives $\partial\varphi^A/\partial x^\alpha$.

components of $\delta x^\alpha$ (i.e., $\delta x^a$ for $a = 1, 2, \cdots, n$), which are now *independent* of one another. So given (2.29) and (2.30), it follows that

$$\sum_{\alpha=1}^{3N} \delta x^\alpha \left( F_\alpha^{(c)} - \sum_A \lambda_A C_\alpha^A \right) = 0\,, \tag{2.31}$$

where the summation over $A$ doesn't change the vanishing of the summation over $\alpha$. The $\lambda_A$ are $M$ *undetermined* coefficients, called **Lagrange multipliers**. Note that we can expand the summation as

$$\sum_{a=1}^{n} \delta x^a \left( F_a^{(c)} - \sum_A \lambda_A C_a^A \right) + \sum_{\beta=n+1}^{n+M} \delta x^\beta \left( F_\beta^{(c)} - \sum_A \lambda_A C_\beta^A \right) = 0\,. \tag{2.32}$$

Now using the freedom in the choice of these $M$ multipliers, we can set the coefficients multiplying the last $M$ components of $\delta x^\beta$ equal to zero—i.e.,

$$F_\beta^{(c)} - \sum_A \lambda_A C_\beta^A = 0\,, \qquad \beta = n+1, n+2, \cdots, n+M\,. \tag{2.33}$$

This reduces (2.32) to

$$\sum_{a=1}^{n} \delta x^a \left( F_a^{(c)} - \sum_A \lambda_A C_a^A \right) = 0\,. \tag{2.34}$$

But since the components $\delta x^a$, $a = 1, 2, \cdots, n$, that appear in this summation are independent of one another, it follows that

$$F_a^{(c)} - \sum_A \lambda_A C_a^A = 0\,, \qquad a = 1, 2, \cdots, n\,. \tag{2.35}$$

Together, (2.33) and (2.35) give us the desired result

$$F_\alpha^{(c)} - \sum_A \lambda_A C_\alpha^A = 0\,, \qquad \alpha = 1, 2, \cdots, 3N\,, \tag{2.36}$$

which is a simple rewrite of (2.28). □

**Exercise 2.7** Find the maximum area of a rectangle inscribed in the ellipse $(x/a)^2 + (y/b)^2 = 1$ using two different methods.

(a) Use the method of Lagrange multipliers to solve this problem by maximizing the area function

$$A(x, y) \equiv 4xy \tag{2.37}$$

subject to the constraint that a corner of the rectangle lies on the ellipse

$$\varphi(x, y) \equiv \left(\frac{x}{a}\right)^2 + \left(\frac{y}{b}\right)^2 - 1 = 0\,, \tag{2.38}$$

as shown in Fig. 2.3. What values do you find for the Lagrange multiplier $\lambda$ and the area of the rectangle?

(b) Repeat the above maximization problem, but this time first solve the constraint equation for $y$ in terms of $x$,

$$y = y(x) \equiv b\left[1 - \left(\frac{x}{a}\right)^2\right]^{1/2}\,, \tag{2.39}$$

and then maximize the function

$$\bar{A}(x) = A(x, y)|_{y=y(x)}\,. \tag{2.40}$$

Do you get the same result for the area of the rectangle?

## 2.5 D'Alembert's Principle

D'Alembert's principle is simply an application of the principle of virtual work to the dynamical equations of Newtonian mechanics (2.2) written in the form

$$\dot{\mathbf{p}}_I - \mathbf{F}_I^{(a)} - \mathbf{F}_I^{(c)} = 0\,. \tag{2.41}$$

Since (2.24) shows that the total virtual work of the forces of constraint is zero, it follows that $\sum_I \delta\mathbf{r}_I \cdot (\dot{\mathbf{p}}_I - \mathbf{F}_I^{(a)}) = 0$ for all virtual displacements. Thus, we have the implication

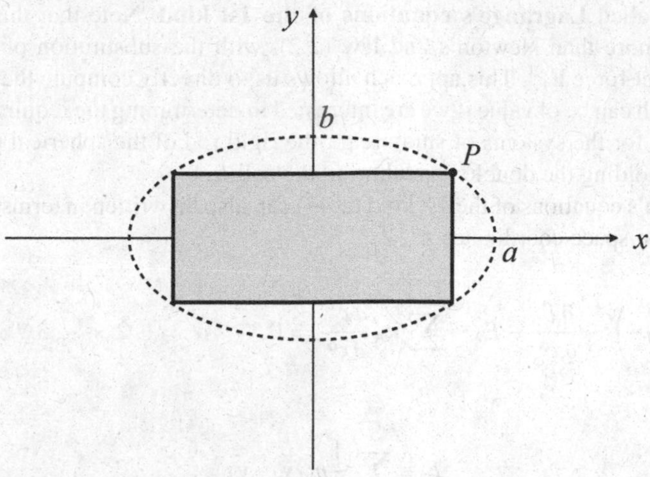


**Fig. 2.3** Rectangle inscribed in an ellipse. The point $P$ defines the size of the rectangle

If $\delta \mathbf{r}_I$ is a virtual displacement, i.e., (2.22) or (2.23), then

$$\sum_I \delta \mathbf{r}_I \cdot \left( \dot{\mathbf{p}}_I - \mathbf{F}_I^{(a)} \right) = 0 . \tag{2.42}$$

This statement constitutes **d'Alembert's principle**. The beauty of d'Alembert's principle is that the constraint *forces* $\mathbf{F}_I^{(c)}$ have dropped out of the problem. The constraints are present only in the form of the kinematical conditions for virtual displacements, (2.22) or (2.23).

## 2.6 Lagrange's Equations of the 1st Kind

Given d'Alembert's principle, we can obtain the equations of motion for a system of particles in two different ways, depending in part on the form of the constraints. The first approach is to apply the method of Lagrange multipliers to (2.42) following the same procedure that we used for the principle of virtual work. The result, for holonomic constraints, is[5]

$$\dot{\mathbf{p}}_I - \mathbf{F}_I^{(a)} - \sum_A \lambda_A \nabla_I \varphi^A = 0 , \qquad I = 1, 2, \cdots , N , \tag{2.43}$$

[5]The result for non-holonomic constraints is the same as (2.43) with $\nabla_I \varphi^A$ replaced by $\mathbf{C}_I^A$.

which are called **Lagrange's equations of the 1st kind**. Note that this equation is nothing more than Newton's 2nd law (2.2), with the substitution of (2.27) for the constraint force $\mathbf{F}_I^{(c)}$. This approach allows us to directly compute the constraint forces, which can be of value if we are interested in determining the required strength of materials for the systems of study (e.g., the rigid rod of the spherical pendulum, or the bolt holding the double pendulum to the ceiling).

Lagrange's equations of the 1st kind (2.43) can also be written in terms of general configuration space coordinates $x^\alpha$,

$$\frac{\mathrm{d}}{\mathrm{d}t}\left(\frac{\partial T}{\partial \dot{x}^\alpha}\right) - \frac{\partial T}{\partial x^\alpha} - F_\alpha - \sum_A \lambda_A \frac{\partial \varphi^A}{\partial x^\alpha} = 0\,, \qquad \alpha = 1, 2, \cdots, 3N\,, \quad (2.44)$$

where

$$T \equiv \sum_I \frac{1}{2} m_I \mathbf{v}_I \cdot \mathbf{v}_I \quad (2.45)$$

is the total kinetic energy of the system and

$$F_\alpha \equiv \sum_I \frac{\partial \mathbf{r}_I}{\partial x^\alpha} \cdot \mathbf{F}_I^{(a)}\,, \qquad \alpha = 1, 2, \cdots, 3N\,, \quad (2.46)$$

are the components of the so-called **generalized force**. This form of Lagrange's equation of the 1st kind is convenient for Hamilton's principle, which we shall discuss in detail in the next chapter. A proof of (2.44) is sketched in the following exercise.

**Exercise 2.8** To derive (2.44), first show that for $\mathbf{r}_I = \mathbf{r}_I(x^1, x^2, \cdots, x^{3N})$, the velocity $\mathbf{v}_I$ is given by

$$\mathbf{v}_I \equiv \frac{\mathrm{d}\mathbf{r}_I}{\mathrm{d}t} = \sum_\alpha \frac{\partial \mathbf{r}_I}{\partial x^\alpha} \dot{x}^\alpha\,, \quad (2.47)$$

which implies

$$\frac{\partial \mathbf{v}_I}{\partial \dot{x}^\alpha} = \frac{\partial \mathbf{r}_I}{\partial x^\alpha} \quad \text{and} \quad \frac{\mathrm{d}}{\mathrm{d}t}\left(\frac{\partial \mathbf{r}_I}{\partial x^\alpha}\right) = \frac{\partial \mathbf{v}_I}{\partial x^\alpha} \quad (2.48)$$

for $I = 1, 2, \cdots, N$. Then use these results to show that

$$\sum_I \frac{\partial \mathbf{r}_I}{\partial x^\alpha} \cdot \dot{\mathbf{p}}_I = \frac{\mathrm{d}}{\mathrm{d}t}\left(\frac{\partial T}{\partial \dot{x}^\alpha}\right) - \frac{\partial T}{\partial x^\alpha}\,. \quad (2.49)$$

**Exercise 2.9** Show that the total kinetic energy of the system can be written as

$$T = \frac{1}{2}\sum_{\alpha,\beta} T_{\alpha\beta}\dot{x}^\alpha \dot{x}^\beta\,, \qquad T_{\alpha\beta} \equiv \sum_I m_I \frac{\partial \mathbf{r}_I}{\partial x^\alpha}\cdot\frac{\partial \mathbf{r}_I}{\partial x^\beta}\,. \tag{2.50}$$

### 2.6.1 Solving Lagrange's Equations of the 1st Kind

To solve Lagrange's equations of the 1st kind (2.43), one typically performs the following steps:

(i) differentiate each of the constraint equations $\varphi^A = 0$ twice with respect to $t$ to get expressions involving $\ddot{\mathbf{r}}_I$;
(ii) substitute for $\ddot{\mathbf{r}}_I$ in these equations using (2.43);
(iii) solve the resulting *algebraic* equations for the Lagrange multipliers $\lambda_A$ in terms of $\mathbf{r}_I$ and $\dot{\mathbf{r}}_I$;
(iv) substitute the solutions for $\lambda_A$ back into (2.43);
(v) finally, solve the resulting equations of motion using standard methods for solving 2nd-order ordinary differential equations.

Note that it is always the case that the Lagrange multipliers can be solved for *algebraically*, while the equations of motion for $\mathbf{r}_I(t)$ are still ordinary differential equations.

***Example 2.3*** In this example, we will consider a bead of mass $m$ constrained to slide without friction on a hoop of wire of radius $R$, which is rotating about a vertical diameter in a uniform gravitational field $-g\hat{\mathbf{z}}$ with constant angular velocity $\omega$, as shown in Fig. 2.4. (See Dutta and Ray (2011) for a complete description of the problem.) Later in this chapter, we will introduce alternative methods for solving this problem, but for now we will demonstrate how the procedure outlined above produces the equations of motion as well as the forces of constraint.

The basic symmetry of the problem encourages us to use spherical coordinates. In these coordinates, the constraint equations are simply

$$\begin{aligned} \varphi^1 &\equiv r - R = 0\,, \\ \varphi^2 &\equiv \phi - \omega t = 0\,. \end{aligned} \tag{2.51}$$

Following the steps outlined above, for step (i), we take the second derivative of each of these constraint equations to find equations involving $\ddot{r}$ and $\ddot{\phi}$. Again, these equation are extraordinarily simple,

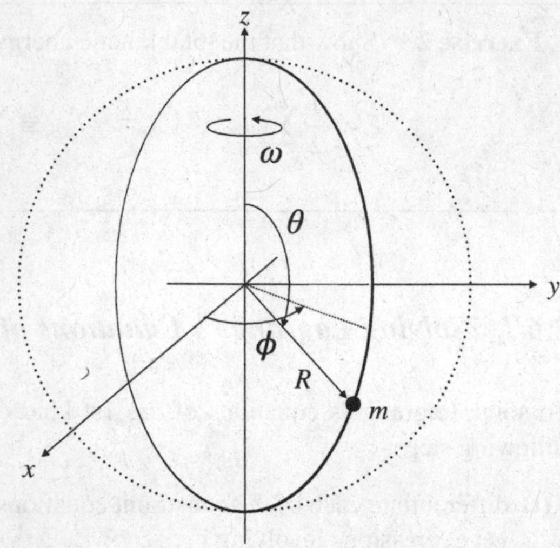

**Fig. 2.4** A bead of mass $m$ constrained to slide on a frictionless hoop of radius $R$ that is rotating with constant angular velocity $\omega$ about a vertical diameter

$$\begin{aligned}\ddot{\varphi}^1 &= \ddot{r} = 0\,,\\ \ddot{\varphi}^2 &= \ddot{\phi} = 0\,.\end{aligned} \tag{2.52}$$

Similar to the procedure outlined in Chap. 1 for obtaining (1.109) (See also Problem 1.1), we find the acceleration in spherical coordinates to be

$$\begin{aligned}\ddot{\mathbf{r}} &= \left(\ddot{r} - r\dot{\phi}^2 \sin^2\theta - r\dot{\theta}^2\right)\hat{\mathbf{r}}\\ &+ \left(r\ddot{\theta} + 2\dot{r}\dot{\theta} - r\dot{\phi}^2 \sin\theta\cos\theta\right)\hat{\boldsymbol{\theta}}\\ &+ \left(r\ddot{\phi}\sin\theta + 2\dot{r}\dot{\phi}\sin\theta + 2r\dot{\phi}\dot{\theta}\cos\theta\right)\hat{\boldsymbol{\phi}}\,,\end{aligned} \tag{2.53}$$

which gives us $\dot{\mathbf{p}} = m\ddot{\mathbf{r}}$. The applied force in this example is simply the force of gravity, so

$$\mathbf{F}^{(a)} = -mg\hat{\mathbf{z}} = -mg\cos\theta\,\hat{\mathbf{r}} + mg\sin\theta\,\hat{\boldsymbol{\theta}}\,. \tag{2.54}$$

To complete the evaluation of the $\dot{\mathbf{p}}$ equation needed for step (ii), we need to determine the gradients of the constraints:

$$\begin{aligned}\nabla\varphi^1 &= \frac{\partial}{\partial r}(r - R)\,\hat{\mathbf{r}} = \hat{\mathbf{r}}\,,\\ \nabla\varphi^2 &= \frac{1}{r\sin\theta}\frac{\partial}{\partial\phi}(\phi - \omega t)\,\hat{\boldsymbol{\phi}} = \frac{1}{r\sin\theta}\hat{\boldsymbol{\phi}}\,.\end{aligned} \tag{2.55}$$

Now, the $\hat{\mathbf{r}}$, $\hat{\boldsymbol{\theta}}$, $\hat{\boldsymbol{\phi}}$ components of $\dot{\mathbf{p}} - \mathbf{F}^{(a)} - \sum\lambda_A\nabla\varphi^A = 0$ give the three equations

$$\begin{aligned}
0 &= m\ddot{r} - mr\dot{\phi}^2 \sin^2\theta - mr\dot{\theta}^2 + mg\cos\theta - \lambda_1\,, \\
0 &= mr\ddot{\theta} + 2m\dot{r}\dot{\theta} - mr\dot{\phi}^2 \sin\theta\cos\theta - mg\sin\theta\,, \\
0 &= mr\ddot{\phi}\sin\theta + 2m\dot{r}\dot{\phi}\sin\theta + 2mr\dot{\phi}\dot{\theta}\cos\theta - \frac{\lambda_2}{r\sin\theta}\,.
\end{aligned} \tag{2.56}$$

Solving for $\ddot{r}$ and $\ddot{\phi}$ in (2.56) and substituting into (2.52) yields two very simple algebraic equations for $\lambda_1$ and $\lambda_2$, which we solve in step (iii) to find

$$\begin{aligned}
\lambda_1 &= mg\cos\theta - mr\dot{\theta}^2 - mr\dot{\phi}^2\sin^2\theta\,, \\
\lambda_2 &= 2mr\dot{r}\dot{\phi}\sin^2\theta + 2mr^2\dot{\phi}\dot{\theta}\sin\theta\cos\theta\,.
\end{aligned} \tag{2.57}$$

With the Lagrange multipliers safely in hand, we can now compute the constraint forces acting on the bead:

$$\begin{aligned}
\mathbf{F}^{(c)} &= \lambda_1 \nabla\varphi^1 + \lambda_2 \nabla\varphi^2 \\
&= m\left(g\cos\theta - r\dot{\theta}^2 - r\dot{\phi}^2\sin^2\theta\right)\hat{\mathbf{r}} + 2m\left(\dot{r}\dot{\phi}\sin\theta + r\dot{\phi}\dot{\theta}\cos\theta\right)\hat{\boldsymbol{\phi}}\,.
\end{aligned} \tag{2.58}$$

The $\hat{\mathbf{r}}$ term contains the centripetal force needed to keep the bead traveling in a circle as well as any force necessary to oppose the radial component of the gravitational force. The $\hat{\boldsymbol{\phi}}$ term is the force necessary to keep the bead moving at the same rate as the wire hoop.

In step (iv) we can now substitute the constraint forces back into the $\dot{\mathbf{p}}$ equation to obtain the equations of motion. The radial equation of motion becomes quite simple,

$$m\ddot{r} = 0\,, \tag{2.59}$$

and can be solved by $\dot{r}$ = constant. The first derivative of $\varphi^1$ indicates that this constant is 0, so finally we have the expected solution of $r = R$. With this solution, we can now look at the azimuthal equation of motion

$$mR\ddot{\phi}\sin\theta = 0\,. \tag{2.60}$$

This is identically solved from the second derivative of $\varphi^2$ given in (2.52). If we look at the first derivative of $\varphi^2$, we see that $\dot{\phi} = \omega$, so $\phi = \omega t + \phi_0$. Finally, we have the $\theta$-equation, which becomes

$$\ddot{\theta} = \sin\theta\left(\omega^2\cos\theta + \frac{g}{R}\right)\,. \tag{2.61}$$

This equation must be solved numerically in the general case, but there are several special cases that can be solved analytically (one of which you will solve in Exercise 2.10). □

**Exercise 2.10** For a given value of $\omega$, it is possible to find a stationary solution with constant $\theta$.

(a) Determine the value of $\omega$ for a stationary solution $\theta = \theta_0$.
(b) Determine the limiting value of $\theta_0$ for $\omega \to \infty$ and find the range of allowed values for $\theta_0$.
(c) What happens if $\theta$ is outside this range?

**Exercise 2.11** Use the solution (2.58) for the constraint forces for the uniformly rotating hoop to show that the work done by the constraint forces as the bead moves from $\theta_1$ to $\theta_2$ is

$$W = mR^2\omega^2\left(\sin^2\theta_2 - \sin^2\theta_1\right) . \tag{2.62}$$

*Hint*: Since the rotation is uniform, $\mathrm{d}\phi = \omega \mathrm{d}t$.

***Example 2.4*** It is interesting to note that the results of Exercise 2.11 show that the work done by the constraint forces is independent of the path taken by the bead. This seems to suggest that the constraint force can be described by the gradient of a potential. This is not true, but the effect of the work done by the constraint force can be expressed in terms of an **effective potential**. To see this, we can obtain a conservation law from (2.61) by first multiplying by $\dot{\theta}$ to obtain:

$$\dot{\theta}\ddot{\theta} = \omega^2\dot{\theta}\sin\theta\cos\theta + \frac{g}{R}\dot{\theta}\sin\theta , \tag{2.63}$$

which can be expressed as a total time derivative:

$$\frac{\mathrm{d}}{\mathrm{d}t}\left(\frac{1}{2}\dot{\theta}^2 - \frac{1}{2}\omega^2\sin^2\theta + \frac{g}{R}\cos\theta\right) = 0 . \tag{2.64}$$

This indicates that

$$\frac{1}{2}mR^2\dot{\theta}^2 - \frac{1}{2}mR^2\omega^2\sin^2\theta + mgR\cos\theta = C \tag{2.65}$$

is conserved. Note that this is not the energy because the sign of the azimuthal component of the kinetic energy is reversed. But recalling the work-energy theorem,

$$\begin{aligned} 0 = \Delta T + \Delta U - W = \frac{1}{2} m R^2 \left( \dot{\theta}_2^2 - \dot{\theta}_1^2 \right) + \frac{1}{2} m R^2 \omega^2 \left( \sin^2 \theta_2 - \sin^2 \theta_1 \right) \\ + m g R \left( \cos \theta_2 - \cos \theta_1 \right) - m R^2 \omega^2 \left( \sin^2 \theta_2 - \sin^2 \theta_1 \right) , \end{aligned} \tag{2.66}$$

we recover the statement that $C$ is conserved. Thus, we can express the energy of the system as

$$T + U + U_{\text{eff}} = \frac{1}{2} m R^2 \left( \dot{\theta}^2 + \omega^2 \sin^2 \theta \right) + m g R \cos \theta - m R^2 \omega^2 \sin^2 \theta , \tag{2.67}$$

where the effective potential is $U_{\text{eff}} = -m R^2 \omega^2 \sin^2 \theta$. The source of this potential energy is, of course, the motor that is driving the hoop with constant angular velocity $\omega$. When the bead slides along the hoop, it changes its value of $\theta$, which changes its azimuthal tangential velocity $\omega R \sin \theta$. This change in velocity is supplied by the motor; as $\theta \to \pi/2$, the motor must supply energy to the system to speed the bead up. Although the force that does the work to provide this energy points in the $\hat{\boldsymbol{\phi}}$ direction, it results in a change in the constraint force that leaves a residual force in the $\hat{\boldsymbol{\theta}}$ direction. This is the force that appears as a gradient of the effective potential. □

## 2.7 Lagrange's Equations of the 2nd Kind

For systems subject to holonomic constraints, there is a second approach for obtaining the equations of motion from d'Alembert's principle. In this approach, one eliminates the constraints altogether by working directly in terms of the generalized coordinates $q^a$, which parameterize the constraint surface $Q$. Note that this approach does not apply to systems subject to non-holonomic constraints, since such constraints do not define a constraint surface $Q$.

This method eliminates the need to solve for the constraint forces, and allows us to *directly* obtain the equations of motion for the generalized coordinates. In this approach, the generalized coordinates $q^a$ are related to the position vectors $\mathbf{r}_I$ via the embedding equations

$$\mathbf{r}_I = \mathbf{r}_I(q^1, q^2, \cdots, q^n, t) , \qquad I = 1, 2, \cdots, N , \tag{2.68}$$

where we have included a possible dependence on $t$ to allow for a *time-dependent* constraint surface, such as the varying-length pendulum described in Fig. 2.2. Given (2.68), it follows that

$$\mathbf{v}_I \equiv \frac{\mathrm{d}\mathbf{r}_I}{\mathrm{d}t} = \sum_a \frac{\partial \mathbf{r}_I}{\partial q^a} \dot{q}^a + \frac{\partial \mathbf{r}_I}{\partial t}, \tag{2.69}$$

$$\delta \mathbf{r}_I = \sum_a \frac{\partial \mathbf{r}_I}{\partial q^a} \delta q^a, \tag{2.70}$$

for $I = 1, 2, \cdots, N$. Note that the expression for the virtual displacement $\delta \mathbf{r}_I$ does not involve a $\delta t$ term since virtual displacements are (by definition) *fixed-time* displacements that are consistent with the constraints. Using the above expression for the velocity $\mathbf{v}_I$, it is also easy to show that

$$\frac{\partial \mathbf{v}_I}{\partial \dot{q}^a} = \frac{\partial \mathbf{r}_I}{\partial q^a} \quad \text{and} \quad \frac{\mathrm{d}}{\mathrm{d}t}\left(\frac{\partial \mathbf{r}_I}{\partial q^a}\right) = \frac{\partial \mathbf{v}_I}{\partial q^a}, \tag{2.71}$$

similar to Exercise 2.8. Thus, expanding the summation in d'Alembert's principle, (2.42), we find

$$\begin{aligned} 0 &= \sum_I \delta \mathbf{r}_I \cdot \left(\dot{\mathbf{p}}_I - \mathbf{F}_I^{(a)}\right) \\ &= \sum_I \sum_a \delta q^a \frac{\partial \mathbf{r}_I}{\partial q^a} \cdot \left(\dot{\mathbf{p}}_I - \mathbf{F}_I^{(a)}\right) \\ &= \sum_a \delta q^a \left( \sum_I \frac{\partial \mathbf{r}_I}{\partial q^a} \cdot \dot{\mathbf{p}}_I - \sum_I \frac{\partial \mathbf{r}_I}{\partial q^a} \cdot \mathbf{F}_I^{(a)} \right). \end{aligned} \tag{2.72}$$

The last term in the parentheses defines the components of the **generalized force**

$$F_a \equiv \sum_I \frac{\partial \mathbf{r}_I}{\partial q^a} \cdot \mathbf{F}_I^{(a)}, \qquad a = 1, 2, \cdots, n, \tag{2.73}$$

while the first term can be simplified by integrating by parts, and then using the two expressions in (2.71):

$$\begin{aligned} \sum_I \frac{\partial \mathbf{r}_I}{\partial q^a} \cdot \dot{\mathbf{p}}_I &= \sum_I \left[ \frac{\mathrm{d}}{\mathrm{d}t}\left(\frac{\partial \mathbf{r}_I}{\partial q^a} \cdot \mathbf{p}_I\right) - \frac{\mathrm{d}}{\mathrm{d}t}\left(\frac{\partial \mathbf{r}_I}{\partial q^a}\right) \cdot \mathbf{p}_I \right] \\ &= \sum_I \left[ \frac{\mathrm{d}}{\mathrm{d}t}\left(\frac{\partial \mathbf{v}_I}{\partial \dot{q}^a} \cdot m_I \mathbf{v}_I\right) - \frac{\partial \mathbf{v}_I}{\partial q^a} \cdot m_I \mathbf{v}_I \right] \\ &= \sum_I \left[ \frac{\mathrm{d}}{\mathrm{d}t}\left(\frac{\partial}{\partial \dot{q}^a}\left(\frac{1}{2} m_I \mathbf{v}_I \cdot \mathbf{v}_I\right)\right) - \frac{\partial}{\partial q^a}\left(\frac{1}{2} m_I \mathbf{v}_I \cdot \mathbf{v}_I\right) \right] \\ &= \frac{\mathrm{d}}{\mathrm{d}t}\left(\frac{\partial T}{\partial \dot{q}^a}\right) - \frac{\partial T}{\partial q^a}, \end{aligned} \tag{2.74}$$

where

$$T \equiv \sum_I \frac{1}{2} m_I \mathbf{v}_I \cdot \mathbf{v}_I \tag{2.75}$$

is the total kinetic energy of the system. Finally, combining (2.73) and (2.74), and using the fact that the $\delta q^a$ are *unconstrained* variations, we can conclude that

$$\frac{\mathrm{d}}{\mathrm{d}t}\left(\frac{\partial T}{\partial \dot{q}^a}\right) - \frac{\partial T}{\partial q^a} - F_a = 0\,, \qquad a = 1, 2, \cdots, n\,. \tag{2.76}$$

These equations are called **Lagrange's equations of the 2nd kind.**

**Exercise 2.12** Show that the total kinetic energy for a system of particles can be written in terms of the generalized coordinates $q^a$ and generalized velocities $\dot{q}^a$ as

$$T = \frac{1}{2}\left(\sum_{a,b} T_{ab}\, \dot{q}^a \dot{q}^b + 2\sum_a T_{a0}\, \dot{q}^a + T_{00}\right), \tag{2.77}$$

where

$$\begin{aligned} T_{ab} &\equiv \sum_I m_I \frac{\partial \mathbf{r}_I}{\partial q^a} \cdot \frac{\partial \mathbf{r}_I}{\partial q^b}\,, \\ T_{a0} &\equiv \sum_I m_I \frac{\partial \mathbf{r}_I}{\partial q^a} \cdot \frac{\partial \mathbf{r}_I}{\partial t}\,, \\ T_{00} &\equiv \sum_I m_I \frac{\partial \mathbf{r}_I}{\partial t} \cdot \frac{\partial \mathbf{r}_I}{\partial t}\,. \end{aligned} \tag{2.78}$$

Note that if the constraint surface is *independent of time*, then $T$ is simply given by $T = \frac{1}{2}\sum_{a,b} T_{ab}\, \dot{q}^a \dot{q}^b$, which is a homogeneous, quadratic function of the generalized velocities $\dot{q}^a$.

**Exercise 2.13** Using the results of Exercise 2.12, calculate the total kinetic energy for:

(a) the simple planar pendulum of Problem 1.3,
(b) the planar double pendulum of Problem 1.4,
(c) the bead-on-a-rotating-hoop from Example 2.3.

## 2.8 Generalized Potentials

Lagrange's equations of the 1st and 2nd kind, written in terms of the kinetic energy $T$ and the components of the generalized force $F_\alpha$ or $F_a$, are the most general forms of these equations. They can be simplified somewhat if the applied forces $\mathbf{F}_I^{(a)}$ can be written as gradients of scalar potentials. For example, if

$$\mathbf{F}_I^{(a)} = -\nabla_I U(\mathbf{r}_1, \mathbf{r}_2, \cdots, \mathbf{r}_N, t)\,, \tag{2.79}$$

then

$$F_\alpha = -\frac{\partial U}{\partial x^\alpha} \quad \text{and} \quad F_a = -\frac{\partial U}{\partial q^a}\,, \tag{2.80}$$

where $U$ on the right-hand sides of these equations are thought of as functions of the $x^\alpha$ or the $q^a$ using $\mathbf{r}_I = \mathbf{r}_I(x^1, \cdots, x^{3N})$ or $\mathbf{r}_I = \mathbf{r}_I(q^1, \cdots, q^n, t)$. Substituting these expressions into (2.44) and (2.76) and defining the **Lagrangian** to be

$$L \equiv T - U\,, \tag{2.81}$$

we obtain

**Lagrange's equations of the 1st and 2nd kind for a generalized potential:**

$$\frac{\mathrm{d}}{\mathrm{d}t}\left(\frac{\partial L}{\partial \dot{x}^\alpha}\right) - \frac{\partial L}{\partial x^\alpha} - \sum_A \lambda_A \frac{\partial \varphi^A}{\partial x^\alpha} = 0\,, \quad \alpha = 1, 2, \cdots, 3N\,, \tag{2.82a}$$

$$\frac{\mathrm{d}}{\mathrm{d}t}\left(\frac{\partial L}{\partial \dot{q}^a}\right) - \frac{\partial L}{\partial q^a} = 0\,, \qquad a = 1, 2, \cdots, n\,, \tag{2.82b}$$

where we used $\partial U/\partial \dot{x}^\alpha = 0$ and $\partial U/\partial \dot{q}^a = 0$ to get the first term in these equations. For non-holonomic constraints, we simply replace $\partial \varphi^A/\partial x^\alpha$ by $C^A_\alpha$ in (2.82a).

More generally, if there exists a function

$$U = U(\mathbf{r}_1, \mathbf{r}_2, \cdots, \mathbf{r}_N, \dot{\mathbf{r}}_1, \dot{\mathbf{r}}_2, \cdots, \dot{\mathbf{r}}_N, t) \tag{2.83}$$

for which

$$\mathbf{F}_I^{(a)} = -\nabla_I U + \frac{\mathrm{d}}{\mathrm{d}t}\left(\frac{\partial U}{\partial \dot{\mathbf{r}}_I}\right), \tag{2.84}$$

then

$$F_\alpha = -\frac{\partial U}{\partial x^\alpha} + \frac{\mathrm{d}}{\mathrm{d}t}\left(\frac{\partial U}{\partial \dot{x}^\alpha}\right) \quad \text{and} \quad F_a = -\frac{\partial U}{\partial q^a} + \frac{\mathrm{d}}{\mathrm{d}t}\left(\frac{\partial U}{\partial \dot{q}^a}\right), \tag{2.85}$$

and we again have (2.82a) and (2.82b), with $L = T - U$. The function $U$ is called a **generalized potential** for the applied forces, since it depends on the velocities ($\dot{x}^\alpha$ or $\dot{q}^a$) in addition to the coordinates ($x^\alpha$ or $q^a$) and time $t$. Following Lanczos (1949), we will call generalized forces derivable from a generalized potential **monogenic**, since there is a *single* function $U$ from which it can be derived. Likewise, we will call generalized forces *not* derivable from a generalized potential **polygenic**. Generalized potentials arise in the context of electromagnetism as illustrated by the following exercise.

**Exercise 2.14** Show that the Lorentz force

$$\mathbf{F} = q(\mathbf{E} + \mathbf{v} \times \mathbf{B}) \tag{2.86}$$

for a charged particle $q$ moving in an electromagnetic field is derivable from the generalized potential

$$U(\mathbf{r}, \dot{\mathbf{r}}, t) = q\,[\Phi(\mathbf{r}, t) - \mathbf{A}(\mathbf{r}, t) \cdot \dot{\mathbf{r}}]\,. \tag{2.87}$$

Recall that $\mathbf{E} = -\boldsymbol{\nabla}\Phi - \partial\mathbf{A}/\partial t$ and $\mathbf{B} = \boldsymbol{\nabla} \times \mathbf{A}$.

Finally, to close out this chapter, we show in the following example under what conditions the total mechanical energy of a system is conserved. We will revisit conservation laws more generally in the context of the Lagrangian and Hamiltonian formulations of mechanics in Sect. 3.3.

***Example 2.5*** Show that if: (i) the constraint surface is independent of time, (ii) the impressed forces are derivable from a potential $U = U(q^1, q^2, \cdots, q^n)$ that depends only on the generalized coordinates, and (iii) all the masses $m_I$ are constant, then the total mechanical energy $E \equiv T + U$ is conserved, i.e., $\mathrm{d}E/\mathrm{d}t = 0$.

*Proof* Given the above assumptions and the results of Exercise 2.12, we can write

$$T = \frac{1}{2}\sum_{a,b} T_{ab}\dot{q}^a\dot{q}^b\,, \qquad T_{ab} \equiv \sum_I m_I \frac{\partial \mathbf{r}_I}{\partial q^a} \cdot \frac{\partial \mathbf{r}_I}{\partial q^b}\,. \tag{2.88}$$

Note that $T_{ab}$ is a function only of the generalized coordinates and not time, since the masses are assumed to be constant. Thus, it follows that

$$\frac{\partial T}{\partial \dot{q}^a} = \sum_b T_{ab}\,\dot{q}^b\,, \qquad \frac{\mathrm{d}T}{\mathrm{d}t} = \sum_a \left(\frac{\partial T}{\partial q^a}\dot{q}^a + \frac{\partial T}{\partial \dot{q}^a}\ddot{q}^a\right). \tag{2.89}$$

In addition,

$$\frac{\mathrm{d}U}{\mathrm{d}t} = \sum_a \frac{\partial U}{\partial q^a}\dot{q}^a \,, \tag{2.90}$$

since $U$ also depends only on the generalized coordinates. Now take Lagrange's equations of the 2nd kind in the form of (2.82b), multiply by $\dot{q}^a$, and sum over the index $a$:

$$\begin{aligned} 0 &= \sum_a \dot{q}^a \left[\frac{\mathrm{d}}{\mathrm{d}t}\left(\frac{\partial L}{\partial \dot{q}^a}\right) - \frac{\partial L}{\partial q^a}\right] \\ &= \sum_a \dot{q}^a \left[\frac{\mathrm{d}}{\mathrm{d}t}\left(\frac{\partial T}{\partial \dot{q}^a}\right) - \frac{\partial T}{\partial q^a} + \frac{\partial U}{\partial q^a}\right] \\ &= \sum_a \left[\frac{\mathrm{d}}{\mathrm{d}t}\left(\dot{q}^a \frac{\partial T}{\partial \dot{q}^a}\right) - \ddot{q}^a \frac{\partial T}{\partial \dot{q}^a} - \dot{q}^a \frac{\partial T}{\partial q^a} + \dot{q}^a \frac{\partial U}{\partial q^a}\right]. \end{aligned} \tag{2.91}$$

Using (2.89), the first term can be written as

$$\sum_a \frac{\mathrm{d}}{\mathrm{d}t}\left(\dot{q}^a \frac{\partial T}{\partial \dot{q}^a}\right) = \frac{\mathrm{d}}{\mathrm{d}t}\left(\sum_a \dot{q}^a \sum_b T_{ab}\dot{q}^b\right) = 2\frac{\mathrm{d}T}{\mathrm{d}t} \,, \tag{2.92}$$

while the middle two terms are equal to $-\mathrm{d}T/\mathrm{d}t$. From (2.90), the last term is $\mathrm{d}U/\mathrm{d}t$. Combining these results, we have

$$0 = 2\frac{\mathrm{d}T}{\mathrm{d}t} - \frac{\mathrm{d}T}{\mathrm{d}t} + \frac{\mathrm{d}U}{\mathrm{d}t} = \frac{\mathrm{d}}{\mathrm{d}t}(T+U) = \frac{\mathrm{d}E}{\mathrm{d}t} \,, \tag{2.93}$$

where $E \equiv T + U$. Thus, the sum of the total kinetic energy $T$ and the potential $U$ is conserved. □

## Suggested References

*Full references are given in the bibliography at the end of the book.*

Dutta and Ray (2011): This article on the arXiv provides a full and in-depth treatment of the problem of a bead on a rotating circular hoop, which we discussed in Example 2.3.

Flannery (2005): A very readable article about the subtleties associated with non-holonomic constraints.

Lanczos (1949): Provides a thorough discussion of holonomic and non-holonomic constraints, the method of Lagrange multipliers, d'Alembert's principle, Lagrange's equations, etc.

## Additional Problems

**Problem 2.1** (*Adapted from Kuchăr (1995).*) A *tether ball* is part of a playground game, consisting of a ball (mass $m$) attached to a string (length $\ell$) that wraps around a pole (cylinder, radius $R$), as shown in Fig. 2.5. For the following problem, treat the ball as a point and the string as ideal (i.e., massless and inextendible), and assume that as the string wraps around the pole, the portion of the string between the position of the ball and the string's last point of contact with the pole, $P$, is *taut* (i.e., described by a segment of a straight line). For this particular problem, you don't need to consider the gravitational force on the ball.

(a) Write down the embedding equations

$$x = x(\theta, \phi), \quad y = y(\theta, \phi), \quad z = z(\theta, \phi), \tag{2.94}$$

for the position of the ball as a function of the generalized coordinates $(\theta, \phi)$, defined in panel (b) of Fig. 2.5.

(b) Express the kinetic energy $T$ of the ball in terms of $(\theta, \phi)$.

You should find:

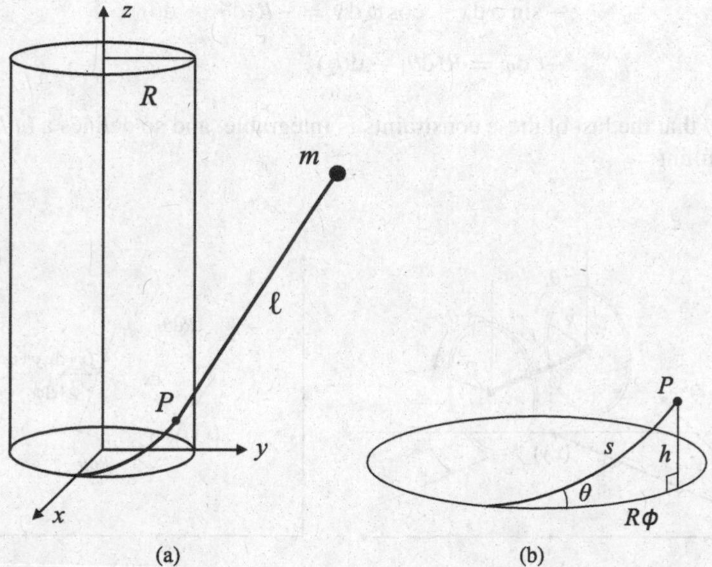


**Fig. 2.5** Panel (a) A tether ball (mass $m$) is attached to a string of length $\ell$ that wraps around a pole of radius $R$. Panel (b) A zoom-in on the bottom portion of the tether ball geometry, with angles $\theta$ and $\phi$ defined. Here $s$ is the arc length of the portion of the string in contact with the pole, and $h$ is $z$-component of point $P$, the string's last point of contact with the pole

$$x = R(\cos\phi + \phi\sin\phi) - \ell\cos\theta\sin\phi\,,$$
$$y = R(\sin\phi - \phi\cos\phi) + \ell\cos\theta\cos\phi\,, \tag{2.95}$$
$$z = \ell\sin\theta\,,$$

and

$$T = \frac{1}{2}m\left[\ell^2\dot{\theta}^2 + (R\phi - \ell\cos\theta)^2\dot{\phi}^2\right]. \tag{2.96}$$

**Problem 2.2** (*Adapted from Problem 1.5 in Goldstein et al. (2002).*) Two wheels of radius $R$ are mounted at the ends of a common axle of length $\ell$, and are free to rotate independently of one another. The system rolls without slipping on a horizontal two-dimensional surface, as shown in Fig. 2.6. Let $x$ and $y$ denote the $x$ and $y$ components of the center of mass of the system, which is located at the midpoint of the axle; $\phi$ denote the angle that the axle makes with the $x$-axis; and $\theta_1$, $\theta_2$ denote the angular positions of fixed points on the two wheels relative to the vertical, as shown in panel (a) of Fig. 2.6. Panel (b) of Fig. 2.6 shows the changes in the coordinates induced by an infinitesimal displacement ds of the system.

(a) Show that these five coordinates are related by three constraints

$$\cos\phi\,\mathrm{d}x + \sin\phi\,\mathrm{d}y = 0\,,$$
$$-\sin\phi\,\mathrm{d}x + \cos\phi\,\mathrm{d}y = \frac{1}{2}R(\mathrm{d}\theta_1 + \mathrm{d}\theta_2)\,, \tag{2.97}$$
$$-\ell\,\mathrm{d}\phi = R(\mathrm{d}\theta_1 - \mathrm{d}\theta_2)\,.$$

(b) Show that the last of these constraints is integrable, and so defines a *holonomic* constraint

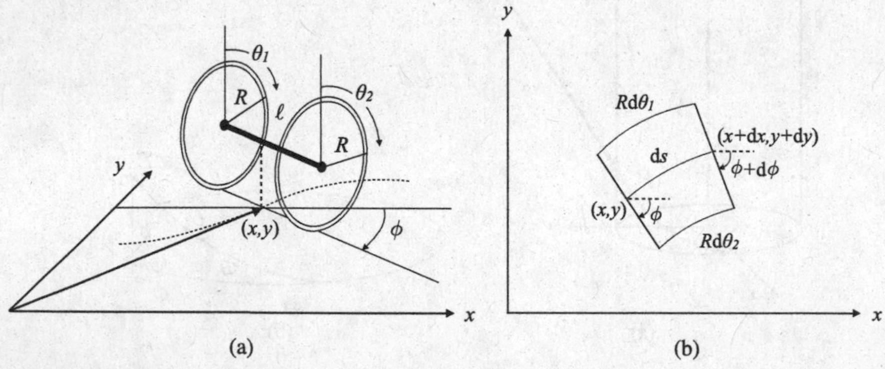


**Fig. 2.6** (a) Two rolling wheels mounted on an axle of length $\ell$. The position of the center of mass of the axle is given by the coordinates $(x,\ y)$. The orientation of the axle with respect to the $x$-axis is given by $\phi$, and the rotation angle of each wheel is given by $\theta_1$ and $\theta_2$. Each wheel is free to rotate at different rates. (b) Changes in the coordinates $\{x, y, \phi, \theta_1, \theta_2\}$ induced by an infinitesimal displacement ds of the system

$$\ell\phi + R(\theta_1 - \theta_2) = \text{const}. \tag{2.98}$$

(c) Show that the three constraints can be combined to yield

$$dx + \frac{1}{2}\ell \sin\phi \, d\phi + R \sin\phi \, d\theta_1 = 0 \tag{2.99}$$

on the coordinate differentials $dx$, $d\phi$, $d\theta_1$.

(d) Show that this constraint cannot be integrated, and hence the set of constraints on the system is non-holonomic.

**Problem 2.3** Consider the rotating hoop problem from Example 2.3, but now let it be accelerating with constant angular acceleration $\alpha$. Repeat the steps of that example to determine the equations of motion and the forces of constraint.

**Problem 2.4** Using Lagrange's equations of the 1st kind, solve for the motion and the constraint force for the simple planar pendulum described in Problem 1.3.

**Problem 2.5** (*Adapted from Kuchǎr (1995).*) A bucket of mass $m$ slides without friction along a string of length $\ell_1 + \ell_2 = 2a$, supported at points $(-a/\sqrt{2}, 0)$ and $(a/\sqrt{2}, 0)$ in a uniform gravitational field $\mathbf{g}$. (See Fig. 2.7.) Assume that the string is ideal—i.e., massless and inextendible.

(a) Explicitly write down the constraint $\ell_1 + \ell_2 = 2a$ in terms of the coordinates $(x, y)$ of the bucket. Show that the constraint can be simplified to

$$\varphi(x, y) \equiv x^2 + 2y^2 - a^2 = 0, \tag{2.100}$$

which is equivalent to the bucket moving along an arc of the ellipse

$$\left(\frac{x}{a}\right)^2 + \left(\frac{y}{b}\right)^2 = 1, \tag{2.101}$$

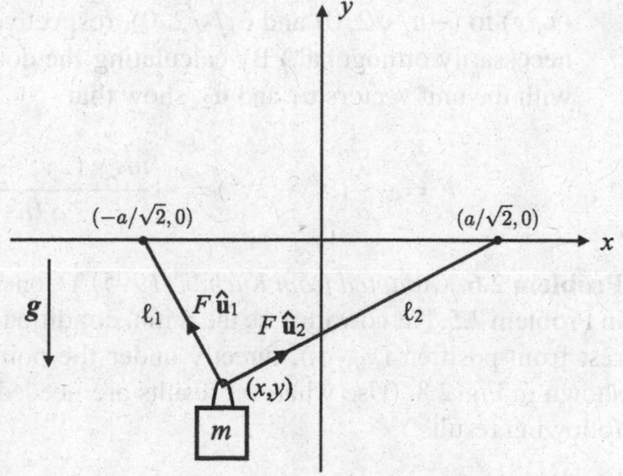


**Fig. 2.7** A sliding bucket in a uniform gravitational field $\mathbf{g}$, at an arbitrary location $(x, y)$ in its motion along the string. Here, $F$ denotes the tension in the string, and $\hat{\mathbf{u}}_1$ and $\hat{\mathbf{u}}_2$ are unit vectors pointing from the position of the bucket to the two points of support

where $b \equiv a/\sqrt{2}$.

(b) Show that Lagrange's equations of the 1st kind for this problem are

$$m\ddot{x} - 2\lambda x = 0\,, \qquad m\ddot{y} + mg - 4\lambda y = 0\,, \tag{2.102}$$

where $\lambda$ is a Lagrange multiplier (to be determined below).

(c) Show that by multiplying the first of the above equations by $\dot{x}$, the second by $\dot{y}$, and then adding the two together using $x\dot{x} + 2y\dot{y} = 0$ (which is the time derivative of the constraint), one obtains

$$\frac{\mathrm{d}}{\mathrm{d}t}\left[\frac{1}{2}m(\dot{x}^2 + \dot{y}^2) + mgy\right] = 0\,, \tag{2.103}$$

which is simply the statement that the total mechanical energy

$$E \equiv \frac{1}{2}m(\dot{x}^2 + \dot{y}^2) + mgy \tag{2.104}$$

is conserved.

(d) Solve for the Lagrange multiplier $\lambda$ by differentiating the constraint (2.100) twice with respect to $t$, and then substituting for $\ddot{x}$ and $\ddot{y}$ in this expression using (2.102). By further eliminating the $\dot{x}$ and $\dot{y}$ terms in favor of $y$ (using conservation of energy and the constraint), show that

$$\lambda = \frac{mgy\left(2y^2 + 3a^2\right) - 2Ea^2}{\left(a^2 + 2y^2\right)^2}\,. \tag{2.105}$$

(e) Calculate the tension in the string by noting that the constraint force $\mathbf{F}^{(c)} = \lambda\boldsymbol{\nabla}\varphi$ can also be written as

$$\mathbf{F}^{(c)} = F\left(\hat{\mathbf{u}}_1 + \hat{\mathbf{u}}_2\right)\,, \tag{2.106}$$

where $\hat{\mathbf{u}}_1$ and $\hat{\mathbf{u}}_2$ are unit vectors, which point from the position of the bucket $(x, y)$ to $(-a/\sqrt{2}, 0)$ and $(a/\sqrt{2}, 0)$, respectively. (Note that $\hat{\mathbf{u}}_1$ and $\hat{\mathbf{u}}_2$ are not necessarily orthogonal.) By calculating the dot product of $\mathbf{F}^{(c)}$ with itself, and with the unit vectors $\hat{\mathbf{u}}_1$ and $\hat{\mathbf{u}}_2$, show that

$$F = -\frac{\lambda}{a}(a^2 + 2y^2) = -\frac{mgy\left(2y^2 + 3a^2\right) - 2Ea^2}{a\left(a^2 + 2y^2\right)}\,. \tag{2.107}$$

**Problem 2.6** (*Adapted from Kuchǎr (1995).*) Consider the sliding bucket described in Problem 2.5, but now impose the initial conditions that the bucket is released from rest from position $(x_0, y_0)$, directly under the point of support at $(-a/\sqrt{2}, 0)$, as shown in Fig. 2.8. (Use whatever results are needed from Problem 2.5 to derive the following results.)

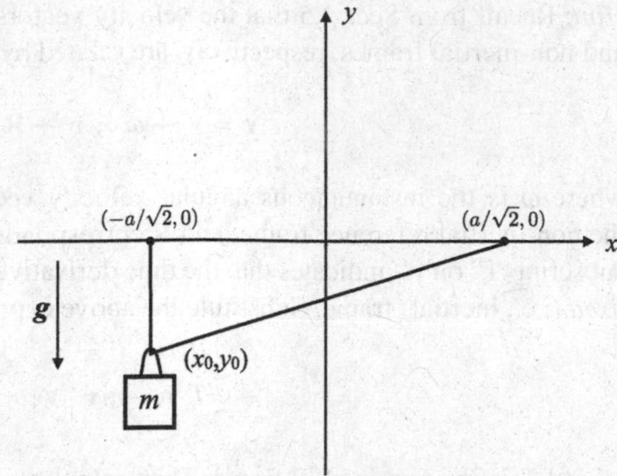


**Fig. 2.8** The sliding bucket from Problem 2.5 is released from rest from position $(x_0, y_0)$, directly under the point of support at $(-a/\sqrt{2}, 0)$

(a) For these initial conditions show that $y_0 = a/2$ and $E = -mga/2$, implying

$$F = -\frac{mg}{a}\frac{(2y^3 + 3ya^2 + a^3)}{(a^2 + 2y^2)}. \tag{2.108}$$

(b) Show that when the bucket is at the lowest point of the ellipse, the tension is given by

$$F_{\text{lowest}} = \frac{mg}{\sqrt{2}}\left(2 - \frac{1}{\sqrt{2}}\right). \tag{2.109}$$

Compare this value to the tension that the string would have if the bucket were just sitting at rest at the lowest point? Which tension is larger?

(c) By using the constraint equation to eliminate $y$ and $\dot{y}$ in terms of $x$ and $\dot{x}$, obtain a 1st-order ordinary differential equation for $x = x(t)$ which can be solved via the method of quadratures, giving

$$t - t_0 = \pm\frac{1}{2}\int_{x_0}^{x} \mathrm{d}x \sqrt{\frac{m(2a^2 - x^2)}{(a^2 - x^2)\left(E + mg\sqrt{\frac{a^2 - x^2}{2}}\right)}}. \tag{2.110}$$

Note that $y < 0$ in this problem, so $y = -\sqrt{y^2}$.

**Problem 2.7** (*Adapted from Kuchǎr (1995).*) Write down an expression for the kinetic energy $T$ of a point mass $m$ in a non-inertial reference frame. Then show that Lagrange's equations of the 2nd kind (2.76) include all of the relevant fictitious force terms. Identify the parts of $T$ that contribute to: (a) the angular acceleration force, (b) the Coriolis force, (c) the centrifugal force, and (d) the translational acceleration force.

*Hint*: Recall from Sect. 1.5 that the velocity vectors $\mathbf{v}$ and $\mathbf{v}'$, as seen in the inertial and non-inertial frames, respectively, are related by

$$\mathbf{v} = \mathbf{v}' + \boldsymbol{\omega} \times \mathbf{r}' + \dot{\mathbf{R}}_{\mathrm{f}} , \tag{2.111}$$

where $\boldsymbol{\omega}$ is the instantaneous angular velocity vector of the rotational motion of the non-inertial reference frame, and $\mathbf{R}$ corresponds to its translational motion. The subscript "f" on $\dot{\mathbf{R}}_{\mathrm{f}}$ indicates that the time derivative is calculated with respect to the *fixed* (i.e., inertial) frame. Substitute the above expression into

$$T \equiv \frac{1}{2} m \mathbf{v} \cdot \mathbf{v} , \tag{2.112}$$

to find $T$ in the non-inertial frame. Then calculate

$$\frac{\mathrm{d}}{\mathrm{d}t}\left(\frac{\partial T}{\partial \dot{q}^a}\right) - \frac{\partial T}{\partial q^a} - F_a = 0 , \tag{2.113}$$

where $q^a$ are the components of $\mathbf{r}'$, and $F_a$ are the components of the applied force in the non-inertial frame. Here $\dot{q}^a$ refers to the time derivative of $q^a$ (i.e., the components of $\mathbf{v}'$) as seen in the non-inertial frame.

**Problem 2.8** Redo the rotating hoop problem from Example 2.3, but this time using Lagrange's equations of the 2nd kind. Of course, by using this method, you will not be interested in solving for the constraint forces, but only in finding the equations of motion for the generalized coordinates describing the position of the bead.

**Problem 2.9** Redo the simple planar pendulum problem described in Problem 1.3, but this time using Lagrange's equations of the 2nd kind. In particular: (a) obtain the equation of motion for the pendulum bob, (b) derive an expression for total energy of the system (which is conserved), and (c) determine the minimum initial velocity for the pendulum bob to make a complete loop-the-loop, requiring that the centripetal force equal the weight of the bob at the top of the loop.

**Problem 2.10** Redo the planar double pendulum problem described in Problem 1.4, but this time using Lagrange's equations of the 2nd kind. Show that you recover the equations of motion given in (1.126).

# Chapter 3
# Hamilton's Principle and Action Integrals

In the previous chapter, we derived Lagrange's equations of motion as a consequence of the principle of virtual work. Here we show that for systems of particles subject to holonomic constraints and forces derivable from a generalized potential, Lagrange's equations can also be obtained by finding the stationary values of a functional, called the "action." By doing so, we convert the problem of finding the equations of motion to a problem in the calculus of variations. We also introduce the Hamiltonian for a system of particles, which can be obtained from the Lagrangian by performing a Legendre transformation. This process leads to Hamilton's equations of motion, which are 1st-order differential equations for the generalized coordinates and momenta. We shall see that the Hamiltonian formulation is particularly well-suited for illustrating the intimate connection between continuous symmetries of the system (if they exist) and conserved quantities. The value of the Lagrangian and Hamiltonian formalisms introduced in this and the previous chapter extend well beyond classical mechanics and are the fundamental underpinnings of all of modern physics.

## 3.1 Hamilton's Principle

If you have already taken a math or physics class where you've studied the calculus of variations, Lagrange's equations of the 2nd kind for forces derivable from a generalized potential,

$$\frac{\mathrm{d}}{\mathrm{d}t}\left(\frac{\partial L}{\partial \dot{q}^a}\right) - \frac{\partial L}{\partial q^a} = 0\,, \qquad a = 1, 2, \cdots, n\,, \tag{3.1}$$



M.J. Benacquista and J.D. Romano, *Classical Mechanics*, Undergraduate Lecture Notes in Physics, https://doi.org/10.1007/978-3-319-68780-3_3

should look very familiar. (If they don't look familiar, you can refer to Appendix C for a quick refresher.) Indeed, they are just the **Euler equations** obtained by finding the stationary values of the definite integral

$$S[q^1, q^2, \cdots, q^n] \equiv \int_{t_1}^{t_2} dt\, L(q^1, q^2, \cdots q^n, \dot{q}^1, \dot{q}^2, \cdots \dot{q}^n, t), \qquad (3.2)$$

for variations of the generalized coordinates $q^a$ having fixed endpoints—i.e.,

$$\delta q^a\big|_{t_1} = 0, \qquad \delta q^a\big|_{t_2} = 0, \qquad a = 1, 2, \cdots, n. \qquad (3.3)$$

The integral of the Lagrangian $L \equiv T - U$ is called the **action** $S$. The equations of motion are obtained from **Hamilton's principle of stationary action**, which can be expressed as:

"Of all possible paths which a system may follow in going from a specified configuration at time $t_1$ to another specified configuration at time $t_2$, the path followed is a stationary point of the action integral."

This is a rather remarkable statement. One introduces a quantity with dimensions of energy×time—the integrated difference between the kinetic and potential energy—and it happens to have a stationary value for the correct trajectory chosen by nature.[1] Hamilton's principle recasts the standard *local* formulation of mechanics (written in terms of differential equations) into a *global* minimization problem over possible trajectories of the particles.

Note that we will often use the shorthand notation

$$S[q] = \int_{t_1}^{t_2} dt\, L(q, \dot{q}, t) \qquad (3.4)$$

in place of (3.2), by letting $q$ and $\dot{q}$ stand for all the generalized coordinates $(q^1, q^2, \cdots, q^n)$ and all the generalized velocities $(\dot{q}^1, \dot{q}^2, \cdots, \dot{q}^n)$, respectively. The equations of motion (3.1), when obtained by the varying the action $S$, are called the **Euler-Lagrange equations**, indicating that these are Euler equations in the context of mechanics.

[1] If you've had any exposure to quantum mechanics, you may recall another quantity with the same dimensions as the action. Planck's constant $h = 6.626 \times 10^{-34}$ J · s, which appears in quantum mechanics, is also called the "quantum of action".

**Exercise 3.1** Consider a single particle of mass $m$ moving in response to an external force $\mathbf{F} = -\nabla U(\mathbf{r}, t)$.

(a) Write down the Lagrangian $L$ and the corresponding Euler-Lagrange equations in Cartesian coordinates $(x, y, z)$.
(b) Repeat part (a) for spherical polar coordinates $(r, \theta, \phi)$.

For part (b) you should find

$$L = \frac{1}{2}m(\dot{r}^2 + r^2\dot{\theta}^2 + r^2 \sin^2\theta\, \dot{\phi}^2) - U(r, \theta, \phi, t)\,, \tag{3.5}$$

and the corresponding Euler-Lagrange equations

$$\begin{aligned}
\frac{\mathrm{d}}{\mathrm{d}t}(m\dot{r}) &= -\frac{\partial U}{\partial r} + mr(\dot{\theta}^2 + \sin^2\theta\, \dot{\phi}^2)\,, \\
\frac{\mathrm{d}}{\mathrm{d}t}(mr^2\dot{\theta}) &= -\frac{\partial U}{\partial \theta} + mr^2 \sin\theta \cos\theta\, \dot{\phi}^2\,, \\
\frac{\mathrm{d}}{\mathrm{d}t}(mr^2 \sin^2\theta\, \dot{\phi}) &= -\frac{\partial U}{\partial \phi}\,.
\end{aligned} \tag{3.6}$$

### 3.1.1 *Proof of Hamilon's Principle*

In Appendix C, we show that $\delta S = 0$ implies (3.1). Here we prove the implication in the opposite direction (i.e., that (3.1) implies $\delta S = 0$), which is more in line with Hamilton's original derivation. We show this for the case of unconstrained systems (or at least systems for which all the variables are independent). This will lead us to a better understanding of the conditions under which Hamilton's principle is valid when there are constraints.

*Proof* So let's start with (3.1) and multiply by a virtual displacement $\delta q^a$, and then sum over $a$,

$$\sum_a \delta q^a \left[ \frac{\mathrm{d}}{\mathrm{d}t}\left(\frac{\partial L}{\partial \dot{q}^a}\right) - \frac{\partial L}{\partial q^a} \right] = 0\,. \tag{3.7}$$

Recall that a virtual displacement $\delta q^a$ maps a configuration $q^a$ at each time $t$ to a new configuration $\overline{q}^a$ at the *same* time,

$$\overline{q}^a(t) = q^a(t) + \delta q^a(t)\,, \qquad a = 1, 2, \cdots, n\,. \tag{3.8}$$

Integrating (3.7) with respect to time yields

$$\begin{aligned} 0 &= \int_{t_1}^{t_2} \mathrm{d}t \sum_a \delta q^a \left[ \frac{\mathrm{d}}{\mathrm{d}t}\left(\frac{\partial L}{\partial \dot{q}^a}\right) - \frac{\partial L}{\partial q^a} \right] \\ &= \int_{t_1}^{t_2} \mathrm{d}t \sum_a \left[ \frac{\mathrm{d}}{\mathrm{d}t}\left(\delta q^a \frac{\partial L}{\partial \dot{q}^a}\right) - \frac{\partial L}{\partial \dot{q}^a}\frac{\mathrm{d}}{\mathrm{d}t}(\delta q^a) - \frac{\partial L}{\partial q^a}\delta q^a \right] \\ &= \sum_a \left(\frac{\partial L}{\partial \dot{q}^a}\right)\delta q^a \Bigg|_{t_1}^{t_2} - \int_{t_1}^{t_2} \mathrm{d}t \sum_a \left[ \frac{\partial L}{\partial \dot{q}^a}\delta \dot{q}^a + \frac{\partial L}{\partial q^a}\delta q^a \right], \end{aligned} \tag{3.9}$$

where we integrated by parts and used the relation that the time-derivative and variation associated with a virtual displacement commute,

$$\frac{\mathrm{d}}{\mathrm{d}t}(\delta q^a) = \delta\left(\frac{\mathrm{d}q^a}{\mathrm{d}t}\right) = \delta \dot{q}^a\,. \tag{3.10}$$

Note that the first term on the right-hand side of (3.9) vanishes if we require that the system be at specified configurations at both $t_1$ and $t_2$, since the variations $\delta q^a$ must then vanish at these times. The last two terms in (3.9) can be written as

$$\int_{t_1}^{t_2} \mathrm{d}t \sum_a \left[ \frac{\partial L}{\partial \dot{q}^a}\delta \dot{q}^a + \frac{\partial L}{\partial q^a}\delta q^a \right] = \int_{t_1}^{t_2} \mathrm{d}t\, \delta L = \delta \int_{t_1}^{t_2} \mathrm{d}t\, L = \delta S\,, \tag{3.11}$$

where $\delta L$ is the infinitesimal change in $L \equiv L(q, \dot{q}, t)$ induced by the virtual displacement $\delta q^a$. Thus, (3.1) implies $\delta S = 0$ as claimed. □

**Exercise 3.2** Show that the Euler-Lagrange equations are unchanged if one adds a *total time derivative* to the Lagrangian—i.e.,

$$L(q, \dot{q}, t) \to L'(q, \dot{q}, t) \equiv L(q, \dot{q}, t) + \frac{\mathrm{d}\Lambda(q, t)}{\mathrm{d}t} \tag{3.12}$$

doesn't change the Euler-Lagrange equations.

**Exercise 3.3** Show that the form of the Euler-Lagrange equations is preserved under an invertible transformation of the generalized coordinates from $q^a$ to $Q^a$,

$$Q^a \equiv Q^a(q,t) \quad \Leftrightarrow \quad q^a \equiv q^a(Q,t)\,, \qquad a = 1, 2, \cdots, n\,. \tag{3.13}$$

*Hint*: The transformed Lagrangian is

$$L'(Q,\dot{Q},t) \equiv L(q,\dot{q},t)|_{q=q(Q,t),\,\dot{q}=\dot{q}(Q,\dot{Q},t)}\,, \tag{3.14}$$

where

$$\dot{q}^a = \sum_b \frac{\partial q^a}{\partial Q^b}\dot{Q}^b + \frac{\partial q^a}{\partial t}\,. \tag{3.15}$$

You should find

$$\frac{\mathrm{d}}{\mathrm{d}t}\left(\frac{\partial L}{\partial \dot{q}^a}\right) - \frac{\partial L}{\partial q^a} = 0 \quad \Leftrightarrow \quad \frac{\mathrm{d}}{\mathrm{d}t}\left(\frac{\partial L'}{\partial \dot{Q}^a}\right) - \frac{\partial L'}{\partial Q^a} = 0\,. \tag{3.16}$$

## 3.2 Constrained Variations

In the previous section, we proved Hamilton's principle for a system of particles expressed in terms of *generalized coordinates* $q^a$, where $a = 1, 2, \cdots, n$. Recall that generalized coordinates represent the *independent* degrees of freedom of the system, since any constraints on the system were effectively eliminated by working in terms of these coordinates. Forces of constraint, for example, do not appear in this formulation of mechanics.

But can Hamilton's principle also be formulated in terms of the *full* set of coordinates $x^\alpha$, $\alpha = 1, 2, \cdots, 3N$ on the configuration space? If there are constraints, then the variations $\delta x^\alpha$ are not independent of one another, but are subject to auxiliary conditions. As usual, these conditions can be incorporated into the formalism by using the method of Lagrange multipliers. But as we shall see below, finding an action functional whose stationary values yield the equations of motion for the system is possible only for a restricted set of constraints and forces. The constraints must be *holonomic* and the forces must be *monogenic* (i.e., derivable from a generalized potential) in order for an action functional to exist. If the generalized force is *polygenic* or if the constraints are *non-holonomic*, then the equations of motion are *not* derivable from the variation of any action functional.

### 3.2.1 Holonomic Constraints

For holonomic constraints, we showed in (2.44) that Lagrange's equations of the 1st kind could be written as

$$\frac{\mathrm{d}}{\mathrm{d}t}\left(\frac{\partial T}{\partial \dot{x}^\alpha}\right) - \frac{\partial T}{\partial x^\alpha} - F_\alpha - \sum_A \lambda_A \frac{\partial \phi^A}{\partial x^\alpha} = 0\,,. \quad \alpha = 1, 2, \cdots, 3N\,, \tag{3.17}$$

where $T$ is the total kinetic energy, $F_\alpha$ are the components of the generalized force, and $\lambda_A$ are Lagrange multipliers. Let's assume, to begin with, that the generalized force $F_\alpha$ is monogenic. (For holonomic constraints with a polygenic generalized force, see Problem 3.3.) Remember that this means that the generalized force can be obtained from a single function $U$, so that

$$F_\alpha = -\frac{\partial U}{\partial x^\alpha} + \frac{\mathrm{d}}{\mathrm{d}t}\left(\frac{\partial U}{\partial \dot{x}^\alpha}\right) \tag{3.18}$$

and

$$\frac{\mathrm{d}}{\mathrm{d}t}\left(\frac{\partial L}{\partial \dot{x}^\alpha}\right) - \frac{\partial L}{\partial x^\alpha} - \sum_A \lambda_A \frac{\partial \phi^A}{\partial x^\alpha} = 0\,, \quad \alpha = 1, 2, \cdots, 3N\,, \tag{3.19}$$

where $L \equiv T - U$. If the generalized force were not monogenic, we would not be able to absorb that term into a Lagrangian involving $U$. If we now follow a similar procedure to that used in Sect. 3.1.1, where we first multiply the above equation by a virtual displacement $\delta x^\alpha$, sum over $\alpha$, then integrate over $t$, and using

$$\sum_\alpha \frac{\partial \phi^A}{\partial x^\alpha} \delta x^\alpha = \delta \phi^A\,, \tag{3.20}$$

we obtain

$$0 = \sum_\alpha \left(\frac{\partial L}{\partial \dot{x}^\alpha}\right) \delta x^\alpha \Bigg|_{t_1}^{t_2} - \delta \int_{t_1}^{t_2} \mathrm{d}t\, L - \int_{t_1}^{t_2} \mathrm{d}t \sum_A \lambda_A \delta \phi^A\,. \tag{3.21}$$

Now the first term on the right-hand side of this equation vanishes for variations with fixed endpoints, while the last term can be written as

$$\begin{aligned} -\int_{t_1}^{t_2} \mathrm{d}t \sum_A \lambda_A \delta \phi^A &= -\int_{t_1}^{t_2} \mathrm{d}t \sum_A \delta(\lambda_A \phi^A) + \int_{t_1}^{t_2} \mathrm{d}t \sum_A \delta \lambda_A \phi^A \\ &= -\delta \int_{t_1}^{t_2} \mathrm{d}t \sum_A \lambda_A \phi^A\,, \end{aligned} \tag{3.22}$$

where we used the constraint equations $\phi^A = 0$, where $A = 1, 2, \cdots, M$, and the linearity of the virtual displacement operation $\delta$ to get the last equality. So we are left with

$$0 = -\delta \int_{t_1}^{t_2} \mathrm{d}t\, L - \delta \int_{t_1}^{t_2} \mathrm{d}t \sum_A \lambda_A \phi^A = -\delta \int_{t_1}^{t_2} \mathrm{d}t\, [L + \sum_A \lambda_A \phi^A] . \tag{3.23}$$

Thus, for monogenic forces, the equations of motion (3.19) and the constraint equations $\phi^A(x, t) = 0$, where $A = 1, 2, \cdots, M$, are the stationary values $\delta\bar{S} = 0$ of the modified action functional

$$\bar{S}[x, \lambda] \equiv \int_{t_1}^{t_2} \mathrm{d}t \left[ L(x, \dot{x}, t) + \sum_A \lambda_A \phi^A(x, t) \right] , \tag{3.24}$$

where $x$, $\dot{x}$, and $\lambda$ are shorthand for all the coordinates $(x^1, x^2, \cdots, x^{3N})$, velocities $(\dot{x}^1, \dot{x}^2, \cdots, \dot{x}^{3N})$, and Lagrange multipliers $(\lambda_1, \lambda_2, \cdots, \lambda_M)$. Note that variation of $\bar{S}$ with respect to $x^\alpha$ yields (3.19), while variation of $\bar{S}$ with respect to the Lagrange multipliers $\lambda_A$ yields the constraint equations $\phi^A(x, t) = 0$, where $A = 1, 2, \cdots, M$. Thus, the constraints do not need to be imposed as *additional conditions*, but arise from the variation of the modified action functional $\bar{S}$ with respect to the Lagrange multipliers.

### 3.2.2 Non-holonomic Constraints

For non-holonomic constraints Lagrange's equation of the 1st kind are

$$\frac{\mathrm{d}}{\mathrm{d}t}\left(\frac{\partial T}{\partial \dot{x}^\alpha}\right) - \frac{\partial T}{\partial x^\alpha} - F_\alpha - \sum_A \lambda_A C^A_\alpha = 0 , \quad \alpha = 1, 2, \cdots, 3N . \tag{3.25}$$

which is simply (2.44) with $\partial\phi^A/\partial x^\alpha$ replaced by $C^A_\alpha$. But even for generalized forces derivable from a potential $U$, for which the above equations simplify to

$$\frac{\mathrm{d}}{\mathrm{d}t}\left(\frac{\partial L}{\partial \dot{x}^\alpha}\right) - \frac{\partial L}{\partial x^\alpha} - \sum_A \lambda_A C^A_\alpha = 0 , \qquad \alpha = 1, 2, \cdots, 3N , \tag{3.26}$$

where $L \equiv T - U$, these equations are *not* the stationary values of an action functional. Multiplying (3.26) by a virtual displacement $\delta x^\alpha$, summing over $\alpha$, and integrating over time, leads to

$$0 = -\delta S - \int_{t_1}^{t_2} \mathrm{d}t \sum_\alpha \sum_A \lambda_A C_\alpha^A \delta x^\alpha \,, \tag{3.27}$$

where $S = \int_{t_1}^{t_2} \mathrm{d}t\, L$. But there is no way to write the second term as the variation of a single object, since the non-holonomic constraints $\sum_\alpha C_\alpha^A \delta x^\alpha = 0$, where $A = 1, 2, \cdots, M$, are *not integrable* and hence cannot be written in the form $\delta\phi^A = 0$ for any set of functions $\phi^A(x, t)$ (See Sects. 2.2.2 and 2.2.3). *Thus, we cannot define an action functional for non-holonomic constraints.*

## 3.3 Conservation Laws Revisited

In Sect. 1.4, we discussed conservation laws for systems of particles subject to external and internal forces. The three fundamental conservation laws were:

I. **Conservation of Linear Momentum:** If the net external force on a system is zero, then the total linear momentum $\mathbf{P} \equiv \sum_I m_I \mathbf{v}_I$ is conserved.
II. **Conservation of Angular Momentum:** If the net external torque on a system is zero and the *strong form* of Newton's 3rd law holds (i.e., the interparticle forces $\mathbf{F}_{JI}$ are directed along the line connecting particles $I$ and $J$), then the total angular momentum $\mathbf{L} \equiv \sum_I \mathbf{r}_I \times \mathbf{p}_I$ is conserved.
III. **Conservation of Mechanical Energy:** If both the external forces and interparticle forces are expressible as gradients of scalar potentials, then the total mechanical energy of the system $E \equiv T + U$ is conserved.

Note that I and II are *vector* conservation laws, which means that if a particular component of the total external force (or torque) is zero, then the corresponding component of the total linear (or angular) momentum is conserved. These conservation laws were proved in Chap. 1 using Newton's laws of motion, assuming either the weak or strong form of Newton's 3rd law, as needed.

Alternatively, one can arrive at similar conservation laws working within the context of the Lagrangian formulation of mechanics. The conserved quantities turn out to be related to *symmetries* of the Lagrangian $L \equiv T - U$, i.e., the total linear momentum, angular momentum, and energy of a system are conserved if the Lagrangian changes by at most a total time derivative (Exercise 3.2) under spatial translations, rotations, and time translations. This result is treated in much more detail in the optional chapter on fields in Sect. 10.4. The results, originally due to Emmy Noether (Noether 1918, 1971), show that conservation laws I and II can be combined into one law[2]:

[2]The underbar on the index $\underline{a}$ in conservation law I/II indicates that this is a particular (i.e., *single*) value of the index; it should not be thought of as a placeholder for *all* possible values, as the index $a$ without an underbar usually represents.

I/II. If the Lagrangian is independent of a particular generalized coordinate $q^{\underline{a}}$, then the corresponding **generalized momentum**

$$p_{\underline{a}} \equiv \frac{\partial L}{\partial \dot{q}^{\underline{a}}} \tag{3.28}$$

is conserved.

III. If the Lagrangian does not depend explicitly on time, then

$$h(q, \dot{q}, t) \equiv \sum_a p_a \dot{q}^a - L(q, \dot{q}, t) \tag{3.29}$$

is conserved (i.e., $dh/dt = 0$) for solutions to the equations of motion.

The function $h$ is called the **energy function**, for reasons that will become more clear below.

**Exercise 3.4** Verify I/II, and III using the Euler-Lagrange equations (3.1).

We see how the conservation theorems I and II are combined in the Lagrangian formulation as the generalized momentum $p_{\underline{a}}$ can correspond either to a component of linear momentum (if $q^{\underline{a}}$ is a translational coordinate, such as $x$) or angular momentum (if $q^{\underline{a}}$ is an angular coordinate, such as $\phi$). Moreover, conservation theorem I/II is valid even when Newton's 3rd law does not hold—e.g., for an electric charge moving in an electromagnetic field. For example, if the scalar and vector potentials $\Phi$ and $\mathbf{A}$ from Exercise 2.14 are independent of $x$, then $p_x = m\dot{x} + qA_x$ is conserved. The momentum of the electromagnetic field is included in the expression for the generalized momentum.

Finally, if the embedding equations $\mathbf{r}_I = \mathbf{r}_I(q^1, q^2, \cdots, q^n, t)$ are independent of time, and the generalized potential $U$ is also independent of the generalized velocities $\dot{q}^a$, where $a = 1, 2, \cdots, n$, then $h$ defined in conservation law III is equal to the total mechanical energy $E \equiv T + U$. To see that this is indeed the case, recall from Exercise 2.12 that if $\mathbf{r}_I = \mathbf{r}_I(q^1, q^2, \cdots, q^n)$, then the total kinetic energy $T = \frac{1}{2}\sum_{a,b} T_{ab}\dot{q}^a\dot{q}^b$, where $T_{ab}$ depends only on the generalized coordinates $q^a$, where $a = 1, 2, \cdots, n$. Thus, $T$ is a homogeneous, quadratic function in the generalized velocities, which implies

$$\sum_a \frac{\partial T}{\partial \dot{q}^a}\dot{q}^a = \sum_a \left(\sum_b T_{ab}\dot{q}^b\right)\dot{q}^a = 2T\,. \tag{3.30}$$

But since $U$ is independent of the generalized velocities, we also have

$$p_a \equiv \frac{\partial L}{\partial \dot{q}^a} = \frac{\partial T}{\partial \dot{q}^a}\,. \tag{3.31}$$

Thus,

$$h \equiv \sum_a p_a\dot{q}^a - L = \sum_a \frac{\partial T}{\partial \dot{q}^a}\dot{q}^a - L = 2T - (T - U) = T + U = E\,. \tag{3.32}$$

Note that the conditions for $h =$ const and $h = E$ are different and need not be satisfied together. For example, for the above derivation, $U$ may depend explicitly on $t$. If that's the case, then $\mathrm{d}h/\mathrm{d}t \neq 0$ even though $h = E$.

## 3.4 Hamilton's Equations

The Euler-Lagrange equations (3.1) are $n$ 2nd-order ordinary differential equations for the generalized coordinates $q^a = q^a(t)$, where $a = 1, 2, \cdots, n$. To solve these equations, it is often convenient to convert this system of equations to $2n$ *1st-order* ordinary differential equations. This can be done using a mathematical technique called a **Legendre transform**, which transforms the Lagrangian $L = L(q, \dot{q}, t)$ to the so-called **Hamiltonian** $H = H(q, p, t)$, where $p_a \equiv \partial L/\partial \dot{q}^a$. As we shall see below, this transformation replaces the $n$ 2nd-order Euler-Lagrange equations for the generalized coordinates $q^a$ by $2n$ 1st-order equations, called **Hamilton's equations**, for the generalized coordinates and momenta $(q^a, p_a)$. But first we will digress a bit and review the Legendre transform in some detail.

### 3.4.1 *Legendre Transform*

Suppose we are given a function $F = F(u, v)$. We will define

$$w \equiv \frac{\partial F}{\partial v}\,, \tag{3.33}$$

and assume that

$$\frac{\partial w}{\partial v} = \frac{\partial}{\partial v}\left(\frac{\partial F}{\partial v}\right) \neq 0 \tag{3.34}$$

so that we can invert (3.33) to obtain $v = v(u, w)$. (This follows from the *implicit function theorem*—see e.g., Boas 2006 for a description of this theorem.) Then

$$G(u, w) \equiv (wv - F(u, v))|_{v=v(u,w)} \tag{3.35}$$

is said to be the the **Legendre transform** of $F$ with respect to $v$. To find the partial derivatives of $G$ with respect to $u$ and $w$, we take the total derivative

$$\mathrm{d}G = v\,\mathrm{d}w + w\,\mathrm{d}v - \mathrm{d}F = v\,\mathrm{d}w + w\,\mathrm{d}v - \frac{\partial F}{\partial u}\mathrm{d}u - \frac{\partial F}{\partial v}\mathrm{d}v = v\,\mathrm{d}w - \frac{\partial F}{\partial u}\mathrm{d}u\,, \tag{3.36}$$

where we used (3.33) for $w$ to get the last equality. Thus,

$$\frac{\partial G}{\partial u} = -\frac{\partial F}{\partial u}\,, \qquad \frac{\partial G}{\partial w} = v\,. \tag{3.37}$$

Although we have been treating $u$ and $v$ as if they were single variables, the above defintions can be trivially extended to the case where $u$ and $v$ are replaced by sets of $n$ variables, $u \equiv (u^1, u^2, \cdots, u^n)$ and $w \equiv (v^1, v^2, \cdots, v^n)$. Then[3]

$$w_i \equiv \frac{\partial F}{\partial v^i}\,, \qquad \det\left(\frac{\partial^2 F}{\partial v^i \partial v^j}\right) \neq 0\,, \tag{3.38}$$

and

$$G(u, w) \equiv \left.\left(\sum_i w_i v^i - F(u, v)\right)\right|_{v=v(u,w)}\,, \tag{3.39}$$

with

$$\frac{\partial G}{\partial u^i} = -\frac{\partial F}{\partial u^i}\,, \qquad \frac{\partial G}{\partial w_i} = v^i\,, \tag{3.40}$$

[3]We are abusing notation slightly in (3.38), writing the determinant of the matrix of 2nd partial derivatives of $F$ with respect to the $v^i$ as the determinant of the matrix components. We will occasionally do this whenever the abstract matrix notation is more cumbersome or less informative than the matrix component notation.

where $i = 1, 2, \cdots, n$. Given this general introduction to Legendre transforms, we are now ready to apply it to the Lagrangian.

### 3.4.2 Hamiltonian

Given the Lagrangian $L(q, \dot{q}, t)$, where $q$ and $\dot{q}$ are shorthand for the set of generalized coordinates and generalized velocities, we define the momenta $p_a$ via

$$p_a \equiv \frac{\partial L}{\partial \dot{q}^a}, \qquad a = 1, 2, \cdots, n. \tag{3.41}$$

Note that $p_a$ plays the role of $w$ in (3.33). The **Hamiltonian** $H(q, p, t)$ is then given by the Legendre transform[4]

$$H(q, p, t) = \left( \sum_a p_a \dot{q}^a - L(q, \dot{q}, t) \right) \Bigg|_{\dot{q} = \dot{q}(q, p, t)}. \tag{3.42}$$

The condition for the inversion of (3.41) is

$$\det \left( \frac{\partial^2 L}{\partial \dot{q}^a \partial \dot{q}^b} \right) \neq 0. \tag{3.43}$$

Note that the Hamiltonian $H$ is numerically equal to the energy function $h$, which was defined by (3.29) in the context of conservation laws. The only difference between $h$ and $H$ is their functional dependence, $h = h(q, \dot{q}, t)$ versus $H = H(q, p, t)$. The $2n$-dimensional space of all possible generalized coordinates and generalized momenta $(q, p) \equiv (q^1, q^2, \cdots, q^n; p_1, p_2, \cdots, p_n)$ on which $H$ is defined is called the **phase space** of the system—as distinguished from the *configuration space* that we have used earlier. When needed, we will denote the phase space by $\Gamma$.

Using (3.40), which relates the partial derivatives for a Legendre transform, and the Euler-Lagrange equations (3.1), it is a relatively simple matter to obtain **Hamilton's equations**

$$\dot{q}^a = \frac{\partial H}{\partial p_a}, \qquad \dot{p}_a = -\frac{\partial H}{\partial q^a}, \qquad a = 1, 2, \cdots, n. \tag{3.44}$$

The first equation is simply $v = \partial G / \partial w$ from (3.40), written in terms of $\dot{q}^a$, $H$, and $p_a$. The second equation is $\partial G / \partial u = -\partial F / \partial u$ written in terms of $H$, $q^a$, and $L$,

[4]Basically $F$, $G$, $u$, $v$, $w$ defined in the previous section are replaced by $L$, $H$, $q$, $\dot{q}$, $p$; and the possible explicit dependence on the time $t$ just goes along for the ride.

and then using (3.1) and (3.41) to write it in terms of $\dot{p}_a$:

$$\frac{\partial H}{\partial q^a} = -\frac{\partial L}{\partial q^a} = -\frac{\mathrm{d}}{\mathrm{d}t}\left(\frac{\partial L}{\partial \dot{q}^a}\right) = -\dot{p}_a \,. \tag{3.45}$$

Since you can't get something for nothing, note that there are now $2n$ 1st-order Hamilton equations as compared to the $n$ 2nd-order Euler-Lagrange equations. Thus, we still require the same number of initial conditions.

***Example 3.1*** Using Hamilton's equations, it is also simple to show that

$$\frac{\mathrm{d}H}{\mathrm{d}t} = \frac{\partial H}{\partial t} = -\frac{\partial L}{\partial t} \,. \tag{3.46}$$

This means that $H$ is conserved if and only if $H$ (or $L$) does not depend explicitly on $t$. Given the numerical equivalence of $H$ and $h$, it follows that $H$ is conserved or equals the total mechanical energy $E$ if and only if $h$ also has these properties.

*Proof* We can prove the above result by applying the chain rule so that

$$\begin{aligned}\frac{\mathrm{d}H}{\mathrm{d}t} &= \frac{\partial H}{\partial t} + \sum_a \left[\frac{\partial H}{\partial q^a}\dot{q}^a + \frac{\partial H}{\partial p_a}\dot{p}_a\right] \\ &= \frac{\partial H}{\partial t} + \sum_a \left[\frac{\partial H}{\partial q^a}\frac{\partial H}{\partial p_a} - \frac{\partial H}{\partial p_a}\frac{\partial H}{\partial q^a}\right] = \frac{\partial H}{\partial t} \,.\end{aligned} \tag{3.47}$$

where we used Hamilton's equations (3.44) to get the second equality. In addition,

$$\begin{aligned}\frac{\partial H}{\partial t} &= \frac{\partial}{\partial t}\left[\left(\sum_a p_a\dot{q}^a - L(q,\dot{q},t)\right)\Big|_{\dot{q}=\dot{q}(q,p,t)}\right] \\ &= \left(\sum_a p_a\frac{\partial \dot{q}^a}{\partial t} - \sum_a \frac{\partial L}{\partial \dot{q}^a}\frac{\partial \dot{q}^a}{\partial t} - \frac{\partial L}{\partial t}\right)\Big|_{\dot{q}=\dot{q}(q,p,t)} \\ &= -\frac{\partial L}{\partial t}\Big|_{\dot{q}=\dot{q}(q,p,t)} \,,\end{aligned} \tag{3.48}$$

where we first used the definition (3.42) of $H$, and then $p_a \equiv \partial L/\partial \dot{q}^a$ to cancel out the first two terms on the second line. □

**Exercise 3.5** Consider a single particle of mass $m$ moving in the presence of a conservative force $\mathbf{F} = -\nabla U$. (a) Starting from the Lagrangian

$$L = \frac{1}{2}mv^2 - U(\mathbf{r}), \tag{3.49}$$

show that the Hamiltonian is given by

$$H = \frac{p^2}{2m} + U(\mathbf{r}). \tag{3.50}$$

(b) Then show that Hamilton's equations $\dot{x}^i = \partial H/\partial p_i$ and $\dot{p}_i = -\partial H/\partial x^i$ recover the definition of momentum $\mathbf{p} = m\dot{\mathbf{r}}$ and Newton's 2nd law $\dot{\mathbf{p}} = -\nabla U$.

### 3.4.3 *1st-Order Action for Hamilton's Equations*

Just as the Euler-Lagrange equations (3.1) can be derived by finding the stationary values of the action $S[q] \equiv \int_{t_1}^{t_2} \mathrm{d}t\, L(q, \dot{q}, t)$, so too can Hamilton's equations (3.44) be derived by finding the stationary values of the *1st-order* action

$$S[q, p] \equiv \int_{t_1}^{t_2} \mathrm{d}t \left[ \sum_a p_a \dot{q}^a - H(q, p, t) \right]. \tag{3.51}$$

Note that the integrand of $S$ is numerically equal to the Lagrangian, where we think of applying the Legendre transform (3.42) in the opposite direction, i.e., from $H(q, p, t)$ to $L(q, \dot{q}, t)$ with $\dot{q}^a \equiv \partial H/\partial p_a$.

To show that the variation of (3.51) does, indeed, yield Hamilton's equations (3.44), let's first consider variations of $S$ induced by variations of the generalized coordinates,

$$\delta S = \int_{t_1}^{t_2} \mathrm{d}t \sum_a \left[ p_a \delta\dot{q}^a - \frac{\partial H}{\partial q^a}\delta q^a \right]. \tag{3.52}$$

Since $\delta\dot{q}^a = \frac{\mathrm{d}}{\mathrm{d}t}\delta q^a$, we can integrate by parts to free up $\delta q^a$, giving

$$\delta S = \sum_a p_a \delta q^a \bigg|_{t_1}^{t_2} - \int_{t_1}^{t_2} \mathrm{d}t \sum_a \delta q^a \left[ \dot{p}_a + \frac{\partial H}{\partial q^a} \right]. \tag{3.53}$$

Since the first term vanishes for variations of the generalized coordinates that have fixed endpoints (i.e., $\delta q^a|_{t_1} = 0$, $\delta q^a|_{t_2} = 0$), we see that $\delta S = 0$ implies

$$\dot{p}_a = -\frac{\partial H}{\partial q^a}, \qquad a = 1, 2, \cdots, n. \tag{3.54}$$

Similary, for variations of the generalized momenta, we obtain

$$\delta S = \int_{t_1}^{t_2} \mathrm{d}t \sum_a \delta p_a \left[ \dot{q}^a - \frac{\partial H}{\partial p_a} \right], \tag{3.55}$$

for which $\delta S = 0$ immediately implies

$$\dot{q}^a = \frac{\partial H}{\partial p_a}, \qquad a = 1, 2, \cdots, n. \tag{3.56}$$

Note that although the variations $\delta p_a$ *need not vanish* at $t_1$ and $t_2$ to obtain this last set of equations, one will sometimes restrict the variations $\delta p_a$ to vanish at the endpoints in order to make this 1st-order action formulation symmetric with respect to variations of $q^a$ and $p_a$.

***Example 3.2*** It is also possible to show that the 1st-order action (3.51) for Hamilton's equations can be obtained directly using the method of Lagrange multipliers (See, e.g., Lanczos (1949), Appendix II). To do this, we first start with the standard 2nd-order action

$$S[q] = \int_{t_1}^{t_2} \mathrm{d}t\, L(q, \dot{q}, t), \tag{3.57}$$

and replace the generalized velocities $\dot{q}^a$ with new variables $v^a$, which we will want to vary *independently* of the $q^a$:

$$S[q, v] = \int_{t_1}^{t_2} \mathrm{d}t\, L(q, v, t). \tag{3.58}$$

In order to obtain the same equations of motions that we originally had from $S[q]$, we impose the *semi-holonomic*[5] auxiliary conditions

$$\phi^a(\dot{q}, v) \equiv \dot{q}^a - v^a = 0, \qquad a = 1, 2, \cdots, n. \tag{3.59}$$

This can be done using (3.141) from Problem 3.2 to obtain a modified action

$$\bar{S}[q, v, p] = \int_{t_1}^{t_2} \mathrm{d}t \left[ L(q, v, t) + \sum_a p_a(\dot{q}^a - v^a) \right], \tag{3.60}$$

[5]Semi-holonomic constraints are described in more detail in Problem 3.2.

where the Lagrange multipliers are denoted here by $p_a$ for reasons that will become apparent shortly.

Since the integrand of (3.60) does not involve any of the time derivatives $\dot{v}^a$, it is possible to eliminate all the $v^a$ from the problem by using the results of Simplification (3), discussed in Appendix C.7.3. That is, we solve the Euler equations for $v^a$ in terms of the remaining variables

$$\frac{\partial L}{\partial v^a} - p_a = 0 \quad \Rightarrow \quad v^a = v^a(q, p, t) \tag{3.61}$$

and obtain the reduced action

$$\underline{S}[q, p] = \int_{t_1}^{t_2} \mathrm{d}t \left[ L(q, v, t) + \sum_a p_a(\dot{q}^a - v^a) \right] \Bigg|_{v=v(q,p,t)} . \tag{3.62}$$

But notice that the integrand of $\underline{S}[q, p]$ can be written as

$$\sum_a p_a \dot{q}^a - \left( \sum_a p_a v^a - L(q, v, t) \right) \Bigg|_{v=v(q,p,t)} = \sum_a p_a \dot{q}^a - H(q, p, t) , \tag{3.63}$$

which is the precisely the integrand of the 1st-order action (3.51). □

## 3.5 Poisson Brackets and Canonical Transformations

In this section, we define a mathematical structure[6] on phase space, called **Poisson brackets**. Hamilton's equations have a very simple representation when expressed in terms of Poisson brackets. We also define an associated class of transformations on phase space, called **canonical transformations**, which preserve the form of the Poisson brackets. Poisson brackets and canonical transformations are powerful mathematical tools that will allow us to understand more fully the connection between continuous *symmetries* of the system (as specified by a 1-parameter family of canonical transformations) and *conserved quantities*. A few such applications of canonical transformations will be discussed in Sect. 3.6.

[6] By "mathematical structure" we simply mean an operation or rule acting on the elements of a set. For example, the dot product of two vectors is an additional mathematical structure on the space of vectors.

### 3.5.1 *Poisson Brackets*

Given two functions on phase space, $f(q, p)$ and $g(q, p)$, we define their **Poisson bracket** $\{f, g\}$ to be

$$\{f, g\} \equiv \sum_a \left( \frac{\partial f}{\partial q^a} \frac{\partial g}{\partial p_a} - \frac{\partial f}{\partial p_a} \frac{\partial g}{\partial q^a} \right) . \tag{3.64}$$

Note that Poisson brackets satisfy the following three properties:

(i) Anti-symmetry:

$$\{f, g\} = -\{g, f\} . \tag{3.65}$$

(ii) Linearity:

$$\{f, g + a\,h\} = \{f, g\} + a\{f, h\} , \tag{3.66}$$

where $a$ is a constant.

(iii) Jacobi identity:

$$\{f, \{g, h\}\} + \{g, \{h, f\}\} + \{h, \{f, g\}\} = 0 . \tag{3.67}$$

**Exercise 3.6** Prove the above three properties, (3.65), (3.66), and (3.67), of Poisson brackets.

**Exercise 3.7** Show that Poisson brackets also satisfy the *product rule*

$$\{f, gh\} = \{f, g\}h + g\{f, h\} , \tag{3.68}$$

and *chain rule*

$$\{f, g(h)\} = \frac{\mathrm{d}g}{\mathrm{d}h}\{f, h\} , \tag{3.69}$$

for the product and composition of two functions.

The Poisson brackets of the generalized coordinates and momenta are particularly simple:

$$\{q^a, q^b\} = 0 , \qquad \{p_a, p_b\} = 0 , \qquad \{q^a, p_b\} = \delta^a_b . \tag{3.70}$$

In addition, Hamilton's equations (3.44) can be written in terms of Poisson brackets as

$$\dot{q}^a = \{q^a, H\}, \qquad \dot{p}_a = \{p_a, H\}, \qquad a = 1, 2, \cdots, n. \tag{3.71}$$

Using these last results, it follows that the total time derivative of *any* function $f(q, p)$, evaluated along a curve $q^a = q^a(t)$, $p_a = p_a(t)$ satisfying the equations of motion, is given by

$$\frac{\mathrm{d}f}{\mathrm{d}t} = \sum_a \left( \frac{\partial f}{\partial q^a}\dot{q}^a + \frac{\partial f}{\partial p_a}\dot{p}_a \right) = \sum_a \left( \frac{\partial f}{\partial q^a}\frac{\partial H}{\partial p_a} - \frac{\partial f}{\partial p_a}\frac{\partial H}{\partial q^a} \right) = \{f, H\}. \tag{3.72}$$

If $f$ also depends explicitly on time, i.e., $f = f(q, p, t)$, then

$$\frac{\mathrm{d}f}{\mathrm{d}t} = \{f, H\} + \frac{\partial f}{\partial t}. \tag{3.73}$$

Thus, we see that the time evolution of a function defined on phase space is *generated* by the Hamiltonian $H$. (See Sect. 3.6.1 and, in particular, the discussion at the end of that section for a more precise meaning of a *generator* of a transformation.) The condition then for $f(q, p)$ to be a conserved quantity is that its Poisson bracket with the Hamiltonian vanish, i.e.,

$$f(q, p) \text{ is conserved} \quad \Leftrightarrow \quad \{f, H\} = 0. \tag{3.74}$$

**Exercise 3.8** Show that if both $f(q, p)$ and $g(q, p)$ are conserved quantities, then their Poisson bracket $\{f, g\}$ is also conserved.

**Exercise 3.9** Using Poisson brackets, show that for a single particle of mass $m$ moving in the presence of a *central* potential $U = U(r)$, the components of the angular momentum vector $\boldsymbol{\ell} \equiv \mathbf{r} \times \mathbf{p}$ are conserved.

### 3.5.2 *Canonical Transformations*

A **canonical transformation** is a (possibly time-dependent) mapping of the phase space variables $(q^a, p_a)$ to a new set of variables

$$Q^a = Q^a(q, p, t), \qquad P_a = P_a(q, p, t), \qquad a = 1, 2, \cdots, n, \tag{3.75}$$

that also satisfy Hamilton's equations

$$\dot{Q}^a = \frac{\partial H'}{\partial P_a}, \quad \dot{P}_a = -\frac{\partial H'}{\partial Q_a}, \quad a = 1, 2, \cdots, n, \tag{3.76}$$

for some (possibly new) Hamiltonian $H' = H'(Q, P, t)$. Since phase space is $2n$-dimensional, there is considerably more freedom in transforming the $q$'s and the $p$'s than there is in transforming just the generalized coordinates $q$ in configuration space. But contrary to the fact that *all* invertible transformations of the $q$'s preserve the form of the Euler-Lagrange equations (See Exercise 3.3), *not all* invertible transformations of the $q$'s and $p$'s lead to new variables satisfying Hamilton's equations. Thus, canonical transformations form a special *subset* of all possible transformations on phase space. The following calculations classify the form of canonical transformations in terms of what are called **generating functions**.

#### 3.5.2.1 Generating Functions for Canonical Transformations

Recall from Sect. 3.4.3 that Hamilton's equations for $(q^a, p_a)$ can be derived from a variational calculation

$$\delta \int_{t_1}^{t_2} \mathrm{d}t \left[ \sum_a p_a \dot{q}^a - H(q, p, t) \right] = 0. \tag{3.77}$$

So, similarly, we must have

$$\delta \int_{t_1}^{t_2} \mathrm{d}t \left[ \sum_a P_a \dot{Q}^a - H'(Q, P, t) \right] = 0 \tag{3.78}$$

for the new phase space variables $(Q^a, P_a)$ to satisfy Hamilton's equations (3.76). But for the above variational equations to hold, the integrands of the two integrals can differ at most[7] by a total time derivative (Exercise 3.2), so that

$$\sum_a p_a \dot{q}^a - H(q, p, t) = \sum_a P_a \dot{Q}^a - H'(Q, P, t) + \frac{\mathrm{d}F}{\mathrm{d}t}. \tag{3.79}$$

This is the condition that the transformation from $(q^a, p_a)$ to $(Q^a, P_a)$ be canonical. Note that each canonical transformation is associated with a specific function $F$, which is called the **generating function** of that canonical transformation. Different generating functions can have different functional forms depending on which set of the old and new canonical variables are *independent* of one another.

[7] We will not consider transformations that simply *rescale* the integrand, such as $Q^a = q^a$, $P_a = \lambda p_a$, where $\lambda$ is a constant, which also preserve Hamilton's equations with $H' = \lambda H$. See Goldstein et al. (2002) for more information about such scaling transformations.

For example, let's assume that all of old and new coordinates, $q^a$ and $Q^a$, are independent of one another, and then rewrite (3.79) by multiplying through by $\mathrm{d}t$ and solving for $\mathrm{d}F$:

$$\mathrm{d}F = \sum_a \left(p_a\,\mathrm{d}q^a - P_a\,\mathrm{d}Q^a\right) + (H' - H)\,\mathrm{d}t\,. \tag{3.80}$$

Taking $F \equiv F_1(q, Q, t)$, it immediately follows that

$$p_a = \frac{\partial F_1}{\partial q^a}\,, \qquad P_a = -\frac{\partial F_1}{\partial Q^a}\,, \qquad H' = H + \frac{\partial F_1}{\partial t}\,. \tag{3.81}$$

It is customary to call such a transformation a **Type I** canonical transformation, with generating function $F_1(q, Q, t)$. **Type II**, **III**, and **IV** canonical transformations are similarly defined in terms of generating functions that depend on $(q, P, t)$, $(p, Q, t)$, and $(p, P, t)$, respectively. So, for a Type II transformation, we rewrite (3.80) as

$$\mathrm{d}F = \sum_a \left(p_a\,\mathrm{d}q^a - \mathrm{d}(P_a Q^a) + Q^a\,\mathrm{d}P_a\right) + (H' - H)\,\mathrm{d}t\,, \tag{3.82}$$

and then rearrange terms to get

$$\mathrm{d}\left(F + \sum_a P_a Q^a\right) = \sum_a \left(p_a\,\mathrm{d}q^a + Q^a\,\mathrm{d}P_a\right) + (H' - H)\,\mathrm{d}t\,. \tag{3.83}$$

Then taking $F_2 \equiv F + \sum_a P_a Q^a$ to be a function $(q, P, t)$, we have

$$p_a = \frac{\partial F_2}{\partial q^a}\,, \qquad Q^a = \frac{\partial F_2}{\partial P_a}\,, \qquad H' = H + \frac{\partial F_2}{\partial t}\,. \tag{3.84}$$

The corresponding transformation equations for Type III and Type IV canonical transformations are similar to (3.81) and (3.84). You are asked to work out explicit expressions for these transformation equations in Problem 3.10.

Note that a *time-independent* canonical transformation is one whose generating function does not explicitly depend on $t$. For such a transformation, the equations simplify a bit as the new Hamiltonian $H'(Q, P, t)$ is equal to the old Hamiltonian $H(q, p, t)$, with $q$ and $p$ replaced by $q = q(Q, P)$ and $p = p(Q, P)$. In other words, the Hamiltonian transforms as a *scalar* function for a time-independent canonical transformation; its value at each point in phase space is unchanged by the transformation, although its functional form in terms of $(Q, P)$ may differ from that in terms of $(q, p)$.

**Exercise 3.10** Show that

$$Q^a = p_a\,, \qquad P_a = -q^a\,, \qquad a = 1, 2, \cdots, n\,, \tag{3.85}$$

is a canonical transformation from $(q, p)$ to $(Q, P)$ with generating function $F_1(q, Q) = \sum_a q^a Q^a$. Thus, there is nothing special about the distinction between coordinates and momenta; you can swap the two, provided you introduces a *minus* sign.

***Example 3.3*** A simple example of a canonical transformation is a so-called **point transformation**

$$Q^a = Q^a(q, t) \qquad \Leftrightarrow \qquad q^a = q^a(Q, t)\,, \tag{3.86}$$

which corresponds to a change of coordinates on configuration space (See also Exercise 3.3). For a concrete example, think of translations or rotations of the generalized coordinates $q^a$. Such a transformation of coordinates induces the following transformation of the velocities,

$$\dot{Q}^a = \sum_b \frac{\partial Q^a}{\partial q^b}\dot{q}^b + \frac{\partial Q^a}{\partial t} \qquad \Leftrightarrow \qquad \dot{q}^a = \sum_b \frac{\partial q^a}{\partial Q^b}\dot{Q}^b + \frac{\partial q^a}{\partial t}\,. \tag{3.87}$$

Since the new Lagrangian is simply the old Lagrangian expressed as a function of the new coordinates and velocities, we have

$$L'(Q, \dot{Q}, t) = L\,(q, \dot{q}, t)|_{q=q(Q,t),\, \dot{q}=\dot{q}(Q,\dot{Q},t)}\,, \tag{3.88}$$

with new momenta

$$P_a \equiv \frac{\partial L'}{\partial \dot{Q}^a} = \sum_b \frac{\partial L}{\partial \dot{q}^b}\frac{\partial \dot{q}^b}{\partial \dot{Q}^a} = \sum_b p_b \frac{\partial q^b}{\partial Q^a}\,. \tag{3.89}$$

Thus,

$$Q^a = Q^a(q, t)\,, \qquad P_a = \sum_b p_b \frac{\partial q^b}{\partial Q^a}\,, \tag{3.90}$$

is the representation of a point transformation in phase space. It is easy to see that this is a Type II canonical transformation with generating function

$$F_2(q, P, t) = \sum_a P_a Q^a(q, t) . \tag{3.91}$$

To verify this explicitly, we calculate

$$\begin{aligned} \frac{\partial F_2}{\partial q^a} &= \sum_b P_b \frac{\partial Q^b}{\partial q^a} = \sum_b \sum_c p_c \frac{\partial q^c}{\partial Q_b} \frac{\partial Q^b}{\partial q^a} = \sum_c p_c \delta^c_a = p_a , \\ \frac{\partial F_2}{\partial P_a} &= Q^a , \end{aligned} \tag{3.92}$$

and

$$\begin{aligned} H' &= \sum_a P_a \dot{Q}^a - L' = \sum_a \sum_b p_b \frac{\partial q^b}{\partial Q^a} \left( \sum_c \frac{\partial Q^a}{\partial q^c} \dot{q}^c + \frac{\partial Q^a}{\partial t} \right) - L \\ &= \sum_b \sum_c p_b \delta^b_c \dot{q}^c + \sum_a P_a \frac{\partial Q^a}{\partial t} - L = \left( \sum_b p_b \dot{q}^b - L \right) + \sum_a P_a \frac{\partial Q^a}{\partial t} \\ &= H + \frac{\partial F_2}{\partial t} , \end{aligned} \tag{3.93}$$

which agree with (3.84). As a trivial example, the *identity* transformation, $Q^a = q^a$, $P_a = p_a$, is a time-independent point transformation with $F_2(q, P) = \sum_a P_a q^a$. □

***Example 3.4*** We can also construct canonical transformations of *mixed type*, which have generating functions that depend on some combination of $n$ old and $n$ new canonical variables, provided they are independent of one another. A simple example of such a mixed type canonical transformation (for $n = 2$ dimensions) is given by

$$\begin{aligned} Q^1 &= q^1 , \qquad P_1 = p_1 , \\ Q^2 &= p_2 , \qquad P_2 = -q^2 . \end{aligned} \tag{3.94}$$

Note that this is just the identity transformation for $(q^1, p_1) \to (Q^1, P_1)$, but a swap of the coordinate and momentum (with the requisite minus sign) for $(q^2, p_2) \to (Q^2, P_2)$. The generating function for this transformation is

$$F(q^1, q^2, P_1, Q^2) = P_1 q^1 + q^2 Q^2 , \tag{3.95}$$

which is the sum of a (1-dimensional) Type II generating function $F_2(q^1, P_1) \equiv P_1 q^1$ for the identity component, and a (1-dimensional) Type I generating function $F_1(q^2, Q^2) \equiv q^2 Q^2$ for the swapping component. Performing the relevant partial derivatives in (3.84) and (3.81), we find

$$
\begin{aligned}
p_1 &= \frac{\partial F}{\partial q^1} = P_1 , \qquad Q^1 = \frac{\partial F}{\partial P_1} = q^1 , \\
p_2 &= \frac{\partial F}{\partial q^2} = Q^2 , \qquad P_2 = -\frac{\partial F}{\partial Q^2} = -q^2 ,
\end{aligned}
\tag{3.96}
$$

which agree with (3.94) as they should. □

Note that the above example can easily be extended to $n$-dimensions, where we swap only *one* pair of canonical variables, e.g.,

$$
Q^{\underline{a}} = p_{\underline{a}} , \qquad P_{\underline{a}} = -q^{\underline{a}} , \tag{3.97}
$$

and use the identity transformation for all the others:

$$
Q^a = q^a , \qquad P_a = p_a , \qquad a \neq \underline{a} . \tag{3.98}
$$

These are called *elementary canonical transformations*. The importance of such transformations shows up in **Carathéodory's theorem**, which we state here without proof. Namely, *any canonical transformation can be composed of elementary canonical transformations followed by a Type I canonical transformation.* In other words, elementary canonical transformations are the building blocks of arbitrary canonical transformations. See e.g., Kuchăr (1995) and Goldstein et al. (2002) for more details.

#### 3.5.2.2 Invariance of Poisson Brackets Under a Canonical Transformation

It is also possible to show that a canonical tranformation preserves the fundamental Poisson bracket relations

$$
\{Q^a, Q^b\}_{(q,p)} = 0 , \qquad \{P_a, P_b\}_{(q,p)} = 0 , \qquad \{Q^a, P_b\}_{(q,p)} = \delta^a_b , \tag{3.99}
$$

where we have included subscripts on the Poisson brackets to explicitly indicate which canonical variables are being differentiated with respect to. The above Poisson brackets are special cases of the more general relation

$$
\{f, g\}_{(q,p)} = \{f, g\}_{(Q,P)} , \tag{3.100}
$$

where $f$ and $g$ are expressed in terms of the appropriate set of canonical variables for the Poisson bracket being evaluated. (See Problem 3.11 for a proof of this relation.) Thus, *canonical transformations preserve the Poisson bracket structure on phase space.*

Mathematically, the above result is analogous to orthogonal transformations in ordinary 3-dimensional space preserving the form of the Euclidean metric when written in terms of Cartesian coordinates $x^i \equiv (x, y, z)$:

$$\boldsymbol{\partial}_i \cdot \boldsymbol{\partial}_j = \delta_{ij}\,, \tag{3.101}$$

where the partial derivative operators $\partial_i \equiv \partial/\partial x^i$ are thought of as coordinate basis vectors $\boldsymbol{\partial}_i$ (See e.g., Appendix A.4.1). Basically, the Poisson bracket of functions in phase space plays the role of the inner product (or dot product) of vectors in Euclidean space (noting, of course, that Poisson brackets are anti-symmetric while the inner product of vectors is symmetric). Just as the inner product of Cartesian coordinate basis vectors can be written in terms of the Kronecker delta symbol, (3.101), so too can the Poission brackets of canonical coordinates and momenta in phase space be written in terms of the Kronecker delta, (3.99). Again, see Problem 3.11 for more details.

## 3.6 Applications of Canonical Transformations

Although canonical transformations have the potential to simplify the equations of motion through a judicious choice of coordinates in phase space, their true value lies in elucidating the deeper meaning behind the Hamiltonian formulation of mechanics. In this section, we will look at a few applications of canonical transformations.

### *3.6.1 Infinitesimal Canonical Transformations*

In the Hamiltonian framework, the instantaneous state of a mechanical system is completely described by a *point* $(q^a, p_a)$ in the $2n$-dimensional phase space $\Gamma$. As the system evolves in time, the point traces out a trajectory in phase space, which is just a 1-dimensional curve $\gamma(t) \equiv (q^a(t), p_a(t))$, as shown in Fig. 3.1 for the case $n = 1$. By introducing the concept of a 1-parameter family of canonical transformations, and considering parameter values sufficiently small that the transformation differs infinitesimally from the identify, then one can show that *time evolution is a continuous family of canonical transformations whose infinitesimal generator is the Hamiltonian* $H$. (To simplify our analysis, we will assume in this and the following subsection on continuous symmetries and conserved quantities that the Hamiltonian $H$ does not depend explicitly on time, i.e., $H \equiv H(q, p)$.)

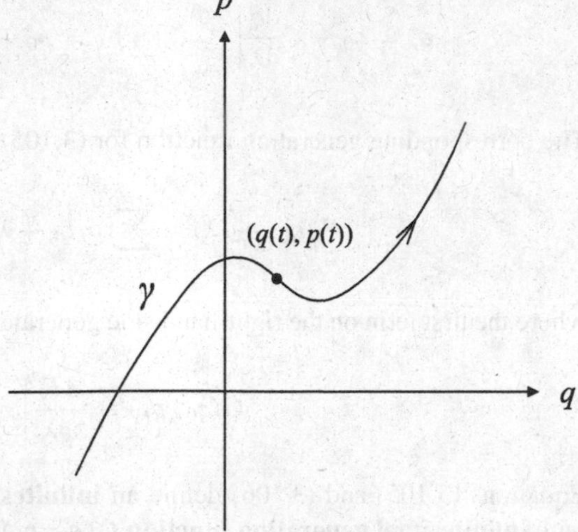


**Fig. 3.1** Hypothetical trajectory in phase space, corresponding to the time evolution of a 1-dimensional system

*Proof* To show this, consider a 1-parameter family of canonical transformations, labeled by a continuous parameter $\lambda$:

$$(q_0, p_0) \quad \rightarrow \quad (q_\lambda, p_\lambda)\,, \tag{3.102}$$

where

$$q_\lambda \equiv q(q_0, p_0, \lambda)\,, \qquad p_\lambda \equiv p(q_0, p_0, \lambda)\,. \tag{3.103}$$

Note that to simplify the notation, we are not including the $a = 1, 2, \cdots, n$ indices on the $q$'s and $p$'s here. If we need to include the $a$ indices, then it is best to write something like $q^a(\lambda)$ and $p_a(\lambda)$ for the transformed variables. Note also that since the discussion at this stage is completely general, the parameter $\lambda$ need *not* correspond to the time $t$.

We will further assume that when $\lambda = 0$, the 1-parameter family of canonical transformations is simply the *identity* transformation

$$q(q_0, p_0, \lambda)|_{\lambda=0} = q_0\,, \qquad p(q_0, p_0, \lambda)|_{\lambda=0} = p_0\,. \tag{3.104}$$

Then making a Taylor series expansion for small values of $\lambda$, we can write[8]

[8]In (3.105) and in all subsequent relevant equations, we ignore terms that are 2nd-order or higher in $\lambda$. Also, we write the derivatives of $q$ and $p$ as *ordinary* derivatives with respect to $\lambda$ (and not partial derivatives), as we are treating $q_0$ and $p_0$ as *fixed* parameters in the expressions $q(q_0, p_0, \lambda)$ and $p(q_0, p_0, \lambda)$.

$$q_\lambda = q_0 + \lambda \left. \frac{dq}{d\lambda} \right|_{\lambda=0}, \qquad p_\lambda = p_0 + \lambda \left. \frac{dp}{d\lambda} \right|_{\lambda=0}. \tag{3.105}$$

The corresponding generating function for (3.105) is

$$F_2(q_0, p_\lambda, \lambda) = \sum p_\lambda q_0 + \lambda G(q_0, p_\lambda), \tag{3.106}$$

where the first term on the right-hand side generates the identity transformation, and

$$G(q_0, p_\lambda) \equiv \left. \frac{\partial F_2}{\partial \lambda} \right|_{\lambda=0}. \tag{3.107}$$

Equations (3.105) and (3.106) define an **infinitesimal canonical transformation** with **infinitesimal generating function** $G(q_0, p_\lambda)$.

Using the transformation (3.84) for a Type II transformation, we have

$$p_0 = \frac{\partial F_2}{\partial q_0} = p_\lambda + \lambda \frac{\partial G}{\partial q_0} = p_0 + \lambda \left. \frac{dp}{d\lambda} \right|_{\lambda=0} + \lambda \frac{\partial G}{\partial q_0}, \tag{3.108}$$

where we used (3.105) to obtain the last equality. Cancelling $p_0$ on both sides of the above equation, we see that

$$\left. \frac{dp}{d\lambda} \right|_{\lambda=0} = -\frac{\partial G}{\partial q_0}. \tag{3.109}$$

Similarly,

$$q_\lambda = \frac{\partial F_2}{\partial p_\lambda} = q_0 + \lambda \frac{\partial G}{\partial p_\lambda} = q_\lambda - \lambda \left. \frac{dq}{d\lambda} \right|_{\lambda=0} + \lambda \frac{\partial G}{\partial p_0}, \tag{3.110}$$

where we used (3.105) and

$$\lambda \frac{\partial G}{\partial p_\lambda} = \lambda \frac{\partial G}{\partial p_0} + O\left(\lambda^2\right) \tag{3.111}$$

to obtain the last equality. Thus, canceling $q_\lambda$ on both sides of (3.110), we obtain

$$\left. \frac{dq}{d\lambda} \right|_{\lambda=0} = \frac{\partial G}{\partial p_0}. \tag{3.112}$$

To summarize, the infinitesimal generating function $G$ induces a change in the phase space variables

$$q_0 \to q_0 + \lambda \left.\frac{\mathrm{d}q}{\mathrm{d}\lambda}\right|_{\lambda=0}, \qquad p_0 \to p_0 + \lambda \left.\frac{\mathrm{d}p}{\mathrm{d}\lambda}\right|_{\lambda=0}, \tag{3.113}$$

with

$$\left.\left(\frac{\mathrm{d}q}{\mathrm{d}\lambda}, \frac{\mathrm{d}p}{\mathrm{d}\lambda}\right)\right|_{\lambda=0} = \left(\frac{\partial G}{\partial p_0}, -\frac{\partial G}{\partial q_0}\right). \tag{3.114}$$

And since there is nothing special about $(q_0, p_0)$, we can drop the 0 subscripts and use this relation for *all* $(q, p)$ in phase space.

Note that (3.109) and (3.112) have the same form as Hamilton's equations (3.44), with the parameter $\lambda$ playing the role of the time $t$, and the infinitesimal generating function $G$ playing the role of the Hamiltonian $H$. In addition, the values of $q$ and $p$ at $\lambda = 0$ can be thought of as their *initial* values. Thus, as mentioned at the start of this subsection, *time evolution in phase space can be interpreted as a continuous family of canonical transformations whose infinitesimal generator is the Hamiltonian* $H$. □

***Example 3.5*** As a simple example of a continuous point transformation, consider a single particle in 3-dimensions, and an (active) rotation about the $\hat{\mathbf{z}}$ axis through an adjustable angle $\theta$ (which corresponds to the parameter $\lambda$ for this example). This is a point transformation (3.90) with

$$\begin{bmatrix} x(\theta) \\ y(\theta) \\ z(\theta) \end{bmatrix} = \begin{bmatrix} \cos\theta & -\sin\theta & 0 \\ \sin\theta & \cos\theta & 0 \\ 0 & 0 & 1 \end{bmatrix} \begin{bmatrix} x(0) \\ y(0) \\ z(0) \end{bmatrix}, \tag{3.115}$$

and

$$p_a(\theta) = \sum_b p_b(0) \frac{\partial x^b(0)}{\partial x^a(\theta)}. \tag{3.116}$$

This last set of equations can also be written in matrix form

$$\begin{bmatrix} p_x(\theta) \\ p_y(\theta) \\ p_z(\theta) \end{bmatrix} = \begin{bmatrix} \cos\theta & -\sin\theta & 0 \\ \sin\theta & \cos\theta & 0 \\ 0 & 0 & 1 \end{bmatrix} \begin{bmatrix} p_x(0) \\ p_y(0) \\ p_z(0) \end{bmatrix}, \tag{3.117}$$

which has the same form as (3.115). If we take $\theta$ to be small, then we obtain an infinitesimal canonical transformation with

$$x(0) \to x(0) - \theta\, y(0)\,, \qquad y(0) \to y(0) + \theta\, x(0)\,, \qquad z(0) \to z(0)\,, \tag{3.118}$$

and similar expressions for the components of the momentum. □

**Exercise 3.11** In this problem, you will show that the components of linear momentum $\mathbf{p}$ and angular momentum $\boldsymbol{\ell} = \mathbf{r} \times \mathbf{p}$ are the infinitesimal generating functions for translations and rotations.

(a) Show that $G \equiv \hat{\mathbf{n}} \cdot \mathbf{p}$ generates infinitesimal translations in the direction of $\hat{\mathbf{n}}$:

$$\left.\frac{\mathrm{d}\mathbf{r}}{\mathrm{d}\lambda}\right|_{\lambda=0} = \hat{\mathbf{n}}\,, \qquad \left.\frac{\mathrm{d}\mathbf{p}}{\mathrm{d}\lambda}\right|_{\lambda=0} = 0\,, \tag{3.119}$$

so that

$$\mathbf{r} \to \mathbf{r} + \lambda\hat{\mathbf{n}}\,, \qquad \mathbf{p} \to \mathbf{p}\,. \tag{3.120}$$

(b) Similarly, show that $G \equiv \hat{\mathbf{n}} \cdot \boldsymbol{\ell}$ generates infinitesimal rotations about the axis $\hat{\mathbf{n}}$ through a counter-clockwise angle $\theta$:

$$\left.\frac{\mathrm{d}\mathbf{r}}{\mathrm{d}\theta}\right|_{\theta=0} = \hat{\mathbf{n}} \times \mathbf{r}\,, \qquad \left.\frac{\mathrm{d}\mathbf{p}}{\mathrm{d}\theta}\right|_{\theta=0} = \hat{\mathbf{n}} \times \mathbf{p}\,, \tag{3.121}$$

so that

$$\mathbf{r} \to \mathbf{r} + \theta\hat{\mathbf{n}} \times \mathbf{r}\,, \qquad \mathbf{p} \to \mathbf{p} + \theta\hat{\mathbf{n}} \times \mathbf{p}\,. \tag{3.122}$$

**Exercise 3.12** A 1-dimensional simple harmonic oscillator of mass $m$ and spring constant $k$ has Hamiltonian

$$H(q, p) = \frac{p^2}{2m} + \frac{1}{2}kq^2\,. \tag{3.123}$$

Show that the trajectory in phase space corresponding to the time evolution of the oscillator is an *ellipse* (traversed clockwise) with semi-major and semi-minor axes $q_0$ and $m\omega q_0$, where $q_0$ is the maximum displacement and $\omega \equiv \sqrt{k/m}$ is the angular frequency of the oscillation. See Fig. 3.2 for a plot of the trajectory.

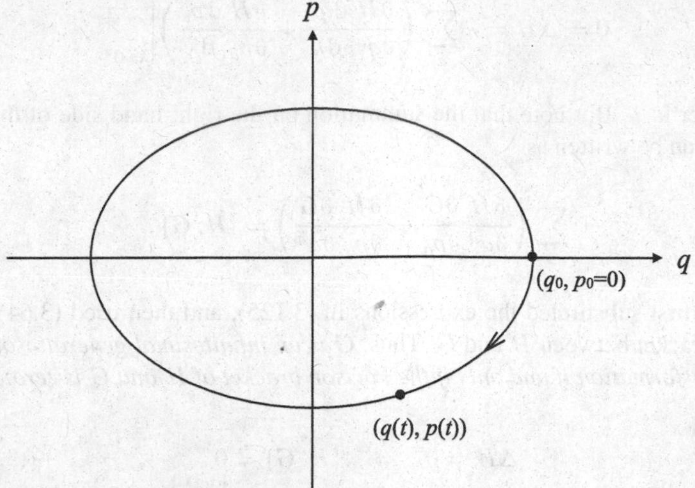


**Fig. 3.2** Trajectory in phase space corresponding to the time evolution of a 1-dimensional simple harmonic oscillator with $q(0) = q_0$, $\dot{q}(0) = 0$

### *3.6.2 Symmetries and Conserved Quantities*

Given the above discussion of infinitesimal canonical transformations, we are now in a position to demonstrate the intimate connection between continuous *symmetries* of the system (if they exist) and *conserved quantities*. We shall see that *infinitesimal generators of symmetry transformations and conserved quantities are one in the same.*

In the Hamiltonian framework, a **continuous symmetry** of the system is a canonical transformation that leaves the functional form of the Hamiltonian *invariant*. In other words, not only must the transformation preserve the *form* of Hamilton's equations, but the Hamiltonian itself must be unchanged by the transformation. For an infinitesimal canonical transformation:

$$q^a \to q^a + \lambda \left.\frac{\mathrm{d}q^a}{\mathrm{d}\lambda}\right|_{\lambda=0}, \qquad p_a \to p_a + \lambda \left.\frac{\mathrm{d}p_a}{\mathrm{d}\lambda}\right|_{\lambda=0}, \tag{3.124}$$

with infinitesimal generator $G$:

$$\left.\frac{\mathrm{d}q^a}{\mathrm{d}\lambda}\right|_{\lambda=0} = \frac{\partial G}{\partial p_a}, \qquad \left.\frac{\mathrm{d}p_a}{\mathrm{d}\lambda}\right|_{\lambda=0} = -\frac{\partial G}{\partial q^a}, \tag{3.125}$$

invariance of the Hamiltonian means

$$H\left(q + \lambda \left.\frac{\mathrm{d}q}{\mathrm{d}\lambda}\right|_{\lambda=0}, p + \lambda \left.\frac{\mathrm{d}p}{\mathrm{d}\lambda}\right|_{\lambda=0}\right) = H(q, p), \tag{3.126}$$

or, equivalently,

$$0 = \Delta H = \lambda \sum_a \left( \frac{\partial H}{\partial q^a} \frac{\mathrm{d}q^a}{\mathrm{d}\lambda} + \frac{\partial H}{\partial p_a} \frac{\mathrm{d}p_a}{\mathrm{d}\lambda} \right) \Big|_{\lambda=0} \tag{3.127}$$

to 1st-order in $\lambda$. But note that the summation on the right-hand side of the above equation can be written as

$$\sum_a \left( \frac{\partial H}{\partial q^a} \frac{\partial G}{\partial p_a} - \frac{\partial H}{\partial p_a} \frac{\partial G}{\partial q^a} \right) = \{H, G\}, \tag{3.128}$$

where we first substituted the expressions in (3.125), and then used (3.64) for the Poisson bracket between $H$ and $G$. Thus, *$G$ is an infinitesimal generator of a symmetry transformation if and only if the Poisson bracket of $H$ and $G$ is zero:*

$$\Delta H = 0 \quad \Leftrightarrow \quad \{H, G\} = 0 . \tag{3.129}$$

But recall from (3.74) that a function $f(q, p)$ on phase space is *conserved* if and only if its Poisson bracket with the Hamiltonian is zero. Thus, *$G$ is an infinitesimal generator of a symmetry transformation if and only if $G$ is a conserved quantity.*

**Exercise 3.13** Recast the conservation laws given in (3.28) and (3.29) as statements about the infinitesimal generators of spatial translations, rotations and time translations.

## 3.7 Transition to Quantum Mechanics

To end this chapter on the Hamiltonian formulation of mechanics, we briefly discuss how one can transition from classical mechanics to quantum theory, starting from Poisson brackets and the classical Hamiltonian. We will assume that the reader is familiar with the basic ideas of quantum mechanics at the level of a modern physics course—i.e., that they have some recollection of wave functions, the Schrödinger equation, quantum operators, etc. For more details, see Griffiths (2005).

In the transition from classical mechanics to quantum mechanics, Poisson brackets are replaced by the **commutator**[9] [ , ] of operators according to the prescription

$$[\hat{f}, \hat{g}] = i\hbar\widehat{\{f, g\}} , \tag{3.130}$$

[9]Recall that the commutator of two operators $\hat{A}$ and $\hat{B}$ is defined by $[\hat{A}, \hat{B}] \equiv \hat{A}\hat{B} - \hat{B}\hat{A}$. In general, operators do not commute, i.e., $\hat{A}\hat{B} \neq \hat{B}\hat{A}$.

where $\hat{f} \equiv f(\hat{q}, \hat{p}, t)$ is the quantum operator corresponding to $f$, keeping in mind issues related to the ordering of the operators in $\hat{f}$. In addition, the classical Hamiltonian $H(q, p, t)$ becomes a quantum operator, $\hat{H} \equiv H(\hat{q}, \hat{p}, t)$, which appears in the **Schrödinger equation**

$$i\hbar \frac{\partial |\Psi(t)\rangle}{\partial t} = \hat{H} |\Psi(t)\rangle \,. \tag{3.131}$$

You may recall that the Schrödinger equation determines the time-evolution of the state vector $|\Psi(t)\rangle$. In the *configuration representation*, $|\Psi(t)\rangle$ is represented by the wave function $\Psi(q, t)$, where $q$ is shorthand for the generalized coordinates on the configuration space $Q$.

***Example 3.6*** As a simple example, consider a single particle of mass $m$ moving in in response to an external force $\mathbf{F} = -\nabla U(\mathbf{r}, t)$. Then in the configuration representation, we have

$$|\Psi(t)\rangle = \Psi(\mathbf{r}, t) \,, \qquad \hat{x} = x \,, \qquad \hat{p}_x = \frac{\hbar}{i}\frac{\partial}{\partial x} \,, \qquad \text{etc.} \,, \tag{3.132}$$

for which

$$i\hbar \frac{\partial \Psi}{\partial t} = -\frac{\hbar^2}{2m} \nabla^2 \Psi + U \Psi \,. \tag{3.133}$$

□

Alternatively, we can express the time evolution of a quantum system in terms of the time derivative of the expectation values of quantum operators representing different classical observables. More explicitly, let $\hat{A}$ denote such a quantum operator corresponding to a classical observable $A$, and denote by $\langle \hat{A} \rangle$ the expectation value of $\hat{A}$ in the state $|\Psi\rangle$. For example, in the configuration representation, $\langle \hat{A} \rangle$ is given by the following integral in configuration space,

$$\langle \hat{A} \rangle = \int \mathrm{d}^n q \; \Psi^* \hat{A} \Psi \,. \tag{3.134}$$

Then one can show using (3.131) that

$$\frac{\mathrm{d}}{\mathrm{d}t} \langle \hat{A} \rangle = \frac{1}{i\hbar} \langle [\hat{A}, \hat{H}] \rangle + \langle \frac{\partial \hat{A}}{\partial t} \rangle \,. \tag{3.135}$$

This is the quantum version of (3.73), which is consistent with the prescription of (3.130).

**Exercise 3.14** Show that the quantum operators defined in (3.132) for the position and momentum variables $x$, $p_x$, etc. satisfy the Poisson bracket-commutator relation (3.130).

**Exercise 3.15** **Ehrenfest's theorem**: Consider a single particle of mass $m$ moving in a potential $U(\mathbf{r}, t)$. Show that

$$m\frac{\mathrm{d}}{\mathrm{d}t}\langle\hat{x}\rangle = \langle\hat{p}_x\rangle\,, \qquad \frac{\mathrm{d}}{\mathrm{d}t}\langle\hat{p}_x\rangle = -\langle\frac{\partial\hat{U}}{\partial x}\rangle\,. \qquad \text{etc.}\,, \tag{3.136}$$

which mimic the classical equations of motion.

## Suggested References

*Full references are given in the bibliography at the end of the book.*

Feynman et al. (1964): Chapter 19 is devoted to Hamilton's principle and action integrals.

Flannery (2005): A very readable article about the subtleties associated with non-holonomic constraints. The article points out an error in the discussion of semi-holonomic constraints given in Goldstein et al. (2002), pp. 46–48. (See also Problem 3.2.)

Griffiths (2005): Provides a more thorough discussion of the transition from classical mechanics to quantum mechanics.

Lanczos (1949): An extremely useful text about the use of variational principles in mechanics. Example 3.2 is from Appendix II of this book.

Landau and Lifshitz (1976): A classic text on the theoretical aspects of classical mechanics, appropriate for graduate students. Chapter VII is devoted to the Hamiltonian formulation of mechanics, including very clear and concise discussions of canonical transformations.

## Additional Problems

**Problem 3.1** An Atwood machine, shown in Fig. 3.3, consists of two masses $m_1$ and $m_2$ connected by an ideal (massless) string of length $\ell$, which hangs over an ideal (massless) pulley. There is a uniform gravitational field $\mathbf{g}$ pointing downward.

(a) Write down the Lagrangian for this system in terms of the single (unconstrained) degree of freedom.

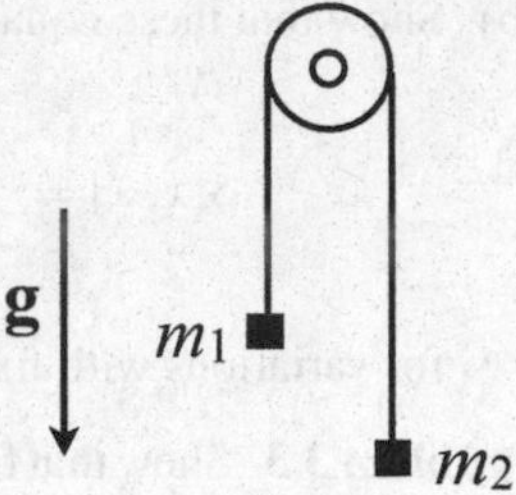


**Fig. 3.3** Atwood machine

(b) Obtain the equation of motion for this system from the Euler-Lagrange equation.
(c) Solve the equation for the motion for the two masses.
(d) Show that the result obtained above agrees with that calculated using standard "freshman physics" techniques.

**Problem 3.2** (See Flannery (2005), who points out an error in the discussion of semi-holonomic constraints given in Goldstein et al. (2002), pp. 46–48.) **Semi-holonomic constraints** are defined to be auxiliary conditions of the form

$$\phi^A(x^1, \cdots, x^{3N}, \dot{x}^1, \cdots, \dot{x}^{3N}, t) = 0, \quad A = 1, 2, \cdots, M, \tag{3.137}$$

where $\phi^A = \mathrm{d}F^A/\mathrm{d}t$ for some set of functions $F^A \equiv F^A(x^1, \cdots, x^{3N}, t)$. Since the $\phi^A$ are total time derivatives, the constraints depend *linearly* on the velocities $\dot{x}^\alpha$:

$$\sum_\alpha \frac{\partial F^A}{\partial x^\alpha}\dot{x}^\alpha + \frac{\partial F^A}{\partial t} = 0. \tag{3.138}$$

By multiplying this last equation by $\mathrm{d}t$ and then restricting to virtual displacements $\delta x^\alpha$ (for which the $\partial F^A/\partial t$ term drops out), we have the following condition on the coordinate differentials

$$\delta F^A \equiv \sum_\alpha \frac{\partial F^A}{\partial x^\alpha}\delta x^\alpha = 0, \tag{3.139}$$

which is exact. Thus, semi-holonomic constraints are trivially integrable, and hence behave like holonomic constraints in regard to the existence of a constraint surface and obtaining the equations of motion from a variational principle.

(a) Show that for semi-holonomic constraints and forces derivable from a generalized potential $U$, Lagrange's equations of the 1st-kind become

$$\frac{\mathrm{d}}{\mathrm{d}t}\left(\frac{\partial L}{\partial \dot{x}^\alpha}\right) - \frac{\partial L}{\partial x^\alpha} - \sum_A\left[\lambda_A\left(\frac{\partial \phi^A}{\partial x^\alpha} - \frac{\mathrm{d}}{\mathrm{d}t}\left(\frac{\partial \phi^A}{\partial \dot{x}^\alpha}\right)\right) - \frac{\mathrm{d}\lambda_A}{\mathrm{d}t}\frac{\partial \phi^A}{\partial \dot{x}^\alpha}\right] = 0, \tag{3.140}$$

where $L \equiv T - U$ is the Lagrangian and $\lambda_A = \lambda_A(t)$ are Lagrange multipliers.

(b) Show that these equations are also derivable from the modified action

$$\bar{S}[x, \lambda] = \int_{t_1}^{t_2} dt \left[ L(x, \dot{x}, t) + \sum_A \lambda_A \phi^A(x, \dot{x}, t) \right] \tag{3.141}$$

for variations with fixed endpoints, i.e., $\delta x^\alpha|_{t_1} = 0$, $\delta x^\alpha|_{t_2} = 0$.

**Problem 3.3** Show that for holonomic constraints and a polygenic generalized force, the closest one can come to obtaining an action whose variation would lead to the equations of motion (3.17) is

$$0 = \delta \int_{t_1}^{t_2} dt \left[ T + \sum_A \lambda_A \phi^A \right] + \int_{t_1}^{t_2} dt \sum_\alpha F_\alpha \delta x^\alpha . \tag{3.142}$$

But this is not the variation of a functional on account of the last term. Thus, the equations of motion for polygenic forces are *not* obtainable from the variation of any action integral.

**Problem 3.4** Show that the Hamiltonian for a particle moving in a potential $U(\mathbf{r}, t)$ in spherical polar coordinates is given by

$$H(r, \theta, \phi, p_r, p_\theta, p_\phi, t) = \frac{1}{2m}\left( p_r^2 + \frac{p_\theta^2}{r^2} + \frac{p_\phi^2}{r^2 \sin^2\theta} \right) + U(r, \theta, \phi, t) \tag{3.143}$$

**Problem 3.5** Show that the Hamiltonian for a point charge $q$ moving in the electromagnetic potential

$$U(\mathbf{r}, \dot{\mathbf{r}}, t) = q\left[\Phi(\mathbf{r}, t) - \mathbf{A}(\mathbf{r}, t) \cdot \dot{\mathbf{r}}\right] \tag{3.144}$$

is given by

$$H(x, y, z, p_x, p_y, p_z, t) = \frac{1}{2m} |\mathbf{p} - q\mathbf{A}|^2 + q\Phi . \tag{3.145}$$

**Problem 3.6** Write down the Hamiltonian and Hamilton's equations of motion for a particle of mass $m$ constrained to move on a cylinder of radius $R$, subject to a force $\mathbf{F} = -k\mathbf{r}$. Solve the equations of motion for the generalized coordinates $z(t)$ and $\phi(t)$, which describe the location of the particle on the cylinder. You should find that $\phi(t)$ increases linearly with $t$, and that $z(t)$ oscillates sinusoidally around $z = 0$ with angular frequency $\omega \equiv \sqrt{k/m}$.

**Problem 3.7** Write down the Hamiltonian and Hamilton's equations of motion for a simple pendulum. Show that you recover the standard equation of motion $\ddot{\theta} = -(g/l) \sin\theta$ for the angular displacement $\theta(t)$.

**Problem 3.8** Consider a single particle of mass $m$ moving in the presence of a time-independent potential $U \equiv U(\mathbf{r})$.

(a) From the Lagrangian $L = T - U$, calculate the momenta $p_a = (p_x, p_y, p_z)$ associated with Cartesian coordinates $q^a = (x, y, z)$.
(b) Do the same for the momenta $P_a = (P_r, P_\theta, P_\phi)$ associated with spherical polar coordinates $Q^a = (r, \theta, \phi)$.
(c) Show explicitly that

$$\sum_a p_a \, dq^a = \sum_a P_a \, dQ^a \,, \tag{3.146}$$

where $dq^a = \dot{q}^a \, dt$, etc., as to be expected for a point transformation.
(d) What does the differential $\sum_a p_a \, dq^a$ correspond to physically?

**Problem 3.9** Use Poisson brackets to determine the time evolution of each component of the angular momentum vector $\boldsymbol{\ell} \equiv \mathbf{r} \times \mathbf{p}$ and recover

$$\frac{d\boldsymbol{\ell}}{dt} = \mathbf{r} \times \mathbf{F} \,. \tag{3.147}$$

**Problem 3.10** (a) Show that a Type III canonical transformation has

$$q^a = -\frac{\partial F_3}{\partial p_a} \,, \qquad P_a = -\frac{\partial F_3}{\partial Q^a} \,, \qquad H' = H + \frac{\partial F_3}{\partial t} \,, \tag{3.148}$$

with $F_3 \equiv F - \sum_a p_a q^a$ expressed as a function of $(p, Q, t)$. (b) Similarly, show that a Type IV canonical transformation has

$$q^a = -\frac{\partial F_4}{\partial p_a} \,, \qquad Q^a = \frac{\partial F_4}{\partial P_a} \,, \qquad H' = H + \frac{\partial F_4}{\partial t} \,, \tag{3.149}$$

with $F_4 \equiv F + \sum_a (P_a Q^a - p_a q^a)$ expressed as a function of $(p, P, t)$.

**Problem 3.11** In this problem you will show that Poisson brackets are preserved by canonical transformations, i.e., $\{f, g\}_{(q,p)} = \{f, g\}_{(Q,P)}$. For simplicity, we will consider time-independent canonical transformations, for which the generating function does not explicitly depend on time, so $H' = H$.

To do so, let $y^\alpha$, where $\alpha = 1, 2, \cdots, 2n$, denote *any* set of coordinates on the $2n$-dimensional phase space. We then define

$$\Omega_{\alpha\beta} \equiv \Big(\sum_a dp_a \wedge dq^a\Big)_{\alpha\beta} \equiv \sum_a \left( \frac{\partial p_a}{\partial y^\alpha} \frac{\partial q^a}{\partial y^\beta} - \frac{\partial p_a}{\partial y^\beta} \frac{\partial q^a}{\partial y^\alpha} \right) , \tag{3.150}$$

which involves the **wedge product**, $dp_a \wedge dq^a$, of the total differentials $dp_a$ and $dq^a$ on phase space. (See Appendix B for a more general discussion of wedge product in the context of differential forms.) Since the wedge product defined above is *anti-symmetric* under interchange of its components $\alpha$ and $\beta$, $\Omega_{\alpha\beta}$ are the components of an anti-symmetric $2n \times 2n$ matrix. ($\Omega$ is called a **symplectic structure** on phase space, analogous to a metric on Euclidean space.)

(a) Show that if we take $y^\alpha = (q^a, p_a)$, then

$$\Omega_{\alpha\beta} = \begin{bmatrix} 0_{n\times n} & -1_{n\times n} \\ 1_{n\times n} & 0_{n\times n} \end{bmatrix}, \tag{3.151}$$

with inverse

$$\left(\Omega^{-1}\right)^{\alpha\beta} = \begin{bmatrix} 0_{n\times n} & 1_{n\times n} \\ -1_{n\times n} & 0_{n\times n} \end{bmatrix}. \tag{3.152}$$

(b) Show that (3.64) for the Poisson brackets of $f(q,p)$ and $g(q,p)$ can be written in terms of the above expression for $(\Omega^{-1})^{\alpha\beta}$ as

$$\{f,g\}_{(q,p)} = \sum_{\alpha,\beta} \left(\Omega^{-1}\right)^{\alpha\beta} \frac{\partial f}{\partial y^\alpha}\frac{\partial g}{\partial y^\beta}. \tag{3.153}$$

(c) Show that the condition for a Type I time-independent canonical transformation can be written as

$$\mathrm{d}F = \sum_a (p_a\,\mathrm{d}q^a - P_a\,\mathrm{d}Q^a) = \sum_\beta \sum_a \left( p_a \frac{\partial q^a}{\partial y^\beta} - P_a \frac{\partial Q^a}{\partial y^\beta}\right)\mathrm{d}y^\beta. \tag{3.154}$$

where we have dropped the $\mathrm{d}t$ term from (3.80) since $H' = H$ for a time-independent canonical transformation.

(d) Show that commutativity of partial derivatives,

$$\frac{\partial^2 F}{\partial y^\alpha \partial y^\beta} = \frac{\partial^2 F}{\partial y^\beta \partial y^\alpha}, \tag{3.155}$$

implies

$$\sum_a \left( \frac{\partial p_a}{\partial y^\alpha}\frac{\partial q^a}{\partial y^\beta} - \frac{\partial p_a}{\partial y^\beta}\frac{\partial q^a}{\partial y^\alpha}\right) = \sum_a \left( \frac{\partial P_a}{\partial y^\alpha}\frac{\partial Q^a}{\partial y^\beta} - \frac{\partial P_a}{\partial y^\beta}\frac{\partial Q^a}{\partial y^\alpha}\right). \tag{3.156}$$

Thus, taking $y^\alpha = (q^a, p_a)$ or $y^\alpha = (Q^a, P_a)$ leads to the same expressions, (3.151) and (3.152), for $\Omega_{\alpha\beta}$ and $(\Omega^{-1})^{\alpha\beta}$. Hence, $\{f,g\}_{(q,p)} = \{f,g\}_{(Q,P)}$ as desired.

**Problem 3.12** Recast the previous problem in the language of differential forms (See Appendix B), obtaining the invariance of the symplectic structure

$$\Omega = \sum_a \mathrm{d}p_a \wedge \mathrm{d}q^a = \sum_a \mathrm{d}P_a \wedge \mathrm{d}Q^a \tag{3.157}$$

under a canonical transformation.

**Problem 3.13** Show that

$$\{\ell_i, \ell_j\} = \sum_k \varepsilon_{ijk}\ell_k\,, \tag{3.158}$$

for the components of the angular momentum vector $\boldsymbol{\ell} = \mathbf{r} \times \mathbf{p}$.

**Problem 3.14** Given a solution for the *actual motion* of a system, we can evaluate the action $S[q] = \int_{t_1}^{t_2} \mathrm{d}t\, L(q, \dot{q}, t)$ along this path, obtaining a function that depends on the configurations and times at the two limits of integration $t_1$ and $t_2$:

$$S(q_2, t_2, q_1, t_1) \equiv \int_{t_1}^{t_2} \mathrm{d}t\, L(q, \dot{q}, t)\Big|_{q(t)=\text{actual motion}}\,. \tag{3.159}$$

This function is called **Hamilton's principal function**. From its construction, Hamilton's principal function implicitly contains all of the information about the actual motion of the system.

(a) Show that Hamilton's principal function for a free particle of mass $m$ moving in one dimension is

$$S(x_2, t_2, x_1, t_1) = \frac{m}{2}\frac{(x_2 - x_1)^2}{(t_2 - t_1)}\,. \tag{3.160}$$

(b) Show that Hamilton's principal function for a 1-dimensional simple harmonic oscillator of mass $m$ and angular frequency $\omega$ is

$$S(q_2, t_2, q_1, t_1) = \frac{m\omega}{2\sin[\omega(t_2 - t_1)]}\left[(q_1^2 + q_2^2)\cos[\omega(t_2 - t_1)] - 2q_1q_2\right]. \tag{3.161}$$

**Problem 3.15** Consider Hamilton's principal function introduced in Problem 3.14, but write it as

$$S \equiv S(q, t, q_0, t_0)\,, \tag{3.162}$$

where we *fix* the initial configuration and initial time, $q_0, t_0$, but let the final configuration and final time, $q, t$, be *variable*. Hamilton's principal function $S$ then becomes a function of just the coordinates and the time at the upper limit of integration for the action integral evaluated along the actual motion of the system; see Fig. 3.4 for the case of a 1-dimensional configuration space.

(a) Show that the partial derivatives of Hamilton's principal function $S$ with respect to the $q$'s and $t$ are given by

$$\frac{\partial S}{\partial q^a} = p_a\,, \qquad \frac{\partial S}{\partial t} = -H\,, \tag{3.163}$$

where $a = 1, 2, \cdots, n$.

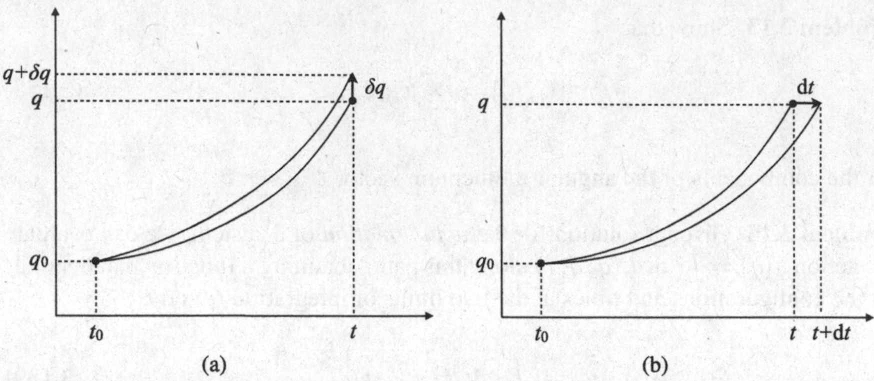


**Fig. 3.4** Independent variations in (a) the final configuration $q$ and (b) the final time $t$, for the action integral evaluated along the actual motion of the system. For both cases, the initial configuration $q_0$ and initial time $t_0$ are fixed

*Hints*: (i) To evaluate $\partial S/\partial q^a$, recall that for an arbitrary variation of the path:

$$\delta S = \sum_a \left(\frac{\partial L}{\partial \dot{q}^a}\right) \delta q^a \Bigg|_{t_0}^{t} + \int_{t_0}^{t} d\bar{t} \sum_a \left[\frac{\partial L}{\partial q^a} - \frac{d}{dt}\left(\frac{\partial L}{\partial \dot{q}^a}\right)\right] \delta q^a . \tag{3.164}$$

Then use the facts that the actual motion of the system satisfies the Euler-Lagrange equations, and that we are fixing the initial configuration at $t_0$. (ii) To evaluate $\partial S/\partial t$, note that the total time derivative of $S$ evaluated along the actual motion of the system is just the Lagrangian, so $dS/dt = L$. Then use the chain rule to evaluate $dS/dt$ for $S \equiv S(q, t, q_0, t_0)$, and recall that $L = \sum_a p_a \dot{q}^a - H$.

(b) Combine the results of part (a) into a *single* equation

$$0 = H\left(q, \frac{\partial S}{\partial q}, t\right) + \frac{\partial S}{\partial t} . \tag{3.165}$$

The above equation is called the **Hamilton-Jacobi equation**; it is a partial differential equation for Hamilton's prinicipal function $S$ with respect to the independent variables $q^a$ and $t$. It turns out that solving the Hamilton-Jacobi equation for $S$ is an *alternative* method of solving the equation of motion for a mechanical system. See e.g., Landau and Lifshitz (1976) and Lanczos (1949) for details.

# Chapter 4
# Central Force Problems

One of the most ubiquitous forces in physics is the central force. In any case where a particle possesses a “charge” that couples to a “field”, the field outside the charge is nearly always a radial field if the particle can be considered to be a dimensionless point. Examples include the classical Newtonian law of gravitation, the electrostatic field, and the Yukawa model of the strong nuclear force. The common feature in these examples is a spherically-symmetric potential that depends only on the radial separation from the point-particle source of the field. In this chapter, we will focus mostly on gravitationally-bound systems, and discuss unbound systems and the scattering problem in Chap. 5.

## 4.1 General Formalism

For central forces that can be described by a potential, $\mathbf{F} = -\nabla U$, the most convenient coordinates are spherical coordinates $(r, \theta, \phi)$. In these coordinates, $U \equiv U(r)$ and the kinetic energy for a particle of mass $m$ is

$$T \equiv \frac{1}{2}m|\dot{\mathbf{r}}|^2 = \frac{1}{2}m\left(\dot{r}^2 + r^2\dot{\theta}^2 + r^2\sin^2\theta\dot{\phi}^2\right). \tag{4.1}$$

The equations of motion for the particle are found from the Euler-Lagrange equations

$$\frac{\mathrm{d}}{\mathrm{d}t}\left(\frac{\partial L}{\partial \dot{q}^a}\right) - \frac{\partial L}{\partial q^a} = 0, \qquad a = 1, 2, 3, \tag{4.2}$$

where

$$L = \frac{1}{2}m\left(\dot{r}^2 + r^2\dot{\theta}^2 + r^2\sin^2\theta\dot{\phi}^2\right) - U(r), \tag{4.3}$$

 
M.J. Benacquista and J.D. Romano, *Classical Mechanics*, Undergraduate Lecture Notes in Physics, https://doi.org/10.1007/978-3-319-68780-3_4

and $q^a \equiv (r, \theta, \phi)$. Since the potential $U$ is a function of $r$ only, the Lagrangian is independent of $\phi$, so we immediately have a conserved quantity

$$p_\phi \equiv \frac{\partial L}{\partial \dot{\phi}} = mr^2 \sin^2 \theta \dot{\phi} = \text{const} . \tag{4.4}$$

As shown in Exercise 4.1, this conserved quantity is simply the $z$-component of the angular momentum vector $\boldsymbol{\ell} \equiv \mathbf{r} \times \mathbf{p}$. But because the problem is spherically symmetric, we are free to choose the orientation of our coordinate system so that the $z$-axis points along *any* direction $\hat{\mathbf{n}}$. Thus, $\hat{\mathbf{n}} \cdot \boldsymbol{\ell} = \text{const}$ for any *any* $\hat{\mathbf{n}}$, which means that the *vector* $\boldsymbol{\ell}$ itself is constant.[1] The motion of a particle subject to a central force is thus restricted to the *two-dimensional* plane perpendicular to $\boldsymbol{\ell}$. By choosing the $z$-axis to point along $\boldsymbol{\ell}$, the motion lies the *equatorial plane* $z = 0$ which has $\theta = \pi/2$.

**Exercise 4.1** Show that $p_\phi = mr^2 \sin^2 \theta \dot{\phi}$ is equal to the $z$-component of the angular momentum vector $\boldsymbol{\ell} \equiv \mathbf{r} \times \mathbf{p}$.

For motion in the equatorial plane, the Euler-Lagrange equations for $(r, \theta, \phi)$ reduce to those for the two-dimensional Lagrangian (Exercise 4.2)

$$L = \frac{1}{2} m \left( \dot{r}^2 + r^2 \dot{\phi}^2 \right) - U(r) . \tag{4.5}$$

As before, the $\phi$ equation of motion gives conservation of angular momentum

$$\ell \equiv mr^2 \dot{\phi} = \text{const} , \tag{4.6}$$

while the radial equation of motion is

$$m\ddot{r} - mr\dot{\phi}^2 + \frac{\partial U}{\partial r} = 0 . \tag{4.7}$$

Since the Lagrangian does not depend explicitly on time, the total energy

$$E = \frac{1}{2} m \left( \dot{r}^2 + r^2 \dot{\phi}^2 \right) + U(r) \tag{4.8}$$

is also conserved.

Note that $\dot{\phi}$ can be eliminated from these last two equations using (4.6). For the radial equation we have

[1] Alternatively, we can deduce that $\boldsymbol{\ell}$ is constant by simply noting that for a central potential $U(r)$, the force $\mathbf{F} = -\nabla U$ points the radial direction, and hence the torque on the particle about the origin is zero. (See Sect. 1.4, conservation law II.).

$$m\ddot{r} - \frac{\ell^2}{mr^3} + \frac{\partial U}{\partial r} = 0\,. \tag{4.9}$$

The term involving the angular momentum can be written as the gradient of an effective radial potential,

$$-\frac{\ell^2}{mr^3} = \frac{\partial}{\partial r}\left(\frac{\ell^2}{2mr^2}\right), \tag{4.10}$$

so that

$$m\ddot{r} = -\frac{\partial U_{\text{eff}}}{\partial r}\,, \qquad U_{\text{eff}}(r) \equiv \frac{\ell^2}{2mr^2} + U(r)\,. \tag{4.11}$$

Similarly, the conservation of energy equation becomes

$$E = \frac{1}{2}m\dot{r}^2 + U_{\text{eff}}(r)\,. \tag{4.12}$$

The quantity $\ell^2/(2mr^2)$ that appears in the effective potential is called the *angular momentum barrier*.

**Exercise 4.2** Show explicitly that the equations of motion for (4.3) reduce to those for (4.5) if one restricts to motion in the equatorial plane $\theta = \pi/2$. (*Hint*: You will need to show that the $\theta$ equation for the first Lagrangian is identically satisfied when $\theta = \pi/2$.)

### *4.1.1 Orbit Equation*

We can obtain an equation for the orbit $r = r(\phi)$ by using the conservation of energy equation (4.12) together with conservation of angular momentum (4.6) to convert time derivatives to derivatives with respect to $\phi$. The differential equation is simplest when expressed in terms of the variable $u \equiv 1/r$. Making this change of variables, and using

$$\frac{\mathrm{d}}{\mathrm{d}t} = \frac{\mathrm{d}\phi}{\mathrm{d}t}\frac{\mathrm{d}}{\mathrm{d}\phi} = \frac{\ell}{mr^2}\frac{\mathrm{d}}{\mathrm{d}\phi} = \frac{\ell}{m}u^2\frac{\mathrm{d}}{\mathrm{d}\phi}\,, \tag{4.13}$$

we obtain

$$u' = \mp\sqrt{\frac{2m}{\ell^2}\left[E - U(1/u)\right] - u^2}\,, \tag{4.14}$$

where $' \equiv \mathrm{d}/\mathrm{d}\phi$. This is a 1st-order separable differential equation, which can be solved via quadratures as

$$\phi - \phi_0 = \int_{u_0}^{u} \frac{\mathrm{d}\bar{u}}{\sqrt{\frac{2m}{\ell^2}\left[E - U(1/\bar{u})\right] - \bar{u}^2}}, \tag{4.15}$$

where we have used $\bar{u}$ to describe the dummy variable in the integration. Note, however, that for an arbitrary potential, it is not guaranteed that the integral can be solved analytically in terms of known functions.

### 4.1.2 Integrable Solutions of the Orbit Equation

One class of potentials that has been extensively studied are the power-law potentials of the form

$$U(r) = Ar^{\alpha} = Au^{-\alpha}. \tag{4.16}$$

For these potentials, (4.15) becomes

$$\phi - \phi_0 = \int_{u_0}^{u} \frac{\mathrm{d}\bar{u}}{\sqrt{\frac{2m}{\ell^2}E - \frac{2mA}{\ell^2}\bar{u}^{-\alpha} - \bar{u}^2}}. \tag{4.17}$$

This integral can be solved in terms of simple trigonometric functions for $\alpha = -2$, $-1$, 2. The class of special functions known as **elliptic integrals** are solutions to integrals of the form

$$\int R(x, \sqrt{y(x)})\,\mathrm{d}x, \tag{4.18}$$

where

$$y(x) = a_0x^4 + a_1x^3 + a_2x^2 + a_3x + a_4. \tag{4.19}$$

Thus, the solution to (4.17) is expressible in terms of basic trig functions and elliptic functions if $\alpha = -4$ and $-3$. (A brief description of elliptical integrals is found in Appendix E.6, while a more complete description can be found in Abramowitz and Stegun 1972.) Additional variable substitutions can expand the number of integrable power-law solutions to include $\alpha = -6$, $+4$, and $+6$. These substitutions can be found in Goldstein et al. (2002).

In the following several sections, we will concentrate on **inverse-square-law forces**, such as Newtonian gravity, which have a $1/r$ potential (so $\alpha = -1$). As we shall see, this particular form of potential allows for a relatively simple analytical treatment.

## 4.2 Kepler's Laws

The orbital motion of the planets in the solar system was studied by Johannes Kepler, who began working with Tycho Brahe in 1600. After Brahe's death in 1601, Kepler had access to Brahe's astrometric data on the motion of the planets, and he began to formulate laws governing planetary motion. This work predated the birth of Newton by about 40 years, and so the laws of Kepler were *phenomenological* in nature. The first two of Kepler's laws describe the shape and motion of individual planetary orbits, while the third law gives the relationship between the orbital periods and sizes of different orbits.

Kepler's 1st law notes that planets move on elliptical orbits with the Sun at one focus of the ellipse (Fig. 4.1). This was based mostly on the observations of the orbit of Mars. Since the Sun is so much more massive than Mars, the center of mass of the Mars-Sun system lies very close to the center of the Sun, and so a more general form of Kepler's 1st law is:

**Kepler's 1st Law:** Gravitational orbits follow ellipses, with the center of mass at one focal point.

Kepler's 2nd law describes the motion of the planet as it moves about its elliptical orbit. The 2nd law is sometimes known as the **equal-area law** (Fig. 4.2).

**Kepler's 2nd Law:** The line joining a planet to the center of mass sweeps out equal areas in equal times throughout the orbit. Thus, the planet moves faster in its orbit when it is closer to the Sun.

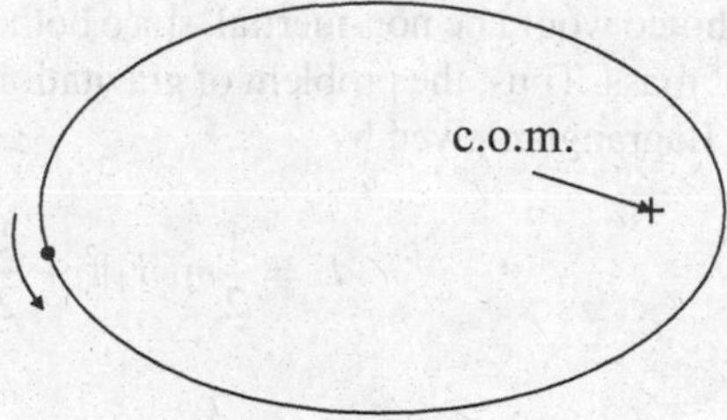


**Fig. 4.1** Kepler's 1st law. The center of mass of the system is located at the focus of an ellipse. The orbiting body moves along the ellipse

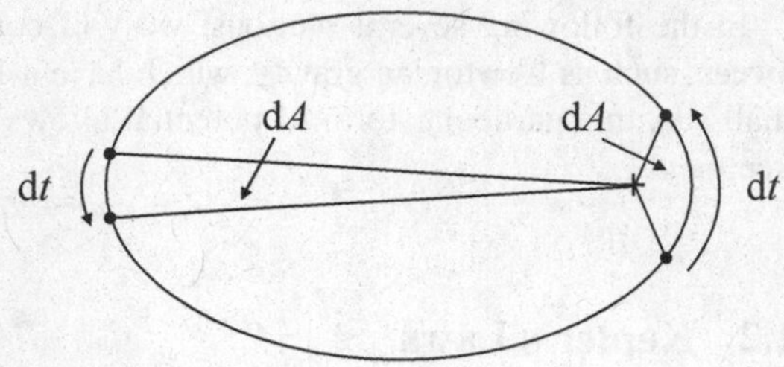

**Fig. 4.2** Kepler's 2nd law. The area d$A$ is the same for all points along the orbit that have equal values of d$t$

Kepler's 3rd law was expressed about ten years later in a five-volume treatise on the harmony of the universe. In this law, Kepler related the orbital period to the size of the orbit for *different* planets. The size of the orbit is measured in terms of the semi-major axis of the ellipse defining the orbit.

**Kepler's 3rd Law:** For two different orbits, the ratio of the squares of the orbital periods is equal to the ratio of the cubes of their semimajor axes,

$$P^2 \propto a^3 , \tag{4.20}$$

where $P$ is the orbital period and $a$ is the semi-major axis.

Using Newton's law of gravitation, we can obtain all of Kepler's laws as a result of solving the central force problem.

## 4.3 Gravitation

The Newtonian gravitational potential

$$U_g = -\frac{Gm_1m_2}{|\mathbf{r}_1 - \mathbf{r}_2|} , \tag{4.21}$$

would correspond to a central force if we chose coordinates centered on one of the masses (say, $m_1$). However, unless $m_1$ were infinitely massive, this coordinate choice would be non-inertial, since both $m_1$ and $m_2$ orbit about their common center of mass. Thus, the problem of gravitational orbits is at first a *two-body* problem with a Lagrangian given by

$$L = \frac{1}{2}m_1|\dot{\mathbf{r}}_1|^2 + \frac{1}{2}m_2|\dot{\mathbf{r}}_2|^2 + \frac{Gm_1m_2}{|\mathbf{r}_1 - \mathbf{r}_2|} . \tag{4.22}$$

But it turns out that this problem can be simplified considerably by working in the *center-of-mass frame*,[2] for which the origin of coordinates lies at the center of mass of the system:

$$m_1\mathbf{r}_1 + m_2\mathbf{r}_2 = \mathbf{0}. \tag{4.23}$$

If we define

$$\mathbf{r} \equiv \mathbf{r}_1 - \mathbf{r}_2 , \tag{4.24}$$

which is the relative separation vector between the two masses, then using the above two equations, it is relatively easy to show that

$$\mathbf{r}_1 = \frac{m_2}{m_1+m_2}\mathbf{r}, \quad \mathbf{r}_2 = -\frac{m_1}{m_1+m_2}\mathbf{r}. \tag{4.25}$$

Substituting these expressions for $\mathbf{r}_1$ and $\mathbf{r}_2$ back into the Lagrangian, (4.22), we find

$$L = \frac{1}{2}\mu|\dot{\mathbf{r}}|^2 + \frac{GM\mu}{r}, \tag{4.26}$$

where

$$M \equiv m_1 + m_2 , \quad \mu \equiv \frac{m_1 m_2}{m_1+m_2}. \tag{4.27}$$

The two masses, $M$ and $\mu$, are the *total mass* and **reduced mass** of the system. Note that the Lagrangian (4.26) now has the same general form as that discussed in Sect. 4.1. We interpret (4.26) as describing the motion of a single body of mass $\mu$ around a *fixed* center of mass $M$. Since $U = -GM\mu/r$ is a central potential, we can conclude again that the angular momentum of the system $\boldsymbol{\ell}$ is conserved, and that the motion takes place in a plane perpendicular to $\boldsymbol{\ell}$. Using plane polar coordinates $(r, \phi)$ to describe the relative separation vector $\mathbf{r}$, we have

$$L = \frac{1}{2}\mu\left(\dot{r}^2 + r^2\dot{\phi}^2\right) + \frac{GM\mu}{r}. \tag{4.28}$$

This is the so-called **effective one-body** (or **reduced-mass**) form of the Lagrangian for the original two-body problem. Once we have a solution of the equations of motion for $\mathbf{r} \equiv (r, \phi)$ for the effective one-body problem, we obtain solutions for $\mathbf{r}_1$ and $\mathbf{r}_2$ for the original two-body problem by simply rescaling $\mathbf{r}$ using (4.25).

**Exercise 4.3** Fill in the steps leading from (4.22)–(4.28).

[2] Since there are no external forces acting on the system, the center-of-mass frame for this problem is an inertial frame.

### 4.3.1 Effective One-Body Problem

By varying the effective one-body Lagrangian (4.28) with respect to the two coordinates, $r$ and $\phi$, we obtain two equations of motion for this system. The angular equation gives conservation of angular momentum (as expected, since the Lagrangian is independent of $\phi$), so we have

$$\mu r^2\dot{\phi} = \ell = \text{const}\,. \tag{4.29}$$

We can recover Kepler's 2nd law from this equation by noting that the infinitesimal area swept out by the line joining the orbiting body to the origin over the time interval $dt$ is

$$\mathrm{d}A = \frac{1}{2}r^2\mathrm{d}\phi = \frac{1}{2}r^2\dot{\phi}\mathrm{d}t = \frac{\ell}{2\mu}\mathrm{d}t\,, \tag{4.30}$$

from which we see that

$$\frac{\mathrm{d}A}{\mathrm{d}t} = \frac{\ell}{2\mu} = \text{const}\,. \tag{4.31}$$

Therefore at any point in the orbit, the line joining the orbiting body to the origin sweeps out the same area in a given time interval. So, Kepler's 2nd law is really just a consequence of conservation of angular momentum. Note that this result is independent of the form of the $U(r)$—i.e., it is valid for *any* central force.

The radial equation of motion is

$$\mu\ddot{r} - \mu r\dot{\phi}^2 + \frac{GM\mu}{r^2} = 0\,. \tag{4.32}$$

It can be written entirely as an equation in $r$ by writing $\dot{\phi}$ in terms of $\ell$ using (4.29):

$$\mu\ddot{r} - \frac{\ell^2}{\mu r^3} + \frac{GM\mu}{r^2} = 0\,. \tag{4.33}$$

We will put off solving this equation explicitly for the time-dependent motion until Sect. 4.3.3. For now we will look into the shape of the orbit. First, we can recover the conserved mechanical energy of the system by multiplying (4.33) by $\dot{r}$ and noting that each term can then be written as a total time derivative,

$$\frac{\mathrm{d}}{\mathrm{d}t}\left\{\frac{1}{2}\mu\dot{r}^2 + \frac{\ell^2}{2\mu r^2} - \frac{GM\mu}{r}\right\} = 0\,. \tag{4.34}$$

The expression in curly brackets is just the total mechanical energy $E \equiv T + U$ of the system

$$E = \frac{1}{2}\mu \dot{r}^2 + \frac{\ell^2}{2\mu r^2} - \frac{GM\mu}{r}, \tag{4.35}$$

with the tangential component $\frac{1}{2}\mu r^2\dot{\phi}^2$ of the kinetic energy expressed in terms of $\ell$. Written in this form, the total mechanical energy looks like that for simple *one-dimensional* motion with an additional $1/r^2$ potential. (This extra potential term is the angular momentum barrier, cf. (4.11).) The effective potential

$$U_{\text{eff}}(r) \equiv \frac{\ell^2}{2\mu r^2} - \frac{GM\mu}{r} \tag{4.36}$$

is shown in Fig. 4.3. Because of the positive $1/r^2$ contribution of the angular momentum barrier, there is a minimum in the effective potential. The minimum value is determined by the angular momentum of the system. For a given angular momentum, there is therefore a minimum allowed total energy of the system at which $E = U_{\min}$. These systems correspond to bound, circular orbits. If the total energy is less than zero, but greater than $U_{\min}$, the orbit is bounded by a minimum radius $r_{\min}$ (**periapsis**) and a maximum radius $r_{\max}$ (**apapsis**). We shall show later that these orbits are the ellipses of Kepler's 1st law. If the total energy is greater than zero, then the system is unbound and there is a point of minimum approach, but no maximum radius. Thus, a particle comes in from infinity and swings back out to infinity. We will determine the shape of these orbits in the next section.

**Exercise 4.4** In a circular orbit, the total energy is equal to the minimum of the effective potential. Use (4.36) to find the radius of a circular orbit for a given angular momentum $\ell$.

### *4.3.2 Classification of Orbits*

To classify the different orbits allowed in Newtonian gravity, we will work with the general orbit equation (4.15) with $m$ replaced by the reduced mass $\mu$ and the potential $U(1/u)$ given by $-GM\mu u$:

$$\phi - \phi_0 = \int_{u_0}^{u} \frac{d\bar{u}}{\sqrt{\frac{2\mu}{\ell^2}\left[E + GM\mu\bar{u}\right] - \bar{u}^2}}. \tag{4.37}$$

The quantity inside the square root is a quadratic function of $\bar{u}$,

$$\frac{2\mu E}{\ell^2} + \frac{2GM\mu^2}{\ell^2}\bar{u} - \bar{u}^2. \tag{4.38}$$

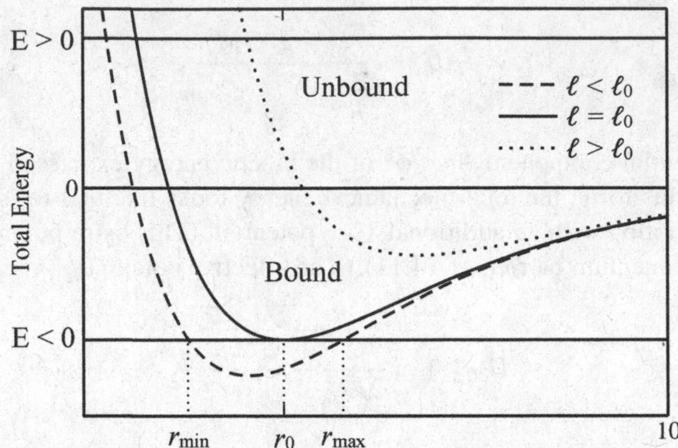


**Fig. 4.3** The effective potential $U_{\text{eff}}(r)$ for the radial coordinate in a $1/r$ potential. The effective potential rises at small radii due to the angular momentum barrier. Particles with positive total energy are on unbound orbits since there is no upper limit to the radius. The point of closest approach in an unbound orbit is the radius at which the energy is equal to the effective potential. If the total energy is negative, then the particle is in a bound orbit. For a given angular momentum, the minimum energy orbit is a circular orbit with radius $r_0$. Other orbits will have a point of closest approach ($r_{\min}$) and a point of farthest approach ($r_{\max}$)

This expression can be simplified by making the following series of substitutions. First, we define

$$\alpha \equiv \frac{\ell^2}{GM\mu^2}, \tag{4.39}$$

so (4.38) can be written as

$$\frac{2E}{GM\mu\alpha} + \frac{2}{\alpha}\bar{u} - \bar{u}^2 = \frac{1}{\alpha^2}\left[\frac{2E\alpha}{GM\mu} + 2\alpha\bar{u} - \alpha^2\bar{u}^2\right]. \tag{4.40}$$

Next, by *completing the square* inside the square brackets, we obtain

$$\frac{1}{\alpha^2}\left[\frac{2E\alpha}{GM\mu} + 1 - (\alpha\bar{u} - 1)^2\right]. \tag{4.41}$$

Finally, by defining

$$e^2 \equiv \frac{2E\alpha}{GM\mu} + 1, \tag{4.42}$$

and making the change of variables $\bar{x} \equiv \alpha\bar{u} - 1$, the orbit equation (4.37) simplifies to

$$\phi - \phi_0 = \int_{x_0}^{x} \frac{d\bar{x}}{\sqrt{e^2 - \bar{x}^2}} . \tag{4.43}$$

The integral on the right-hand side can be easily evaluated. The solution is

$$\phi - \phi_0 = \arcsin\left(\frac{x}{e}\right) - \arcsin\left(\frac{x_0}{e}\right) . \tag{4.44}$$

We can choose the coordinate system so that $\phi_0 = 0$ when $x_0 = e$, and we have

$$x = e \sin(\phi + \pi/2) = e \cos\phi . \tag{4.45}$$

Reversing all the substitutions gives the parametric equation for an ellipse in polar coordinates, with the origin located at one focus:

$$r = \frac{\alpha}{1 + e\cos\phi} . \tag{4.46}$$

You are probably more familiar with the equation for an ellipse in Cartesian coordinates, with the origin located at the *center* of the ellipse:

$$\frac{x^2}{a^2} + \frac{y^2}{b^2} = 1 , \tag{4.47}$$

where $a$ is the **semi-major axis** and $b$ is the **semi-minor axis** of the ellipse. The eccentricity of the ellipse is given by $e$. Since $e$ is defined in terms of $E$ and $\ell$, we

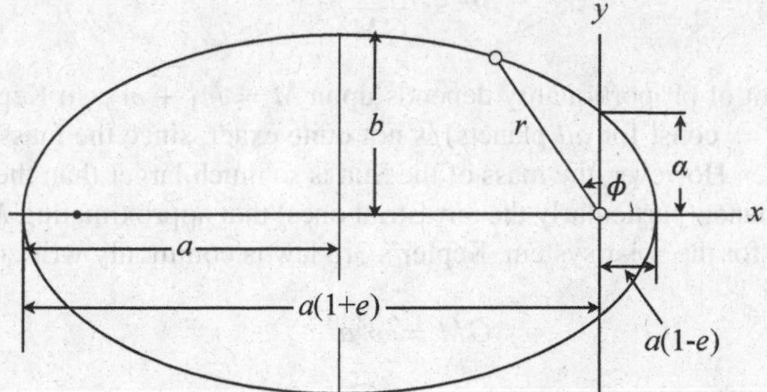


**Fig. 4.4** The ellipse and its properties: $a$ and $b$ are the semi-major and semi-minor axes of the ellipse, $e$ is its eccentricity, $\alpha$ is the latus rectum, and $(r, \phi)$ are the polar coordinates for a point on the ellipse, as measured with respect to $x$ and $y$ axes with the origin at one of the focal points

see that the eccentricity is a function of the energy and angular momentum of the orbit. The relationship between the two descriptions is shown in Fig. 4.4.

**Exercise 4.5** (a) Verify that (4.46) with $0 \le e < 1$ satisfies

$$\frac{(x - x_0)^2}{a^2} + \frac{y^2}{b^2} = 1\,, \tag{4.48}$$

with

$$x_0 = -\frac{\alpha e}{1 - e^2}\,, \qquad a = \frac{\alpha}{1 - e^2}\,, \qquad b = \frac{\alpha}{\sqrt{1 - e^2}}\,. \tag{4.49}$$

(b) Show that the last two relations can also be obtained directly from the definitions of $\alpha$, $a$, and $e$ given in Fig. 4.4. *Hint*: Recall that an ellipse can also be defined as the locus of points $P$ such that the sum of the distances to $P$ from the two focal points is a constant.

Using the result of the previous problem, we recover Kepler's 1st law that the orbits of planets are ellipses with the Sun at one focus, with the orbit given by

$$r = \frac{a\left(1 - e^2\right)}{1 + e\cos\phi}\,. \tag{4.50}$$

**Exercise 4.6** Use (4.31) to find a relation between the orbital period $P$ and the total area of the ellipse. Knowing that the area of an ellipse is $\pi ab$, you can now use the equation for the orbit along with the definition of $\alpha$ to obtain Kepler's 3rd law. Show that

$$P^2 = \frac{4\pi^2}{GM}a^3\,. \tag{4.51}$$

The constant of proportionality depends upon $M = m_1 + m_2$, so Kepler's 3rd law ($P^2/a^3 =$ const for *all* planets) is not quite exact, since the masses of the planets differ. However, the mass of the Sun is so much larger than the mass of any other planet (particularly the terrestrial ones) that approximating $M \sim M_\odot$ works well for the solar system. Kepler's 3rd law is commonly written as

$$GM = \omega^2 a^3\,, \tag{4.52}$$

where $\omega \equiv 2\pi/P$.

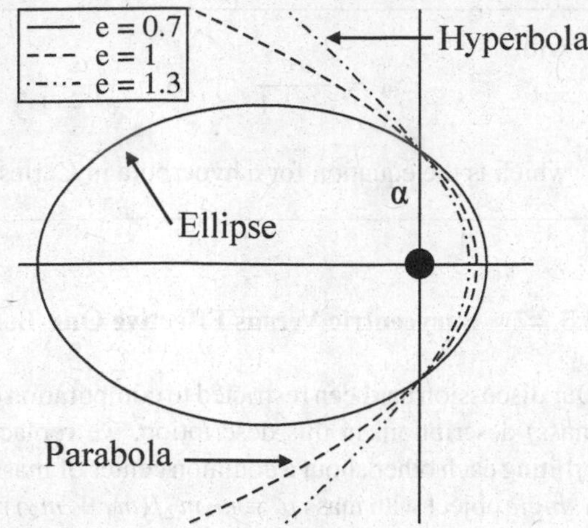


**Fig. 4.5** The orbits for bound ($e < 1$), critical ($e = 1$), and unbound ($e > 1$) systems. Bound orbits are ellipses with the central mass at one focus. Critical orbits are parabolas with the central mass at the focus. The kinetic energy of a critical orbit is zero at infinity. Unbound orbits are hyperbolae with the central mass at one focus. The kinetic energy of an unbound orbit is positive at infinity. The angular momentum for each orbit is $\ell = \sqrt{GM\mu^2\alpha}$

#### 4.3.2.1 Critical and Unbound Orbits

For bound orbits, $E < 0$ and $e < 1$. Since the method of solution for $r(\phi)$ outlined above is independent of whether the orbit is bound or not, we can also explore the properties of unbound orbits. For the critical case of $E = 0$, (4.42) requires that $e = 1$. The parametric equation then reads

$$r = \frac{\alpha}{1 + \cos\phi}, \tag{4.53}$$

which is the equation for a parabola with a point of closest approach of $\alpha/2$, as shown in Fig. 4.5. If the angular momentum is zero, then $e = 1$ as well. In that case, $\alpha = 0$ and the orbit is simply a straight drop from infinity into the central mass. For zero-angular momentum orbits, the energy can take any value, although negative energy orbits do not start at infinite distance. If the energy is positive, then $e > 1$ and the resulting orbit is hyperbolic.

**Exercise 4.7** (a) Show that (4.46) with $e = 1$ satisfies

$$x = \frac{\alpha}{2} - \frac{1}{2\alpha}y^2, \tag{4.54}$$

which is the equation of a parabola in Cartesian coordinates.
(b) Similarly, show that (4.46) with $e > 1$ satisfies

$$\frac{(x - x_0)^2}{A^2} - \frac{y^2}{B^2} = 1, \tag{4.55}$$

with

$$x_0 = \frac{\alpha e}{e^2 - 1}, \qquad A = \frac{\alpha}{e^2 - 1}, \qquad B = \frac{\alpha}{\sqrt{e^2 - 1}}, \tag{4.56}$$

which is the equation for a hyperbola in Cartesian coordinates.

#### 4.3.2.2 Barycentric Versus Effective One-Body Orbits

Our discussion has been restricted to computation of the effective one-body (reduced-mass) description. In this description, we replaced the true system of two masses orbiting each other about a common center of mass with an idealized system in which a *single* object with mass $\mu \equiv m_1 m_2/(m_1 + m_2)$ moves in an inertial reference frame under the influence of a central $1/r$ potential with a coupling strength given by $GM\mu$, where $M \equiv m_1 + m_2$. The beauty of this description is that there is no second mass $M$ located at the origin—merely a potential. The orbit $\mathbf{r}(t)$ that is calculated for this single particle can be used to obtain the orbits for both masses in the true system. If we choose to use the **barycenter** frame with the origin placed at the center of mass, then we are using an inertial reference frame and we can simply use the rescalings of (4.25) to obtain the orbits $\mathbf{r}_1(t)$ and $\mathbf{r}_2(t)$ for the individual masses directly from $\mathbf{r}(t)$.

When analyzing the orbits of bodies in astronomy, the location of the center of mass is often very difficult to determine. Thus, it is more common to use a coordinate system that is centered on one of the two masses in order to describe the position of the other mass. Unfortunately, this is a non-inertial, body-centered reference frame that will introduce fictitious forces related to the acceleration of the origin relative to the inertial reference frame. For concreteness, let's assume that the origin is placed on $m_2$. In this case, the equation of motion for $m_1$ is

$$m_1 \ddot{\mathbf{r}} = -\frac{G m_1 m_2}{r^3}\mathbf{r} - m_1 \ddot{\mathbf{r}}_2 \tag{4.57}$$

where $\mathbf{r}$ is the same vector that we used in the effective one-body frame, and the last term on the right-hand side is the fictitious force due to the acceleration of the origin about the center of mass. By moving the $-m_1\ddot{\mathbf{r}}_2$ to the left-hand side and using the relationship (4.25) between $\mathbf{r}_2$ and $\mathbf{r}$, we can absorb this fictitious force into a rescaled mass of the orbiting body $m_1$ using

$$m_1 \ddot{\mathbf{r}} + m_1 \ddot{\mathbf{r}}_2 = m_1\left(1 - \frac{m_1}{M}\right)\ddot{\mathbf{r}} = \mu \ddot{\mathbf{r}} = -\frac{G m_1 m_2}{r^3}\mathbf{r} = -\frac{GM\mu}{r^3}\mathbf{r}. \tag{4.58}$$

Thus, we recover the equation of motion obtained within the effective one-body frame. Because of this direct correspondence between adopting a reduced mass in the inertial effective one-body frame and absorbing a fictitious force into a rescaled mass in the non-inertial body-centered frame, it is tempting to equate the two reference frames. However, it is important to remember that the effective one-body frame is a mathematical construct used to obtain the equations of motion from within an inertial reference frame. As long as we restrict ourselves to two-body problems we can use these frames interchangeably, but as soon as an additional (third) particle is introduced, we must either revert to the inertial barycenter frame or introduce fictitious forces.

### 4.3.3 Equations of Motion and Their Solutions

Equations (4.35) and (4.29), for conservation of mechanical energy $E$ and conservation of angular momentum $\ell$, can be written as

$$\begin{aligned} \dot{r}^2 &= \frac{2E}{\mu} - \frac{\ell^2}{\mu^2 r^2} + \frac{2GM}{r}, \\ \dot{\phi} &= \frac{\ell}{\mu r^2}, \end{aligned} \tag{4.59}$$

which are two coupled 1st-order ordinary differential equations in the independent variable $t$. We can use the definition of $\alpha$ and the solution for the orbit $r(\phi)$, to decouple the equations and find the following equation for $\dot{\phi}$:

$$\dot{\phi} = \frac{G^2 M^2 \mu^3}{\ell^3} (1 + e\cos\phi)^2 . \tag{4.60}$$

Since this equation is separable, it can be integrated to give $t(\phi)$, which can (in principle) be inverted to obtain $\phi(t)$. The corresponding $r(t)$ is then obtained from (4.46) by replacing $\phi$ with $\phi(t)$. It turns out, however, that the inversion of $t(\phi)$ can be quite tricky. So we should look for other ways to solve for the time dependence.

### 4.3.4 Kepler's Equation

Another way of obtaining the time dependence of the motion involves **Kepler's equation**, which describes the time dependence of an angle measured from the center of the orbital ellipse. Consider an **auxiliary circle** of radius $a$, centered at point $O$, that circumscribes an ellipse with semi-major axis $a$, as shown in Fig. 4.6. The central mass is located at a focus of the ellipse, designated by $S$ in the figure. The location of the orbiting body is designated by $P$. The major axis of the ellipse is the line $\overline{A\Pi}$, and the **true anomaly**, $\phi$, is $\angle \Pi SP$. Now, consider a line normal to $\overline{A\Pi}$, passing through $P$ on the ellipse, and intersecting the circle at $Q$. We define $R$ to be the point where this line intersects $\overline{A\Pi}$. The **eccentric anomaly** is the angle $\psi$

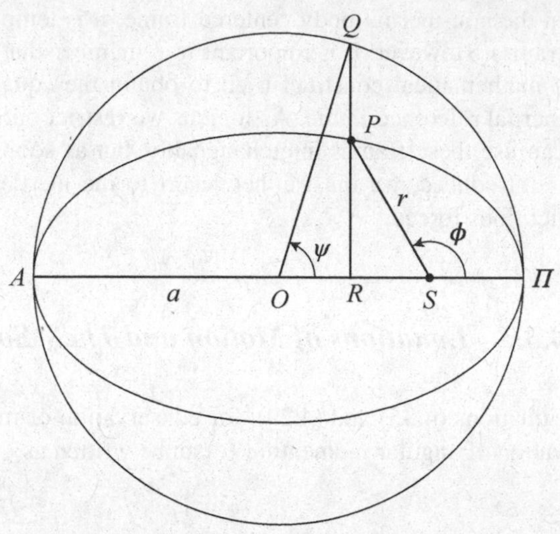


**Fig. 4.6** Configuration of the orbit and the auxiliary circle, showing the true anomaly $\phi$ and the eccentric anomaly $\psi$. The central body is at $S$ and the orbiting body is at $P$. The center of the circle is at $O$, and the periapsis and apapsis are at $\Pi$ and $A$, respectively

defined by $\angle\, \Pi O Q$. (Note that astronomers often use $E$ for the eccentric anomaly, but that is reserved for energy in this book.)

The position of the orbiting body can be described in terms of the eccentric anomaly by

$$\begin{aligned} r\cos\phi &= a\cos\psi - ae\,, \\ r\sin\phi &= a\sin\psi\sqrt{1-e^2}\,, \end{aligned} \tag{4.61}$$

where the last equation is a direct consequence of the relation $\overline{PR}/\overline{QR} = b/a = \sqrt{1-e^2}$ (Exercise 4.8). From these two equations, we find

$$r = a\,(1 - e\cos\psi)\,. \tag{4.62}$$

**Exercise 4.8** Use the equations for a circle and an ellipse in Cartesian coordinates to show that $\overline{PR}/\overline{QR} = b/a = \sqrt{1-e^2}$.

To determine the eccentric anomaly as function of time, $\psi = \psi(t)$, we first need to find a relationship between $\mathrm{d}\phi$ and $\mathrm{d}\psi$. Using (4.61) and (4.62), it follows that

$$\sin\phi = \frac{b\sin\psi}{a(1-e\cos\psi)}\,. \tag{4.63}$$

Taking the differential of both sides gives

$$\cos\phi\,\mathrm{d}\phi = \frac{b}{a}\frac{(\cos\psi(1-e\cos\psi)\,\mathrm{d}\psi - e\sin^2\psi\,\mathrm{d}\psi)}{(1-e\cos\psi)^2}\,, \tag{4.64}$$

which can be solved to yield

$$\mathrm{d}\phi = \frac{b}{a\,(1-e\cos\psi)}\,\mathrm{d}\psi\,. \tag{4.65}$$

Writing conservation of angular momentum (4.29) as

$$\mathrm{d}\phi = \frac{\ell}{\mu r^2}\mathrm{d}t\,, \tag{4.66}$$

and substituting the above expressions for $\mathrm{d}\phi$ and $r$, we obtain the following integrable equation

$$(1-e\cos\psi)\,\mathrm{d}\psi = \frac{\ell}{\mu ab}\,\mathrm{d}t\,. \tag{4.67}$$

Setting $t = 0$ at periapsis implies $\psi(0) = 0$. So integrating the above equation subject to this boundary condition gives

$$\psi - e\sin\psi = \frac{\ell t}{\mu ab}\,. \tag{4.68}$$

This equation can be simplified further if we note that Kepler's 2nd law (4.31) can be written as

$$\frac{\pi ab}{P} = \frac{\ell}{2\mu}\,, \tag{4.69}$$

where $\pi ab$ is the area of the ellipse and $P$ is the orbital period. Thus, $\ell/(\mu ab) = 2\pi/P \equiv \omega$, where $\omega$ is the orbital angular frequency. Making this substitution into (4.68), we arrive at the final form of Kepler's equation

$$\psi - e\sin\psi = \omega t\,. \tag{4.70}$$

Astronomers usually denote $\omega t$ by $M$ and call this the **mean anomaly**. (The mean anomaly is the angle that an object moving with *constant* angular velocity $\omega$ would sweep out in a time interval $t$.) For obvious reasons, we will not use $M$ for the mean anomaly, so we will just use $\omega t$. Kepler's equation is a transcendental equation which can be solved numerically or by Fourier series.

### 4.3.5 Fourier Series Solution to Kepler's Equation

The method of Fourier series solution to Kepler's equation was first applied by Bessel, who proposed a solution of the form

$$\psi = \omega t + \sum_{n=1}^{\infty} b_n(e) \sin(n\omega t) . \tag{4.71}$$

This is a series over the interval $0 \leq \omega t \leq \pi$, so the coefficients $b_n(e)$ are found via

$$b_n(e) = \frac{2}{\pi} \int_0^{\pi} (\psi - \theta) \sin n\theta \, \mathrm{d}\theta , \tag{4.72}$$

where $\theta \equiv \omega t$, using the orthogonality of the set of functions $\{\sin n\theta | n = 1, 2, \cdots\}$ on this interval. Integrating by parts, we find

$$b_n(e) = -\frac{2}{n\pi} \left[ (\psi - \theta) \cos n\theta \Big|_0^{\pi} - \int_0^{\pi} \left( \frac{\mathrm{d}\psi}{\mathrm{d}\theta} - 1 \right) \cos n\theta \, \mathrm{d}\theta \right] . \tag{4.73}$$

The first term is zero because $\psi = 0$ when $\theta = 0$ and $\psi = \pi$ when $\theta = \pi$. Thus, we have

$$b_n(e) = \frac{2}{n\pi} \int_0^{\pi} \cos n\theta \, \mathrm{d}\psi . \tag{4.74}$$

We can now substitute $\theta = \psi - e \sin \psi$ into this integral, and recall the integral representation of the Bessel function (Appendix E.5.1, (E.75)):

$$J_n(z) = \frac{1}{\pi} \int_0^{\pi} \cos(nx - z \sin x) \mathrm{d}x , \tag{4.75}$$

to arrive at the solution

$$b_n(e) = \frac{2}{n} J_n(ne) . \tag{4.76}$$

Thus, the eccentric anomaly is given by

$$\psi = \omega t + \sum_{n=1}^{\infty} \frac{2}{n} J_n(ne) \sin(n\omega t) . \tag{4.77}$$

With this, we can now compute $r \cos \phi$ and $r \sin \phi$ using (4.61).

**Exercise 4.9** We can also determine the time dependence of $r$ and $\phi$ directly from the eccentric anomaly using

$$r = a\,(1 - e\cos\psi)\,, \tag{4.78a}$$

$$\tan\frac{\phi}{2} = \sqrt{\frac{1+e}{1-e}}\,\tan\frac{\psi}{2}\,. \tag{4.78b}$$

Obtain these two expressions using some algebra, trigonometry, and Fig. 4.6.

## 4.4 Virial Theorem

The **virial theorem** relates the time average of the kinetic energy of a system of particles to the time average of the potential energy of the system. It has applications across a wide range of statistical mechanical systems, but we will focus here on its application to the central force problem. To begin with we will consider a system of $N$ particles and then later specialize to the two-body problem.

The virial theorem is obtained by considering the scalar quantity

$$G \equiv \sum_I \mathbf{p}_I \cdot \mathbf{r}_I\,. \tag{4.79}$$

The total time derivative of $G$ is

$$\frac{\mathrm{d}G}{\mathrm{d}t} = \sum_I \mathbf{p}_I \cdot \dot{\mathbf{r}}_I + \sum_I \dot{\mathbf{p}}_I \cdot \mathbf{r}_I\,. \tag{4.80}$$

The first term is simply twice the total kinetic energy, and by Newton's second law, the second term is

$$\sum_I \dot{\mathbf{p}}_I \cdot \mathbf{r}_I = \sum_I \mathbf{F}_I \cdot \mathbf{r}_I\,. \tag{4.81}$$

Thus,

$$\frac{\mathrm{d}G}{\mathrm{d}t} = 2T + \sum_I \mathbf{F}_I \cdot \mathbf{r}_I\,. \tag{4.82}$$

If we now look at the time average of these quantities over some time interval $\Delta t$, we have

$$\frac{1}{\Delta t}\int_0^{\Delta t} \frac{\mathrm{d}G}{\mathrm{d}t}\,\mathrm{d}t = 2\overline{T} + \overline{\sum_I \mathbf{F}_I \cdot \mathbf{r}_I} = \frac{1}{\Delta t}\left(G(\Delta t) - G(0)\right)\,. \tag{4.83}$$

Now, if the values of $\mathbf{p}_I$ and $\mathbf{r}_I$ are such that $G$ is finite at all times (such as is the case in a bound orbit or a bound system of particles), then we can always choose $\Delta t$ to be large enough that the right-hand side of (4.83) is vanishingly small. The virial theorem then reads

$$\overline{T} = -\frac{1}{2}\overline{\sum_I \mathbf{F}_I \cdot \mathbf{r}_I}\,, \tag{4.84}$$

in the limit of large $\Delta t$. The equality is *exact* if the motion is *periodic* and $\Delta t$ is equal to the period. This is the most general form of the virial theorem, which does not depend on the specific form of the forces acting on the individual particles.

If the forces are interparticle central forces, then

$$\sum_I \mathbf{F}_I \cdot \mathbf{r}_I = \sum_I \sum_J \mathbf{F}_{JI} \cdot \mathbf{r}_I = \frac{1}{2}\sum_{I,J} \mathbf{F}_{JI} \cdot \mathbf{r}_{IJ}\,, \tag{4.85}$$

where we have used Newton's 3rd law to obtain the last equality, and where we have defined $\mathbf{F}_{II} \equiv 0$ to allow the sum to run over *all* $I$ and $J$. If the central forces are also conservative, so that $\mathbf{F}_{JI} = -\boldsymbol{\nabla}_{IJ} U_{IJ}(r_{IJ})$, then

$$\sum_I \mathbf{F}_I \cdot \mathbf{r}_I = -\frac{1}{2}\sum_{I,J} \frac{\partial U_{IJ}}{\partial r_{IJ}} r_{IJ}\,. \tag{4.86}$$

Finally, if the central potential is a power law with $U_{IJ}(r_{IJ}) = A_{IJ} r_{IJ}^{\alpha}$ for some $\alpha$ (e.g., the gravitational force, with $A_{IJ} = -G m_I m_J$ and $\alpha = -1$), then

$$\sum_I \mathbf{F}_I \cdot \mathbf{r}_I = -\frac{1}{2}\sum_{I,J} A_{IJ}\alpha r_{IJ}^{\alpha-1} r_{IJ} = -\frac{\alpha}{2}\sum_{I,J} U_{IJ}(r_{IJ}) = -\alpha U\,, \tag{4.87}$$

where $U$ is the total potential energy of the system. Thus, for inverse-square law, conservative, central forces the virial theorem becomes

$$\overline{T} = -\frac{1}{2}\overline{U}\,. \tag{4.88}$$

**Exercise 4.10** For a circular orbit under the gravitational force, the kinetic and potential energies are constants. Show that the virial theorem is satisfied by explicitly calculating $T$ and $U$ for the two-body system.

### 4.4.1 Equations of State

The virial theorem can also be used to determine the equations of state for gases, where the average kinetic energy of the gas particles is related to the internal energy of the gas. For the simple case of a monatomic gas, where there is no mechanism to store energy within the atom, the average kinetic energy is related to the temperature of the gas. Since $T$ is commonly used for temperature in thermodynamics, we will suspend using it for the kinetic energy in this subsection and use $E_{\text{kin}}$ instead. From the definition of temperature, the average kinetic energy for a system of $N$ gas particles is

$$\overline{E}_{\text{kin}} = \frac{3}{2}NkT\,. \tag{4.89}$$

For ideal gases, the particles are widely separated, so that they do not interact. Therefore, the only forces acting on the particles are the forces of constraint that keep the gas enclosed in a volume $V$. These forces act only at the surface of the enclosure and point inward, normal to the surface, $-P\hat{\mathbf{n}}$. Thus,

$$-\frac{1}{2}\sum_I \mathbf{F}_I \cdot \mathbf{r}_I \to \frac{1}{2}\oint_S \mathrm{d}a\, P\hat{\mathbf{n}} \cdot \mathbf{r}\,. \tag{4.90}$$

Using the divergence theorem (A.87), we can convert the integral over the surface to an integral over the volume

$$\oint_S \mathrm{d}a\, \hat{\mathbf{n}} \cdot \mathbf{r} = \int_V \mathrm{d}V\, \boldsymbol{\nabla} \cdot \mathbf{r} = 3V\,. \tag{4.91}$$

So the virial theorem (4.84) gives us

$$\frac{3}{2}NkT = \frac{3}{2}PV\,, \tag{4.92}$$

which is just the **ideal gas law**

$$PV = NkT\,. \tag{4.93}$$

Note that for cases where there are also interparticle (central) interactions derivable from a total potential $U$, the right-hand side of (4.92) needs to be supplemented by a term equal to $\alpha\overline{U}/2$, leading to

$$PV = NkT - \frac{1}{3}\alpha\overline{U}\,. \tag{4.94}$$

## 4.5 Closed Orbits

The effective one-dimensional potential of (4.12) will admit stable, bound orbits if there is a local minimum of

$$U_{\text{eff}}(r) = \frac{\ell^2}{2\mu r^2} + U(r)\,, \tag{4.95}$$

at some $r_0 > 0$ which corresponds to the radius of a circular orbit for a given $\ell$. Setting $\mathrm{d}U_{\text{eff}}/\mathrm{d}r = 0$ at $r_0$ implies

$$\left.\frac{\mathrm{d}U}{\mathrm{d}r}\right|_{r_0} = \frac{\ell^2}{\mu r_0^3}\,, \qquad \text{(equilibrium condition)} \tag{4.96}$$

which indicates that the potential $U$ must have a positive slope at $r_0$, so the central force $F(r) = -\mathrm{d}U/\mathrm{d}r$ is attractive there. The effective potential has a minimum at this point only if the second derivative of $U_{\text{eff}}(r)$ is positive at $r_0$. Thus,

$$\left.\frac{\mathrm{d}^2 U_{\text{eff}}}{\mathrm{d}r^2}\right|_{r_0} = \frac{3\ell^2}{\mu r_0^4} + \left.\frac{\mathrm{d}^2 U}{\mathrm{d}r^2}\right|_{r_0} > 0\,, \tag{4.97}$$

or, equivalently,

$$\left.\frac{\mathrm{d}^2 U}{\mathrm{d}r^2}\right|_{r_0} > -\frac{3}{r_0}\left.\frac{\mathrm{d}U}{\mathrm{d}r}\right|_{r_0}\,, \qquad \text{(stability condition)} \tag{4.98}$$

where we used (4.96) to get the right-hand side. Potentials which admit stable, bound orbits must satisfy this inequality.

***Example 4.1*** If the potential is a power law of the form $U(r) = Ar^\alpha$ in the neighborhood of $r_0$, then it is easy to show that (4.98) reduces to

$$A\alpha\,(\alpha - 1)\, r_0^{\alpha-2} > -3A\alpha r_0^{\alpha-2}\,, \tag{4.99}$$

which implies

$$\alpha > -2\,. \tag{4.100}$$

Note that this requirement holds only in the vicinity of the minimum of $U_{\text{eff}}(r)$. If we require the existence of stable circular orbits at *all* radii, then a power-law potential $U(r)$ must be less steep than $1/r^2$ at *all values* of $r$. □

For a given angular momentum $\ell$, the energy of the circular orbit is the *lowest* possible energy in the neighborhood of $r_0$. The energy is then fixed by $\ell$ and $r_0$ as

$$E_0 = U(r_0) + \frac{\ell^2}{2\mu r_0^2}\,. \tag{4.101}$$

For energies slightly above $E_0$, the orbit will be *nearly circular*, being perturbed slightly away from $r = r_0$. For the case of the gravitational potential, this orbit is an ellipse. For other potentials, it is not necessarily the case that $r(\phi) = r(\phi + 2\pi)$, so that the orbit would come back on itself. For Newtonian gravity, the particle returns back to its original position after one revolution. More generally, a **closed orbit** is one for which the particle returns back to its original position after a *finite number* of revolutions.

### 4.5.1 *Driven Harmonic Oscillator Equation for* $u \equiv 1/r$

To further analyze the requirements on the potential for there to be closed orbits, it is convenient to take the equation of motion

$$\mu\ddot{r} - \frac{\ell^2}{\mu r^3} + \frac{\mathrm{d}U}{\mathrm{d}r} = 0 \tag{4.102}$$

for a particle in orbit about a central force $F = -\mathrm{d}U/\mathrm{d}r$, and rewrite it in terms of the variable $u \equiv 1/r$. Making this substitution for $r$, and using (4.13) to convert derivatives with respect to time to derivatives with respect to $\phi$, we find

$$-\frac{\ell^2}{\mu}u^2\left(u'' + u\right) - F(1/u) = 0\,. \tag{4.103}$$

Multiplying this equation by $-\mu/\ell^2 u^2$ and rearranging terms, we see that the differential equation is the same as that for a **driven harmonic oscillator**:

$$u'' + u = \Lambda(u)\,, \tag{4.104}$$

with driving force

$$\Lambda(u) \equiv -\frac{\mu}{\ell^2 u^2}F(1/u)\,. \tag{4.105}$$

If the orbit is circular with $u_0 = 1/r_0$, then $u'' = 0$, and we have a constraint relating $u_0$ and $\Lambda(u_0)$ (or $F(1/u_0)$), so

$$u_0 = \Lambda(u_0) = -\frac{\mu}{\ell^2 u_0^2} F(1/u_0) \,. \tag{4.106}$$

**Exercise 4.11** Show that in terms of the potential $U$, the driving force $\Lambda(u)$ can be written as

$$\Lambda(u) = -\frac{\mu}{\ell^2}\frac{\mathrm{d}U(1/u)}{\mathrm{d}u} \,. \tag{4.107}$$

### *4.5.2 Nearly Circular Orbits (1st-Order Perturbations)*

We now consider small perturbations to the circular orbit by letting $u \equiv u_0 + \eta$, where $\eta \ll u_0$. In this case, the driven harmonic oscillator equation reads

$$\eta'' + u_0 + \eta = \Lambda(u_0 + \eta) \,. \tag{4.108}$$

We can Taylor expand $\Lambda$ on the right hand side as

$$\Lambda(u_0 + \eta) = \Lambda(u_0) + \eta \left.\frac{\mathrm{d}\Lambda}{\mathrm{d}u}\right|_{u_0} + \frac{1}{2}\eta^2 \left.\frac{\mathrm{d}^2\Lambda}{\mathrm{d}u^2}\right|_{u_0} + \cdots \,. \tag{4.109}$$

Since the perturbation is small, we can discard terms containing $\eta^2$ and higher powers of $\eta$. (We will do this for all subsequent relevant equations.) Noting that $u_0 = \Lambda(u_0)$, we find the perturbed equation of motion to be

$$\eta'' = -\eta\left(1 - \left.\frac{\mathrm{d}\Lambda}{\mathrm{d}u}\right|_{u_0}\right) \,. \tag{4.110}$$

The perturbed orbit will not be stable unless the factor in parentheses is positive definite, so we can write

$$\eta'' = -\beta^2 \eta \,, \tag{4.111}$$

where

$$\beta^2 \equiv 1 - \left.\frac{\mathrm{d}\Lambda}{\mathrm{d}u}\right|_{u_0} \,. \tag{4.112}$$

This is just the standard simple harmonic oscillator equation, which has periodic solutions $\sin(\beta\phi)$ and $\cos(\beta\phi)$. Without loss of generality (since the unperturbed orbit is circular), we can choose the boundary conditions so that

$$\eta(\phi) = \eta_1 \cos(\beta\phi), \qquad (4.113)$$

where the amplitude $\eta_1 \ll u_0$.

**Exercise 4.12** Show that the requirement that $\beta^2$ be positive definite in (4.112) is satisfied by (4.98).

For the orbit to be closed, $\eta$ must return to itself after some integer number of orbits (let's say $q$). This means that

$$\eta(\phi + 2\pi q) = \eta(\phi), \qquad (4.114)$$

or, equivalently,

$$\cos(\beta[\phi + 2\pi q]) = \cos(\beta\phi), \qquad (4.115)$$

which implies $\beta q = p$ for some integer $p$. Hence, $\beta$ must be a rational number

$$\beta = p/q, \qquad \text{(closed 1st order)}. \qquad (4.116)$$

Now, since $\beta$ is related to $\Lambda$ and $\Lambda$ is related to $F$, a nearly circular orbit will be closed only for forces that satisfy (4.112). From the definition of $\Lambda$ given in (4.105), we have

$$\left.\frac{\mathrm{d}\Lambda}{\mathrm{d}u}\right|_{u_0} = \frac{2\mu}{\ell^2 u_0^3} F(1/u_0) - \frac{\mu}{\ell^2 u_0^2} \left.\frac{\mathrm{d}F(1/u)}{\mathrm{d}u}\right|_{u_0}. \qquad (4.117)$$

But the requirement (4.106) for a circular orbit gives

$$\frac{\mu}{\ell^2 u_0^2} = -\frac{u_0}{F(1/u_0)}. \qquad (4.118)$$

Substituting this back into (4.117) gives

$$\left.\frac{\mathrm{d}\Lambda}{\mathrm{d}u}\right|_{u_0} = -2 + \frac{u_0}{F(1/u_0)} \left.\frac{\mathrm{d}F(1/u)}{\mathrm{d}u}\right|_{u_0}, \qquad (4.119)$$

so

$$3 - \beta^2 = \frac{u_0}{F(1/u_0)} \left.\frac{\mathrm{d}F(1/u)}{\mathrm{d}u}\right|_0 = \left.\frac{\mathrm{d}\ln F(1/u)}{\mathrm{d}\ln u}\right|_{u_0}, \qquad (4.120)$$

where we used the fact that $\mathrm{d}\ln x = \mathrm{d}x/x$ to get the last equality.

We have shown above that in order for a nearly circular orbit to be closed, $\beta$ must be a rational number. But since we can smoothly change both the angular momentum and total energy to go from one circular orbit to another, $\beta$ must be the *same* rational number for *all* circular orbits. Otherwise there would be some nearly circular orbits that were not closed. Thus, (4.120) must be true for *all* values of $u_0$, with a constant value for $\beta$, which means that we can drop the subscript "0" from $u$. Equation (4.120) is then a simple separable differential equation which can be solved as

$$\left(3-\beta^2\right) \mathrm{d} \ln u = \mathrm{d} \ln F(1/u) \quad \Rightarrow \quad F(1/u) = A u^{3-\beta^2} = \frac{A}{r^{3-\beta^2}}, \quad (4.121)$$

where $A$ is an arbitrary constant defining the strength of the force, and $\beta$ is a rational number. All forces of this type will have closed orbits that deviate slightly from circular. Examples of such nearly circular orbits are shown in Fig. 4.7 for four different values of $\beta$.

If we insist that *all* orbits be closed (and not just those that deviate slightly from circular), then we must include the *higher-order terms* in the Taylor expansion of $\Lambda$. We shall see in the following subsection that this imposes a further restriction on the form of $\beta^2$ for closed orbits.

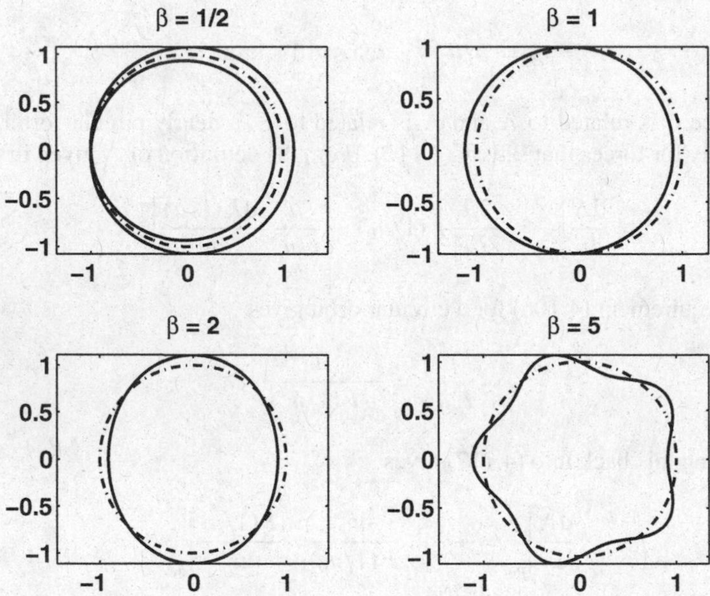


**Fig. 4.7** Closed orbits that deviate slightly from circular orbits. The four plots correspond to $\beta = 1/2$, 1, 2, and 5. The dot-dashed curves shows the original circular orbits and the solid curves show the slightly perturbed orbits

### 4.5.3 *Higher-Order Perturbations*

We start by considering a Fourier expansion of the perturbation

$$\eta(\phi) = \sum_{n=0}^{\infty} \eta_n \cos(n\beta\phi), \tag{4.122}$$

allowing contributions from higher-order harmonics, as well as a constant term $\eta_0$, as compared to (4.113). We will make the assumption that the coefficients $\eta_n$ decrease for larger values of $n$, so that we can truncate the series at some desired level of precision. For example, when deriving the differential equation for $\eta$ at a given order of the perturbation, we need to be *consistent* between the order of the Fourier series expansion of $\eta$ and the Taylor series expansion of $\Lambda(u_0 + \eta)$ around the circular orbit $u = u_0$. We also know that as the orbit approaches circularity, the dominant term will be $\eta_1$, since that is the only surviving term at the 1st-order approximation (4.113) derived above.

#### 4.5.3.1 Second-Order Calculation

Let's begin by considering the next order of approximation by going out to $n = 2$. Thus,

$$\eta(\phi) = \eta_0 + \eta_1 \cos(\beta\phi) + \eta_2 \cos(2\beta\phi), \tag{4.123}$$

where $\beta$ is still given by (4.112). Since this is the next order in the expansion, we assume that the new terms are small compared to $\eta_1$ (i.e., $\eta_0,\ \eta_2 \ll \eta_1$), so we keep terms of order $\eta_0$, $\eta_2$, and $\eta_1^2$. The equation for $\eta$ is now

$$\eta'' + \eta = \eta \left.\frac{\mathrm{d}\Lambda}{\mathrm{d}u}\right|_{u_0} + \frac{1}{2}\eta^2 \left.\frac{\mathrm{d}^2\Lambda}{\mathrm{d}u^2}\right|_{u_0}, \tag{4.124}$$

where we have Taylor expanded $\Lambda(u_0 + \eta)$, keeping terms to 2nd order in $\eta$. Looking at each term in the above equation separately, we have for $\eta''$,

$$\eta'' = -\beta^2 \eta_1 \cos(\beta\phi) - 4\beta^2 \eta_2 \cos(2\beta\phi), \tag{4.125}$$

and for $\eta^2$,

$$\eta^2 = \eta_1^2 \cos^2(\beta\phi) = \frac{1}{2}\eta_1^2 (1 + \cos(2\beta\phi)), \tag{4.126}$$

where we discarded terms proportional to $\eta_0\eta_1$, $\eta_2\eta_1$, $\eta_0^2$, $\eta_2^2$, etc., since they are 3rd-order small or smaller. The first derivative of $\Lambda(u)$ is given in terms of the definition

(4.112) of $\beta^2$,

$$\left.\frac{\mathrm{d}\Lambda}{\mathrm{d}u}\right|_{u_0} = 1 - \beta^2 \,, \tag{4.127}$$

but we also know from (4.105) and (4.121) that

$$\Lambda(u) = -\frac{\mu}{\ell^2 u^2} F(1/u) = -\frac{\mu A}{\ell^2} u^{1-\beta^2} \,, \tag{4.128}$$

so

$$\left.\frac{\mathrm{d}\Lambda}{\mathrm{d}u}\right|_{u_0} = -\frac{\mu A}{\ell^2} \left(1 - \beta^2\right) u_0^{-\beta^2} \,. \tag{4.129}$$

Comparing the right-hand sides of (4.127) and (4.129), we have

$$u_0^{\beta^2} = -\frac{\mu A}{\ell^2} \,, \tag{4.130}$$

which allows us to write

$$\Lambda(u) = u_0^{\beta^2} u^{1-\beta^2} \,. \tag{4.131}$$

From this expression, we can now easily compute the second derivative of $\Lambda(u)$,

$$\left.\frac{\mathrm{d}^2\Lambda}{\mathrm{d}u^2}\right|_{u_0} = -\beta^2 \left(1 - \beta^2\right) u_0^{-1} \,. \tag{4.132}$$

We are now in a position to combine everything together in (4.124). The left-hand and right-hand sides of this equation are

$$\begin{aligned} \mathrm{LHS} &= \eta_0 + \left(1 - \beta^2\right) \eta_1 \cos(\beta\phi) + \left(1 - 4\beta^2\right) \eta_2 \cos(2\beta\phi) \,, \\ \mathrm{RHS} &= \left(1 - \beta^2\right) \left[\eta_0 + \eta_1 \cos(\beta\phi) + \eta_2 \cos(2\beta\phi)\right] \\ &\quad - \frac{1}{2} \frac{\beta^2 \left(1 - \beta^2\right)}{u_0} \frac{1}{2} \eta_1^2 \left(1 + \cos(2\beta\phi)\right) . \end{aligned} \tag{4.133}$$

Grouping terms according to the cosines, we obtain the following three equations:

$$\begin{aligned} \eta_0 &= \left(1 - \beta^2\right) \eta_0 - \frac{1}{4} \frac{\beta^2 \left(1 - \beta^2\right)}{u_0} \eta_1^2 \,, \\ \left(1 - \beta^2\right) \eta_1 &= \left(1 - \beta^2\right) \eta_1 \,, \\ \left(1 - 4\beta^2\right) \eta_2 &= \left(1 - \beta^2\right) \eta_2 - \frac{1}{4} \frac{\beta^2 \left(1 - \beta^2\right)}{u_0} \eta_1^2 \,. \end{aligned} \tag{4.134}$$

The middle equation is a tautology and tells us nothing, while the other two yield expressions for $\eta_0$ and $\eta_2$:

$$\begin{aligned} \eta_0 &= -\frac{1}{4}\frac{(1-\beta^2)}{u_0}\eta_1^2\,, \\ \eta_2 &= +\frac{1}{12}\frac{(1-\beta^2)}{u_0}\eta_1^2\,, \end{aligned} \tag{4.135}$$

which imply $\eta_0 = -3\eta_2$. This confirms that $\eta_0$ and $\eta_2$ are of order $\eta_1^2$ (and thus an order smaller than $\eta_1$), but it doesn't further constrain the nature of the force law in order to guarantee closed orbits. In order to do that, we need to go to 3rd order in the Taylor expansion and include the $\eta_3 \cos(3\beta\phi)$ term in the Fourier expansion.

#### 4.5.3.2 Third-Order Calculation

The calculation to 3rd-order in the perturbation $\eta$ is rather long but conceptually straight-forward, so we leave it as a problem for the reader (Problem 4.3). The result, which we summarize here, is particularly interesting, as it drastically restricts the allowed values of $\beta$—and hence the allowed central forces—that admit closed bound orbits. The result was first derived by Joseph Bertrand in 1873, and is referred to in the literature as **Bertrand's theorem.**

**Bertrand's theorem**: The only central forces that give rise to closed bound orbits are the inverse-square-law force $F \propto r^{-2}$ and Hooke's law for the 3-dimensional harmonic oscillator $F \propto r$.

**Exercise 4.13** The 3-dimensional harmonic oscillator potential for Hooke's law ($F \propto r$) is $U = kr^2/2$, where $k$ is the coupling strength of the potential (think of it as the spring constant). Use Cartesian coordinates to solve for the motion of a particle of mass $m$ under the influence of this potential, subject to the initial conditions that $x(0) = a$, $y(0) = 0$, $\dot{x}(0) = 0$, $\dot{y}(0) = b\omega$, where $\omega \equiv \sqrt{k/m}$. Show that the orbit is an ellipse centered at the origin with

$$\frac{x^2}{a^2} + \frac{y^2}{b^2} = 1\,. \tag{4.136}$$

Thus, the orbits for this Hooke's law potential are *closed*, consistent with Bertrand's theorem for $\beta = 2$.

## 4.6 Another Conserved Quantity for Inverse-Square-Law Forces

From the previous section, we found that only two central potentials allow for all bound orbits to be closed. For potentials such as the gravitational potential with $U \propto r^{-1}$, we have $\beta = 1$, and so the particle returns back to its initial position after one orbit. For the simple harmonic oscillator potential, with $U \propto r^2$, we have $\beta = 2$, and so the particle returns to its original perturbation after a half of an orbit. This means that the orbits for the simple harmonic oscillator have an additional symmetry about the origin (Exercise 4.13).

If we consider the general orbits of these two potentials, we see that the inverse-square-law force results in ellipses with the origin located at a focus, and the harmonic oscillator force results in ellipses with the origin at the center of the ellipse. We can define a preferred line in either system by using the major axis of the ellipse. With the simple harmonic oscillator, both ends of the major axis are equidistant from the origin, so we cannot assign a preferred direction to this line. However, with the inverse-square-law force, we can also assign a direction to this line by having it point in the direction of the periapsis. *Thus, it is possible to define a conserved vector associated with the orbit for an inverse-square-law force.*

Clearly, the length of this vector can be arbitrarily constructed from various other constants of the motion (e.g., $\boldsymbol{\ell}$, $E$, $e$, ...), so all we really care about is constructing a constant vector from $\mathbf{r}$, $\mathbf{p}$, $\boldsymbol{\ell}$, ... that points in the direction of the periapsis. (Recall that for our choice of coordinates, periapsis is in the $+\hat{\mathbf{x}}$-direction.) The proportionality factor, provided it is constant, is not important for this calculation.

### 4.6.1 *Motion of the Momemtum Vector in Momentum Space*

So let's start by highlighting an interesting feature of the momentum of a particle orbiting an inverse-square-law central force. We'll use the case of the Newtonian gravitational force, but the results are easily generalizable to *any* inverse-square-law force. We write the momentum vector of a particle in orbit as

$$\mathbf{p} = \mu\dot{\mathbf{r}} = \mu\left[\left(\dot{r}\cos\phi - r\dot{\phi}\sin\phi\right)\hat{\mathbf{x}} + \left(\dot{r}\sin\phi + r\dot{\phi}\cos\phi\right)\hat{\mathbf{y}}\right]. \tag{4.137}$$

Substituting $\dot{\phi} = \ell/\mu r^2$ and

$$\dot{r} = \dot{\phi}\frac{\mathrm{d}r}{\mathrm{d}\phi} = \frac{\ell}{\mu r}\frac{e\sin\phi}{1+e\cos\phi}, \tag{4.138}$$

we obtain

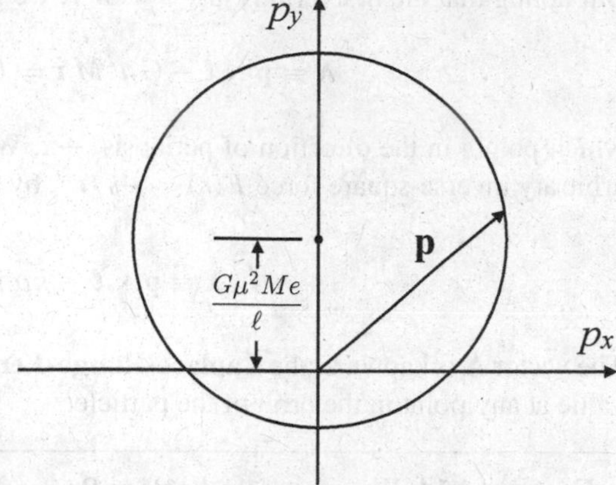


**Fig. 4.8** Motion of the momentum vector $\mathbf{p}$ in momentum space. The momentum traces out a circle whose center is offset in the $p_y$ direction by an amount $G\mu^2 Me/\ell$

$$\mathbf{p} = \frac{\ell}{r}\left[\left(\frac{e\sin\phi\cos\phi}{1+e\cos\phi} - \sin\phi\right)\hat{\mathbf{x}} + \left(\frac{e\sin^2\phi}{1+e\cos\phi} + \cos\phi\right)\hat{\mathbf{y}}\right], \tag{4.139}$$

which simplifies to

$$\mathbf{p} = \frac{\ell}{r(1+e\cos\phi)}\left[-\sin\phi\,\hat{\mathbf{x}} + (\cos\phi + e)\,\hat{\mathbf{y}}\right]. \tag{4.140}$$

Using (4.46), we have $r\,(1+e\cos\phi) = \alpha = \ell^2/\left(G\mu^2 M\right)$, so

$$\mathbf{p} = \frac{G\mu^2 M}{\ell}\left(-\sin\phi\,\hat{\mathbf{x}} + \cos\phi\,\hat{\mathbf{y}}\right) + \frac{G\mu^2 Me}{\ell}\,\hat{\mathbf{y}}. \tag{4.141}$$

Thus, we see that the momentum vector $\mathbf{p}$ traces out a circle of radius $G\mu^2 M/\ell$ in momentum space, where the center of the circle is offset by $G\mu^2 Me/\ell$ in the $p_y$ direction. See Fig. 4.8.

### 4.6.2 Laplace-Runge-Lenz Vector

In order to construct a conserved vector that lies within the orbital plane, we should consider creating it from $\mathbf{r}\times\boldsymbol{\ell}$ or $\mathbf{p}\times\boldsymbol{\ell}$, with the possible addition of vectors proportional to $\mathbf{r}$ or $\mathbf{p}$. Let's look at $\mathbf{p}\times\boldsymbol{\ell}$ first, recalling that $\boldsymbol{\ell} = \ell\,\hat{\mathbf{z}}$. Using (4.141), we see that

$$\mathbf{p}\times\boldsymbol{\ell} = G\mu^2 M\left(\cos\phi\,\hat{\mathbf{x}} + \sin\phi\,\hat{\mathbf{y}}\right) + G\mu^2 Me\,\hat{\mathbf{x}}. \tag{4.142}$$

But noting that the first term is just $G\mu^2 M\,\hat{\mathbf{r}}$, we have found a constant vector

$$\mathbf{A} \equiv \mathbf{p} \times \boldsymbol{\ell} - G\mu^2 M\,\hat{\mathbf{r}} = G\mu^2 M e\,\hat{\mathbf{x}}\,, \tag{4.143}$$

which points in the direction of periapsis, $+\hat{\mathbf{x}}$. We can easily generalize this to an arbitrary inverse-square force $F(r) = -k/r^2$, by simply replacing $GM\mu$ with $k$:

$$\mathbf{A} \equiv \mathbf{p} \times \boldsymbol{\ell} - k\mu\,\hat{\mathbf{r}}\,. \tag{4.144}$$

The vector **A** is known as the **Laplace-Runge-Lenz** vector. Note that **A** has the same value at any point in the orbit of the particle.

**Exercise 4.14** We can use the Laplace-Runge-Lenz vector to quickly calculate the equation of the orbit for a particle. Use the definition (4.143) of **A** and the fact that $\mathbf{r} \cdot \mathbf{A} = rA\cos\phi$ to recover the ellipse solution (4.46).

**Exercise 4.15** Show by direct differentiation that the Laplace-Runge-Lenz vector defined by (4.144) is a conserved quantity—i.e., that $\mathrm{d}\mathbf{A}/\mathrm{d}t = 0$.

## 4.7 Three-Body Problem

We have concentrated on the two-body problem for central forces because it can be solved analytically in a number of cases; it can be reduced to an effective one-body problem and the resulting symmetry makes the problem tractable. This symmetry is lost when going to a larger number of bodies. Even for the simplest case of three bodies, Poincaré showed that there is no analytic solution to the general gravitational three-body problem. It is possible for exact solutions to be obtained for only very special configurations.

One special case that we will look at is appropriate for the analysis of binary stellar evolution and for interplanetary navigation of spacecraft. We restrict ourselves to the case where one of the three bodies is substantially less massive than the other two. In this case, the motion of the two massive bodies is unaffected by the presence of the third body and so their motion is entirely determined by the two-body problem described earlier in this chapter. The third body then moves about in the time-varying potential created by the other two bodies. This is known as the **Roche model**.

Since the motion of the two large bodies is already determined, the Roche model focuses on the motion of the third body. We choose a coordinate system that is co-rotating with the orbit of the binary. For simplicity, we will assume that the orbit is circular. To be specific, we choose coordinates with the binary in the $xy$-plane and the $z$-axis aligned with the orbital angular momentum vector, as shown in Fig. 4.9.

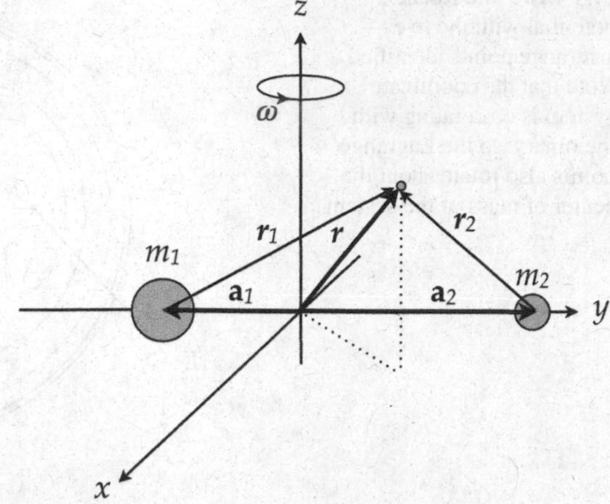


**Fig. 4.9** The co-rotating Cartesian coordinate system for describing the motion of a third body within the potential of a binary system

The origin is co-located with the center of mass of the binary and the two bodies lie along the $y$-axis. We choose the orientation of the $y$-axis such that the more massive object in the binary is placed along the negative $y$-axis. We'll label the most massive object $m_1$, so the positions of the two objects are $\mathbf{a}_1 = -a_1\hat{\mathbf{y}}$ and $\mathbf{a}_2 = a_2\hat{\mathbf{y}}$. The rotational rate is determined from Kepler's law, so the coordinate system rotates with

$$\omega = \sqrt{\frac{G\left(m_1 + m_2\right)}{\left(a_1 + a_2\right)^3}}. \tag{4.145}$$

In this configuration, the potential that the third body experiences is stationary so that energy is conserved. If the orbit of the binary were not circular, this would not be the case and the binary could inject or remove energy from the motion of the third body. In this coordinate system, the Lagrangian for the third body $m$ is

$$L = \frac{1}{2}m\left(\dot{x}^2 + \dot{y}^2 + \dot{z}^2\right) + \frac{1}{2}m\omega^2\left(x^2 + y^2\right) + \frac{Gm_1 m}{r_1} + \frac{Gm_2 m}{r_2}, \tag{4.146}$$

where the second term describes the kinetic energy of rotation due to the non-inertial reference frame, and $r_1 = |\mathbf{r} - \mathbf{a}_1|$ and $r_2 = |\mathbf{r} - \mathbf{a}_2|$. We now group the last three terms in the Lagrangian into an effective potential,

$$U_{\mathrm{R}} = -\frac{1}{2}m\omega^2\left(x^2 + y^2\right) - \frac{Gm_1 m}{r_1} - \frac{Gm_2 m}{r_2}, \tag{4.147}$$

known as the **Roche potential**. There are five stationary points (two maxima and three saddle points) in the orbital plane of the Roche potential. These are known as the **Lagrange points** and are shown in Fig. 4.10.

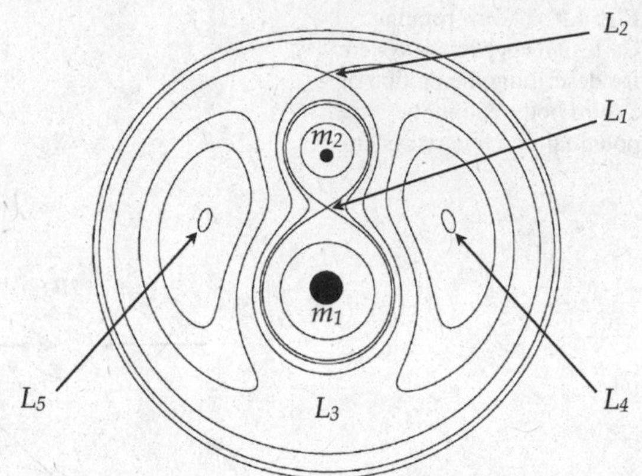


**Fig. 4.10** The Roche potential with the five Lagrange points identified. Note that the coordinate system is co-rotating with the binary, so the Lagrange points also rotate about the center of mass of the system

> **Exercise 4.16** Confirm by direct substitution that the points $L_4$ and $L_5$ each form separate equilateral triangles with the two masses $m_1$ and $m_2$. (*Hint:* Use geometry to find the coordinates of $L_4$ and $L_5$, and then show that these coordinates lie at stationary points (they are actually maxima) of the Roche potential.)

## Suggested References

*Full references are given in the bibliography at the end of the book.*

Arnold (1978): A discussion of the mathematical methods of classical mechanics, written for mathematics students or physicists with a mathematical bent. Chapter 2, Sect. 8 of this book has a series of problems that constitute an alternative proof of Bertrand's theorem.

Benacquista (2013): A similar treatment of the central force problem in an astrophysics context is given in Chap. 2, while the three-body problem in the context of mass-transferring binaries is given in Chap. 13.

Bertrand (1873): Bertrand's original paper, in French.

Goldstein et al. (2002): The second edition has a nice discussion of Bertrand's theorem in the appendix.

Hilditch (2001): This book has a more observational-based treatment of the central force problem and the Roche potential in the context of astrophysics. Chapter 2 covers the central force and Chap. 4 covers the Roche potential.

Santos et al. (2007): An English translation of Bertrand's theorem.

## Additional Problems

**Problem 4.1** (*Adapted from Kuchăr 1995.*) Consider the motion (in the $xy$-plane) of a mass $m$ about two *fixed* mass points $m_1$ and $m_2$ located at $(\ell, 0)$ and $(-\ell, 0)$, respectively. For this problem, define **elliptic coordinates** $(\xi, \eta)$ as

$$x = \ell \cosh \xi \cos \eta \,, \qquad y = \ell \sinh \xi \sin \eta \,. \tag{4.148}$$

(a) Show that curves of constant $\xi$ are *ellipses*, with focal points $(\ell, 0)$ and $(-\ell, 0)$, and semi-major and semi-minor axes

$$a \equiv \ell \cosh \xi \,, \qquad b \equiv \ell |\sinh \xi| \,. \tag{4.149}$$

Note that for $\xi = 0$, the ellipse degenerates to a line connecting the two focal points.

(b) Show that curves of constant $\eta$ are hyperbolae that open to the right and left with focal points $(\ell, 0)$ and $(-\ell, 0)$, and semi-axes

$$a \equiv \ell |\cos \eta| \,, \qquad b \equiv \ell |\sin \eta| \,. \tag{4.150}$$

Note that for $\eta = 0$ and $\eta = \pi$, the hyperbolae degenerate to lines along the $x$-axis, $x \in [\ell, \infty)$ and $x \in (-\infty, -\ell]$, respectively. For $\eta = \pi/2$, the hyperbola is the $y$-axis from $-\infty$ to $\infty$.

(c) Calculate the coordinate basis vectors $\boldsymbol{\partial}_\xi$ and $\boldsymbol{\partial}_\eta$ using the expressions

$$\boldsymbol{\partial}_\xi = \frac{\partial x}{\partial \xi}\hat{\mathbf{x}} + \frac{\partial y}{\partial \xi}\hat{\mathbf{y}} \,, \qquad \boldsymbol{\partial}_\eta = \frac{\partial x}{\partial \eta}\hat{\mathbf{x}} + \frac{\partial y}{\partial \eta}\hat{\mathbf{y}} \,. \tag{4.151}$$

(d) Using the results of the previous part, show that the coordinates $(\xi, \eta)$ are *orthogonal*, i.e.,

$$\boldsymbol{\partial}_\xi \cdot \boldsymbol{\partial}_\eta = 0 \,, \tag{4.152}$$

and that the norm of each coordinate basis vector is given by

$$|\boldsymbol{\partial}_\xi| = |\boldsymbol{\partial}_\eta| = \ell(\cosh^2 \xi - \cos^2 \eta)^{1/2} \,. \tag{4.153}$$

(e) Show that the kinetic energy of the mass $m$ can be written in elliptic coordinates as

$$T = \frac{1}{2} m \ell^2 (\cosh^2 \xi - \cos^2 \eta)(\dot{\xi}^2 + \dot{\eta}^2) \,. \tag{4.154}$$

(f) Show that gravitational potential energy between $m$ and the two fixed mass points $m_1$ and $m_2$ can be written in elliptic coordinates as

$$U = -\frac{Gm}{\ell}\left[\frac{m_1}{(\cosh\xi - \cos\eta)} + \frac{m_2}{(\cosh\xi + \cos\eta)}\right], \tag{4.155}$$

with corresponding gravitational force

$$\begin{aligned}\mathbf{F} = &-\frac{Gm}{\ell^2(\cosh^2\xi - \cos^2\eta)^{1/2}} \\ &\times\left\{\left[\frac{m_1 \sinh\xi}{(\cosh\xi - \cos\eta)^2} + \frac{m_2 \sinh\xi}{(\cosh\xi + \cos\eta)^2}\right]\hat{\boldsymbol{\xi}}\right. \\ &\left.+\left[\frac{m_1 \sin\eta}{(\cosh\xi - \cos\eta)^2} - \frac{m_2 \sin\eta}{(\cosh\xi + \cos\eta)^2}\right]\hat{\boldsymbol{\eta}}\right\}. \end{aligned} \tag{4.156}$$

**Problem 4.2** Rederive the power-law form of the force (4.121) for closed slightly perturbed circular orbits, working primarily with the *potential* $U$ instead of the force $F$.

(a) First, show that in terms of $u \equiv 1/r$,

$$F(1/u) = u^2 \frac{\mathrm{d}U(1/u)}{\mathrm{d}u}, \tag{4.157}$$

and the condition for a circular orbit (4.96) can be written as

$$-\frac{1}{u_0}\left.\frac{\mathrm{d}U(1/u)}{\mathrm{d}u}\right|_{u_0} = \frac{\ell^2}{\mu}. \tag{4.158}$$

(b) Then using the results of Exercise 4.11, show that

$$1 - \beta^2 = \left.\frac{\mathrm{d}\Lambda}{\mathrm{d}u}\right|_{u_0} = \left[-\frac{\mu}{\ell^2}\frac{\mathrm{d}^2U(1/u)}{\mathrm{d}u^2}\right]_{u_0}. \tag{4.159}$$

(c) Eliminate $\mu/\ell^2$ from the above equation using (4.158), thereby obtaining

$$u_0 \left.\frac{\mathrm{d}^2U(1/u)}{\mathrm{d}u^2}\right|_{u_0} - (1-\beta^2)\left.\frac{\mathrm{d}U(1/u)}{\mathrm{d}u}\right|_{u_0} = 0. \tag{4.160}$$

(d) As discussed in the main text, since $\beta$ is a rational number, the above equation must hold for *all* values of $u_0$, allowing us to drop the subscript "0". The resulting differential equation can be written in terms of $G(u) \equiv \mathrm{d}U(1/u)/\mathrm{d}u$ as

$$u\frac{\mathrm{d}G}{\mathrm{d}u} - (1-\beta^2)G(u) = 0. \tag{4.161}$$

Solve this equation, finding

$$G(u) = Au^{1-\beta^2} \quad \Rightarrow \quad U(1/u) = \frac{A}{2-\beta^2}u^{2-\beta^2} + B\,. \tag{4.162}$$

(e) Finally, using (4.157), recover $F(1/u) = Au^{3-\beta^2}$, which is (4.121).

**Problem 4.3** In this problem, you will determine the constraint on $\beta$ when the perturbation on a circular orbit is carried out to 3rd order in the Taylor expansion and to 3rd order in the Fourier expansion.

(a) Carry the perturbation expansion in (4.122) out to 3rd order, including the term $\eta_3 \cos(3\beta\phi)$, and show that

$$\begin{aligned}\eta^2 &= \frac{1}{2}\eta_1^2 + (2\eta_0\eta_1 + \eta_1\eta_2)\cos(\beta\phi) + \frac{1}{2}\eta_1^2\cos(2\beta\phi) + \eta_1\eta_2\cos(3\beta\phi)\,,\\ \eta^3 &= \frac{3}{4}\eta_1^3\cos(\beta\phi) + \frac{1}{4}\eta_1^3\cos(3\beta\phi)\,,\end{aligned} \tag{4.163}$$

and

$$\left.\frac{d^3\Lambda}{du^3}\right|_{u_0} = \beta^2(1-\beta^2)(1+\beta^2)u_0^{-2}\,. \tag{4.164}$$

(b) Show that grouping together the $\cos(n\beta\phi)$ terms results in the four equations:

$$\begin{aligned}\eta_0 &= -\frac{1}{4}\frac{(1-\beta^2)}{u_0}\eta_1^2\,,\\ 0 &= +\frac{1}{12}\frac{\beta^2(1-\beta^2)(4-\beta^2)}{u_0^2}\eta_1^3\,,\\ \eta_2 &= +\frac{1}{12}\frac{(1-\beta^2)}{u_0}\eta_1^2\,,\\ \eta_3 &= -\frac{1}{96}\frac{\beta^2(1-\beta^2)}{u_0^2}\eta_1^3\,.\end{aligned} \tag{4.165}$$

(c) The second of these equations does not constrain $\eta_1$, but it does provide a constraint on the allowed values of $\beta$. The three allowed values are $\beta = 2$, 1, or 0, and so the three allowed force laws are: $F \propto r$, $F \propto r^{-2}$, and $F \propto r^{-3}$. The first two are the familiar 3-dimensional harmonic oscillator of Hooke's law and the inverse-square-law of Newtonian gravity. There is something very strange about the last force law. If $\beta = 0$, then the perturbed orbit is also circular. Determine the effective potential for this force law, and show that the only stable orbits are circular.

**Problem 4.4** In this problem, you will derive expressions for the apapsis of an orbit in terms of the orbital parameters.

(a) Use conservation of angular momentum and energy to derive the following expression for the apapsis $r_a$ of a particle in orbit about a mass $M$:

$$r_a = \frac{r^2 v^2 \sin^2\theta}{GM}\left[1 - \sqrt{1 - \frac{2rv^2\sin^2\theta}{GM} + \frac{r^2 v^4 \sin^2\theta}{G^2M^2}}\right]^{-1}, \quad (4.166)$$

where $r$ and $v$ are the distance and speed of the particle at any point in its orbit, and $\theta$ is the angle between **r** and **v**.

(b) Show that for the case of a circular orbit, you recover the virial theorem.

(c) For the special case of $r = r_p$ and $v = v_p$ being determined at periapsis, show that:

$$r_a = r_p\left[\frac{2GM}{r_p v_p^2} - 1\right]^{-1}. \quad (4.167)$$

**Problem 4.5** The Apollo spacecraft were placed in a nearly circular orbit about the Earth and then given a boost for the **Trans Lunar Injection** maneuver, which would send the spacecraft out to the Moon. Once at the Moon, the spacecraft would perform another burn to slow down and settle into orbit about the Moon (the **Lunar Orbit Insertion**). The actual orbital path was a figure-eight due to the gravitational influence of the Moon. Here, we will estimate the change in velocities required for both maneuvers using a simplified model of the Earth-Moon gravitational potential.

(a) After launch, the spacecraft were placed in a circular orbit with an altitude of 180 nautical miles. This corresponds to an orbital radius of $6.71 \times 10^6$ m. Use the results from Problem 4.4 to determine the required additional velocity (or "*delta-vee*") in order to lift the spacecraft into an orbit with an apogee a little past the distance of the Moon (say, $4.00 \times 10^8$ m). What is the eccentricity of this new orbit about the Earth?

(b) Use Kepler's laws to determine how much time it will take for the spacecraft to go from the Trans Lunar Injection maneuver to apogee.

(c) Assume that you have correctly aimed your orbit so that the Moon will lie along the semimajor axis at a distance of $3.84 \times 10^8$ m from the Earth. What is the separation between the Moon and the spacecraft? (Ignore any effects of the Moon's gravitation on the spacecraft's orbit.)

(d) Let's now turn off the Earth's gravity and turn on the Moon's gravity. What new delta-vee is required to slow down the spacecraft during the Lunar Orbit Insertion to place it into a circular orbit about the Moon?

**Problem 4.6** The **Yukawa Potential**

$$U(r) = -\frac{ke^{-r/a}}{r} \quad (4.168)$$

is used to describe short-range forces, and in quantum mechanics it describes an interaction mediated by a massive scalar field.

(a) Show that for all values of $\ell > 0$, the effective potential, $U_{\text{eff}} > 0$ for $r \to \infty$ and for $r \to 0$.

(b) Show that there is a maximum value of $\ell$ for circular orbits, and that this occurs at $r_0 = a\left(1+\sqrt{5}\right)/2$. (This is the *golden ratio* that appears in mathematics, architecture, and art.)
(c) Is the circular orbit with $r = r_0$ stable?

**Problem 4.7** Closed orbits in the Yukawa potential.

(a) Use the Yukawa potential from Problem 4.6 to determine an expression for $\beta$ in terms of $u_0$ and $a$.
(b) What is the value of $u_0$ for which $\beta = 0$?
(c) For what value of $u_0$ is the orbital path closed after two orbits of the central potential?

**Problem 4.8** In general relativity, the orbit of a particle about a non-spinning, spherically symmetric mass is determined by extremizing the proper time of the particle's world line through the curved, four-dimensional spacetime. The result is a radial equation of motion that can be expressed in terms of an effective potential

$$U_{\text{eff}}(r) = \frac{\ell^2}{2\mu r^2}\left(1 - \frac{2GM}{rc^2}\right) - \frac{GM\mu}{r}, \tag{4.169}$$

where $M$ is the mass of the central object and $\mu$ is the mass of the orbiting object. (Note that in the limit $c \to \infty$, $U_{\text{eff}}$ becomes the familiar effective potential for Newtonian gravity.)

(a) Find an expression for the radius of a stable circular orbit.
(b) Determine the minimum value of $\ell$ for stable circular orbits.
(c) What is the radius of the minimum stable circular orbit?

**Problem 4.9** The **periapse precession** $\omega_{\text{p}}$ is the rate at which the line joining the center of the force to the point of closest approach (the periapse) rotates. (See Fig. 4.11.) It is measured as an angular velocity. Show that the periapse precession for a nearly circular orbit can be expressed as

$$\omega_{\text{p}} = (1-\beta)\frac{\ell}{\mu r_0^2}, \tag{4.170}$$

where $r_0$ is the radius of the perturbed circular orbit.

**Problem 4.10** In general relativity, the radial equation of motion for a planet of mass $\mu$ in orbit around a non-spinning, spherically symmetric object of mass $M$ can be expressed as if the Newtonian gravitational potential were modified as

$$U(r) = -\frac{GM\mu}{r} - \frac{\ell^2 GM}{\mu c^2 r^3}. \tag{4.171}$$

(See (4.169) from Problem 4.8 for the corresponding effective potential $U_{\text{eff}}(r)$.)

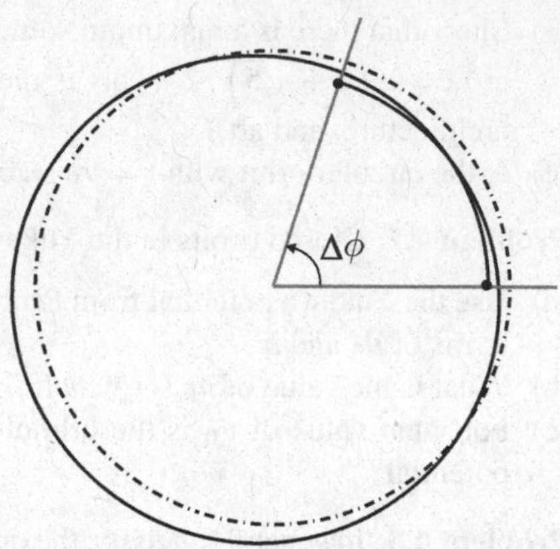


**Fig. 4.11** Illustration of periapse precession, for a single orbit. The perturbed orbit is shown as a solid black line; the unperturbed circular orbit is shown as the dashed-dotted line. For this example $\beta = 5/6$

(a) For nearly circular orbits, show that the precession rate (See Problem 4.9) can be written as

$$\omega_{\mathrm{p}} = \left(1 - \sqrt{1 - \frac{6GM}{r_0 c^2}}\right)\frac{2\pi}{P_{\mathrm{orb}}}, \tag{4.172}$$

where $r_0$ is the semi-major axis (or average orbital radius), and $P_{\mathrm{orb}}$ is the orbital period of the planet.

(b) Use the appropriate values to determine the relativistic precession rate of Mercury. How does your value compare with the observed value of 43″/century?

**Problem 4.11** The **Plummer potential**

$$\Phi(r) = -\frac{GM}{\sqrt{r^2 + a^2}} \tag{4.173}$$

is a gravitational potential that models the equilibrium distribution of stars in a spherical cluster such as a globular cluster. The gravitational potential energy in a volume element $\mathrm{d}V = r^2 \sin\theta \, \mathrm{d}r\mathrm{d}\theta\mathrm{d}\phi$ is then

$$\mathscr{U}(r)\mathrm{d}V = \Phi\rho\mathrm{d}V\,, \tag{4.174}$$

where

$$\rho = \frac{3Ma^2}{4\pi\left(r^2 + a^2\right)^{5/2}} \tag{4.175}$$

is the stellar mass distribution in the equilibrium configuration and $\mathscr{U}$ is the gravitational potential energy density.

(a) Integrate the density $\rho$ over all space to determine the total mass of the cluster.

(b) Determine the mass enclosed within the radius $r = a$.

(c) The virial theorem relates the time-averaged kinetic energy to the time-averaged potential energy over several cycles of the system. But in this case, we can also average over many stars within a shell of radius $r$ and thickness $\mathrm{d}r$, so that

$$\begin{aligned} \overline{\mathscr{U}}(r)\mathrm{d}V &= \rho\Phi\, 4\pi r^2 \mathrm{d}r\,, \\ \overline{\mathscr{T}}(r)\mathrm{d}V &= \frac{1}{2}\rho\sigma_\mathrm{v}^2\, 4\pi r^2 \mathrm{d}r\,, \end{aligned} \tag{4.176}$$

where $\mathscr{T}$ is the kinetic energy density, and $\sigma_\mathrm{v}^2$ is the so-called **velocity dispersion**. (It is the variance of the velocity distribution for the individual stars.) Using the above equation and the virial theorem, show that $\sigma_\mathrm{v}^2(r) = -\Phi(r)$ for the Plummer model of a globular cluster.

**Problem 4.12** Astronomers do not have the luxury of measuring the 3-dimensional properties of a globular cluster. Instead, they measure the 2-dimensional projection of the cluster on the plane of the sky.

(a) Using the properties of the Plummer model developed in Problem 4.11, rewrite the mass distribution $\rho$ in terms of cylindrical coordinates with the $z$-axis pointing along the line of sight to the cluster. Determine the total mass of stars in an infinitely-long cylinder of radius $a$ centered on the cluster.
(b) When measuring the velocity dispersion, astronomers can determine only its *radial* component $\sigma_\mathrm{r}^2$. The full "de-projected" velocity dispersion is then $3\sigma_\mathrm{r}^2$. (But since the measured $\sigma_\mathrm{r}^2$ is integrated along the line of sight through different $\sigma_\mathrm{v}^2(r)$, its measured value will not come from a Gaussian distribution.) Assuming that $3\sigma_\mathrm{r}^2$ can be approximated by the central velocity dispersion $\sigma_\mathrm{v}^2(0)$ from part (c) of Problem 4.11, determine an estimate of the mass of the globular cluster in terms of the measured quantities $\sigma_\mathrm{r}^2$ and $a$. Does this approximation provide an upper or lower bound on the mass of the cluster?

**Problem 4.13** In this problem, you will calculate the Poisson brackets of the components of the Laplace-Runge-Lenz vector.

(a) Using the fundamental Poisson bracket relations

$$\{x^i, x^j\} = 0\,, \qquad \{p_i, p_j\} = 0\,, \qquad \{x^i, p_j\} = \delta^i_j\,, \tag{4.177}$$

evaluate the Poisson brackets

$$\{\ell_i, A_j\}\,, \qquad \{A_i, H\}\,, \qquad \{A_i, A_j\}\,, \tag{4.178}$$

where $\ell_i$ are the components of the angular momentum vector $\boldsymbol{\ell} \equiv \mathbf{r} \times \mathbf{p}$, $A_i$ are the components of the Laplace-Runge-Lenz vector

$$\mathbf{A} \equiv \mathbf{p} \times \boldsymbol{\ell} - k\mu\hat{\mathbf{r}}\,, \tag{4.179}$$

and

$$H = \frac{p^2}{2\mu} - \frac{k}{r} \tag{4.180}$$

is the Hamiltonian for a $1/r$ potential.
You should find

$$\{\ell_i, A_j\} = \sum_k \varepsilon_{ijk} A_k\,, \quad \{A_i, H\} = 0\,, \quad \{A_i, A_k\} = 2\mu|E| \sum_k \varepsilon_{ijk}\ell_k\,, \tag{4.181}$$

where

$$E \equiv -\left(\frac{p^2}{2\mu} - \frac{k}{r}\right) \tag{4.182}$$

is the conserved energy for a bound orbit.

(b) Show that by renormalizing $A_i$ so that $D_i \equiv A_i/\sqrt{2\mu|E|}$, you obtain

$$\{D_i, D_j\} = \sum_k \varepsilon_{ijk}\ell_k\,. \tag{4.183}$$

(c) Interpret the Poisson bracket relations in (4.181) in terms of: (i) the transformation properties of **A** under rotations, (ii) the fact that **A** is a conserved quantity, and (iii) whether it is possible to obtain additional conserved quantities by taking Poisson brackets of the components of **A** with itself.

(d) Recall from Problem 3.13 that the Poisson brackets of the components of the angular momentum vector $\boldsymbol{\ell}$ are

$$\{\ell_i, \ell_j\} = \sum_k \varepsilon_{ijk}\ell_k\,. \tag{4.184}$$

Thus, together with (4.183), we see that the Poisson bracket algebra of the components $\ell_i$ and $D_i$ is

$$\{\ell_i, \ell_j\} = \sum_k \varepsilon_{ijk}\ell_k\,, \quad \{D_i, D_j\} = \sum_k \varepsilon_{ijk}\ell_k\,, \quad \{\ell_i, D_j\} = \sum_k \varepsilon_{ijk} D_k\,, \tag{4.185}$$

which is closed amongst themselves. Show that the 6-dimensional space of anti-symmetric $4 \times 4$ matrices obeys the same algebra, but with Poisson brackets replaced by commutator of matrices. For the mathematically inclined, this is the *Lie algebra* of the group $SO(4)$ of special orthogonal transformations in 4 dimensions.

# Chapter 5
# Scattering

In the last chapter, we focused our attention on bound orbits for a central force. Here we will look in detail at *unbound* systems, for which either the central force is repulsive or the incoming particle has sufficiently large kinetic energy that the total energy of the system is positive. In scattering problems involving two particles, the incoming particle comes in from a great distance (infinity), reaches a point of closest approach to the second (target) particle, and then continues on to infinity in a new direction. In short, the trajectory of the incoming particle is *deflected* by the scattering interaction. The target particle (assuming that it is not a *fixed* scattering center), will also be affected by the interaction. In this chapter, we develop the tools needed to calculate the deflection angles of the two particles as a function of the impact parameter and initial velocity of the incoming particle.

## 5.1 Review of Collisions

Before exploring the details of scattering off of central forces, we will review collisions, which should be familiar from introductory physics courses. For this simplified treatment, the precise nature of the force responsible for the scattering can be ignored.

### *5.1.1 Infinitely-Massive Second Object*

The simplest form of a collision is the rebounding of a low-mass particle off of an infinitely massive, immovable object such as a wall or floor. In this case momentum

 
M.J. Benacquista and J.D. Romano, *Classical Mechanics*, Undergraduate Lecture Notes in Physics, https://doi.org/10.1007/978-3-319-68780-3_5

is not conserved, and the change in momentum of the particle is caused by the impulse delivered by the normal force of the wall or floor during the collision. If the initial kinetic energy of the particle is conserved and not dissipated to other forms of energy such as heat or sound, then the collision is **elastic**. For such a collision, the magnitude of the momentum of the particle is constant, but the direction changes. Since the impulse is normal to the surface, the angle of reflection is equal to the angle of incidence as measured with respect to the normal.

The collision is said to be **inelastic** if some of the kinetic energy is lost to other forms of energy during the collision. The loss of kinetic energy can be described phenomenologically by the **coefficient of restitution**, which parameterizes the reduction in rebound velocity. Although the symbol $e$ is often used for the coefficient of restitution, we will use $\eta$ to avoid confusion with the eccentricity of an orbit. In the case of an object colliding normal to a surface, the fraction of kinetic energy that is retained after the collision is related to the coefficient of restitution by

$$\eta^2 \equiv \frac{K_{\text{final}}}{K_{\text{initial}}}, \tag{5.1}$$

where $\eta$ runs from 0 (perfectly inelastic) to 1 (elastic). For an inelastic collision of a ball with the floor at an angle, there may be different coefficients of restitution for the normal and parallel components of the velocity. Thus, the angle of incidence need not equal the angle of reflection for inelastic collisions (Exercise 5.2).

**Exercise 5.1** A ball dropped onto a floor experiences an inelastic collision with coefficient of restitution $\eta$. Show that if the ball is dropped vertically from a height $h$, then it will come to rest after a time

$$T = \sqrt{\frac{2h}{g}}\left[\frac{1+\eta}{1-\eta}\right]. \tag{5.2}$$

Comment on the extreme cases of $\eta = 0$ and $\eta = 1$.

**Exercise 5.2** Determine the relationship between the angle of incidence and angle of reflection for a ball that bounces off of a floor with coefficient of restitution $\eta$. For this calculation, assume that the coefficient of restitution affects only the component of the velocity that is perpendicular to the floor.

### 5.1.2 Finite-Mass Second Object

If the second object is not infinitely massive, then it will recoil from the collision. In the absence of any net external force, the total momentum of the system will then be conserved in this type of collision. If we only consider two-body collisions and central (or radial) forces, then the motion can always be confined to a properly chosen plane. In a typical problem, the initial velocities of the two particles are known, and we seek to determine the final velocities of the particles after the collision. If we consider the most general problem, there are four unknown quantities needed to describe the final state of the system. Thus, we will either need four equations to determine these unknowns, or we will need to reduce the number of unknowns in the problem.

If the collision is **perfectly inelastic**, then the particles stick together, leaving a single object after the collision. In this case, the number of unknowns is reduced to two and the problem can be solved easily using only the conservation of momentum equation (its two components). With a judicious choice of reference frame, we can simplify the problem. In the **lab frame**, one of the two particles (called the *target particle*) is at rest at the origin prior to the collision. In this frame, the total momentum of the system is equal to the momentum of the other particle (called the *incident particle*), and the problem is reduced to one dimension along the trajectory of the incident particle.

If the collision is elastic, then the number of unknowns required to describe the final state after the collision remains at four, but we only have three equations (two from conservation of momentum and one from conservation of kinetic energy). Thus, one of the four unknowns must be specified in order to solve the problem. Frequently this unknown is chosen to be the angle through which one of the particles is deflected from its original trajectory. In the lab frame, the target particle is at rest, so this angle is taken to be the deflection angle of the incident particle.

### 5.1.3 Barycenter (Center-of-Mass) Frame

A different choice of reference frame can simplify the problem even more. If we consider the **barycenter frame**, in which the center of mass of the system is at rest at the origin, then the total momentum of the system is zero. This reduces the problem for a perfectly inelastic collision to a *triviality* as the final state of the system (i.e., the two particles stuck together) is simply the combined particle at rest at the origin.

**Exercise 5.3** Consider two particles of mass $m_1$ and $m_2$. Particle 1 approaches the origin along the negative $x$-axis with speed $u_1$; particle 2 approaches the origin along the negative $y$-axis with speed $u_2$. They undergo a perfectly inelastic collision at the origin and stick together, leaving the origin with velocity $\mathbf{v}$.

(a) Determine the speed and direction of the combined mass after the collision in terms of the individual masses and initial velocities.
(b) Transform to the lab frame so that particle 2 is at rest at the origin and particle 1 is moving along the new $x'$-axis. Determine the speed and direction of the combined mass after the collision in the lab frame.
(c) Transform to the barycenter frame and show that this frame is moving with respect to the original frame with velocity $\mathbf{v}$.

Using the barycenter frame also simplifies the description of an elastic collision. In this frame, the two particles initially come together along one axis. They collide at the origin and then separate along a different axis. Thus, the angle of deflection for each particle is the same. Furthermore, since the center of mass remains at rest and kinetic energy is conserved, the scattered speed of each particle is identical to its initial speed. Transforming back from the barycenter frame to the lab frame allows us to relate the angle of deflection in the barycenter frame to the angle of deflection in the lab frame.

***Example 5.1*** Consider two particles of mass $m_1$ and $m_2$, which undergo an elastic collision. In the lab frame, $m_1$ is coming in from the left at speed $u_1$, while $m_2$ is at rest at the origin. After the collision, $m_1$ is deflected by an angle $\psi$ above the horizontal with speed $v_1$, and $m_2$ recoils with speed $v_2$ at an angle $\zeta$ below the horizontal, as shown in panel (a) of Fig. 5.1.

In the barycenter frame, the two particles approach the center of mass with speeds $u_1'$ and $u_2'$, respectively. The solution for the speeds after the collision is very simple in the barycenter frame. The initial and final total momenta are both zero, so

$$m_1 u_1' = m_2 u_2' , \qquad m_1 v_1' = m_2 v_2' , \tag{5.3}$$

and the total kinetic energy is conserved, so

$$\frac{1}{2} m_1 {u_1'}^2 + \frac{1}{2} m_2 {u_2'}^2 = \frac{1}{2} m_1 {v_1'}^2 + \frac{1}{2} m_2 {v_2'}^2 . \tag{5.4}$$

It is easy to show that the two particles then scatter with speeds $v_1' = u_1'$ and $v_2' = u_2'$. Particle 1 will be deflected upward and to the right at some angle $\theta$, while particle 2 is deflected downward and to the left at the same angle $\theta$, as shown in panel (b) of Fig. 5.1.

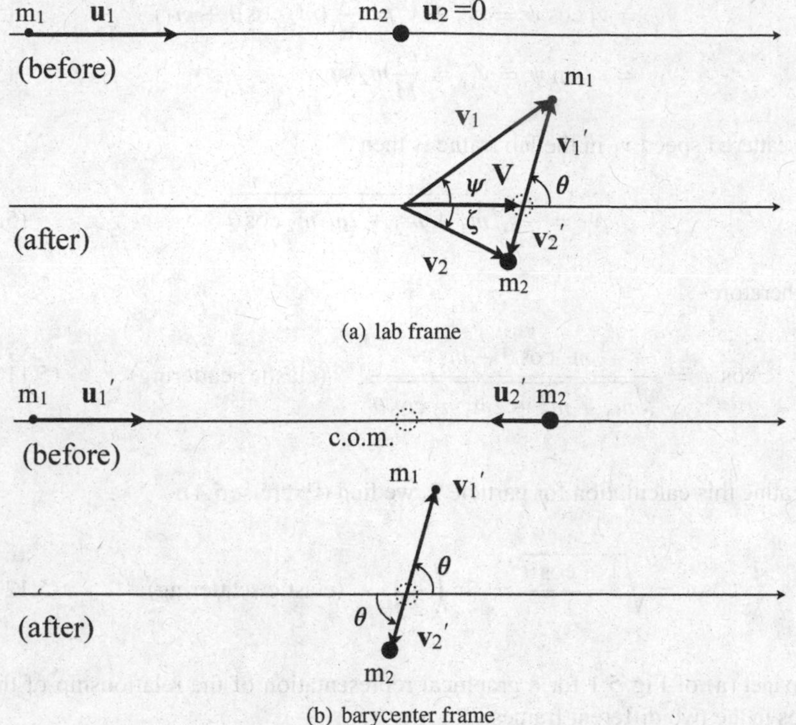


**Fig. 5.1** The scattering event as seen in: (a) the lab frame, and (b) the barycenter frame. "Before" and "after" refer to the configurations before and after the collision. The relationship between the final velocities and scattering angles in the two frames are shown together in panel (**a**)

We can relate the solution in the barycenter frame to the scattered velocities in the lab frame by noting that the barycenter frame is moving with respect to the lab frame with velocity

$$\mathbf{V} = \frac{m_1 \mathbf{u}_1}{M}, \tag{5.5}$$

where $M \equiv m_1 + m_2$. Thus, the velocities in the barycenter frame can be related to the velocities in the lab frame by simply subtracting **V**. Explicitly,

$$v_1' = u_1' = u_1 - V = u_1 \frac{m_2}{M}, \tag{5.6}$$

$$v_2' = u_2' = V = u_1 \frac{m_1}{M}. \tag{5.7}$$

We can now determine the values of the scattered speeds in the lab frame and obtain a relationship between $\theta$ and the angles $\psi$ and $\zeta$. We will do this for particle 1, noting that $\mathbf{v}_1 = \mathbf{v}_1' + \mathbf{V}$. Thus,

$$v_{1x} = v_1 \cos\psi = v'_{1x} + V = \frac{u_1}{M}(m_2 \cos\theta + m_1)\,, \tag{5.8}$$

$$v_{1y} = v_1 \sin\psi = v'_{1y} = \frac{u_1}{M} m_2 \sin\theta\,. \tag{5.9}$$

The scattered speed $v_1$ in the lab frame is then

$$v_1 = \frac{u_1}{M}\sqrt{m_1^2 + m_2^2 + 2m_1 m_2 \cos\theta}\,, \tag{5.10}$$

and therefore

$$\cos\psi = \frac{m_2\cos\theta + m_1}{\sqrt{m_1^2 + m_2^2 + 2m_1 m_2\cos\theta}} \quad \text{(elastic scattering)}\,. \tag{5.11}$$

Repeating this calculation for particle 2, we find (Exercise 5.4):

$$\cos\zeta = \sqrt{\frac{1-\cos\theta}{2}} = \sin\left(\frac{\theta}{2}\right) \quad \text{(elastic scattering)}\,. \tag{5.12}$$

See panel (a) of Fig. 5.1 for a graphical representation of the relationship of these angles in the two different frames. □

**Exercise 5.4** Verify (5.12).

**Exercise 5.5** Show that for the case of equal-mass elastic scattering, the angle between the velocities of the incident and target particles after scattering is always 90° in the lab frame.

In this introductory section, we have ignored the details of the collision itself and have treated that part of the problem as a "black box" in which impulses are delivered to each particle and kinetic energy may be dissipated into other, unaccounted-for forms. In the remaining sections of this chapter, we will look at the details of the scattering forces to better understand and describe the factors that determine the deflection angles of the incident and target particles. We will focus primarily on elastic collisions, discussing a phenomenological treatment of inelastic collisions only briefly at the end of the chapter.

## 5.2 The Hard Sphere

To start with a very simple example of a scattering force, consider the scattering off of a hard sphere of radius $a$ located at the origin. We'll have the incident particle come in from the $-z$ direction. The incoming direction of the particle must be offset from the $z$-axis by a distance less than $a$ in order to collide with the sphere. This offset is known as the **impact parameter** and is usually denoted by $b$. After the collision, the particle will go off on a straight trajectory deflected by an angle $\theta$ with respect to the incoming trajectory. This angle corresponds to the polar angle $\theta$ in spherical coordinates. At the point of impact, the particle reaches $r_{\text{min}}$, its distance of closest approach. For the hard sphere, $r_{\text{min}} = a$. The point of impact can also be described by the polar angle $\phi_{\text{m}}$. The configuration is shown in Fig. 5.2.

We can find a relationship between the impact parameter $b$ and scattering angle $\theta$ for an elastic collision by noting that the magnitude of the particle's momentum must be unchanged during the collision. (We are assuming that the hard sphere is either infinitely massive or "nailed" to the universe, so that it doesn't move at all during the collision—we are not conserving momentum.)

We don't consider rotating particles here, so the scattering force is strictly the normal force between the particle and the sphere, and it is directed radially outward. The impulse of this force is such that it simply reverses the direction of the component of the particle's momentum that is normal to the surface of the sphere at the point of contact. From Fig. 5.2, we see that we can write the initial momentum as $\mathbf{p}_{\text{i}} = p\,\hat{\mathbf{z}}$. The component of momentum normal to the surface is $\left(\mathbf{p}_{\text{i}} \cdot \hat{\mathbf{n}}\right)\hat{\mathbf{n}}$ with $\hat{\mathbf{n}} = \sin\phi_{\text{m}}\,\hat{\mathbf{x}} + \cos\phi_{\text{m}}\,\hat{\mathbf{z}}$. Thus, we can obtain the scattered momentum vector from

$$\mathbf{p}_{\text{r}} = \mathbf{p}_{\text{i}} + \Delta\mathbf{p}\,, \tag{5.13}$$

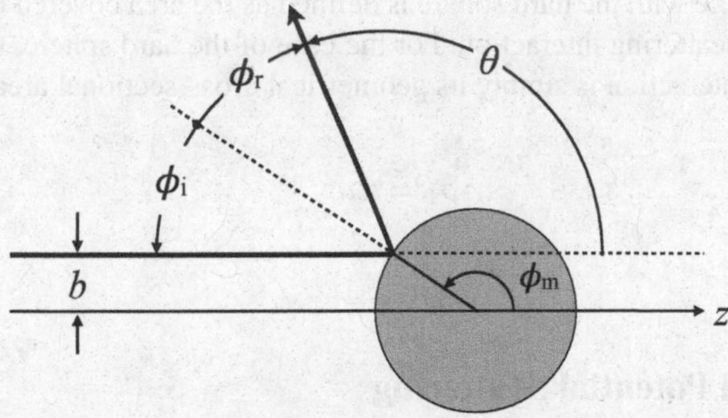


**Fig. 5.2** The configuration for the scattering of a point particle off of a fixed hard sphere of radius $a$. The particle comes in from the left (from the $-z$ direction) with an impact parameter of $b$, and strikes the surface of the sphere at an angle of $\phi_{\text{i}}$ relative to the normal of the surface. The polar angle of the normal is the angle $\phi_{\text{m}}$, which describes the point of closest approach. After scattering, the particle leaves the sphere at an angle $\phi_{\text{r}}$ relative to the normal. The final trajectory is described by the polar angle $\theta$

where

$$\Delta \mathbf{p} = \Delta p\, \hat{\mathbf{n}} , \qquad \Delta p = 2 \left| \mathbf{p}_\mathrm{i} \cdot \hat{\mathbf{n}} \right| = -2p \cos \phi_\mathrm{m} . \tag{5.14}$$

Note that $\Delta p \geq 0$ because $\pi/2 \leq \phi_\mathrm{m} \leq \pi$. With a little bit of algebra and judicious use of trig identities, we find the scattered momentum vector to be

$$\mathbf{p}_\mathrm{r} = -p \left( \sin 2\phi_\mathrm{m}\, \hat{\mathbf{x}} + \cos 2\phi_\mathrm{m}\, \hat{\mathbf{z}} \right) . \tag{5.15}$$

We can see from Fig. 5.2 that $\phi_\mathrm{i} + \phi_\mathrm{m} = \pi$, so $\phi_\mathrm{m} = \pi - \phi_\mathrm{i}$. This means that

$$\sin 2\phi_\mathrm{m} = -\sin 2\phi_\mathrm{i} , \qquad \cos 2\phi_\mathrm{m} = \cos 2\phi_\mathrm{i} , \tag{5.16}$$

for which

$$\mathbf{p}_\mathrm{r} = p \left( \sin 2\phi_\mathrm{i}\, \hat{\mathbf{x}} - \cos 2\phi_\mathrm{i}\, \hat{\mathbf{z}} \right) . \tag{5.17}$$

Thus, $\phi_\mathrm{i} = \phi_\mathrm{r}$ (as expected), and the scattering angle $\theta = \pi - 2\phi_\mathrm{i}$. Again, from the figure, we can see that $b = a \sin \phi_\mathrm{i}$, so

$$\theta = \pi - 2 \arcsin \left( \frac{b}{a} \right) , \tag{5.18}$$

or, equivalently,

$$b = a \cos(\theta/2) . \tag{5.19}$$

For all of the above formulas, we are assuming that $b \leq a$.

Clearly, there is no scattering if $b > a$. The **total cross section** $\sigma_\mathrm{T}$ for the interaction of the particle with the hard sphere is defined as the area covered by values of $b$ that result in a scattering interaction. For the case of the hard sphere, the total cross section for an interaction is simply its geometrical cross-sectional area

$$\sigma_\mathrm{T} = \pi a^2 . \tag{5.20}$$

## 5.3 Central Potential Scattering

Let's now extend the analysis to scattering off of a general central potential. Since there is no hard surface as there was in the case of the hard sphere, the point of closest approach will depend not only on the impact parameter, but also on the initial kinetic energy (and hence the initial momentum) and the functional form of the potential. We can retain the fixed central point by adopting the *effective one-body frame* discussed in Chap. 4, and so many of the features of Fig. 5.2 carry over to this more general case.

Assuming the collision is elastic, the particle's trajectory is symmetric about the line of closest approach. This allows us to use simple geometry to express the scattering angle $\theta$ in terms of the polar angle of closest approach $\phi_m$. For the case of repulsive scattering, the trajectory of the scattered particle follows a curve like that shown in Fig. 5.3. Detail of the angles involved are shown in Fig. 5.4. From the figure, it is easy to find the scattering angle by adding up all the angles along the line of the incoming asymptote. This gives

$$\phi_m - \theta + \phi_m - \theta + \theta = \pi \quad \Rightarrow \quad \theta = 2\phi_m - \pi \,. \tag{5.21}$$

For the case of an attractive central force, the trajectory of the scattered particle follows a curve like that shown in Fig. 5.5. Again, details of the angles involved in the scattering are shown in Fig. 5.6. For this case, we add up the angles along the outgoing asymptote. This gives

$$\phi_m + \phi_m + \theta = \pi \quad \Rightarrow \quad \theta = \pi - 2\phi_m \,. \tag{5.22}$$

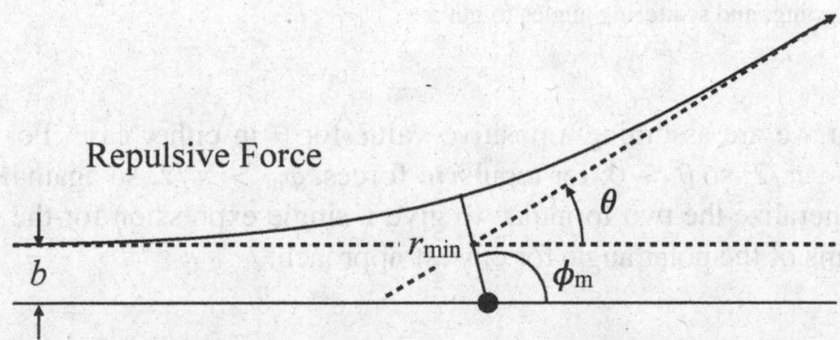


**Fig. 5.3** The trajectory of a particle scattered off of a repulsive potential, showing the impact parameter $b$, the point of closest approach $r_{min}$, and its angle $\phi_m$. The scattered angle is $\theta$

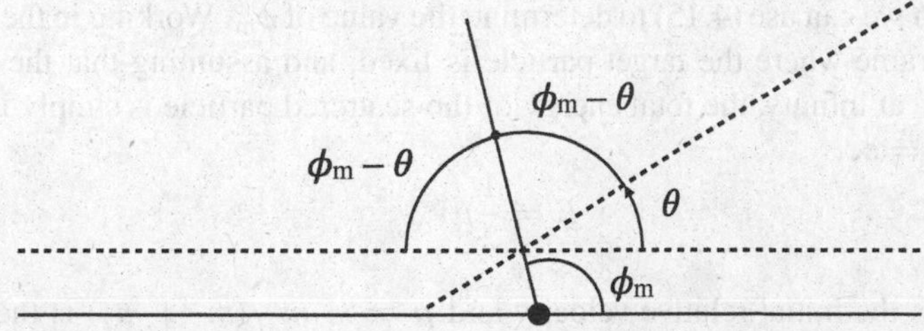


**Fig. 5.4** Relationship between the scattering angle $\theta$ and the polar angle of closest approach $\phi_m$. Since the asymptotic trajectories are symmetric about the line of closest approach, the sum of the angles for the incoming trajectory, outgoing trajectory, and scattering angles add up to $\pi$

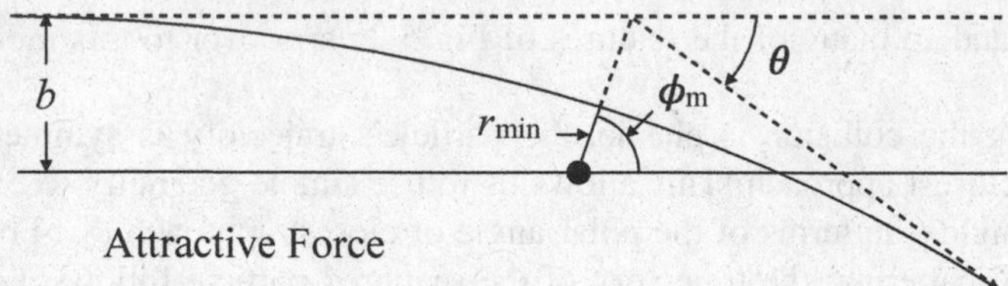


**Fig. 5.5** The trajectory of a particle scattered off of an attractive potential, showing the impact parameter $b$, the point of closest approach $r_{min}$, and its angle $\phi_m$. The scattered angle is $\theta$

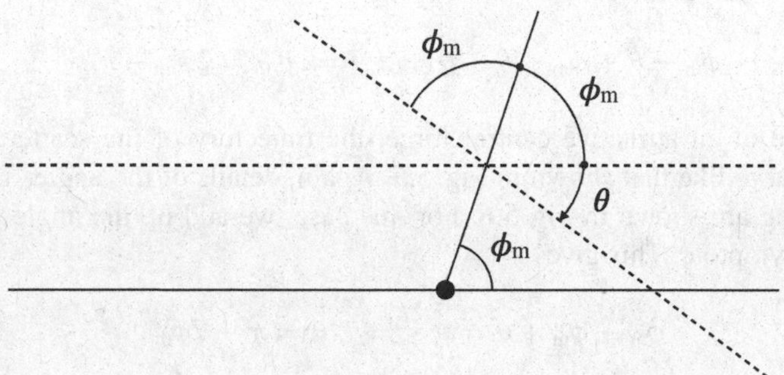


**Fig. 5.6** The angles associated with attractive scattering. As in Fig. 5.4, we sum the angles for the incoming, outgoing, and scattering angles to get $\pi$

Note that we are assuming a positive value for $\theta$ in either case. For attractive forces, $\phi_m < \pi/2$, so $\theta > 0$; for repulsive forces, $\phi_m > \pi/2$, so again $\theta > 0$. We can thus generalize the two formulae to give a single expression for the scattering angle in terms of the polar angle for closest approach:

$$\theta = |\pi - 2\phi_m| . \tag{5.23}$$

This is as far as we can go without specifying the potential. Once the potential is known, then we can use (4.15) to determine the value of $\phi_m$. Working in the effective one-body frame where the target particle is fixed, and assuming that the potential goes to zero at infinity, the total energy of the scattered particle is simply its kinetic energy at $r = \infty$,

$$E = \frac{1}{2}\mu v_\infty^2 , \tag{5.24}$$

where $v_\infty$ is the initial relative velocity and $\mu \equiv m_1 m_2/(m_1 + m_2)$ is the reduced mass of the system. The angular momentum can also be expressed in terms of $v_\infty$ and the impact parameter $b$,

$$\ell = \mu b v_\infty , \tag{5.25}$$

using the fact that $b$ is normal to the incoming trajectory of the scattered particle. At the point of closest approach $\dot{r} = 0$, so we can use the conservation of energy equation (4.12) to solve for $r_{\text{min}}$,

$$2\mu r_{\text{min}}^2 \left(E - U(r_{\text{min}})\right) = \ell^2\,, \tag{5.26}$$

where $E$ and $\ell$ are given by (5.24) and (5.25).

With $r_{\text{min}}$ in hand, we use the usual definition $u \equiv 1/r$ from Chap. 4 and (4.15) to find $\phi_{\text{m}}$. At $r = \infty$, $u_0 = 0$ and $\phi_0 = \pi$, so

$$\phi_{\text{m}} = \pi - \int_0^{u_{\text{min}}} \frac{\mathrm{d}u}{\sqrt{\frac{2\mu}{\ell^2}\left[E - U(1/u)\right] - u^2}}\,, \tag{5.27}$$

where $u_{\text{min}} \equiv 1/r_{\text{min}}$. We now have a formal solution for $\phi_{\text{m}}$ in terms of $v_\infty$ and $b$, which can then be used to find $\theta = |\pi - 2\phi_{\text{m}}|$. In practice, this integral usually cannot be solved analytically, so (5.27) is most often used in setting up a numerical solution.

## 5.4 Differential Cross Section

For potentials that may extend to infinity, it is not necessarily the case that there will be a finite range of impact parameters that result in scattering. Thus, it is unlikely that we will have an easily-defined total cross section such as we found for the hard sphere. However, we can define a **differential cross section** $\mathrm{d}\sigma/\mathrm{d}\Omega$, which relates an infinitesimal range of impact parameters $b$ to $b + \mathrm{d}b$, to scattering through an infinitesimal range of angles $\theta$ to $\theta + \mathrm{d}\theta$.

For central potentials, the scattering event preserves the azimuthal angle, so we can describe the differential cross section as the area of an infinitesimal annulus of radius $b$ and thickness $\mathrm{d}b$,

$$\mathrm{d}\sigma = 2\pi b\, \mathrm{d}b\,. \tag{5.28}$$

Particles entering from within this annulus will then be scattered through a solid angle

$$\mathrm{d}\Omega = 2\pi \sin\theta\, \mathrm{d}\theta\,, \tag{5.29}$$

as shown in Fig. 5.7. The differential cross section for elastic scattering from a central potential is then defined to be the ratio

$$\frac{\mathrm{d}\sigma}{\mathrm{d}\Omega} \equiv \frac{b}{\sin\theta}\left|\frac{\mathrm{d}b}{\mathrm{d}\theta}\right|, \tag{5.30}$$

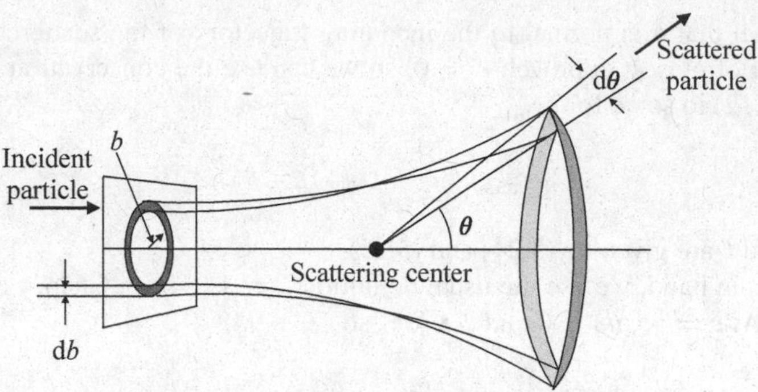


**Fig. 5.7** Configuration for the differential cross section. The particle coming in from the left anywhere within the annulus of radius $b$ and thickness $\mathrm{d}b$ (the dark shaded region) will scatter off the scattering center through the solid angle ring of angle $\theta$ and thickness $\mathrm{d}\theta$ (the light shaded region). Note that for this case the angle of scattering decreases with increasing impact parameter, so $\mathrm{d}b/\mathrm{d}\theta$ is negative

where we have taken the absolute value of $\mathrm{d}b/\mathrm{d}\theta$ in order to ensure a positive differential cross section (as the derivative $\mathrm{d}b/\mathrm{d}\theta$ is usually negative). The total cross section is the integral of the differential cross section over all solid angles,

$$\sigma_{\mathrm{T}} \equiv \int \left(\frac{\mathrm{d}\sigma}{\mathrm{d}\Omega}\right) \mathrm{d}\Omega \,. \tag{5.31}$$

## 5.4.1 Scattering of a Beam of Incident Particles

Usually particle scattering involves a steady stream of incident particles scattering off of an array of target particles. This allows for a different (but equivalent) way of viewing the calculation of the differential cross section. For this case, we will consider a beam of particles (each of mass $\mu$ and energy $E$), incident on an array of target particles (each of mass $M$), fixed in the effective one-body frame. The *intensity* (or flux density) $I_0$ of the incident beam is defined as the number of particles leaving the gun per unit time, per unit area perpendicular to the direction of the beam. Different particles in the beam will have different impact parameters $b$ relative to the target particles, and hence will scatter through different angles $\theta$. A detector placed at an angle $\theta$ with respect to the initial beam path will record the number of particles $\mathrm{d}N$ scattered into the solid angle $\mathrm{d}\Omega$ in a time interval $\mathrm{d}t$. The set-up is shown in Fig. 5.8.

The differential cross section $\mathrm{d}\sigma/\mathrm{d}\Omega$ for scattering of a beam of particles into the solid angle $\mathrm{d}\Omega$ for a particular $\theta$ is thus defined as the ratio

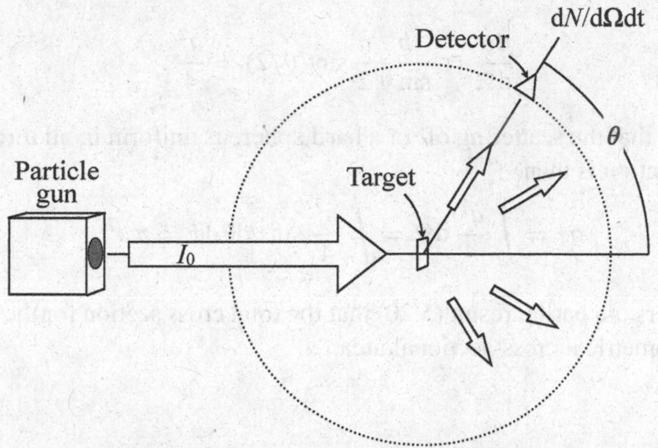


**Fig. 5.8** A particle scattering experiment. The particle gun produces a steady stream of particles with energy $E$. The intensity of the beam is $I_0$. The particles scatter off of the target and are received by a detector placed at angle $\theta$ with respect to the incoming beam

$$\frac{\mathrm{d}\sigma}{\mathrm{d}\Omega} \equiv \frac{\mathrm{d}N}{I_0\,\mathrm{d}\Omega\,\mathrm{d}t}\,, \tag{5.32}$$

which has units of area per solid angle. As mentioned previously, scattering off of a central potential is azimuthally symmetric, so $\mathrm{d}\Omega = 2\pi\sin\theta\,\mathrm{d}\theta$. In additon, the number of scattered particles $\mathrm{d}N$, which appears in the numerator of (5.32), must be the same as the number of incident particles passing through the annular region $2\pi b\,\mathrm{d}b$ in time interval $\mathrm{d}t$—i.e.,

$$\mathrm{d}N = I_0\,2\pi b\,\mathrm{d}b\,\mathrm{d}t\,. \tag{5.33}$$

Making these substitutions into (5.32), we find

$$\frac{\mathrm{d}\sigma}{\mathrm{d}\Omega} \equiv \frac{\mathrm{d}N}{I_0\,\mathrm{d}\Omega\,\mathrm{d}t} = \frac{I_0\,2\pi b\,\mathrm{d}b\,\mathrm{d}t}{I_0\,2\pi\sin\theta\,\mathrm{d}\theta\,\mathrm{d}t} = \frac{b}{\sin\theta}\left|\frac{\mathrm{d}b}{\mathrm{d}\theta}\right|\,, \tag{5.34}$$

which recovers our earlier expression (5.30) for the differential cross section. Again, we have included an absolute value of the derivative $\mathrm{d}b/\mathrm{d}\theta$ in the last equality of (5.34) in order that the differential cross section be positive.

***Example 5.2*** Before going on to specific examples, let's return briefly to the hard sphere to see how the new machinery can be used to recover the results from Sect. 5.2. We will bypass developing a potential for the hard sphere (which would look something like a step function of radius $a$), and use the relation between $b$ and $\theta$ given in (5.19),

$$b = a\cos(\theta/2)\,, \tag{5.35}$$

so that

$$\frac{d\sigma}{d\Omega} = \frac{b}{\sin\theta}\frac{a}{2}\sin(\theta/2) = \frac{a^2}{4}. \tag{5.36}$$

Thus, we see that the scattering off of a hard sphere is uniform in all directions. The total cross section is then

$$\sigma_T = \int \frac{a^2}{4}\,d\Omega = \int \frac{a^2}{4}\sin\theta\, d\theta\, d\phi = \pi a^2, \tag{5.37}$$

which recovers our earlier result (5.20) that the total cross section for the hard sphere is just its geometrical cross-sectional area. □

## 5.5 Gravitational Scattering

For gravitational scattering of a mass $m_1$ off of another mass $m_2$, we will continue working in the effective one-body frame introduced in Chap. 4. Recall that, in this frame, positions and velocities are defined with respect to the second mass, which acts like a *fixed* scattering center with mass $M \equiv m_1 + m_2$, while the first mass acts as if it had a mass equal to the reduced mass of the system $\mu \equiv m_1 m_2/(m_1 + m_2)$. Note that if $m_1 \ll m_2$, then $M \approx m_2$ and $\mu \approx m_1$, and the effective one-body and inertial reference frames will be approximately the same for this case.

In Chap. 4, we derived the solution for the motion of a particle under the influence of the gravitational potential. The main result is given by (4.46),

$$r = \frac{\alpha}{1 + e\cos\phi}, \tag{5.38}$$

where $e \geq 1$ for unbound orbits. In addition, $\alpha$ and $e$ are related to the angular momentum $\ell$ and total energy $E$ of the system via (4.39) and (4.42):

$$\alpha = \frac{\ell^2}{GM\mu^2}, \quad e^2 = 1 + \frac{2E\alpha}{GM\mu}. \tag{5.39}$$

For the case of scattering, we will assume that the incident mass comes in from infinity with (relative) speed $v_\infty$. We can then express $\ell$ and $E$ in terms of the impact parameter $b$ and the velocity at infinity $v_\infty$, as we did in (5.25) and (5.24). With a little bit of algebra the eccentricity can be written as

$$e^2 = 1 + \left(\frac{bv_\infty^2}{GM}\right)^2. \tag{5.40}$$

The orbit equation describes the motion of the particle in a reference frame where the central mass is at the origin and the point of closest approach $r = r_{\min}$ occurs at $\phi = 0$. The scattered particle comes in from from the upper left quadrant and leaves in the lower left quadrant. This is not in the same orientation as the situation described in Fig. 5.5, but it is easy to relate the quantities between both orientations. Again, using the definitions of $\alpha$ and $e$, we can show that

$$r_{\text{min}} = b\sqrt{\frac{e-1}{e+1}}. \tag{5.41}$$

Because the eccentricity $e \geq 1$, (5.38) shows that $r$ goes to infinity for a particular angle $\phi_0$, for which

$$\cos\phi_0 = -\frac{1}{e}. \tag{5.42}$$

This angle is also the angle that the asymptotic trajectory at infinity makes with the vector $\mathbf{r}_{\text{min}}$. We can use this angle to define the polar angle of closest approach, so $\phi_{\text{m}} = \pi - \phi_0$, as shown in Fig. 5.9. (Here we should note that the angles being described are *polar angles*, so they are always positive and lie between 0 and $\pi$.) The angle of scattering is then

$$\theta = \pi - 2\phi_{\text{m}} = -\pi + 2\phi_0 = -\pi + 2\arccos\left(-\frac{1}{e}\right), \tag{5.43}$$

so

$$\cos\left(\frac{\pi}{2} + \frac{\theta}{2}\right) = -\sin\left(\frac{\theta}{2}\right) = -\frac{1}{e}. \tag{5.44}$$

But remember that

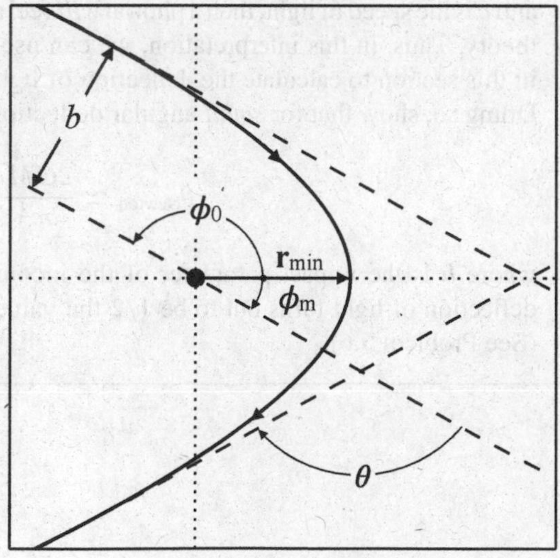


**Fig. 5.9** The relation between the hyperbolic trajectory and the coordinates used in scattering

$$e^2 = 1 + \left(\frac{bv_\infty^2}{GM}\right)^2 . \tag{5.45}$$

To simplify things, we dig up the trig identity

$$\cot y = \sqrt{\csc^2 y - 1} , \tag{5.46}$$

and let $y = \theta/2$, to find

$$\cot\left(\frac{\theta}{2}\right) = \frac{bv_\infty^2}{GM} . \tag{5.47}$$

Finally, by directly differentiating (5.47) and using (5.30), we obtain the differential cross section for gravitational scattering

$$\frac{d\sigma}{d\Omega} = \frac{G^2M^2}{4v_\infty^4} \csc^4(\theta/2) . \tag{5.48}$$

**Exercise 5.6** Verify (5.48).

**Exercise 5.7** In a strict interpretation of Newtonian gravity, a massless particle (e.g., a photon) does not feel a gravitational force. If, however, we treat a photon as an "ordinary" (massive) particle with mass $m \equiv E/c^2$, where $E$ is its energy and $c$ is the speed of light, then a photon *will* feel a gravitational force in Newton's theory. Thus, in this interpretation, we can use the same formalism developed in this section to calculate the deflection of light by a central force of mass $M$. Doing so, show that for *small* angular deflections,

$$\theta_{\text{Newton}} \simeq \frac{2GM}{bc^2} , \tag{5.49}$$

where $b$ is the impact parameter of the incoming photon. This value for the deflection of light turns out to be $1/2$ the value predicted by general relativity (See Problem 5.6).

## 5.6 Rutherford Scattering

Electromagnetism provides another simple well-known central potential describing an inverse-square-law force. This force can be repulsive as well as attractive. Mathematically, the scattering trajectories for two charged particles interacting are identical to the gravitational case, but with $GM\mu$ replaced by $kQq$, where $Q$ and $q$ are the charges of the particles and $k \equiv 1/4\pi\varepsilon_0$ is Coulomb's constant. Thus, using the results of the previous section, we can immediately write down an expression for the scattering angle (in the effective one-body frame) of an incident particle of charge $q$ off of another particle of charge $Q$,

$$\cot\left(\frac{\theta}{2}\right) = \frac{2Eb}{kQq}, \tag{5.50}$$

where we replaced $v_\infty^2$ in the gravitational expression, (5.47), by by $2E/\mu$. Similarly, we can obtain the **Rutherford formula** for scattering

$$\frac{d\sigma}{d\Omega} = \left(\frac{kQq}{4E}\right)^2 \csc^4(\theta/2). \tag{5.51}$$

The Rutherford formula was originally applied to the scattering of $\alpha$-particles (helium nuclei with charge $q = 2e$) off of a target of gold nuclei (with charge $Q = 79e$). The detection of $\alpha$-particles scattered at very large $\theta$ indicated that the gold nuclei acted like point charges down to very small impact parameters. This indicated that the nucleus of an atom was very small compared to the size of the atom.

## 5.7 Example: Gravitational Slingshot

In this section, we discuss a simplified model of the **gravitational slingshot**, which is used in interplanetary travel to increase the velocity of a spacecraft relative to the Sun, thereby changing the spacecraft's heliocentric orbit. For our particular example, we will scatter a small spacecraft off of the gravitational potential of Jupiter in order to take it from one orbit to another orbit with a larger semi-major axis. (Since the mass of the spacecraft is small compared to the mass of Jupiter and to the mass of the Sun, the results that we have calculated in the effective one-body frame apply here.) When setting up this problem, we will be using several tools that we developed in Chap. 4.

So we start with a spacecraft in orbit about the Earth, and we first need to inject it into an orbit that will cross the orbit of Jupiter, as shown in Fig. 5.10. The Jupiter injection orbit is an ellipse that has a perihelion at Earth's orbit and an aphelion just

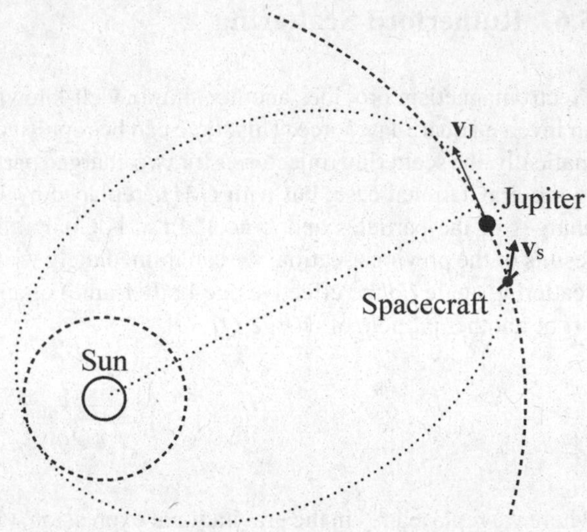


**Fig. 5.10** Gravitational slingshot of a spacecraft off of Jupiter. This is in the heliocentric reference frame where the spacecraft is on the Jupiter injection orbit prior to the scattering event, with initial velocity $\mathbf{v}_S$. Jupiter is on a circular orbit with radius 5.2 AU and tangential velocity $\mathbf{v}_J$

beyond Jupiter's orbit. For the purposes of this example, we will take the aphelion to be 5.3 AU[1] (Jupiter's orbit is nearly circular at 5.2 AU). For this orbit,

$$r_p = a(1 - e) = 1\,\text{AU}\,, \qquad r_a = a(1 + e) = 5.3\,\text{AU}\,. \tag{5.52}$$

Thus, $a = 3.15$ AU and $e = 0.683$.

**Exercise 5.8** Assume that the spacecraft is initially in a circular orbit about the Sun with an orbital radius of 1 AU. Show that it must be given an additional velocity of 9 km/s in order to be placed in the Jupiter injection orbit described above.

When the spacecraft approaches Jupiter's orbit, the scattering event will take place over a short enough time and small enough spatial extent that we can consider the problem to be a simple elastic scattering event in an inertial reference frame that is instantaneously comoving with Jupiter's orbital velocity. In the absence of any scattering, the spacecraft's velocity will make an angle $\alpha$ with respect to the radius of Jupiter's orbit. The geometry of the intersection between the Jupiter injection orbit and Jupiter's orbit is independent of the actual position of Jupiter. Of course, we want Jupiter to be somewhere in the vicinity of the spacecraft, so we can take advantage of the gravitational scattering, but that is a matter of timing of the launch of the spacecraft. This timing is done so that the intersection between the two orbits occurs a distance $d$ behind Jupiter. The relevant quantities in the Sun's reference frame are shown in Fig. 5.11.

[1]The **Astronomical Unit** AU is the mean distance between the Earth and the Sun.

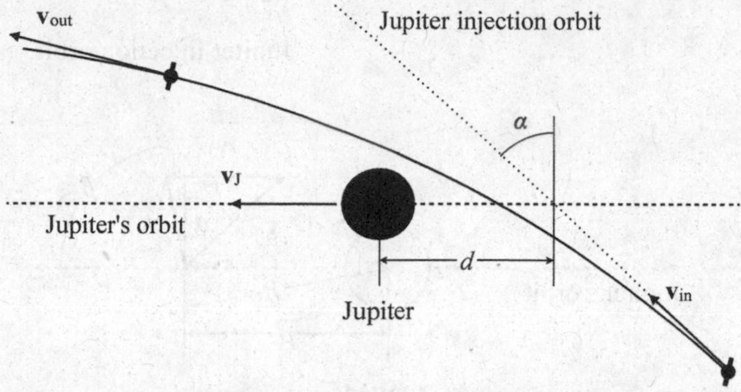


**Fig. 5.11** Scattering in the heliocentric reference frame. The spacecraft comes in from the right with heliocentric velocity $\mathbf{v}_{\text{in}}$ at an angle $\alpha$ relative to the radius of Jupiter's orbit. In the absence of any gravitational scattering by Jupiter, the Jupiter injection orbit would cross Jupiter's orbit a distance $d$ behind the planet. After the scattering event the outgoing heliocentric velocity is $\mathbf{v}_{\text{out}}$

**Exercise 5.9** Using the orbit equation, (4.50), we can find the angle $\alpha$ from

$$\tan\alpha = \frac{v_\phi}{v_r} = \frac{r\,\mathrm{d}\phi/\mathrm{d}t}{\mathrm{d}r/\mathrm{d}t} = r\frac{\mathrm{d}\phi}{\mathrm{d}r}\,. \tag{5.53}$$

Show that

$$\tan\alpha = \sqrt{\frac{1-e^2}{e^2-(1-r/a)^2}}\,. \tag{5.54}$$

We're now going to need the impact parameter $b$ in order to find the deflection angle in the co-moving coordinate system. (Recall from Fig. 5.5 that the impact parameter is the perpendicular distance between the scattering center and the asymptote of the initial velocity of the incoming object.) To transform to the co-moving frame, we simply subtract $\mathbf{v}_{\text{J}}$ from all velocities in the problem. Thus, the initial velocity in this frame is $\mathbf{v}_{\text{i}} = \mathbf{v}_{\text{in}} - \mathbf{v}_{\text{J}}$. The angle between $\mathbf{v}_{\text{i}}$ and the path of Jupiter's orbit is denoted by $\beta$ in Fig. 5.12; it satisfies

$$\tan\beta = \frac{v_{\text{in}}\cos\alpha}{v_{\text{J}} - v_{\text{in}}\sin\alpha}\,. \tag{5.55}$$

Once we know $\beta$, we then have $b = d\sin\beta$.

**Exercise 5.10** Referring to Fig. 5.12, verify (5.55). Show also that

$$v_{\text{i}}^2 = v_{\text{in}}^2 + v_{\text{J}}^2 - 2v_{\text{in}}v_{\text{J}}\sin\alpha\,. \tag{5.56}$$

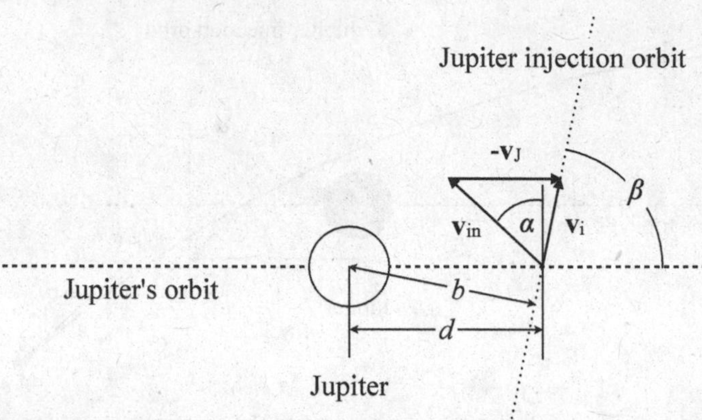


**Fig. 5.12** Scattering as seen with respect to the co-moving reference frame of Jupiter. In this frame, Jupiter is at rest, and the unscattered path of the spacecraft is tilted with respect to the orbital path of Jupiter by the angle $\beta$. The impact parameter, $b$, is the perpendicular distance between Jupiter and the unscattered path

Given the impact parameter $b$, it is now a simple matter to determine the scattering angle in the co-moving frame of Jupiter. Taking $v_\infty = v_i$ and using (5.47), we have

$$\cot\left(\frac{\theta}{2}\right) = \frac{bv_i^2}{GM_J}, \tag{5.57}$$

where $M_J$ is the mass of Jupiter. Since the scattering is elastic, the outgoing speed $v_o$ is the *same* as the incoming speed $v_i$, noting that this is with respect to the co-moving frame. The angle that $\mathbf{v}_o$ makes with the orbital path of Jupiter is $\theta + \beta$, as can be seen in Fig. 5.13. With the magnitude and direction of $\mathbf{v}_o$ known, we can find the scattered velocity of the spacecraft in the heliocentric reference frame, $\mathbf{v}_{out} = \mathbf{v}_o + \mathbf{v}_J$, which has a magnitude *greater* than $v_i$ as expected.

At this point, it is simply a matter of putting together all the different quantities that we have computed throughout this example. The end result is not a simple expression and therefore not very illuminating, so here we will just lay out the equations that will be used. In order to describe the vectors, we will consider two coordinate systems, $O$ and $O'$, which are related by a simple Galilean velocity transformation. In the heliocentric frame $O$, the $x$-axis is chosen to be tangent to Jupiter's orbital path, pointing in the retrograde direction, and the $y$-axis is chosen to lie along a radial line from the Sun. In the co-moving frame $O'$, the $x'$-axis is parallel to the $x$-axis, and the $y'$-axis lies along the line joining the Sun and Jupiter. During the short time of the scattering encounter, $O'$ is simply moving with velocity $\mathbf{v}_J$ in the $-x$ direction of $O$. Thus,

$$\mathbf{v}_{out} = (v_i \cos(\theta + \beta) - v_J)\,\hat{\mathbf{x}} + v_i \sin(\theta + \beta)\hat{\mathbf{y}}, \tag{5.58}$$

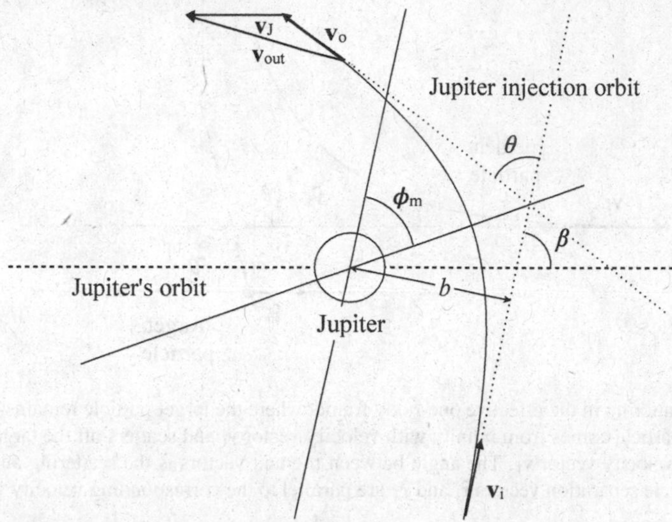


**Fig. 5.13** The configuration after the spacecraft scatters off of Jupiter, as seen with respect to the co-moving reference frame of Jupiter. Within this frame, the scattered velocity is $\mathbf{v}_\mathrm{o}$. When converted to the heliocentric frame, the scattered velocity is $\mathbf{v}_\mathrm{out} = \mathbf{v}_\mathrm{o} + \mathbf{v}_\mathrm{J}$

with $v_\mathrm{i}$ given by (5.56), $\theta$ given by (5.57), and $\beta$ given by (5.55). In order to determine $\beta$, we need to know the angle $\alpha$, which is given by (5.54) with $r$ equal to the radius of Jupiter's orbit, 5.2 AU. In order to determine $\theta$, we need to know the impact parameter $b$, which is given by $b = d \sin\beta$. The initial conditions on this problem are then $a$ (the semi-major axis of the initial orbit), $e$ (its eccentricity), and $d$ (the distance between Jupiter and the point where the spacecraft's orbit crosses Jupiter's orbit).

**Exercise 5.11** Determine the scattered velocity (speed and direction) for a spacecraft that was launched from Earth on an initial orbit that reached an aphelion $r_\mathrm{a} = 5.3$ AU, with (a) $d = 1000 R_\mathrm{J}$, (b) $d = 100 R_\mathrm{J}$, and (c) $d = 10 R_\mathrm{J}$. ($R_\mathrm{J}$ is the equatorial radius of Jupiter.) What is the aphelion distance of the new orbits after scattering for each of the three cases above?

## 5.8 Transformation to the Lab Frame

So far, we have been analyzing scattering as an effective one-body problem, with all motion defined in terms of the interparticle separation $\mathbf{r} \equiv \mathbf{r}_1 - \mathbf{r}_2$ and relative velocity $\mathbf{v} \equiv \mathbf{v}_1 - \mathbf{v}_2$. In this frame, the scattering angle $\theta$ is related to the angle between the asymptotic initial and final separation vectors. To be precise, if $\mathbf{r}_\mathrm{i} \equiv \mathbf{r}_{1\mathrm{i}} - \mathbf{r}_{2\mathrm{i}}$ is

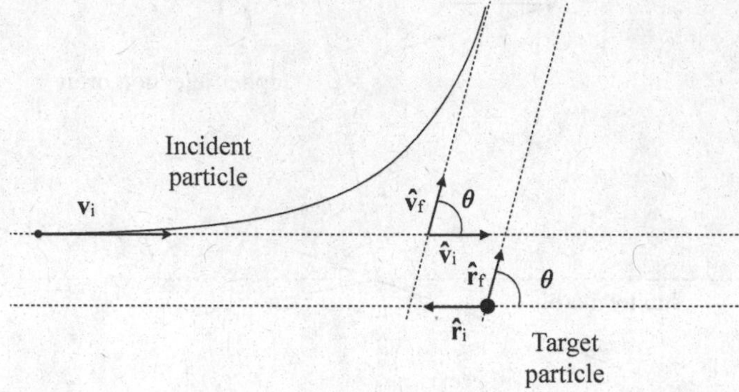


**Fig. 5.14** Scattering in the effective one-body frame, where the target particle remains at rest and the incident particle comes from infinity with velocity vector $\mathbf{v}_i$ and scatters off the target particle, leaving with velocity vector $\mathbf{v}_f$. The angle between the unit vectors is the scattering angle $\theta$. The unit interparticle separation vectors $\hat{\mathbf{r}}_i$ and $\hat{\mathbf{r}}_f$ are parallel to the corresponding velocity vectors

the initial separation vector at $t = -\infty$ and $\mathbf{r}_f \equiv \mathbf{r}_{1f} - \mathbf{r}_{2f}$ is the final separation vector at $t = +\infty$, then $\theta$ is defined by $-\hat{\mathbf{r}}_i \cdot \hat{\mathbf{r}}_f = \cos\theta$ . Since the asymptotic initial and final velocities are parallel to their respective separation vectors, we can also use $\cos\theta = \hat{\mathbf{v}}_i \cdot \hat{\mathbf{v}}_f$ , as can be seen in Fig. 5.14.

Because the relative separation vectors and relative velocity vectors are both defined in terms of the particles themselves and not any coordinate system, the angle between them is independent of the coordinate choice. In the barycenter frame, the origin always lies on the line joining the two particles, so the position vectors of the individual particles in this frame are parallel to the relative separation vector. Thus, the scattering angle will be the same in the barycenter frame as in the effective one-body frame, as shown in Fig. 5.15.

When a beam of particles is scattered off of a target, we don't make our angle measurements from interparticle trajectories. Instead, we use a reference frame that is fixed on the target and we measure the final trajectory of the scattered particle relative to the fixed target. This is the same reference frame as the lab frame discussed in Sect. 5.1. Therefore, in the lab frame, the angle measured between the incoming and outgoing beam is $\psi$, and the recoil angle is $\zeta$ as shown in Fig. 5.16. We measure $\psi$ and a corresponding scattering cross section $\sigma(\psi)$ in our experiments, but we want to compare these results with $\theta$ and $\sigma(\theta)$ which are more easily calculated in the barycenter or effective one-body frames. If we suppress the impact parameter, the relationship between these angles and their respective reference frames is identical to that shown in Fig. 5.1. We will occasionally refer to this figure in the calculations that follow.

As before, we will assume that, prior to scattering, the target particle is at rest in the lab frame. Denoting the position vectors of the two particles by $\mathbf{r}_1$ and $\mathbf{r}_2$, the location of the center of mass in the lab frame is

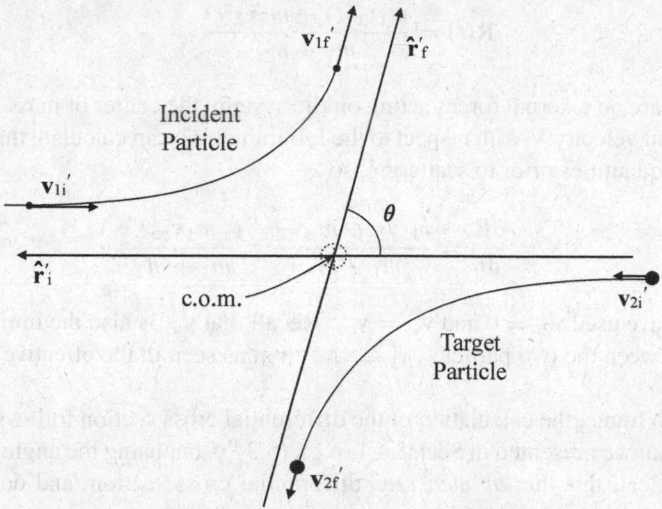


**Fig. 5.15** Scattering in the barycenter frame. The center of mass remains at rest, the target particle comes in from the right, and the incident particle comes in from the left. The interparticle separation vectors $\hat{\mathbf{r}}_i'$ and $\hat{\mathbf{r}}_f'$ are the same as in the effective one-body frame, so the scattering angle $\theta$ is the same. The incident particle's incoming and outgoing velocity vectors $\mathbf{v}_{1i}'$ and $\mathbf{v}_{1f}'$ are parallel to the relative velocity vectors $\mathbf{v}_i$ and $\mathbf{v}_f$ in the effective one-body frame, although the magnitudes of these velocity vectors in the two frames are different from one another

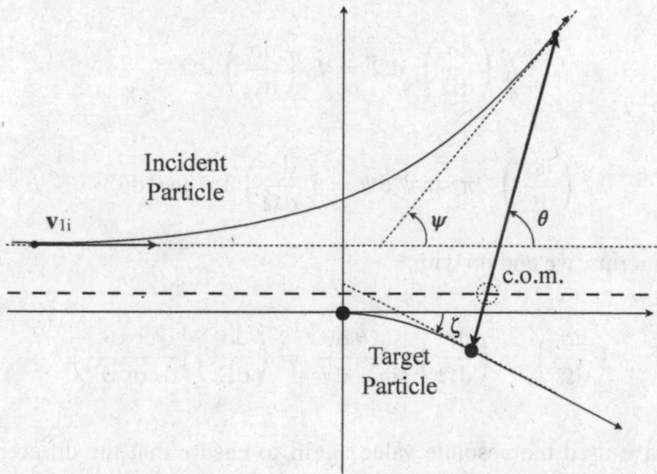


**Fig. 5.16** Scattering as seen in the lab frame, where the target particle is initially at rest and recoils as the incident particle approaches. The center of mass of the system moves to the right in the lab frame with constant velocity, along the dashed horizontal line shown in the figure. The scattering angles for the incident and and target particles are $\psi$ and $\zeta$, respectively. The scattering angle $\theta$ for the two particles as seen in the barycenter frame is also shown

$$\mathbf{R}(t) = \frac{m_1\mathbf{r}_1(t) + m_2\mathbf{r}_2(t)}{m_1 + m_2} . \tag{5.59}$$

Since there are no external forces acting on the system, the center of mass will move with constant velocity $\mathbf{V}$ with respect to the lab frame. We can calculate this velocity in terms of quantities prior to scattering, so

$$\mathbf{V} = \frac{\mathrm{d}\mathbf{R}}{\mathrm{d}t} = \frac{m_1\mathbf{v}_{1\mathrm{i}} + m_2\mathbf{v}_{2\mathrm{i}}}{m_1 + m_2} = \frac{m_1\mathbf{v}_\infty}{m_1 + m_2} , \tag{5.60}$$

where we have used $\mathbf{v}_{2i} = 0$ and $\mathbf{v}_{1i} = \mathbf{v}_\infty$. Recall that $\mathbf{v}_\infty$ is also the initial relative velocity between the two particles, $\mathbf{v}_\mathrm{i} \equiv \mathbf{v}_{1\mathrm{i}} - \mathbf{v}_{2\mathrm{i}}$, as seen in the effective one-body frame.

In the lab frame, the calculation of the differential cross section follows the same argument that we presented in Sect. 5.4.1 (e.g., (5.34)), but using the angle $\psi$ instead of $\theta$ . We'll call this the *lab-measured* differential cross section, and denote it by $(\mathrm{d}\sigma/\mathrm{d}\Omega)'$. It is thus given by

$$\left(\frac{\mathrm{d}\sigma}{\mathrm{d}\Omega}\right)' = \frac{b}{\sin\psi}\left|\frac{\mathrm{d}b}{\mathrm{d}\psi}\right| . \tag{5.61}$$

Since the number of particles entering the detector does not (and should not) depend upon our choice of reference frame, we note that

$$I_0\left(\frac{\mathrm{d}\sigma}{\mathrm{d}\Omega}\right)' \mathrm{d}\Omega' = I_0\left(\frac{\mathrm{d}\sigma}{\mathrm{d}\Omega}\right) \mathrm{d}\Omega , \tag{5.62}$$

so

$$\left(\frac{\mathrm{d}\sigma}{\mathrm{d}\Omega}\right)' 2\pi \sin\psi \, \mathrm{d}\psi = \left(\frac{\mathrm{d}\sigma}{\mathrm{d}\Omega}\right) 2\pi \sin\theta \, \mathrm{d}\theta . \tag{5.63}$$

Rearranging terms, we end up with

$$\left(\frac{\mathrm{d}\sigma}{\mathrm{d}\Omega}\right)' = \left(\frac{\mathrm{d}\sigma}{\mathrm{d}\Omega}\right)\frac{\sin\theta \, \mathrm{d}\theta}{\sin\psi \, \mathrm{d}\psi} = \left(\frac{\mathrm{d}\sigma}{\mathrm{d}\Omega}\right)\left|\frac{\mathrm{d}\cos\theta}{\mathrm{d}\cos\psi}\right| , \tag{5.64}$$

where we have used the absolute value again to ensure that the differential cross section is positive. Thus, to relate the measured differential cross section $(\mathrm{d}\sigma/\mathrm{d}\Omega)'$ with the theoretical or computed differential cross section $\mathrm{d}\sigma/\mathrm{d}\Omega$, we need an expression relating $\cos\theta$ to $\cos\psi$.

### 5.8.1 *Elastic Scattering in the Lab Frame*

We have already determined the relationship between $\cos\theta$ and $\cos\psi$ for the case of elastic scattering (See (5.11), Example 5.1). This expression involves both masses, but it can be written solely in terms of the mass ratio:

$$\cos\psi = \frac{\cos\theta + \rho_1}{\sqrt{1 + 2\rho_1\cos\theta + \rho_1^2}}\,, \quad \text{where} \quad \rho_1 \equiv m_1/m_2\,. \tag{5.65}$$

Taking the appropriate derivative in (5.64), we find

$$\left(\frac{d\sigma}{d\Omega}\right)' = \frac{\left(1 + 2\rho_1\cos\theta + \rho_1^2\right)^{3/2}}{1 + \rho_1\cos\theta}\left(\frac{d\sigma}{d\Omega}\right). \tag{5.66}$$

In the lab frame, the target particles (which are at rest prior to the scattering) recoil and carry away some energy from the incident particles. Thus, even in the case of elastic scattering, the scattered beam will show a *decrease* in energy compared with the incident beam. We can determine this decrease in terms of the mass ratio $\rho_1$ and deflection angle $\theta$. The scattered speed in the lab frame is given by (5.10), with $u_1$ replaced by $v_\infty$. Thus, the ratio of kinetic energies is

$$\frac{T_{1f}}{T_{1i}} = \frac{v_{1f}^2}{v_\infty^2} = \frac{1 + 2\rho_1\cos\theta + \rho_1^2}{(1+\rho_1)^2} < 1\,. \tag{5.67}$$

In practice, the target is often a foil and the actual target particles are the nuclei of that foil. Although the target particles recoil after a scattering event, they remain bound to the foil and the missing kinetic energy of the incident particles shows up as an increase in the internal energy of the foil.

In the special case of *equal-mass* particles, $\rho_1 = 1$ and (5.65) takes on the very simple form

$$\cos\psi = \sqrt{\frac{1+\cos\theta}{2}} = \cos\left(\frac{\theta}{2}\right), \tag{5.68}$$

so

$$\psi = \theta/2 \qquad \text{(equal mass, elastic scattering)}\,. \tag{5.69}$$

This means that in the case of elastic scattering of equal-mass particles, the incident particles cannot scatter by more than 90° in the lab frame.

**Exercise 5.12** Show that

$$\left(\frac{d\sigma}{d\Omega}\right)' = 4\cos\psi\left(\frac{d\sigma}{d\Omega}\right) \tag{5.70}$$

for elastic scattering of equal-mass particles.

### 5.8.2 *Inelastic Scattering in the Lab Frame*

**Inelastic collisions** do not conserve kinetic energy within the barycenter or effective one-body frames. The initial kinetic energy is dissipated during the collision. For hard sphere collisions, the energy is dissipated through deformation of the sphere or excitation of vibrations within the sphere. For stellar scattering, if the encounter is close enough, the gravitational interaction can raise tides on the stars, which can then dissipate energy through couplings with resonances in the star or with the rotational energy in the star. If the scattering particles are molecules with internal structure, the scattering interaction can excite rotations or vibrations in the molecules. In all these cases, a fraction of the initial kinetic energy is converted to some sort of internal energy that can then be dissipated through heat or photon emission through a quantum transition. In these cases, the energy is stored internally during the scattering interaction.

Another form of inelastic scattering involves the emission of radiation during the scattering event. For example, if a charged particle is accelerated, it will emit electromagnetic radiation that carries away some of the energy of the incident particle. If very massive compact objects, such as black holes, scatter off of each other with small impact parameters, then the system will emit a burst of gravitational radiation during the close passage of the objects. In both of these cases, the emitted radiation carries away both energy *and* momentum. Thus, for inelastic scattering through emission of radiation, the barycenter frame will not be inertial unless we also include the radiation field as well. Thus, we will not consider this type of inelastic scattering in this section.

The geometry of the asymptotic velocity vectors and scattering angles for momentum-conserving inelastic scattering remains the same as in panel (a) of Fig. 5.1, only we can no longer equate the initial and final speeds as we did in the case of elastic scattering (See Example 5.1). Decomposing $\mathbf{v}_{1f} = \mathbf{v}'_{1f} + \mathbf{V}$ into its $x$ and $y$ components, we see that

$$\begin{aligned} v_{1f}\cos\psi &= v'_{1f}\cos\theta + V\,, \\ v_{1f}\sin\psi &= v'_{1f}\sin\theta\,. \end{aligned} \tag{5.71}$$

Squaring these equations and adding gives us a relationship between the magnitudes of the velocities

$$v_{1f}^2 = v_{1f}'^2 + 2v_{1f}' V \cos\theta + V^2 , \tag{5.72}$$

which can be used to eliminate $v_{1f}$ from either of the equations in (5.71):

$$\cos\psi = \frac{\cos\theta + \rho_1}{\sqrt{1 + 2\rho_1 \cos\theta + \rho_1^2}} , \quad \text{where} \quad \rho_1 \equiv V/v_{1f}' . \tag{5.73}$$

Notice that this expression is identical to that for elastic scattering (5.65), except that $\rho_1$ is defined differently. Consequently, the differential cross section can still be found from (5.66) using the new definition of $\rho_1$. Repeating this process for the target particle with $\mathbf{v}_{2f} = \mathbf{v}_{2f}' + \mathbf{V}$, we find

$$\cos\zeta = \frac{-\cos\theta + \rho_2}{\sqrt{1 - 2\rho_2 \cos\theta + \rho_2^2}} , \quad \text{where} \quad \rho_2 \equiv V/v_{2f}' . \tag{5.74}$$

**Exercise 5.13** Verify (5.74) for the scattering angle $\zeta$ of the recoiled target particle.

All that's left to do is to express $\rho_1$ and $\rho_2$ in terms of quantities that can be calculated in the effective one-body frame. This can be done using conservation of total linear momentum. Using (5.60), we immediately have $V = m_1 v_\infty/(m_1 + m_2)$. In addition, since $m_1\mathbf{v}_1' + m_2\mathbf{v}_2' = \mathbf{0}$ in the barycenter frame and $\mathbf{v} \equiv \mathbf{v}_1' - \mathbf{v}_2'$, we also have

$$\mathbf{v}_1' = \frac{m_2}{m_1 + m_2}\mathbf{v} , \qquad \mathbf{v}_2' = -\frac{m_1}{m_1 + m_2}\mathbf{v} . \tag{5.75}$$

Thus, $v_{1f}' = m_2 v_f/(m_1 + m_2)$ and $v_{2f}' = m_1 v_f/(m_1 + m_2)$. Putting these results together, we find

$$\rho_1 \equiv \frac{V}{v_{1f}'} = \frac{m_1}{m_2}\frac{v_\infty}{v_f} , \qquad \rho_2 \equiv \frac{V}{v_{2f}'} = \frac{v_\infty}{v_f} , \tag{5.76}$$

which involve the ratio of the initial and final relative velocities.

### 5.8.3 Phenomenological Treatment of Inelastic Scattering

We know from (5.76) that $\rho_1$ depends on the initial and final relative velocities between the incident and target particle. Unless the details of the energy dissipation mechanism are known, we must rely on a more empirical or phenomenological approach. Because we are only considering inelastic collisions that dissipate their energies internally, the linear momentum of the center of mass does not change during the scattering event. Thus, the change in the kinetic energy of the system is entirely due to a change in the kinetic energy of the effective one-body system. We describe this change using a parameter $Q$ (known as the $Q$-**value** of the inelastic collision). Thus,

$$\frac{1}{2}\mu v_{\rm f}^2 = \frac{1}{2}\mu v_\infty^2 + Q\,. \tag{5.77}$$

Clearly from this expression, $Q < 0$. We can solve (5.77) for the ratio $v_{\rm f}/v_\infty$ and obtain

$$\frac{v_{\rm f}}{v_\infty} = \sqrt{1 + \frac{2Q}{\mu v_\infty^2}}\,. \tag{5.78}$$

We can also express this in terms of the energy of the incident particle in the lab frame, $E_{\rm lab} = m_1 v_\infty^2/2$, to get

$$\frac{v_{\rm f}}{v_\infty} = \sqrt{1 + \frac{(m_1 + m_2)}{m_2}\frac{Q}{E_{\rm lab}}}\,, \tag{5.79}$$

which leads us to

$$\rho_1 = \frac{m_1}{m_2}\sqrt{\frac{m_2}{m_2 + Q\,(m_1 + m_2)\,/E_{\rm lab}}}\,. \tag{5.80}$$

Thus, for scattering of an incident beam of energy $E$ off of target particles through an inelastic interaction described by a given $Q$-value, we can compute the differential cross section in the lab frame using (5.66), where $\rho_1$ is given by (5.80).

## Suggested References

*Full references are given in the bibliography at the end of the book.*

Goldstein et al. (2002): Contains a fairly standard treatment of scattering as part of the discussion of central forces.

Landau and Lifshitz (1976): A classic graduate-level text on classical mechanics. Several of the additional problems in this chapter were adapted from problems in the relevant sections of this book.

## Additional Problems

**Problem 5.1** (*Adapted from Kuchǎr (1995).*) For an elastic collision, total kinetic energy is conserved, i.e.,

$$\sum_I \frac{1}{2} m_I v_{I\mathrm{i}}^2 = \sum_I \frac{1}{2} m_I v_{I\mathrm{f}}^2 \,, \tag{5.81}$$

where $\mathbf{v}_{I\mathrm{i}}$ and $\mathbf{v}_{I\mathrm{f}}$ denote the initial and final velocities of particle $I$. The principle of relativity (See Sect. 11.1), which states that the laws of physics should be the same in all inertial reference frames, requires that if the total kinetic energy is conserved in one inertial frame $O$, then it must also be conserved in *any* other inertial frame $O'$, i.e., which moves with *constant* velocity $\mathbf{u}$ relative to $O$. Show that this requirement implies conservation of total momentum,

$$\sum_I m_I \mathbf{v}_{I\mathrm{i}} = \sum_I m_I \mathbf{v}_{I\mathrm{f}} \,. \tag{5.82}$$

**Problem 5.2** The **spontaneous disintegration** of a particle of a mass $M$ into two constituent particles of mass $m_1$ and $m_2$ can be thought of as the *time-reverse* of a perfectly inelastic collision of the two constituent particles. Assume that the internal energies of the particles are $E_\mathrm{i}$ and $E_{1\mathrm{i}}$, $E_{2\mathrm{i}}$, respectively, and that $m_1 + m_2 = M$. (Recall that total mass is conserved in Newtonian mechanics; in special relativity, total mass is conserved only for elastic collisions.)

(a) What constraint is there on the internal energies of the particles?
(b) Show that the final speeds of the two constituent particles as measured in the barycenter frame are given by

$$v_1 = \frac{\sqrt{2\mu \Delta E_\mathrm{i}}}{m_1} \,, \qquad v_2 = \frac{\sqrt{2\mu \Delta E_\mathrm{i}}}{m_2} \,, \tag{5.83}$$

where $\Delta E_\mathrm{i} \equiv E_\mathrm{i} - E_{1\mathrm{i}} - E_{2\mathrm{i}}$ and $\mu \equiv m_1 m_2 / M$ is the reduced mass of the system.

**Problem 5.3** Consider a particle of mass $m$ moving from one region (e.g., $z > 0$) with constant potential $U_1$ to another region ($z < 0$) with constant potential $U_2$. Let $\mathbf{v}_1$, $\mathbf{v}_2$ denote the velocities of the particle in these two regions, and let $\phi_1$, $\phi_2$ denote the angles that the velocity vectors make with the normal to the interface to the two regions, as shown in Fig. 5.17.

(a) Show that the motion lies in a 2-dimensional plane, with

$$v_1 \sin\phi_1 = v_2 \sin\phi_2 \,, \qquad v_2^2 - v_1^2 = \frac{2}{m}(U_1 - U_2) \,. \tag{5.84}$$

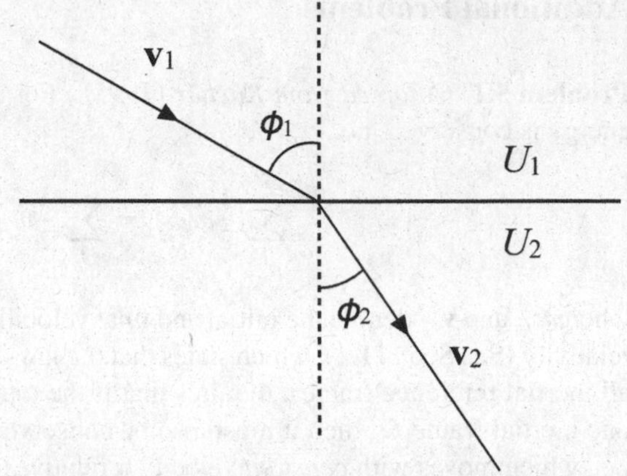


**Fig. 5.17** Particle moving from one region (e.g., $z > 0$) with potential $U_1 = \text{const}$ to another region ($z < 0$) with potential $U_2 = \text{const}$

(b) Rewrite these two formulae as

$$\sin\phi_1 = n\sin\phi_2\,, \qquad n \equiv \sqrt{1 + \frac{2(U_1 - U_2)}{mv_1^2}}\,. \tag{5.85}$$

Thus, the trajectory of the particle "refracts" as it passes from region 1 to region 2, obeying a formula similar to Snell's law for optics.

**Problem 5.4** *(Adapted from Landau and Lifshitz (1976), Sect. 18.)*

(a) Determine the total cross section $\sigma_T$ for a particle to fall into the center of the potential $U = -A/r^2$, with $A > 0$. You should find

$$\sigma_T = \frac{2\pi A}{\mu v_\infty^2}\,. \tag{5.86}$$

*Hint*: For given values of $A$ and $v_\infty$, find the maximum value of the impact parameter $b$ that allows the particle to fall into the center.

(b) Repeat for the potential $U = -A/r^n$, with $A > 0$, where $n > 2$. You should find

$$\sigma_T = \pi n(n-2)^{(2-n)/n}\left(\frac{A}{\mu v_\infty^2}\right)^{2/n}\,. \tag{5.87}$$

*Hint*: For this potential, the particle energy $E$ must be greater than or equal to the maximum value of the effective potential:

$$U_0 \equiv U_{\text{eff}}\big|_{\max} = \frac{1}{2}(n-2)A\left(\frac{\mu b^2 v_\infty^2}{An}\right)^{n/(n-2)}\,. \tag{5.88}$$

So first derive the above expression for $U_0$. Then set $E = U_0$ to determine the maximum value of the impact parameter $b$ that allows the particle to fall into the center.

**Problem 5.5** (*Adapted from Landau and Lifshitz (1976), Sect. 19, Problem 2.*) Calculate the differential cross section $\mathrm{d}\sigma/\mathrm{d}\Omega$ and the total cross section $\sigma_\mathrm{T}$ for scattering off the finite *spherical square-well* potential

$$U(r) = \begin{cases} -U_0\,, & 0 \le r \le a \\ 0\,, & r > a \end{cases} \tag{5.89}$$

where $U_0 > 0$. You should find

$$\frac{\mathrm{d}\sigma}{\mathrm{d}\Omega} = \frac{a^2 n^2}{4\cos(\theta/2)} \frac{(n\cos(\theta/2) - 1)(n - \cos(\theta/2))}{(1 - 2n\cos(\theta/2) + n^2)^2}\,, \tag{5.90}$$

and

$$\sigma_T = \pi a^2\,, \tag{5.91}$$

where

$$n \equiv \sqrt{1 + \frac{2U_0}{\mu v_\infty^2}}\,. \tag{5.92}$$

*Hint*: Similar to Problem 5.3, the trajectory of the incoming particle will be "refracted" as it enters and leaves the scattering potential, as shown in Fig. 5.18. The angles $\alpha$ and $\beta$ are related by the Snell's-law-like formula

$$\sin\alpha = n\sin\beta\,, \tag{5.93}$$

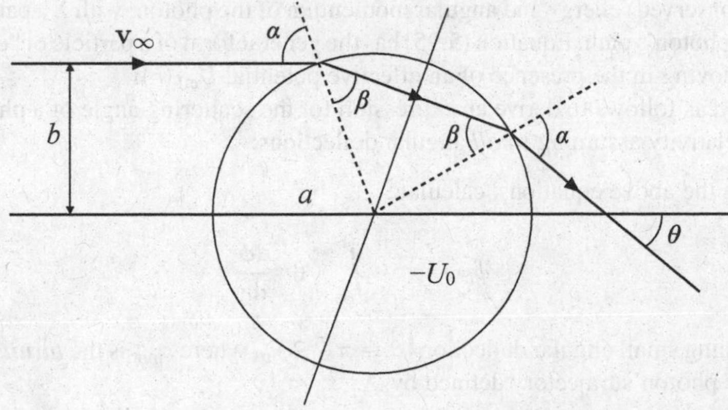


**Fig. 5.18** Geometry for scattering off a spherical square well of radius $a$ and depth $U_0$. The trajectory of the particle refracts as it enters and leaves the well

where $n$ is given above. Second, using geometry you should be able to show that $\theta = 2(\alpha - \beta)$ and $b = a \sin \alpha$, which together with the previous equation allows you to relate the impact parameter $b$ to the scattering angle $\theta$:

$$\left(\frac{a}{b}\right)^2 = \frac{1 - 2n\cos(\theta/2) + n^2}{n^2 \sin^2(\theta/2)} . \tag{5.94}$$

Differentiating this expression to find $\mathrm{d}b/\mathrm{d}\theta$, and using (5.30), leads to the above answer for $\mathrm{d}\sigma/\mathrm{d}\Omega$. Finally, to find $\sigma_T$, integrate $\mathrm{d}\sigma/\mathrm{d}\Omega$ over all solid angles (a cone due to azimuthal symmetry) from $\theta_{\min} = 0$ (corresponding to $b = 0$) to $\theta_{\max} = 2\cos^{-1}(1/n)$ (corresponding to $b = a$). This should give you the expected answer for the geometrical cross-sectional area of the sphere.

**Problem 5.6** To properly describe the motion of a massless particle, e.g., a photon, in the presence of a central force, one needs to use the framework of Einstein's theory of general relativity. Using the equations of general relativity (See, e.g., Hartle 2003), one can show that the motion of a photon is described by the radial equation

$$\frac{1}{2}\left(\frac{\mathrm{d}r}{\mathrm{d}\lambda}\right)^2 + U_{\text{eff}}(r) = \frac{1}{2}\frac{E^2}{c^2} , \tag{5.95}$$

where

$$U_{\text{eff}}(r) \equiv \frac{1}{2}\frac{\ell^2}{r^2}\left(1 - \frac{2GM}{rc^2}\right) . \tag{5.96}$$

In the above equations, $c$ is the speed of light, and

$$E \equiv c^2\left(1 - \frac{2GM}{rc^2}\right)\frac{\mathrm{d}t}{\mathrm{d}\lambda} , \qquad \ell \equiv r^2\frac{\mathrm{d}\phi}{\mathrm{d}\lambda} , \tag{5.97}$$

are the (conserved) energy and angular momentum of the photon, with $\lambda$ a parameter along the photon's path. Equation (5.95) has the general form of a particle of "energy" $E^2/2c^2$ moving in the presence of an effective potential $U_{\text{eff}}(r)$.

Proceed as follows to derive an expression for the scattering angle of a photon in general relativity assuming *small* angular deflections:

(a) Using the above equations, calculate

$$\phi_{\text{m}} \equiv \pi + \int_{\infty}^{r_{\min}} \mathrm{d}r \, \frac{\mathrm{d}\phi}{\mathrm{d}r} , \tag{5.98}$$

assuming small angular deflections $\theta \equiv \pi - 2\phi_{\text{m}}$, where $r_{\min}$ is the *turning point* of the photon's trajectory defined by

$$\left.\frac{\mathrm{d}r}{\mathrm{d}\lambda}\right|_{r=r_{\min}} = 0 \quad \Leftrightarrow \quad \frac{1}{2}\frac{E^2}{c^2} = U_{\text{eff}}(r_{\min}) . \tag{5.99}$$

(*Hint*: Make a change variables from $r$ to $u \equiv r_{\rm min}/r$ in the integral. Also, by assuming small angular deflections so that $GM/r_{\rm min}c^2 \ll 1$, you will be able to simplify the denominator of the integrand.) You should find

$$\phi_{\rm m} \simeq \frac{\pi}{2} - \frac{2GM}{r_{\rm min}c^2}. \tag{5.100}$$

(b) Using $r_{\rm min} \approx b$ for small angular deflections, where $b$ is the impact parameter, show that the deflection angle $\theta \equiv \pi - 2\phi_{\rm m}$ becomes

$$\theta_{\rm Einstein} \simeq \frac{4GM}{bc^2}. \tag{5.101}$$

This is a factor of two larger than the Newtonian value $\theta_{\rm Newton}$ given in Exercise 5.7, (5.49).

(c) Show that for $M$ equal to the mass of the Sun ($M = M_\odot \approx 2 \times 10^{30}$ kg) and $b$ equal to the Sun's radius ($b = R_\odot \approx 7 \times 10^5$ km), the deflection of light grazing the surface of the Sun is given by

$$\theta_{\rm Einstein} \approx 8.5 \times 10^{-6}\ {\rm rad} = 1.75\ {\rm arcsec}. \tag{5.102}$$

This predicition of general relativity was verified by Eddington in 1919 during a solar eclipse expedition.

**Problem 5.7** In general relativity, the radial equation for the motion of a massive particle in the presence of a central force of mass $M$ is given by

$$\frac{1}{2}\mu\left(\frac{{\rm d}r}{{\rm d}\tau}\right)^2 + U_{\rm eff}(r) = \frac{1}{2}\frac{E^2}{\mu c^2}, \tag{5.103}$$

where[2]

$$\begin{aligned} U_{\rm eff}(r) &\equiv \frac{1}{2}\left(\frac{\ell^2}{\mu r^2} + \mu c^2\right)\left(1 - \frac{2GM}{rc^2}\right) \\ &= \frac{1}{2}\mu c^2 - \frac{GM\mu}{r} + \frac{1}{2}\frac{\ell^2}{\mu r^2} - \frac{GM\ell^2}{\mu c^2 r^3}. \end{aligned} \tag{5.104}$$

This is similar to the radial equation for a photon in general relativity as discussed in Problem 5.6. Note that the first term in the effective potential is a constant (which could be absorbed into the "energy" $E^2/2\mu c^2$ on the right-hand side if we wanted

[2]See also Problem 4.8. The expression for $U_{\rm eff}(r)$ given in (5.104) differs from that in (4.169) by the presence of the constant energy term $\mu c^2/2$. This does not change the equations of motion, only the value of the total energy.

to); the second and third terms are the familiar Newtonian potential and angular momentum contribution to the potential; and the fourth term is the correction to the potential that comes from general relativity. The conserved energy and angular momentum of the particle are given by

$$E \equiv \mu c^2 \left(1 - \frac{2GM}{rc^2}\right) \frac{dt}{d\tau}\,, \qquad \ell \equiv \mu r^2 \frac{d\phi}{d\tau}\,, \tag{5.105}$$

with $\tau$ being the *proper time*—i.e., the time as measured by a clock "carried" along the particle's path.

Proceed as follows to derive an expression for the scattering angle of a massive particle in general relativity assuming *small* angular deflections:

(a) Similar to Problem 5.6, use the above equations to calculate

$$\phi_{\rm m} \equiv \pi + \int_{\infty}^{r_{\rm min}} dr\, \frac{d\phi}{dr}\,, \tag{5.106}$$

assuming small angular deflections $\theta \equiv \pi - 2\phi_m$, where $r_{\rm min}$ is the *turning point* of the particle's trajectory. (*Hint*: Make a change of variables from $r$ to $u \equiv r_{\rm min}/r$ in the integral, and use the small angular deflection assumption to simplify the denominator of the integrand.) You should find

$$\phi_{\rm m} \simeq \frac{\pi}{2} - \frac{GM\mu^2 r_{\rm min}}{\ell^2} - \frac{2GM}{r_{\rm min}c^2}\,. \tag{5.107}$$

(b) Using $r_{\rm min} \approx b$ (impact parameter) for small angular deflections, and

$$\ell = \frac{\mu b v_\infty}{\sqrt{1 - v_\infty^2/c^2}}\,, \tag{5.108}$$

show that the deflection angle $\theta \equiv \pi - 2\phi_{\rm m}$ can be written as

$$\theta_{\rm Einstein} \simeq \frac{2GM}{b v_\infty^2}\left(1 + \frac{v_\infty^2}{c^2}\right), \tag{5.109}$$

where we added the "Einstein" subscript to indicate that this is the fully-general relativitistic result.

(c) In Sect. 5.5, we calculated the deflection angle for scattering in Newtonian gravity. Using (5.47), show that in the limit of small deflection angles,

$$\theta_{\text{Newton}} \simeq \frac{2GM}{bv_\infty^2} . \quad (5.110)$$

Thus,

$$\theta_{\text{Einstein}} = \theta_{\text{Newton}} \left(1 + v_\infty^2/c^2\right) , \quad (5.111)$$

implying that the small-angle scattering of a massive particle in general relativity is larger than that in Newtonian gravity by a factor of $(1 + v_\infty^2/c^2)$.

# Chapter 6
# Rigid Body Kinematics

So far, our formulation of classical mechanics has been limited to that of point particles—i.e., *idealized mass points* having no spatial extent. Whenever we described the motion of an extended object, such as a skier accelerating down a ski slope or a car driving along a road, we considered just a single point of the object (usually its center of mass) and ignored its *orientation* about that point. In this and the following chapter, we extend our analysis of motion to **rigid bodies**—i.e., bodies having spatial extent, but which have a *fixed* shape, unchanged by any forces or torques that might act on them. In this chapter we develop the *kinematical framework* needed for describing the complicated translational and rotational motion of a rigid body as it moves through space. In the following chapter, we discuss the forces and torques responsible for such motion, deriving the Euler equations for rigid body motion, which are the equivalent of Newton's law of motion for point particles.

## 6.1 Generalized Coordinates for a Rigid Body

By a **rigid body**, we mean any spatially-distributed mass whose shape does not change during its motion. Thus, as a rigid body moves through space, although the positions $\mathbf{r}_I(t)$ of the constituent mass points can change with time, the distances

$$|\mathbf{r}_I(t) - \mathbf{r}_J(t)| = c_{IJ}\,, \qquad I, J = 1, 2 \ldots, N \tag{6.1}$$

between mass points will be constant (See Fig. 6.1). These constraints reduce the number of degrees of freedom for a rigid body to just 6 as we explain below. To do the counting, note that the $N(N-1)/2$ constraints of (6.1) are not all independent, since once you know the distances between mass point $m_I$ and *three* non-colinear *reference* mass points (e.g., $m_1$, $m_2$, $m_3$), then you've fixed its position relative to all the other mass points. This means that there are $(N-3)(N-4)/2$ *redundant*



M.J. Benacquista and J.D. Romano, *Classical Mechanics*, Undergraduate Lecture Notes in Physics, https://doi.org/10.1007/978-3-319-68780-3_6

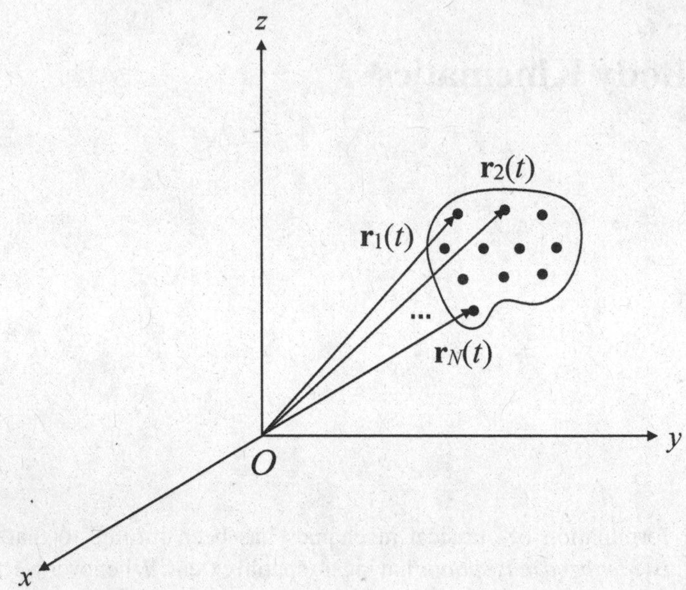


**Fig. 6.1** Individual mass points that constitute a rigid body. As the rigid body moves through space, the distances between the mass points $|\mathbf{r}_I(t) - \mathbf{r}_J(t)|$ do not change

constraints, corresponding to the distances between $m_I$ and the $N-3$ mass points $m_4, m_5, \ldots, m_N$. Thus, the number of *independent* constraints is

$$\frac{1}{2}N(N-1) - \frac{1}{2}(N-3)(N-4) = 3N - 6\,, \tag{6.2}$$

which, when subtracted from the $3N$ degrees of freedom associated with $N$ mass points, gives you the 6 remaining degrees of freedom as claimed.

It is customary to use three parameters to specify the three degrees of freedom associated with the location of a *fixed* point in the body (e.g., the center of mass). The remaining three degrees of freedom then correspond to the orientation of a **body frame** $O'$: $(x', y', z')$ relative to an inertial frame $O$: $(x, y, z)$. (The body frame is fixed in the rigid body and has its origin at the fixed point $O'$, as shown in Fig. 6.2.) Three parameters are then needed to specify the relative orientation of these two frames. In what follows, we will denote the orthonormal basis vectors for the inertial frame $O$ by $\hat{\mathbf{e}}_1, \hat{\mathbf{e}}_2, \hat{\mathbf{e}}_3$. For the body frame $O'$, we will denote the orthonormal basis vectors by $\hat{\mathbf{e}}_{1'}, \hat{\mathbf{e}}_{2'}, \hat{\mathbf{e}}_{3'}$, with primes on the indices.

There are two basic approaches for specifying the orientation of the body frame relative to the inertial frame, which we will describe in the next three sections. **Euler** (or **Tait-Bryan**) **angles** $(\phi, \theta, \psi)$ are three rotation angles that can be used to specify the relationship between the two frames. The motion of the rigid body over time is then given by the motion of the fixed point $\mathbf{R}(t)$ and the changing orientation

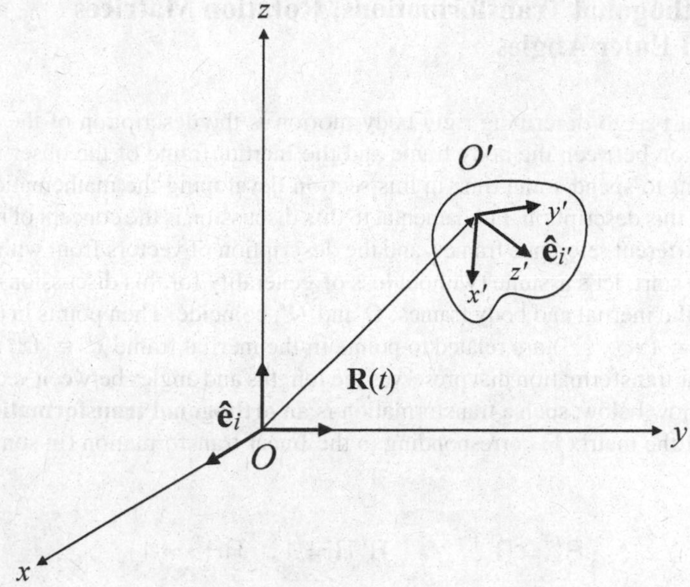


**Fig. 6.2** Inertial and body frames, $O$ and $O'$, used to describe the motion of a rigid body. $\mathbf{R}(t)$ is the position vector of some fixed point in the body, which we will usually take to be at the body's center of mass

angles $(\phi(t), \theta(t), \psi(t))$. An alternative approach is to use the **axis-angle** $(\hat{\mathbf{n}}, \Psi)$ representation of a rotation—a single rotation through the angle $\Psi$ about a specified axis $\hat{\mathbf{n}}$—to relate the two frames. This latter approach is closely related to specifying the instantaneous angular velocity vector $\boldsymbol{\omega}(t)$, in addition to $\mathbf{R}(t)$, to describe the motion of a rigid body.

Finally, in Sect. 6.6, we will describe how unit **quaternions** can also be used to represent rotations. Like the axis-angle representation, the quaternion representation of rotations has an advantage over the more traditional Euler (or Tait-Bryan) angle representation. This is related to a condition called **gimbal lock** (Sect. 6.6.2), which corresponds to a degeneracy[1] in an angular parameterization of rotations. The fact that the quaternion representation avoids the gimbal lock condition has made it the "representation of choice" for the physics engines used by the gaming industry for simulating realistic rigid body motion.

---

[1]This degeneracy is very similar to the degeneracy that exists in the standard $(\theta, \phi)$ parametrization of a 2-sphere at the North and South poles.

## 6.2 Orthogonal Transformations, Rotation Matrices and Euler Angles

An integral part to describing rigid body motion is the description of the changes in orientation between the body frame and the inertial frame of the observer. So it is important to spend some time in this section developing the mathematical tools needed for this description. Fundamental to this discussion is the concept of rotations between different reference frames, and the description of vectors from within those frames. To start, let's assume (without loss of generality for this discussion) that the origins of the inertial and body frames, $O$ and $O'$, coincide. Then points in the body frame $x^{i'} \equiv (x', y', z')$ are related to points in the inertial frame $x^i \equiv (x, y, z)$ via some linear transformation that preserves the lengths and angles between vectors. As we shall show below, such a transformation is an **orthogonal transformation**. This means that the matrix $\mathsf{R}$ corresponding to the linear transformation (in some basis) satisfies

$$\mathsf{R}^T = \mathsf{R}^{-1} \quad \Leftrightarrow \quad \mathsf{R}^T\mathsf{R} = 1\,, \quad \mathsf{R}\mathsf{R}^T = 1\,. \tag{6.3}$$

Here the superscript $T$ stands for the *transpose* operation, which simply swaps rows and columns, and the superscript $-1$ denotes matrix inverse. (See Appendix D if you need a quick refresher on linear algebra.)

***Example 6.1*** To prove the above claim, let's write the length of a vector $\mathbf{v}$ using matrix notation,[2] so

$$\mathbf{v} \cdot \mathbf{v} = \mathsf{v}^T\mathsf{v}\,. \tag{6.4}$$

The transformed vector is $\mathsf{Rv}$, and so its length is

$$(\mathsf{Rv})^T\,(\mathsf{Rv})\,. \tag{6.5}$$

If the transformation is to preserve the lengths, then we have the equation

$$\mathsf{v}^T\mathsf{v} = (\mathsf{Rv})^T\,(\mathsf{Rv}) = \mathsf{v}^T\,\mathsf{R}^T\,\mathsf{Rv}\,. \tag{6.6}$$

Since the equality must hold for all vectors $\mathbf{v}$, we can conclude that

$$\mathsf{R}^T\mathsf{R} = 1\,. \tag{6.7}$$

Taking the determinant of the above equation and using the properties of determinants described in Appendix D.4.3.5 yields

[2] We are working here with *real-valued* vectors and matrices, so the expressions for the inner products above involve only the *transpose* operation $T$ and not Hermitian conjugate † as in (D.78).

$$\det(\mathsf{R}^T \mathsf{R}) = \det \mathsf{R}^T \det \mathsf{R} = \det \mathsf{R} \det \mathsf{R} = 1 \quad \Rightarrow \quad \det \mathsf{R} = \pm 1 \,. \tag{6.8}$$

Thus, $\mathsf{R}$ is an invertible matrix, so (6.7) implies that $\mathsf{R}^{-1} = \mathsf{R}^T$ as desired. □

**Exercise 6.1** In the above example, we showed that an orthogonal transformation preserves the length of a vector $\mathbf{v}$. But besides preserving length, an orthogonal transformation must also preserve the *angle* between two vectors. Verify this by showing that the inner product between two vectors $\mathbf{u}$ and $\mathbf{v}$ can be written solely in terms of their lengths,

$$\mathbf{u} \cdot \mathbf{v} = \frac{1}{2}(|\mathbf{u} + \mathbf{v}|^2 - |\mathbf{u}|^2 - |\mathbf{v}|^2) \,. \tag{6.9}$$

Thus, an orthogonal transformation preserves lengths of vectors if and only if it preserves the inner product, and hence the angle, between any two vectors.

### 6.2.1 *Passive Versus Active Transformations*

For the majority of what we need to do in this and the following chapter, it is most convenient to interpret orthogonal transformations and rotations as **passive transformations**, which transform a reference frame $O$: $(x, y, z)$ to a new reference frame $O'$: $(x', y', z')$, but keep the vectors $\mathbf{A}$ fixed. This is opposed to **active transformations**, which transform the vectors, $\mathbf{A} \rightarrow \mathbf{A}'$, but keep the reference frame fixed. Represented as a matrix, a passive transformation relates the components of the same vector $\mathbf{A}$ with respect to the basis vectors $\hat{\mathbf{e}}_1$, $\hat{\mathbf{e}}_2$, $\hat{\mathbf{e}}_3$ and $\hat{\mathbf{e}}_{1'}$, $\hat{\mathbf{e}}_{2'}$, $\hat{\mathbf{e}}_{3'}$ of the two different reference frames, while an active transformation relates the components of the two vectors $\mathbf{A}$ and $\mathbf{A}'$ with respect to the same frame. For both passive and active rotations about an axis $\hat{\mathbf{n}}$ through some angle $\psi$, we will take positive $\psi$ to correspond to a *counter-clockwise* (CCW) angle as seen looking down onto the rotation axis.[3] With this convention, passive and active transformations are *inverses* of one another—e.g., a passive rotation of a reference frame through the angle $\psi$ is equivalent to an active rotation of vectors through the angle $-\psi$. (Note that a CCW rotation (either passive or active) through the angle $-\psi$ is equivalent to a CW rotation through $\psi$.)

An intuitive way of understanding the difference between active and passive rotations is to consider that you are looking at a tree. If you rotate your head to the left, the tree will appear to rotate to the right in your field of view. That is a passive rotation. For an active rotation, the tree actually falls down to the right while you're looking at it. Note that the active rotation of the tree to the right is mimicked by the passive rotation of your head to the left. In other words, a passive rotation goes to a

[3]This convention is consistent with the right-hand rule, where the thumb of your right hand points along $\hat{\mathbf{n}}$ and your fingers curl around in the direction of a CCW rotation through the angle $\psi$.

non-inertial reference frame and introduces fictitious forces, while an active rotation retains the inertial reference frame and describes real forces and torques.

***Example 6.2*** To demonstrate explicitly the inverse relation between passive and active transformations, let's first consider passive and active rotations around the $z$-axis through a positive angle $\psi$. These rotations are shown graphically in Fig. 6.3. Note that the $(x', y')$ reference frame is rotated CCW by $\psi$ relative to the $(x, y)$ frame. Similarly, the vector $\mathbf{A}'$ is rotated CCW by $\psi$ relative to $\mathbf{A}$.

For the passive transformation, we see that the components of $\mathbf{A}$ in the new coordinate system are

$$\begin{aligned} A_{x'} &= A\cos\phi' = A\cos(\phi-\psi) \\ &= A\cos\phi\cos\psi + A\sin\phi\sin\psi \\ &= A_x\cos\psi + A_y\sin\psi\,, \end{aligned} \tag{6.10}$$

and

$$\begin{aligned} A_{y'} &= A\sin\phi' = A\sin(\phi-\psi) \\ &= A\sin\phi\cos\psi - A\cos\phi\sin\psi \\ &= -A_x\sin\psi + A_y\cos\psi\,, \end{aligned} \tag{6.11}$$

where we used trig identities for the sine and cosine of the difference of two angles. Thus, the linear transformation for the passive rotation is

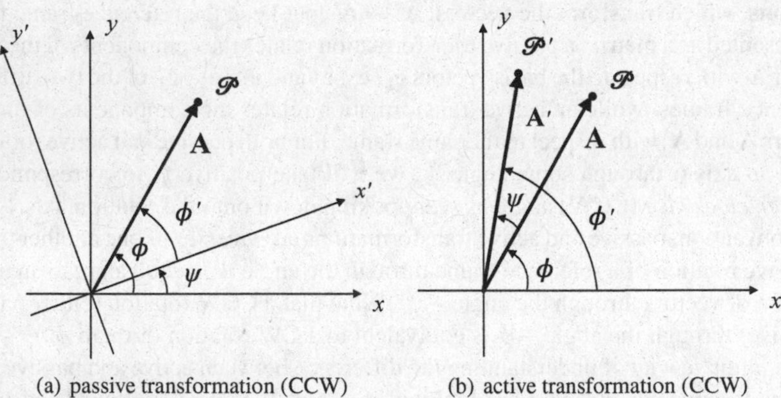


**Fig. 6.3** Passive and active rotations through a counter-clockwise angle $\psi$

$$\mathsf{R}^{\text{passive}}(\psi) = \begin{bmatrix} \cos\psi & \sin\psi \\ -\sin\psi & \cos\psi \end{bmatrix} . \tag{6.12}$$

Similarly, for the active transformation, the components of the new vector $\mathbf{A}'$ are

$$\begin{aligned} A'_x &= A\cos\phi' = A\cos(\phi+\psi) \\ &= A\cos\phi\cos\psi - A\sin\phi\sin\psi \\ &= A_x\,\cos\psi - A_y\,\sin\psi\,, \end{aligned} \tag{6.13}$$

and

$$\begin{aligned} A'_y &= A\sin\phi' = A\sin(\phi+\psi) \\ &= A\sin\phi\cos\psi + A\cos\phi\sin\psi \\ &= A_x\,\sin\psi + A_y\,\cos\psi\,. \end{aligned} \tag{6.14}$$

Thus, the linear transformation for the active rotation is

$$\mathsf{R}^{\text{active}}(\psi) = \begin{bmatrix} \cos\psi & -\sin\psi \\ \sin\psi & \cos\psi \end{bmatrix} . \tag{6.15}$$

There are two useful things to note about these linear transformations. First, the orthonormal basis vectors for the passive transformation are related by

$$\begin{aligned} \hat{\mathbf{e}}_{x'} &= \cos\psi\,\hat{\mathbf{e}}_x + \sin\psi\,\hat{\mathbf{e}}_y\,, \\ \hat{\mathbf{e}}_{y'} &= -\sin\psi\,\hat{\mathbf{e}}_x + \cos\psi\,\hat{\mathbf{e}}_y\,, \end{aligned} \tag{6.16}$$

which is similar in form to the relationship between the components given in (6.10) and (6.11). Second, the matrices $\mathsf{R}^{\text{passive}}(\psi)$ and $\mathsf{R}^{\text{active}}(\psi)$ are inverses of one another so that

$$\mathsf{R}^{\text{active}}(\psi) = \mathsf{R}^{\text{passive}}(-\psi) = [\mathsf{R}^{\text{passive}}(\psi)]^T = [\mathsf{R}^{\text{passive}}(\psi)]^{-1}\,. \tag{6.17}$$

Thus, by simply changing the sign of the angle, one changes a passive rotation matrix into a active rotation matrix, and vice-versa.

This can also be seen graphically by comparing Figs. 6.3 and 6.4. In Fig. 6.4 we show the effect of passive and active transformations for a CW rotation through the angle $\psi$ (which is equivalent to CCW rotation through $-\psi$).

The angle $\phi' = \phi + \psi$ in panel (a) of Fig. 6.4 agrees with that in panel (b) of Fig. 6.3; similarly, the angle $\phi' = \phi - \psi$ in panel (b) of Fig. 6.4 agrees with that in panel (a) of Fig. 6.3. So again we see the inverse relationship between passive and active rotations. □

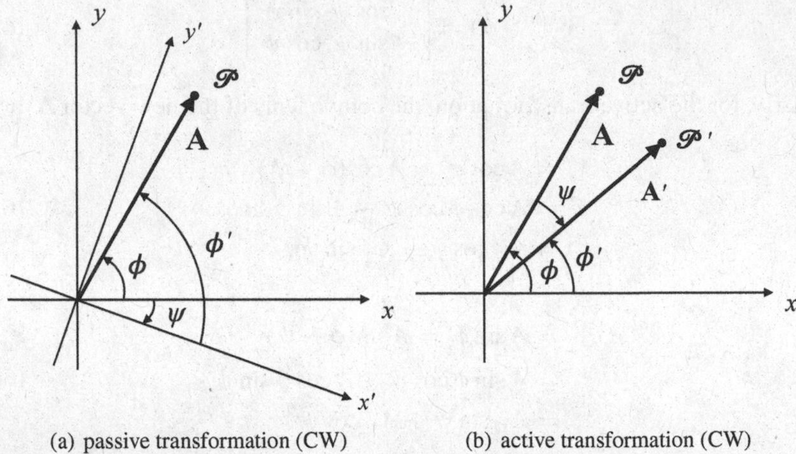


(a) passive transformation (CW) (b) active transformation (CW)

**Fig. 6.4** Passive and active rotations through a clockwise angle $\psi$

The example given above is for a two-dimensional reference frame with rotations taking place within the plane. Thus, an arbitrary orientation of one frame with respect to the other could be described by a single rotation through an angle $\psi$. (We are ignoring for the moment the possibility of picking one frame up into the third dimension and flipping it over—i.e., a *reflection*. That discussion comes later.)

In the more physical case of three dimensions, we can describe the transformation in terms of dot products between the basis vectors in one frame and those in the other frame. Let's consider the matrix components $R_{i'j}$ of an orthogonal transformation, which (as a passive transformation) maps the components of a vector **A** with respect to the inertial frame basis vectors $\hat{\mathbf{e}}_i$ to the components of the same vector with respect to the body frame basis vectors $\hat{\mathbf{e}}_{i'}$. Following the general prescription for a change of basis as discussed in Appendix D.4.2, we can write

$$\hat{\mathbf{e}}_j = \sum_{k'} R_{k'j}\hat{\mathbf{e}}_{k'}\,, \qquad j = 1, 2, 3\,. \tag{6.18}$$

Since $\mathbf{A} = \sum_i A_i\hat{\mathbf{e}}_i = \sum_{i'} A_{i'}\hat{\mathbf{e}}_{i'}$, we also have

$$A_{i'} = \sum_j R_{i'j}A_j\,, \qquad i' = 1', 2', 3'\,. \tag{6.19}$$

Now, if we take the dot product $\hat{\mathbf{e}}_{i'} \cdot \hat{\mathbf{e}}_j$, then using (6.18), we find

$$R_{i'j} = \hat{\mathbf{e}}_{i'} \cdot \hat{\mathbf{e}}_j = \cos\theta_{i'j}\,, \tag{6.20}$$

where $\theta_{i'j}$ is the angle between $\hat{\mathbf{e}}_{i'}$ and $\hat{\mathbf{e}}_j$. The **direction cosines** are $\cos\theta_{i'j}$ and they relate the basis vectors of the two systems. Collecting these 9 numbers as a matrix R, we have

$$\mathsf{R} = \begin{bmatrix} \hat{\mathbf{e}}_{1'} \cdot \hat{\mathbf{e}}_1 & \hat{\mathbf{e}}_{1'} \cdot \hat{\mathbf{e}}_2 & \hat{\mathbf{e}}_{1'} \cdot \hat{\mathbf{e}}_3 \\ \hat{\mathbf{e}}_{2'} \cdot \hat{\mathbf{e}}_1 & \hat{\mathbf{e}}_{2'} \cdot \hat{\mathbf{e}}_2 & \hat{\mathbf{e}}_{2'} \cdot \hat{\mathbf{e}}_3 \\ \hat{\mathbf{e}}_{3'} \cdot \hat{\mathbf{e}}_1 & \hat{\mathbf{e}}_{3'} \cdot \hat{\mathbf{e}}_2 & \hat{\mathbf{e}}_{3'} \cdot \hat{\mathbf{e}}_3 \end{bmatrix} . \tag{6.21}$$

Since an orthogonal transformation satisfies $\mathsf{R}^{-1} = \mathsf{R}^T$ or, equivalently,

$$(R^{-1})_{ji'} = (R^T)_{ji'} = R_{i'j} , \tag{6.22}$$

we can *invert* the relationship (6.18) between the basis vectors as

$$\hat{\mathbf{e}}_{k'} = \sum_j (R^{-1})_{jk'} \hat{\mathbf{e}}_j = \sum_j R_{k'j} \hat{\mathbf{e}}_j , \quad k' = 1', 2', 3' . \tag{6.23}$$

Thus, (6.19) and (6.23), which relate the components of a vector **A** and the two sets of orthonormal basis vectors, have exactly the same form. (This is what we also found for Example 6.2.) Keep in mind, however, that in (6.19) the matrix components $R_{i'j}$ multiply the (scalar) *components* of a vector, while in (6.23) they multiply the unit vectors themselves.

Using (6.22), the orthogonality condition $\mathsf{R}\mathsf{R}^T = \mathsf{R}^T\mathsf{R} = \mathsf{1}$ can be written in component form as either

$$\sum_i R_{j'i} R_{k'i} = \delta_{j'k'} , \tag{6.24}$$

or

$$\sum_{i'} R_{i'j} R_{i'k} = \delta_{jk} . \tag{6.25}$$

These are 6 conditions on the 9 components of the matrix $R_{i'j}$. Thus, only 3 independent parameters are needed to describe an orthogonal transformation, consistent with our discussion in Sect. 6.1 regarding the number of generalized coordinates needed to describe the orientation of a rigid body.

### 6.2.2 *Orthogonal Group and Special Orthogonal Group*

The set of orthogonal transformations in 3-dimensions has the properties of a mathematical **group**. A group is a set of elements, which we'll denote here by $G \equiv \{A_i\}$, together with an operation, typically called *group multiplication*, which we will denote by $*$. In order to be a group, the set must be closed under group multiplication, which requires that if $A_1 * A_2 = C$, then $C$ must also be an element of $G$.

Group multiplication must be associative, so that $A_1 * (A_2 * A_3) = (A_1 * A_2) * A_3$. Furthermore, there must be a special element of the set, $A_0$, called the **identity**, such that $A_0 * A_i = A_i * A_0 = A_i$. With the identity in hand, we arrive at the fourth property of a group. There must exist an inverse for each element $A_i$, such that $A_i^{-1} * A_i = A_i * A_i^{-1} = A_0$. Note that although it is not necessary, in general, for group multiplication to be commutative, the identity element must commute with all the elements of the group.

For the set of orthogonal transformations, group multiplication is successive application of the transformations or, equivalently, multiplication of the corresponding matrices. In this case, the identity element is simply the identity matrix $\mathsf{1}$, and the inverse of $\mathsf{R}$ is its transpose, $\mathsf{R}^{-1} = \mathsf{R}^T$. The particular group associated with orthogonal transformations in three dimensions is called the **orthogonal group** and is denoted by $O(3)$.

**Exercise 6.2** If a matrix $\mathsf{R}$ is an element of $O(3)$, then $\mathsf{R}^T = \mathsf{R}^{-1}$. Show that if $\mathsf{R}_1$ and $\mathsf{R}_2$ are elements of $O(3)$, then $\mathsf{R}_3 \equiv \mathsf{R}_1\mathsf{R}_2$ is also an element of $O(3)$.

We know that matrix multiplication is not commutative, so that

$$\mathsf{R}_1\mathsf{R}_2 \neq \mathsf{R}_2\mathsf{R}_1 \quad \text{(in general)}\,. \tag{6.26}$$

Thus, successive application of two orthogonal transformations does not necessarily commute. To demonstrate this, take a rectangular block—like a textbook—and apply two successive rotations to it: first rotate the block by 90° counter-clockwise around the $z'$ axis, and then by 90° counter-clockwise around the *transformed* $y'$-axis, where the $x'$, $y'$, $z'$ axes are attached to the block. After returning the block to its original orientation, apply the two rotations again, but this time in the *opposite* order. The final configuration of the block will be different for the two different sequences of rotations as illustrated graphically in Fig. 6.5.

In (6.8), we showed that the determinant of an orthogonal matrix is either $\pm 1$. An example of an orthogonal transformation with determinant $-1$ is the **inversion** (or **parity**) **transformation** $\mathsf{P} = \text{diag}(-1, -1, -1)$. It changes a right-handed system of coordinates to a left-handed system as shown in Fig. 6.6. In two dimensions, a parity transformation is equivalent to picking the frame up out of two dimensions and flipping it over—i.e., reflecting about some axis. If the determinant of an orthogonal transformation $\mathsf{R}$ is equal to 1, then $\mathsf{R}$ is said to be an element of the **special orthogonal group** of transformations, denoted $SO(3)$. As the transformations involved with the motion of a rigid body are all continously deformable to the identity transformation (which has unit determinant), rigid body transformations must also have unit determinant and thus are elements of $SO(3)$.

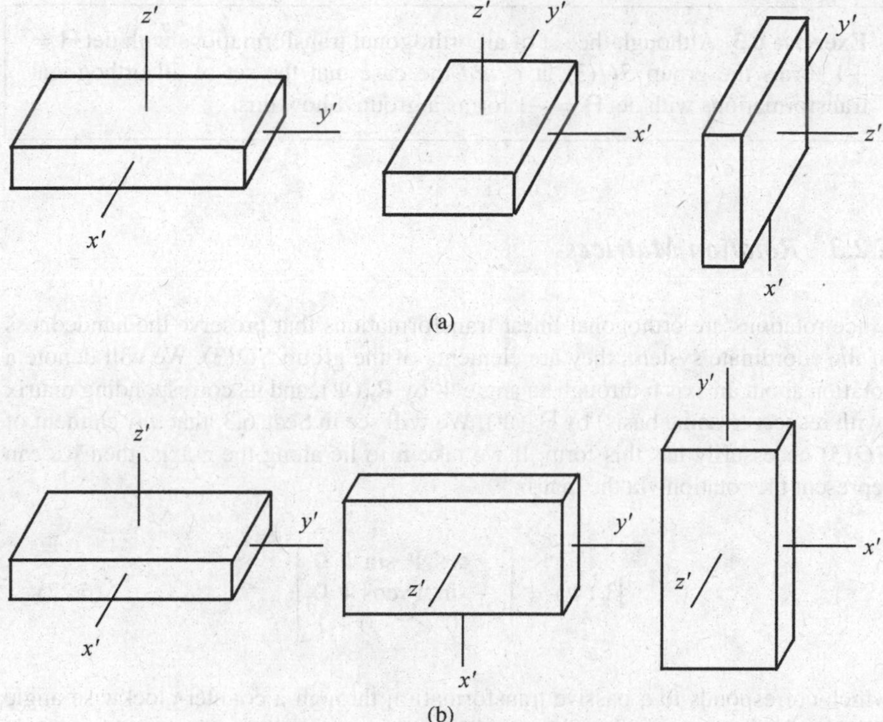


**Fig. 6.5** Non-commutating rotations applied to a rectangular block. Panel (a) First rotate around the $z'$-axis by 90°, then around the transformed $y'$-axis by 90°. Panel (b) First rotate around the $y'$-axis by 90°, then around the transformed $z'$-axis by 90°. The final configurations of the rectangular block are different

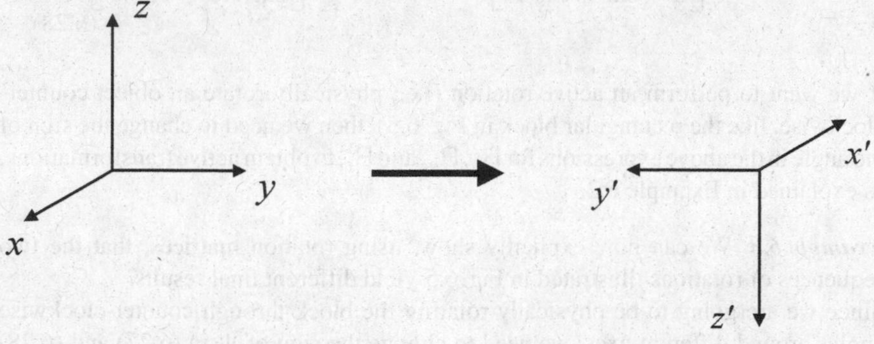


**Fig. 6.6** An inversion maps a right-handed coordinate system into a left-handed coordinate system

**Exercise 6.3** Although the set of all orthogonal transformations with $\det \mathsf{R} = +1$ forms the group $SO(3)$, it is *not* the case that the set of all orthogonal transformations with $\det \mathsf{R} = -1$ forms a group. Show this.

### 6.2.3 Rotation Matrices

Since rotations are orthogonal linear transformations that preserve the handedness of the coordinate system, they are elements of the group $SO(3)$. We will denote a rotation about an axis $\hat{\mathbf{n}}$ through an angle $\Psi$ by $\mathbf{R}_{\hat{\mathbf{n}}}(\Psi)$, and its corresponding matrix (with respect to some basis) by $\mathsf{R}_{\hat{\mathbf{n}}}(\Psi)$. We will see in Sect. 6.3, that *any* element of $SO(3)$ necessarily has this form. If we take $\hat{\mathbf{n}}$ to lie along the $z$ axis, then we can represent the rotation via the matrix

$$\mathsf{R}_z(\Psi) = \begin{bmatrix} \cos\Psi & \sin\Psi & 0 \\ -\sin\Psi & \cos\Psi & 0 \\ 0 & 0 & 1 \end{bmatrix}, \tag{6.27}$$

which corresponds to a passive transformation through a counter-clockwise angle $\Psi$. Note the similarity with (6.12). Similarly, for rotations about the $x$ and $y$ axes we have

$$\mathsf{R}_x(\Psi) = \begin{bmatrix} 1 & 0 & 0 \\ 0 & \cos\Psi & \sin\Psi \\ 0 & -\sin\Psi & \cos\Psi \end{bmatrix}, \quad \mathsf{R}_y(\Psi) = \begin{bmatrix} \cos\Psi & 0 & -\sin\Psi \\ 0 & 1 & 0 \\ \sin\Psi & 0 & \cos\Psi \end{bmatrix}. \tag{6.28}$$

If we want to perform an active rotation (i.e., physically rotate an object counter-clockwise, like the rectangular block in Fig. 6.5), then we need to change the sign of the angle in the above expressions for $\mathsf{R}_x$, $\mathsf{R}_y$, and $\mathsf{R}_z$, to obtain active transformations, as explained in Example 6.2.

***Example 6.3*** We can now explicitly show, using rotation matrices, that the two sequences of rotations illustrated in Fig. 6.5 yield different final results.
Since we are going to be physically rotating the block through counter-clockwise angles around different axes, we need to change the sign of $\Psi$ in (6.27) and (6.28) for the rotation matrices $\mathsf{R}_x$, $\mathsf{R}_y$, $\mathsf{R}_z$. To simplify the notation a bit, we will define

$$\mathsf{Y} = \mathsf{R}_y(-90°) = \begin{bmatrix} 0 & 0 & 1 \\ 0 & 1 & 0 \\ -1 & 0 & 0 \end{bmatrix}, \quad \mathsf{Z} = \mathsf{R}_z(-90°) = \begin{bmatrix} 0 & -1 & 0 \\ 1 & 0 & 0 \\ 0 & 0 & 1 \end{bmatrix}, \tag{6.29}$$

and choose a fixed reference frame $(x, y, z)$ that agrees with the initial orientation of the $(x', y', z')$ frame attached to the rectangular block. It is important to remember that the transformations $\mathsf{Y}$ and $\mathsf{Z}$ are *active* rotations that are moving the basis vectors associated with $(x', y', z')$ within the fixed $(x, y, z)$ frame, and that the axes of rotation implied by $\mathsf{Y}$ and $\mathsf{Z}$ are the axes within that fixed $(x, y, z)$ frame.

For the sequence of rotations illustrated in panel (a) of Fig. 6.5, we begin by rotating the block counter-clockwise around the $z'$-axis, by 90°. The matrix which performs this first rotation is just $\mathsf{Z}$, since the $z$ and $z'$ axes intially line up. Since the second rotation is around the *transformed* $y'$ axis, we can't simply represent it by the matrix $\mathsf{Y}$, since that performs rotations around the $y$-axis of the fixed frame. What we need to do is first *undo* the $\mathsf{Z}$ rotation using $\mathsf{Z}^{-1}$ (this brings the two coordinate frames back together again); then apply $\mathsf{Y}$; and then reapply $\mathsf{Z}$ to bring the block into its final state. If we denote this counter-clockwise rotation by 90° around the transformed $y'$ axis by $\mathsf{Y}'$, then

$$\mathsf{Y}' = \mathsf{Z}\mathsf{Y}\mathsf{Z}^{-1}\,. \tag{6.30}$$

This particular combination of operations is called **conjugation** of $\mathsf{Y}$ by $\mathsf{Z}$. The result of the two rotations $\mathsf{Z}$ and $\mathsf{Y}'$ applied in succession is then

$$\mathsf{Y}'\mathsf{Z} = \mathsf{Z}\mathsf{Y}\mathsf{Z}^{-1}\mathsf{Z} = \mathsf{Z}\mathsf{Y}\,, \tag{6.31}$$

where we used $\mathsf{Z}^{-1}\mathsf{Z} = \mathsf{1}$. So the final result of rotating the block first around $z = z'$ and then around the transformed $y'$ axis is equivalent to first rotating the block around $y$ and then around $z$, where $y$ and $z$ are the directions specified by the fixed frame. (The intermediate stages are different, however.) The matrices $\mathsf{Y}'$ and $Z$ on the left-hand side of (6.31) are called **intrinsic rotation matrices**, since they are defined with respect to axes intrinsic to the body; while $\mathsf{Z}$ and $\mathsf{Y}$ on the right-hand side are called **extrinsic rotation matrices**, since they are defined with respect to axes fixed in space—i.e., extrinsic to the body.

Repeating the above argument for the second sequence of transformations, which is shown in panel (b) of Fig. 6.5, we have

$$\mathsf{Z}'\mathsf{Y} = \mathsf{Y}\mathsf{Z}\,. \tag{6.32}$$

It is now a simple matter to compare the results of multiplying the two matrices, $\mathsf{Y}$ and $\mathsf{Z}$, in opposite orders. We find

$$\mathsf{Z}\mathsf{Y} = \begin{bmatrix} 0 & -1 & 0 \\ 0 & 0 & 1 \\ -1 & 0 & 0 \end{bmatrix}, \quad \mathsf{Y}\mathsf{Z} = \begin{bmatrix} 0 & 0 & 1 \\ 1 & 0 & 0 \\ 0 & 1 & 0 \end{bmatrix}, \tag{6.33}$$

which disagree, thus confirming that these two sequences of rotations produce different results. In short, *rotations do not commute in general.* □

**Exercise 6.4** Show that the simple result $\mathsf{Y'Z} = \mathsf{ZY}$ derived above generalizes to an arbitrary number of transformations applied in succession

$$\cdots \mathsf{D'''C''B'A} = \mathsf{ABCD}\cdots, \tag{6.34}$$

where $'$, $''$, $'''$, etc. indicate that the relevant transformations are to be applied to the object after it has already undergone the 1st, 2nd, 3rd, $\cdots$ transformations. Note that the rotation angles for these transformations need not be equal to 90°, and all three basic extrinsic rotations can be involved.

#### 6.2.3.1 Euler Angle Representation

The **Euler angle representation** of an orthogonal transformation $\mathsf{R}$ relates the orientation of the $(x, y, z)$ and $(x', y', z')$ coordinate frames shown in Fig. 6.7. In this representation, $(\theta, \phi)$ are the standard polar and azimuthal angles on the 2-sphere corresponding to the direction of $\hat{\mathbf{z}}'$, and $\psi$ is the angle needed to rotate the standard 2-sphere basis vectors $\hat{\boldsymbol{\theta}}$, $\hat{\boldsymbol{\phi}}$ into $\hat{\mathbf{x}}'$, $\hat{\mathbf{y}}'$. The complete rotation is the equivalent of a product of three rotations

$$\mathsf{R}(\phi, \theta, \psi) = \mathsf{R}_z(\psi)\mathsf{R}_y(\theta)\mathsf{R}_z(\phi)\,. \tag{6.35}$$

The matrix $\mathsf{R}(\phi, \theta, \psi)$ represents a passive transformation, so it maps the components of a vector $\mathbf{A}$ with respect to the $(x, y, z)$ frame to the components of the same vector with respect to the $(x', y', z')$ frame.

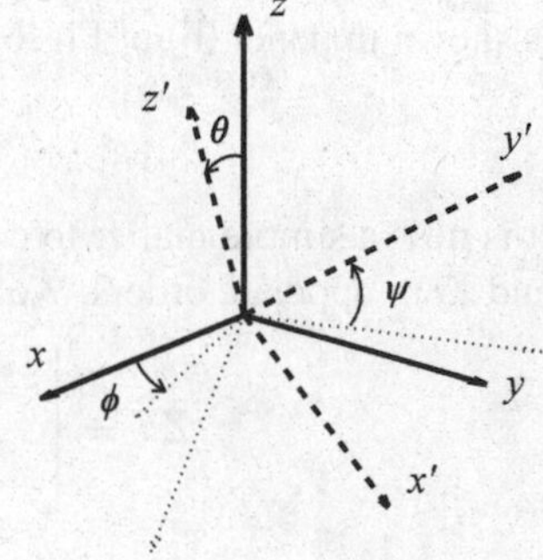


**Fig. 6.7** Relationship between the $(x, y, z)$ and $(x', y', z')$ coordinate frames in terms of the Euler angles $(\phi, \theta, \psi)$ using the $zyz$ convention

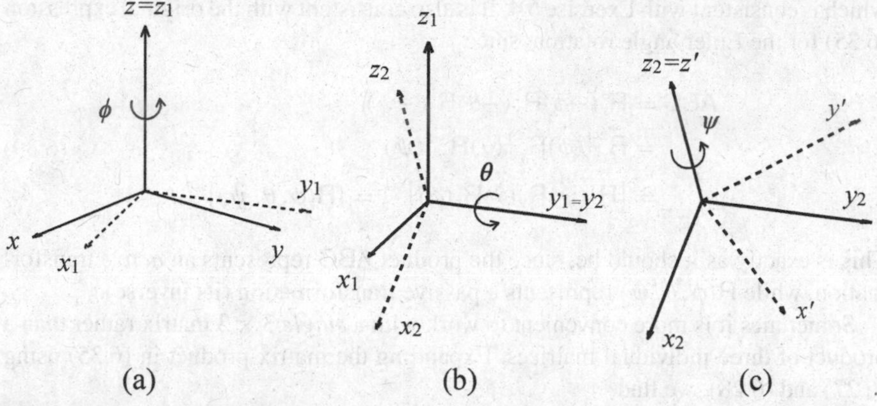


**Fig. 6.8** Successive rotations that map the $(x, y, z)$ coordinate frame to the $(x', y', z')$ coordinate frame, using the $zyz$ convention for Euler angles $(\phi, \theta, \psi)$

To get to the final coordinate frame $(x', y', z')$ starting from $(x, y, z)$, one first rotates the system counter-clockwise around the $z$-axis by $\phi$; then around the transformed $y$-axis (which we denote by $y_1$) by $\theta$; and then around the transformed $z$-axis (which we denote by $z_2$) by $\psi$. These successive transformations are shown in Fig. 6.8. But given our discussion in Example 6.3, we know that these successive transformations are not simply $\mathsf{R}_z(\phi)$, $\mathsf{R}_y(\theta)$, and $\mathsf{R}_z(\psi)$, since they are *passive* transformations defined by the *extrinsic* $(x, y, z)$ coordinate frame, and not by the successive transformed axes. So if we define

$$\mathsf{A} \equiv \mathsf{R}_z(-\phi)\,, \qquad \mathsf{B} \equiv \mathsf{R}_y(-\theta)\,, \qquad \mathsf{C} \equiv \mathsf{R}_z(-\psi)\,, \tag{6.36}$$

then following the discussion given in Example 6.3 and Exercise 6.4, the three active transformations illustrated in Fig. 6.8 are

$$\begin{aligned} &\mathsf{A}\,, \\ &\mathsf{B}' = \mathsf{A}\mathsf{B}\mathsf{A}^{-1}\,, \\ &\mathsf{C}'' = (\mathsf{B}'\mathsf{A})\mathsf{C}(\mathsf{B}'\mathsf{A})^{-1} = \mathsf{A}\mathsf{B}\mathsf{C}(\mathsf{A}\mathsf{B})^{-1}\,. \end{aligned} \tag{6.37}$$

This convention of performing the rotations is called the $zyz$ **convention**, since the second rotation is about the transformed $y$-axis, which is often called the **line of nodes**. One nice property of the $zyz$-convention is that the $z'$-axis points in the direction of $(\theta, \phi)$, which has the standard interpretation in terms of spherical coordinates for the fixed $(x, y, z)$ coordinate frame.

Concatenating the above transformations yields

$$\mathsf{C}''\mathsf{B}'\mathsf{A} = \mathsf{A}\mathsf{B}\mathsf{C}\,, \tag{6.38}$$

which is consistent with Exercise 6.4. It is also consistent with the original expression (6.35) for the Euler angle rotation, since

$$\begin{aligned} \mathsf{ABC} &= \mathsf{R}_z(-\phi)\mathsf{R}_y(-\theta)\mathsf{R}_z(-\psi) \\ &= \mathsf{R}_z^{-1}(\phi)\mathsf{R}_y^{-1}(\theta)\mathsf{R}_z^{-1}(\psi) \\ &= \left[\mathsf{R}_z(\psi)\mathsf{R}_y(\theta)\mathsf{R}_z(\phi)\right]^{-1} = [\mathsf{R}(\phi,\theta,\psi)]^{-1} . \end{aligned} \tag{6.39}$$

This is exactly as it should be, since the product ABC represents an *active* transformation, while $\mathsf{R}(\phi, \theta, \psi)$ represents a passive transformation (its inverse).

Sometimes it is more convenient to work with a *single* $3 \times 3$ matrix rather than a product of three individual matrices. Expanding the matrix product in (6.35) using (6.27) and (6.28), we find

$$\mathsf{R}(\phi,\theta,\psi) = \begin{bmatrix} c\theta c\phi c\psi - s\phi s\psi & c\theta s\phi c\psi + c\phi s\psi & -s\theta c\psi \\ -c\theta c\phi s\psi - s\phi c\psi & -c\theta s\phi s\psi + c\phi c\psi & s\theta s\psi \\ s\theta c\phi & s\theta s\phi & c\theta \end{bmatrix} , \tag{6.40}$$

where we introduced the notation $c\theta \equiv \cos\theta$, $s\theta \equiv \sin\theta$, etc., to save some space with the writing. To convert (6.40) to a form appropriate for an *active* transformation, we can simply take its transpose, using the fact that $\mathsf{R}^T = \mathsf{R}^{-1}$ for an orthogonal transformation.

**Exercise 6.5** Verify (6.40) starting with the matrix representation for $\mathsf{R}(\phi, \theta, \psi)$ given in (6.35).

#### 6.2.3.2 Tait-Bryan Angle Representation

If the angles $(\phi, \theta, \psi)$ describing the rotations are associated with three *different* axes (e.g., $z, y, x$ instead of $z, y, z$), then the angles are called **Tait-Bryan angles** and the product matrix

$$\mathsf{R} \equiv \mathsf{R}_x(\psi)\mathsf{R}_y(\theta)\mathsf{R}_z(\phi) \tag{6.41}$$

is called the $xyz$ **convention** for the Tait-Bryan angles. This matrix relates the components of a vector **A** with respect to the $(x, y, z)$ and $(x', y', z')$ coordinate frames shown in the top panel of Fig. 6.9. The successive rotations that map the $(x, y, z)$ coordinate frame to the $(x', y', z')$ coordinate frame for this particular representation are shown in the bottom three panels of the same figure. The Tait-Bryan angles $(\phi, \theta, \psi)$ are most often used to describe the orientation of an aircraft (or similar object), and in such a context go by the names **yaw**, **pitch**, and **roll**, or **heading**, **elevation**, and **bank**. Yaw or heading specifies the left-right direction of the aircraft; pitch or elevation specifies whether the aircraft is ascending or descending; and roll

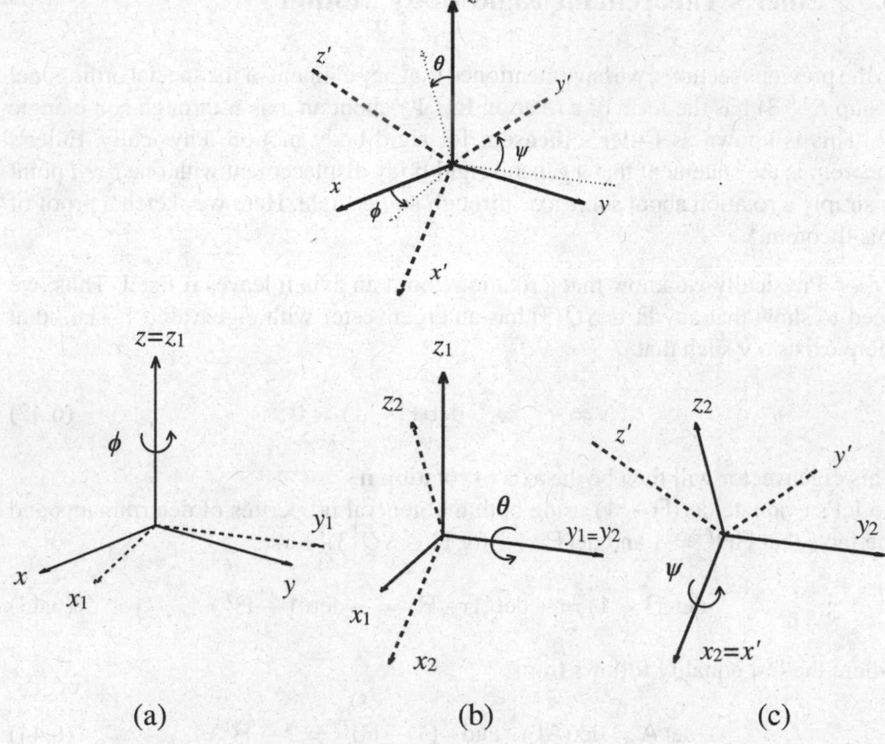


**Fig. 6.9** Same as Figs. 6.7 and 6.8, but using the Tait-Bryan $xyz$ convention for the angles $(\phi, \theta, \psi)$

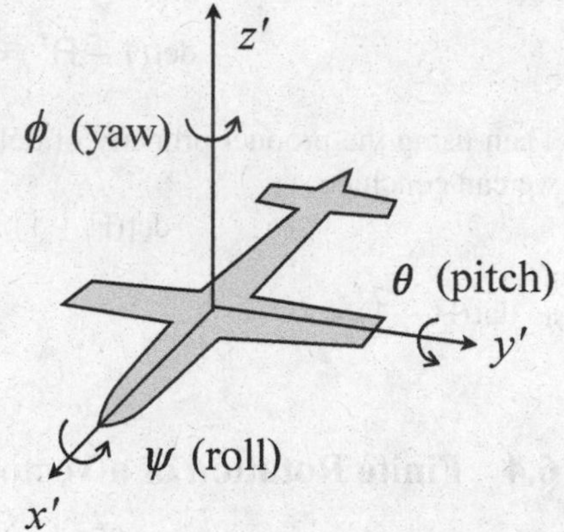


**Fig. 6.10** The Tait-Bryan angles $(\phi, \theta, \psi)$ using the $xyz$ convention are often called yaw, pitch, and roll in the context of describing, e.g., the orientation of an aircraft

or bank specifies if the aircraft has any rotational motion (spin) about its direction of motion. This is illustrated graphically in Fig. 6.10.

## 6.3 Euler's Theorem for Rigid Body Motion

In the previous sections, we have mentioned that any element of the special orthogonal group $SO(3)$ has the form of a rotation $\mathbf{R}_{\hat{\mathbf{n}}}(\Psi)$ about an axis $\hat{\mathbf{n}}$ through some angle $\Psi$. This is known as **Euler's theorem** for rigid body motion. Physically, Euler's theorem is the statement that a general rigid body displacement with one *fixed* point is simply a rotation about some axis through some angle. Here we sketch a proof of this theorem.[4]

*Proof* Physically we know that a rotation about an axis $\hat{\mathbf{n}}$ leaves $\hat{\mathbf{n}}$ fixed. Thus, we need to show that any $\mathsf{R} \in SO(3)$ has an eigenvector with eigenvalue 1—i.e., that there exists a $\mathsf{v}$ such that

$$\mathsf{R}\mathsf{v} = \mathsf{v} \quad \Leftrightarrow \quad \det(\mathsf{R} - \mathsf{1}) = 0 . \tag{6.42}$$

This eigenvector will then be the axis of rotation $\hat{\mathbf{n}}$.
So let's evaluate $\det(\mathsf{R} - \mathsf{1})$ using both the general properties of determinants and the facts that $\mathsf{R}\mathsf{R}^T = \mathsf{1}$ and $\det \mathsf{R} = 1$ for $\mathsf{R} \in SO(3)$. First,

$$\det(\mathsf{R} - \mathsf{1}) = -\det(\mathsf{1} - \mathsf{R}) = -\det(\mathsf{1} - \mathsf{R}^T) , \tag{6.43}$$

where the last equality follows from

$$\det \mathsf{A} = \det(\mathsf{A}^T) \quad \text{and} \quad (\mathsf{1} - \mathsf{R})^T = \mathsf{1} - \mathsf{R}^T . \tag{6.44}$$

But since $\mathsf{R}\mathsf{R}^T = \mathsf{1}$, we can also write

$$-\det(\mathsf{1} - \mathsf{R}^T) = -\det((\mathsf{R} - \mathsf{1})\mathsf{R}^T) . \tag{6.45}$$

Then using the product property of determinants again and $\det(\mathsf{R}^T) = \det \mathsf{R} = 1$, we can conclude

$$\det(\mathsf{R} - \mathsf{1}) = -\det(\mathsf{R} - \mathsf{1}) , \tag{6.46}$$

so $\det(\mathsf{R} - \mathsf{1}) = 0$. □

## 6.4 Finite Rotation of a Vector

We can obtain an explicit matrix representation of $\mathbf{R}_{\hat{\mathbf{n}}}(\Psi)$ in the $(x, y, z)$ coordinate frame, which will allow us to coveniently switch between the axis-angle and Euler

[4]Since this proof and the following discussion will make use of eigenvectors, eigenvalues, similarity transformations, etc., please see the appropriate sections in Appendix D or some other reference if you need a refresher on any of these topics.

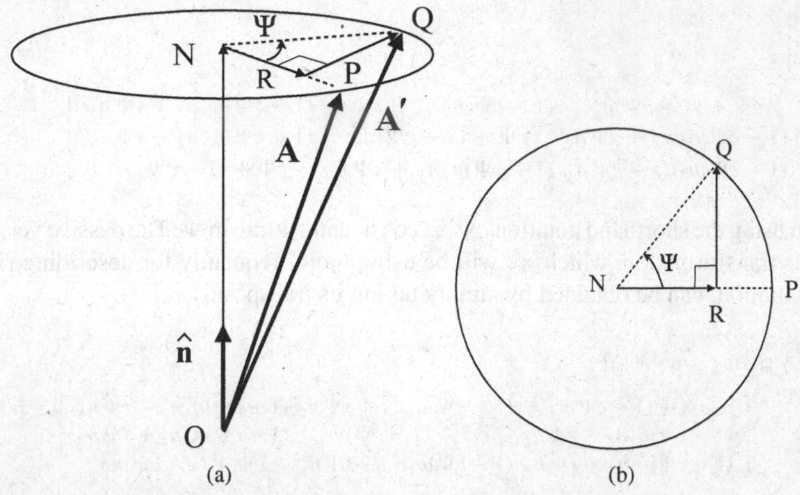


**Fig. 6.11** Geometrical set-up for the finite rotation of a vector, $\mathbf{A} \mapsto \mathbf{A}'$, under an active rotation about $\hat{\mathbf{n}}$ by $\Psi$. Panel (a) Perspective view. Panel (b) Top view

angle representations of rotations. To do this we need to know what $\mathbf{R}_{\hat{\mathbf{n}}}(\Psi)$ does to the orthonormal basis vectors $\hat{\mathbf{e}}_1, \hat{\mathbf{e}}_2, \hat{\mathbf{e}}_3$, or, equivalently, to an arbitrary vector $\mathbf{A}$. So in the discussion that follows, it is simplest to consider an *active* rotation of a vector $\mathbf{A}$ around $\hat{\mathbf{n}}$ through a counter-clockwise angle $\Psi$. To make things concrete, let $O$ denote the origin, $N$ be the point along $\hat{\mathbf{n}}$ such that the length of $ON$ equals $\mathbf{A} \cdot \hat{\mathbf{n}}$, $P$ be the tip of $\mathbf{A}$, $Q$ be the tip of $\mathbf{A}'$, and $R$ be the point along $NP$ for which a perpendicular drawn from $R$ intersects $Q$. (See Fig. 6.11.) Then in terms of these quantities we have

$$\mathbf{A}' = \mathbf{ON} + \mathbf{NR} + \mathbf{RQ}\,. \tag{6.47}$$

Using geometry, we can also show

$$\mathbf{ON} = (\mathbf{A} \cdot \hat{\mathbf{n}})\hat{\mathbf{n}}\,, \quad \mathbf{NR} = \cos\Psi\,(\mathbf{A} - (\mathbf{A} \cdot \hat{\mathbf{n}})\hat{\mathbf{n}})\,, \quad \mathbf{RQ} = \sin\Psi\,(\hat{\mathbf{n}} \times \mathbf{A})\,. \tag{6.48}$$

Thus,

$$\mathbf{A}' = \cos\Psi\,\mathbf{A} + (1 - \cos\Psi)(\mathbf{A} \cdot \hat{\mathbf{n}})\,\hat{\mathbf{n}} + \sin\Psi\,(\hat{\mathbf{n}} \times \mathbf{A})\,. \tag{6.49}$$

If we define $\mathsf{A} \equiv [A_x, A_y, A_z]^T$ and $\mathsf{A}' \equiv [A'_x, A'_y, A'_z]^T$ to represent the vectors $\mathbf{A}$ and $\mathbf{A}'$, then we can write (6.49) as a matrix equation

$$\mathsf{A}' = \mathsf{R}^{\text{active}}\mathsf{A}\,, \tag{6.50}$$

where

$$\mathsf{R}^{\text{active}} = \begin{bmatrix} c\Psi + (1-c\Psi)n_x^2 & (1-c\Psi)n_x n_y - s\Psi\, n_z & (1-c\Psi)n_x n_z + s\Psi\, n_y \\ (1-c\Psi)n_y n_x + s\Psi\, n_z & c\Psi + (1-c\Psi)n_y^2 & (1-c\Psi)n_y n_z - s\Psi\, n_x \\ (1-c\Psi)n_z n_x - s\Psi\, n_y & (1-c\Psi)n_z n_y + s\Psi\, n_x & c\Psi + (1-c\Psi)n_z^2 \end{bmatrix}, \tag{6.51}$$

again using the shorthand notation $c\Psi \equiv \cos\Psi$ and $s\Psi \equiv \sin\Psi$. The passive version of this transformation, which we will be using more frequently for describing rigid body motion, can be obtained by simply taking its transpose,

$$\mathsf{R}_{\hat{\mathbf{n}}}(\Psi) \equiv [\mathsf{R}^{\text{active}}]^T = \begin{bmatrix} c\Psi + (1-c\Psi)n_x^2 & (1-c\Psi)n_x n_y + s\Psi\, n_z & (1-c\Psi)n_x n_z - s\Psi\, n_y \\ (1-c\Psi)n_y n_x - s\Psi\, n_z & c\Psi + (1-c\Psi)n_y^2 & (1-c\Psi)n_y n_z + s\Psi\, n_x \\ (1-c\Psi)n_z n_x + s\Psi\, n_y & (1-c\Psi)n_z n_y - s\Psi\, n_x & c\Psi + (1-c\Psi)n_z^2 \end{bmatrix}. \tag{6.52}$$

By inspection we see that

$$\begin{bmatrix} n_x \\ n_y \\ n_z \end{bmatrix} = \frac{1}{2\sin\Psi} \begin{bmatrix} R_{y'z} - R_{z'y} \\ R_{z'x} - R_{x'z} \\ R_{x'y} - R_{y'x} \end{bmatrix}. \tag{6.53}$$

In terms of the Euler-angle representation (6.40), the right-hand side becomes

$$\begin{bmatrix} n_x \\ n_y \\ n_z \end{bmatrix} = \frac{1}{2\sin\Psi} \begin{bmatrix} \sin\theta(\sin\psi - \sin\phi) \\ \sin\theta(\cos\psi + \cos\phi) \\ (1+\cos\theta)\sin(\phi+\psi) \end{bmatrix}. \tag{6.54}$$

Now, since $\hat{\mathbf{n}}$ is a unit vector, the normalization condition $n_x^2 + n_y^2 + n_z^2 = 1$ allows us to derive a relationship between $\Psi$ and the Euler angles $(\phi, \theta, \psi)$. After some relatively straightforward algebra, which takes advantage of the trig identity $1 + \cos x = 2\cos^2(x/2)$, one can show

$$\cos\left(\frac{\Psi}{2}\right) = \cos\left(\frac{\theta}{2}\right)\cos\left(\frac{\phi+\psi}{2}\right). \tag{6.55}$$

This solution has the property that when $\theta = \phi = 0$, a rotation around $\hat{\mathbf{n}} = \hat{\mathbf{z}}$ by $\Psi$ is the same as an Euler angle rotation around $\hat{\mathbf{n}} = \hat{\mathbf{z}}$ by $\psi$. Alternatively, one can derive this expression for $\Psi$ by taking the trace of $\mathsf{R}(\phi, \theta, \psi)$, given by (6.40), and equating it to the trace of $\mathsf{S}\mathsf{R}_{\hat{\mathbf{n}}}(\Psi)\mathsf{S}^{-1}$, where $\mathsf{S}$ is a similarity transformation to a new set of coordinates $(x', y', z')$ with $\hat{\mathbf{z}}'$ aligned along the rotation axis $\hat{\mathbf{n}}$ (Problem 6.3).

In summary, (6.52) gives the matrix representation of a passive rotation around $\hat{\mathbf{n}}$ by $\Psi$, and (6.54) and (6.55) give the components of the axis $\hat{\mathbf{n}}$ and the rotation angle $\Psi$ in terms of the Euler angles $(\phi, \theta, \psi)$ in the $zyz$ convention. As we move to continuous rotations and the dynamics of rigid bodies, we will find the $(\hat{\mathbf{n}}, \Psi)$ representation useful in describing the instantaneous angular velocity of the body.

**Exercise 6.6** Verify (6.52), (6.54), and (6.55).

**Exercise 6.7** Calculate the eigenvectors having eigenvalue 1 for the two matrices $\mathsf{ZY}$ and $\mathsf{YZ}$ in Example 6.3. What are the rotation angles $\Psi$ corresponding to these two matrices? Do the eigenvectors (rotation axes $\hat{\mathbf{n}}$) and rotation angles $\Psi$ agree physically with what you expect from Fig. 6.5?

## 6.5 Infinitesimal Orthogonal Transformations

Throw a rigid body into the air and watch it move. Relative to some fixed point in the body, e.g., its center of mass, the body is undergoing complicated rotational motion about an instantaneous axis of rotation, which is changing its direction and magnitude from one instant to the next. To describe the motion of such an object, we need to be able to work with rotations defined over infinitesimally small time intervals $dt$. Such rotations differ only slightly from the identity transformation and hence are examples of **infinitesimal transformations**. In matrix form we write them as

$$\mathsf{R} = 1 + \varepsilon\,, \tag{6.56}$$

where $\varepsilon$ is a $3 \times 3$ matrix, all of whose elements are small compared to 1. Thus, when working with infinitesimal transformations, we can ignore terms that are *2nd-order* in $\varepsilon$, which simplifies many of the calculations. For example, the inverse $\mathsf{R}^{-1}$ of an infinitesimal transformation is given to leading order by

$$\mathsf{R}^{-1} = 1 - \varepsilon\,, \tag{6.57}$$

which follows from

$$(1 - \varepsilon)(1 + \varepsilon) = 1 - \varepsilon + \varepsilon - \varepsilon\varepsilon = 1 + O(\varepsilon^2)\,. \tag{6.58}$$

In addition,

$$\mathsf{R}_1\mathsf{R}_2 = 1 + \varepsilon_1 + \varepsilon_2 + O(\varepsilon^2) = \mathsf{R}_2\mathsf{R}_1\,, \tag{6.59}$$

so infinitesimal transformations commute. (Order $\varepsilon^2$ and higher-order terms are responsible for the non-commutativity of *finite* transformations.)

Since an infinitesimal rotation is an element of $SO(3)$, we also have $\mathsf{R}^T = \mathsf{R}^{-1}$. This condition implies

$$\varepsilon^T = -\varepsilon\,, \tag{6.60}$$

which means that $\varepsilon$ is an *anti-symmetric* matrix. Since an anti-symmetric matrix in 3-dimensions has three independent components, we can associate a vector $\mathrm{d}\Psi \equiv \left[\mathrm{d}\Psi_x\,,\mathrm{d}\Psi_y\,,\mathrm{d}\Psi_z\right]^T$ with an infinitesimal active rotation via

$$\varepsilon = \begin{bmatrix} 0 & -\mathrm{d}\Psi_z & \mathrm{d}\Psi_y \\ \mathrm{d}\Psi_z & 0 & -\mathrm{d}\Psi_x \\ -\mathrm{d}\Psi_y & \mathrm{d}\Psi_x & 0 \end{bmatrix}. \tag{6.61}$$

We choose this definition so that an infinitesimal active rotation applied to a vector gives

$$\mathbf{A}' = \mathbf{A} + \mathrm{d}\boldsymbol{\Psi} \times \mathbf{A}\,. \tag{6.62}$$

This is consistent with (6.49) for a finite active rotation about $\hat{\mathbf{n}}$ restricted to the infinitesimal angle $\mathrm{d}\Psi$:

$$\mathbf{A}' = \mathbf{A} + (\hat{\mathbf{n}} \times \mathbf{A})\,\mathrm{d}\Psi\,, \tag{6.63}$$

with the identification $\mathrm{d}\boldsymbol{\Psi} = \hat{\mathbf{n}}\,\mathrm{d}\Psi$.

**Exercise 6.8** Note that (6.61) can be written as

$$\varepsilon = \mathrm{d}\Psi_x\,\mathsf{L}_x + \mathrm{d}\Psi_y\,\mathsf{L}_y + \mathrm{d}\Psi_z\,\mathsf{L}_z\,, \tag{6.64}$$

where

$$\mathsf{L}_x \equiv \begin{bmatrix} 0 & 0 & 0 \\ 0 & 0 & -1 \\ 0 & 1 & 0 \end{bmatrix}, \quad \mathsf{L}_y \equiv \begin{bmatrix} 0 & 0 & 1 \\ 0 & 0 & 0 \\ -1 & 0 & 0 \end{bmatrix}, \quad \mathsf{L}_z \equiv \begin{bmatrix} 0 & -1 & 0 \\ 1 & 0 & 0 \\ 0 & 0 & 0 \end{bmatrix}. \tag{6.65}$$

Show that

$$[\mathsf{L}_x, \mathsf{L}_y] = \mathsf{L}_z\,, \quad [\mathsf{L}_y, \mathsf{L}_z] = \mathsf{L}_x\,, \quad [\mathsf{L}_z, \mathsf{L}_x] = \mathsf{L}_y\,, \tag{6.66}$$

where

$$[\mathsf{A}, \mathsf{B}] \equiv \mathsf{AB} - \mathsf{BA} \tag{6.67}$$

is the **commutator** of the two matrices $\mathsf{A}$ and $\mathsf{B}$. Given their relationship to $\varepsilon$, the matrices $\mathsf{L}_x$, $\mathsf{L}_y$, $\mathsf{L}_z$ are said to be *generators* of infinitesimal rotations about the $x$, $y$, and $z$ axes. (See also Exercise 3.11.)

### 6.5.1 Instantaneous Angular Velocity Vector

The **instantaneous angular velocity vector** $\boldsymbol{\omega}$ describes an infinitesimal rotation in the time interval $\mathrm{d}t$. It is related to the infinitesimal rotation vector $\hat{\mathbf{n}}\,\mathrm{d}\Psi$ via the formula

$$\boldsymbol{\omega}\,\mathrm{d}t = \hat{\mathbf{n}}\,\mathrm{d}\Psi = \hat{\mathbf{n}}_\phi\,\mathrm{d}\phi + \hat{\mathbf{n}}_\theta\,\mathrm{d}\theta + \hat{\mathbf{n}}_\psi\,\mathrm{d}\psi\,, \tag{6.68}$$

where $\hat{\mathbf{n}}_\phi$, $\hat{\mathbf{n}}_\theta$, $\hat{\mathbf{n}}_\psi$ point along the axes of the three Euler angle rotations through the infinitesimal angles $\mathrm{d}\phi$, $\mathrm{d}\theta$, $\mathrm{d}\psi$, respectively (See Fig. 6.8):

$$\begin{aligned}
\hat{\mathbf{n}}_\phi &= \hat{\mathbf{z}}\,,\\
\hat{\mathbf{n}}_\theta &= -\sin\phi\,\hat{\mathbf{x}} + \cos\phi\,\hat{\mathbf{y}}\,,\\
\hat{\mathbf{n}}_\psi &= \sin\theta\cos\phi\,\hat{\mathbf{x}} + \sin\theta\sin\phi\,\hat{\mathbf{y}} + \cos\theta\,\hat{\mathbf{z}}\,.
\end{aligned} \tag{6.69}$$

Note that regardless of the orientation of the rigid body, $\hat{\mathbf{n}}_\phi$ always points along the inertial frame $z$-axis, and $\hat{\mathbf{n}}_\theta$ always lies in the inertial frame $xy$-plane. (For Tait-Bryan angles using the $xyz$-convention, $\hat{\mathbf{n}}_\psi = \cos\theta\cos\phi\,\hat{\mathbf{x}} + \cos\theta\sin\phi\,\hat{\mathbf{y}} - \sin\theta\,\hat{\mathbf{z}}$.) Using these expressions, we can then write $\boldsymbol{\omega}$ in terms of either the fixed (inertial) frame $(x, y, z)$:

$$\begin{bmatrix}\omega_x\\ \omega_y\\ \omega_z\end{bmatrix} = \begin{bmatrix}\sin\theta\cos\phi\,\dot\psi - \sin\phi\,\dot\theta\\ \sin\theta\sin\phi\,\dot\psi + \cos\phi\,\dot\theta\\ \cos\theta\,\dot\psi + \dot\phi\end{bmatrix}, \tag{6.70}$$

or the body frame $(x', y', z')$:

$$\begin{bmatrix}\omega_{x'}\\ \omega_{y'}\\ \omega_{z'}\end{bmatrix} = \begin{bmatrix}-\sin\theta\cos\psi\,\dot\phi + \sin\psi\,\dot\theta\\ \sin\theta\sin\psi\,\dot\phi + \cos\psi\,\dot\theta\\ \cos\theta\,\dot\phi + \dot\psi\end{bmatrix}, \tag{6.71}$$

where in both cases the time derivatives of the Euler angles are with respect to the inertial frame. As we shall see below, the instantaneous angular velocity vector $\boldsymbol{\omega}$ is the most important quantity for describing the rotational state of motion of a rigid body.

**Exercise 6.9** Derive the above expressions for $\boldsymbol{\omega}$. (*Hint*: Given the components of $\boldsymbol{\omega}$ with respect to the inertial frame, you can multiply them by the transformation matrix $\mathsf{R}(\phi, \theta, \psi)$ given in (6.40) to obtain the components with respect to the body frame.)

### 6.5.2 Velocity and Acceleration in the Inertial and Body Frames

In Sect. 1.5, we discussed the motion of a particle as seen from the perspective of a non-inertial reference frame. There we derived expressions that relate the velocity and acceleration of a particle as seen in a *fixed* (i.e., inertial) frame to the same quantities calculated in a non-inertial frame, which could have both translational and rotational motion. The key results, which we take from that section, are

$$\mathbf{v} = \mathbf{v}' + \boldsymbol{\omega} \times \mathbf{r}' + \dot{\mathbf{R}}, \tag{6.72}$$

and

$$\mathbf{a} = \mathbf{a}' + \dot{\boldsymbol{\omega}} \times \mathbf{r}' + 2\boldsymbol{\omega} \times \mathbf{v}' + \boldsymbol{\omega} \times (\boldsymbol{\omega} \times \mathbf{r}') + \ddot{\mathbf{R}}, \tag{6.73}$$

where primes denote quantities calculated in the non-inerial frame. In the above equations $\mathbf{R}(t)$ denotes the position vector of the origin $O'$ of the non-inertial frame relative to the origin $O$ of the inertial frame. (See Fig. 1.5.) Its time derivatives, $\dot{\mathbf{R}}$ and $\ddot{\mathbf{R}}$, are calculated with respect to the inertial frame. Since the time derivative of the instantaneous angular velocity vector $\boldsymbol{\omega}$ is the same when calculated in either the inertial or non-inertial reference frame, see (1.71), we can write $\mathrm{d}\boldsymbol{\omega}/\mathrm{d}t \equiv \dot{\boldsymbol{\omega}}$ without any ambiguity.

For describing rigid body motion, the body frame $O'$: $(x', y', z')$ is *attached* to the rigid body, so the individual mass points comprising the body have zero velocity and zero acceleration with respect to the body frame. Thus, the above equations simplify for rigid body motion:

$$\mathbf{v} = \boldsymbol{\omega} \times \mathbf{r}' + \dot{\mathbf{R}}, \qquad \mathbf{a} = \dot{\boldsymbol{\omega}} \times \mathbf{r}' + \boldsymbol{\omega} \times (\boldsymbol{\omega} \times \mathbf{r}') + \ddot{\mathbf{R}}. \tag{6.74}$$

It also turns out that the angular velocity $\boldsymbol{\omega}$ of a rigid body is *independent* of the choice of body frame. To see this, let $O''$: $(x'', y'', z'')$ denote a new body frame with origin $O''$ located at $\mathbf{S} \equiv \mathbf{R} + \mathbf{d}$. (The body frame axes for $O''$ can also be oriented differently than those for $O'$.) With respect to this new frame, a point $\wp$ in the rigid body is described by position vector $\mathbf{r}''$, which is related to $\mathbf{r}'$ via $\mathbf{r}' = \mathbf{r}'' + \mathbf{d}$. Making this substitution into the first equation of (6.74), we have

$$\mathbf{v} = \boldsymbol{\omega} \times (\mathbf{r}'' + \mathbf{d}) + \dot{\mathbf{R}} = \boldsymbol{\omega} \times \mathbf{r}'' + \left(\boldsymbol{\omega} \times \mathbf{d} + \dot{\mathbf{R}}\right). \tag{6.75}$$

But in the new frame, we can write down

$$\mathbf{v} = \boldsymbol{\Omega} \times \mathbf{r}'' + \dot{\mathbf{S}}, \tag{6.76}$$

where $\boldsymbol{\Omega}$ denotes its angular velocity. Equating these last two expressions for $\mathbf{v}$ for all points in the body, we can conclude that

$$\dot{\mathbf{S}} = \boldsymbol{\omega} \times \mathbf{d} + \dot{\mathbf{R}}\,, \qquad \boldsymbol{\Omega} = \boldsymbol{\omega}\,. \tag{6.77}$$

Thus, the translational component of $\mathbf{v}$ changes if we change the body frame, but the rotational component is unaffected by such a change. In other words, the angular velocity $\boldsymbol{\omega}$ is a property of the motion of rigid body, and doesn't depend on the choice of body frame.

Finally, if there is a fixed point in the body that we can take as the origin of coordinates for *both* the inertial and body frames (so $\mathbf{R}(t) = \mathbf{0}$), then the equations for $\mathbf{v}$ and $\mathbf{a}$ simplify even more:

$$\mathbf{v} = \boldsymbol{\omega} \times \mathbf{r}'\,, \qquad \mathbf{a} = \dot{\boldsymbol{\omega}} \times \mathbf{r}' + \boldsymbol{\omega} \times (\boldsymbol{\omega} \times \mathbf{r}')\,. \tag{6.78}$$

Alternatively, we can interpret the velocities and accelerations in (6.78) as the rotational velocity and acceleration of the particles of the rigid body after we have subtracted off the translational motion of the body itself. The above expressions will be used repeatedly when discussing rigid body dynamics in Chap. 7.

## 6.6 Quaternion Representation of Rotations

To end this chapter on rigid body kinematics, we will discuss a representation of rotations that uses unit *quaternions* in place of Euler (or Tait-Bryan) angles to describe a general rotation. As we shall see below, the quaternion representation is similar to the axis-angle representation discussed in Sects. 6.3 and 6.4, but it doesn't require the use of *matrices* such as (6.52). Quaternions also avoid the problem of *gimbal lock* (Sect. 6.6.2), which is associated with a degeneracy in the Euler (or Tait-Bryan) parameterization of rotations. This has made it a particularly useful representation for video game designers who need to simulate realistic rigid body motion.

### *6.6.1 Quaternions*

**Quaternions** are effectively a generalization of complex numbers $z = x + \mathrm{i}y$, which have *three* "imaginary" components $\mathbf{i}$, $\mathbf{j}$, $\mathbf{k}$ that obey the rules of ordinary algebra and satisfy

$$\mathbf{i}^2 = \mathbf{j}^2 = \mathbf{k}^2 = \mathbf{ijk} = -1\,. \tag{6.79}$$

A general quaterion $q$ can be written as the sum

$$q = w + x\mathbf{i} + y\mathbf{j} + z\mathbf{k}, \tag{6.80}$$

where $(w, x, y, z)$ are real variables. From (6.79) one can show that

$$\mathbf{ij} = \mathbf{k}, \quad \mathbf{jk} = \mathbf{i}, \quad \mathbf{ki} = \mathbf{j}. \tag{6.81}$$

But one can also show

$$\mathbf{ji} = -\mathbf{k}, \quad \mathbf{kj} = -\mathbf{i}, \quad \mathbf{ik} = -\mathbf{j}, \tag{6.82}$$

so unlike multiplication of real numbers or complex numbers, multiplication of quaternions is *not* commutative.

**Exercise 6.10** Show that composition of rotations in 2-dimensions, e.g.,

$$\mathsf{R}(\theta) = \begin{bmatrix} \cos\theta & \sin\theta \\ -\sin\theta & \cos\theta \end{bmatrix}, \tag{6.83}$$

obeys exactly the same properties as multiplication of complex numbers, $z = x + \mathrm{i}y$, having unit magnitude. What's the mapping between these two spaces?

### 6.6.1.1 Vectors as Quaternions

For reasons that will become clear shortly, it is convenient to use a vector symbol, e.g., $\mathbf{v}$, to denote the pure "imaginary" component of a quaternion,

$$\mathbf{v} \equiv x\mathbf{i} + y\mathbf{j} + z\mathbf{k}, \tag{6.84}$$

so that a general quaternion can be written as

$$q = w + x\mathbf{i} + y\mathbf{j} + z\mathbf{k} \equiv w + \mathbf{v}. \tag{6.85}$$

In terms of this notation we can multiply two vector quaternions $\mathbf{u}$ and $\mathbf{v}$, using (6.79) and its consequences, (6.81) and (6.82). The result is

$$\mathbf{uv} = -\mathbf{u}\cdot\mathbf{v} + \mathbf{u}\times\mathbf{v}, \tag{6.86}$$

where

$$\begin{aligned} \mathbf{u}\cdot\mathbf{v} &= u_x v_x + u_y v_y + u_z v_z, \\ \mathbf{u}\times\mathbf{v} &= (u_y v_z - u_z v_y)\,\mathbf{i} + (u_z v_x - u_x v_z)\,\mathbf{j} + (u_x v_y - u_y v_x)\,\mathbf{k}, \end{aligned} \tag{6.87}$$

are the dot product (A.4) and cross product (A.5) of ordinary 3-dimensional vectors. (Note that **i**, **j**, **k** are playing the role here of the standard orthonormal basis vectors, which we have been denoting by $\hat{\mathbf{x}}$, $\hat{\mathbf{y}}$, $\hat{\mathbf{z}}$.) It is then fairly easy to extend (6.86) to general quaternions $q_1 \equiv s + \mathbf{u}$ and $q_2 \equiv t + \mathbf{v}$, which have "scalar" components $s$ and $t$:

$$(s + \mathbf{u})(t + \mathbf{v}) = st + s\mathbf{v} + t\mathbf{u} - \mathbf{u} \cdot \mathbf{v} + \mathbf{u} \times \mathbf{v} . \tag{6.88}$$

**Exercise 6.11** (a) Verify (6.86) and (6.88). (b) Show that (6.88) reduces to the familiar product of ordinary complex numbers if $q_1 = a + b\mathbf{i}$ and $q_2 = c + d\mathbf{i}$.

#### 6.6.1.2 Unit Quaternions

*Unit* quaternions $q = w + x\mathbf{i} + y\mathbf{j} + z\mathbf{k}$ are simply quaternions with unit norm,

$$|q|^2 \equiv w^2 + x^2 + y^2 + z^2 = 1 . \tag{6.89}$$

The set of all points $(w, x, y, z)$ satisfying (6.89) defines the **3-sphere** $S^3$. The 3-sphere is a generalization[5] of the 2-sphere $S^2$, which is defined by $x^2 + y^2 + z^2 = 1$. As we shall see below, it is convenient to parametrize the 3-sphere as follows:

$$w = \cos(\Psi/2) , \qquad (x, y, z) = \sin(\Psi/2)\ (n_x, n_y, n_z) , \tag{6.90}$$

where

$$\Psi \in [0, 2\pi] , \qquad n_x^2 + n_y^2 + n_z^2 = 1 . \tag{6.91}$$

This parametrization is illustrated graphically in Fig. 6.12, where one dimension is necessarily suppressed. The lines of "latitude", which look like circles in perspective, are actually 2-dimensional spheres of radius $\sin(\Psi/2)$, with a point on the 2-sphere corresponding to $\sin(\Psi/2)$ times the unit vector $\hat{\mathbf{n}} \equiv (n_x, n_y, n_z)$. The North pole corresponds to a rotation by 0°, while the equator corresponds to a rotation by 180°.

[5] A 2-sphere is the two-dimensional surface of a solid sphere (or ball) in three dimensions, while a 3-sphere is the three-dimensional surface of a solid ball in four dimensions.

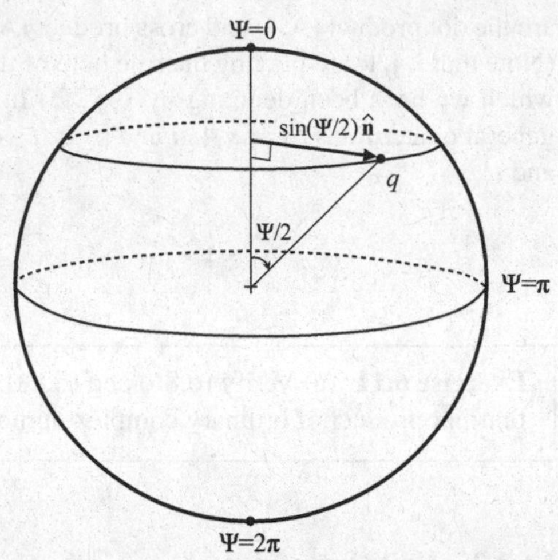


**Fig. 6.12** Parametrization of the 3-sphere $S^3$ (the space of unit quaternions) in terms of $\Psi \in [0, 2\pi]$ and a unit vector $\hat{\mathbf{n}}$ (i.e., a point on the unit 2-sphere $S^2$). Note that lines of "latitude", corresponding to a fixed value of $\Psi$, are actually 2-dimensional spheres of radius $\sin(\Psi/2)$, and not circles in perspective as they appear in the figure

#### 6.6.1.3 Unit Quaternions as Rotations

If the above parametrization of unit quaternions reminds you of the axis-angle $(\hat{\mathbf{n}}, \Psi)$ representation of rotations (Sects. 6.3 and 6.4), then you probably won't be too surprised to find out that unit quaternions can actually be thought of as *rotation operators*, which rotate vectors about $\hat{\mathbf{n}}$ through the angle $\Psi$. To demonstrate this explicity, we calculate the product $q\mathbf{v}q^{-1}$, where

$$q = \cos(\Psi/2) + \sin(\Psi/2)\,(n_x\mathbf{i} + n_y\mathbf{j} + n_z\mathbf{k}), \tag{6.92}$$

and

$$\mathbf{v} = v_x\mathbf{i} + v_y\mathbf{j} + v_z\mathbf{k}, \tag{6.93}$$

with $n_x^2 + n_y^2 + n_z^2 = 1$. Note that $q^{-1}$, which is the inverse of $q$, is given by (6.92) with $\Psi$ replaced by $-\Psi$. After a couple of lines of quaternion algebra, we find

$$q\mathbf{v}q^{-1} = \cos\Psi\,\mathbf{v} + (1 - \cos\Psi)\,(\mathbf{v}\cdot\mathbf{n})\hat{\mathbf{n}} + \sin\Psi\,(\hat{\mathbf{n}} \times \mathbf{v}), \tag{6.94}$$

which has *exactly* the same form as (6.49). So $\mathbf{v}' \equiv q\mathbf{v}q^{-1}$ is the new vector obtained from $\mathbf{v}$ under an active rotation about $\hat{\mathbf{n}}$ by $\Psi$.

Although it's not necessary to do so, we can construct an ordinary $3 \times 3$ rotation matrix in terms of the components of a unit quaternion. In terms of $\hat{\mathbf{n}}$ and $\Psi$, this matrix is identical to (6.51). In terms of $(w, x, y, z)$, one can show that

$$\mathsf{R}^{\text{active}} = \begin{bmatrix} w^2+x^2-y^2-z^2 & 2(xy-wz) & 2(xz+wy) \\ 2(yx+wz) & w^2-x^2+y^2-z^2 & 2(yz-wx) \\ 2(zx-wy) & 2(zy+wx) & w^2-x^2-y^2+z^2 \end{bmatrix}. \quad (6.95)$$

Since (6.95) depends quadratically on the components $(w, x, y, z)$ of $q$, we obtain the same rotation matrix if we use $-q$ in place of $q$. (One can also see this by noting that $q\mathbf{v}q^{-1}$ is independent of the sign of $q$.) Thus, $q$ and its **antipodal point** $-q$ correspond to the same rotation. Said another way, the unit quaternions $S^3$ are a *double-cover* of the set of rotations $SO(3)$. This will be described in more detail in the following subsection.

**Exercise 6.12** (a) Verify (6.94). (b) Also, verify that the transpose of (6.95) is the same as taking its inverse. [*Hint*: If $\Psi \rightarrow -\Psi$, then $w \rightarrow w$, while $(x, y, z) \rightarrow (-x, -y-z)$.]

### 6.6.2 Gimbal Lock and Parametrizations of $SO(3)$

One of the consequences of Euler's theorem (Sect. 6.3) is that the special orthogonal group $SO(3)$ is 3-dimensional: any element of $SO(3)$ has the form of a rotation about some axis $\hat{\mathbf{n}}$ through some angle $\Psi$, and hence can be characterized by 3 numbers. Using Euler (or Tait-Bryan) angles $(\phi, \theta, \psi)$ is one way of parametrizing this space, similar to the standard $(\theta, \phi)$ parametrization of the 2-dimensional sphere $S^2$. But like $(\theta, \phi)$, the Euler (or Tait-Bryan) angles are *degenerate*—i.e., different values of these angles don't always define different points in the space (similar to the North pole of a 2-sphere having $\theta = 0$ and $\phi$ equal to anything.) In the context of rotations and rigid body motion, such a degeneracy is called **gimbal lock**.

Physically, a *gimbal* is a pivoted support that allows rotational motion about a single axis. The classic example of a three-axis gimbal system is a *gyroscope*, a schematic of which is shown in Fig. 6.13. The three gimbals are the dark, medium, and light gray rings, with their corresponding rotation axes indicated by dashed lines. The gimbals fit one inside the other and rotate around the short "posts" that connect neighboring rings. Using Tait-Bryan angles $(\phi, \theta, \psi)$ in the $xyz$-convention to parameterize the yaw, pitch, and roll orientation of an object—e.g., the arrow shown in the figure—effectively imposes this gimbal structure on the allowed motion of the object. That is, we should imagine the arrow that is shown in the figure as *fixed* to the innermost (i.e., roll) gimbal of the gyroscope. Note also that different rotations of the gyroscope have a *hierarchy* associated with them: (i) A rotation around the yaw axis $\hat{\mathbf{n}}_\phi$ (which always points in the $z$-direction of the fixed inertial frame) carries the yaw, pitch, and roll gimbals with it; (ii) a rotation around the pitch axis $\hat{\mathbf{n}}_\theta$ (which always lies in the $xy$-plane of the inertial frame) carries the pitch and roll gimbals with it, but not the yaw gimbal; and (iii) a rotation about the roll axis $\hat{\mathbf{n}}_\psi$ (which can point in *any* direction), carries only the roll gimbal with it.

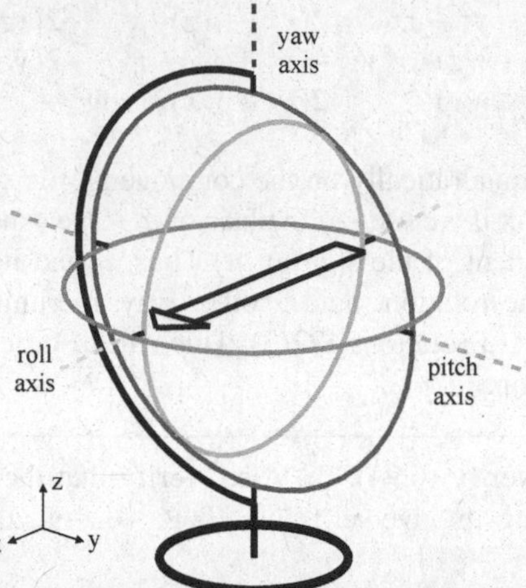


**Fig. 6.13** Schematic illustration of a gyroscope, which is an example of a 3-axis gimbal system. The gimbals are shown as three rings (dark, medium, and light gray), with their corresponding rotation axes indicated by dashed lines. The initial orientation of the gyroscope has the roll, pitch, and yaw axes directed along the $x$, $y$, and $z$ axes of the fixed inertial frame. The arrow object should be thought of as fixed to the innermost gimbal

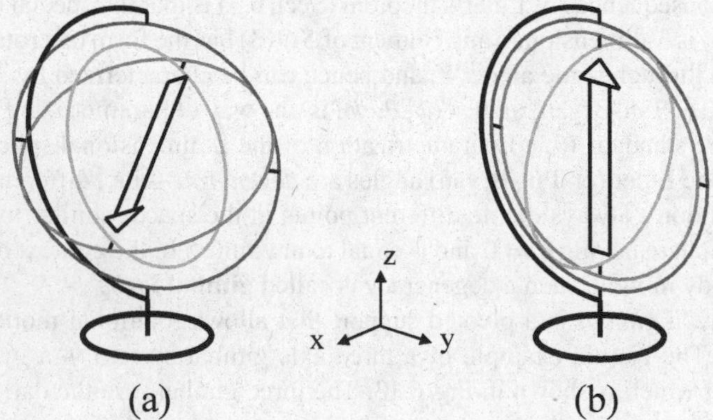


**Fig. 6.14** Normal and gimbal lock configurations of the gyroscope and arrow. Panel (a) In the "normal" configuration, the yaw, pitch, and roll rotation axes all point in different directions. Panel (b) In the gimbal lock configuration, the yaw and roll axes both point in the $z$-direction, corresponding to the loss of one degree of freedom

Figure 6.14 shows the result of two different rotations of the gyroscope and arrow from its initial configuration shown in Fig. 6.13. Panel (a) shows the final result of a sequence of yaw, pitch, and roll rotations by $\phi = 20°$, $\theta = 30°$, and $\psi = 30°$, respectively. This is a *normal* configuration in the sense that the yaw, pitch, and roll axes all point in *different* directions. Panel (b) show the result of a single pitch rotation by $\theta = -\pi/2$. For this case, both the yaw and roll axes are *aligned*—they point in the

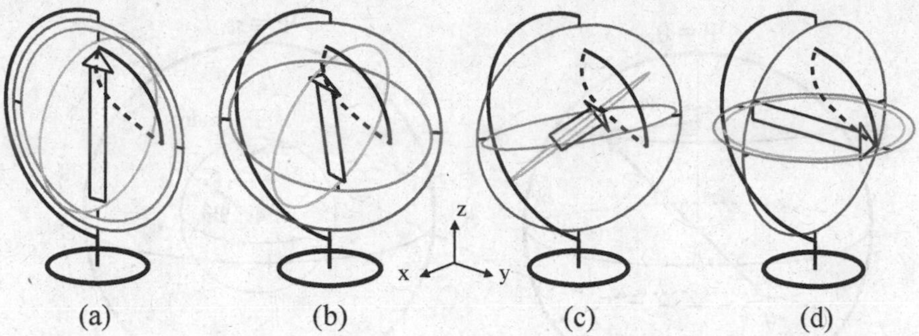


**Fig. 6.15** Sequence of gyroscope orientations for a rotation that takes the arrow from its initial vertical orientation, shown in panel (a), to its final horizontal orientation, shown in panel (d). The solid curve, which is an arc of a great circle, is the path the arrow head would take if one could rotate the arrow directly about the $x$-axis. The dashed curve is the actual path that the arrow head takes as determined by a sequence of yaw, pitch, and roll rotations

$z$-direction. In this orientation, subsequent yaw and roll rotations are *degenerate*—i.e., they produce the same rotation of the arrow. This condition is called **gimbal lock.** Although nothing is physically locked in this orientation, one degree of freedom has been lost; no rotation can change the yaw (heading) of the arrow. To get out of the gimbal lock state, one must first do a pitch rotation to move the roll axis away from vertical. Then a yaw rotation can change the heading of the arrow.

One problem with gimbal lock is that the rotation of an object out of a gimbal lock state is quite different than it would be in the absence of the gimbal structure imposed by the Tait-Bryan (or Euler) angle parametrization. To illustrate this, let's suppose that we want to rotate the arrow from the gimbal lock configuration described above (shown again in panel (a) of Fig. 6.15) so that the arrow ultimately points along the $y$-axis as shown in panel (d) of Fig. 6.15. Although it would be simplest to perform this rotation by rotating the arrow around the $x$-axis by $-90°$ (shown in the panels of Fig. 6.15 by the solid curve, which is an arc of a great circle), such a rotation axis is not available given the initial orientation of the gyroscope and its associated gimbal structure. Instead, one must perform a sequence of yaw, pitch, and roll rotations, which leads to the motion shown by the dashed curve in Fig. 6.15. This "unexpected" motion of the arrow from panel (a) to (d) is why gimbal lock is problematic for video-game developers who want to simulate realistic rigid body motion. An axis-angle or quaternion parametrization of the orientation of the arrow would not be straight-jacketed by the implicit gimbal structure imposed by the Tait-Bryan (or Euler) angle parametrization.

#### 6.6.2.1 Different Parametrizations of $SO(3)$

To avoid gimbal lock is relatively simple: one should use a different set of coordinates in the neighborhood of the original coordinate degeneracies. This can be done

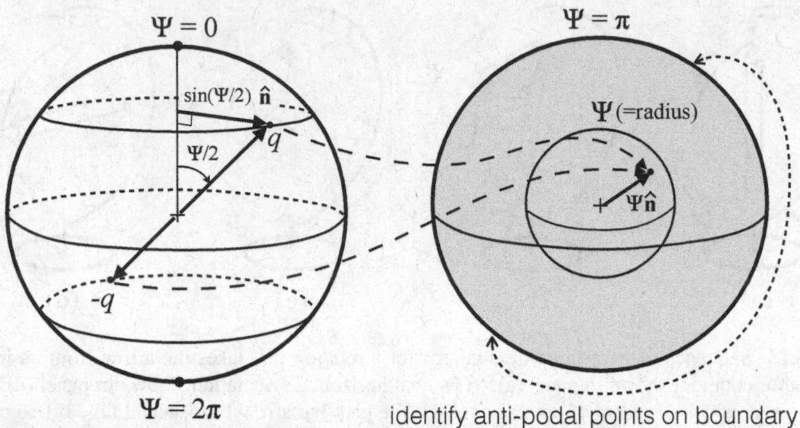


**Fig. 6.16** The set of unit quaternions $S^3$ is a double cover of the space of rotations $SO(3)$. The antipodal points $q$ and $-q$ in $S^3$ map to the same rotation $\Psi\,\hat{\mathbf{n}}$ in $SO(3)$. See text for more details

physically by adding a *redundant* degree of freedom (i.e., an additional gimbal) that one "activates" whenever the original gimbal lock condition is approached. Mathematically, this corresponds to using a set of coordinates that *covers* the relevant space. For example, for the 2-sphere, we can avoid the $\theta = 0$ and $\theta = \pi$ degeneracies at the North and South poles by working with Cartesian coordinates $(x, y, z)$ that satisfy the constraint

$$x^2 + y^2 + z^2 = 1\,. \tag{6.96}$$

Then the North pole is given uniquely by $(0, 0, 1)$ and the South pole by $(0, 0, -1)$. The same can be done for the space of rotations $SO(3)$. That is, instead of working with the Euler or Tait-Bryan angular coordinates $(\phi, \theta, \psi)$, one should work with Cartesian coordinates $(w, x, y, z)$ that satisfy the constraint

$$w^2 + x^2 + y^2 + z^2 = 1\,. \tag{6.97}$$

Recall that such coordinates define the space of unit quaternions, where $w$ is associated with the rotation angle. This is the same as the 3-sphere $S^3$ (Sect. 6.6.1.2).

As mentioned at the end of Sect. 6.6.1.3, the set of unit quaternions $S^3$ is actually a **double-cover** of the space of rotations $SO(3)$. This is illustrated graphically in Fig. 6.16. On the left of the figure is the 3-sphere $S^3$, where one dimension has necessarily been suppressed (See also Fig. 6.12). On the right of the figure is the space of rotations $SO(3)$, represented by a solid *ball* in 3 dimensions of radius $\pi$, and where antipodal points on the boundary are identified. The radii of the 2-spheres foliating this 3-d ball correspond to the angle of rotation $\Psi \in [0, \pi]$, and the unit vector $\hat{\mathbf{n}}$, which points from the center of the ball out to some point in the 3-d ball, corresponds to the axis of rotation. Since antipodal points $q$ and $-q$ in $S^3$ correspond to the same physical rotation (a rotation around $\hat{\mathbf{n}}$ by $\Psi$ is the same as a rotation around

$-\hat{\mathbf{n}}$ by $2\pi - \Psi$) they map to the same point in $SO(3)$. This is shown in the figure by the two dashed lines connecting $q$ and $-q$ to $\Psi\hat{\mathbf{n}}$.

## Suggested References

*Full references are given in the bibliography at the end of the book.*

Goldstein et al. (2002): One of the classic texts on classical mechanics. Our presentation of rigid body motion in this and the following chapter follows the basic structure of Chaps. 4 and 5 in Goldstein. Note that Goldstein uses the $zxz$ convention for Euler angle rotations, as opposed to the $zyz$ convention, which we have chosen to use.

Griffiths (2005): An excellent introduction to quantum mechanics appropriate for both undergraduate and beginning graduate students. Chapters 4 and 5 have a more detailed description of Pauli spin matrices and the spin-statistic theorem, which are briefly mentioned in Problems 6.5 and 6.6.

Kuipers (1999): An detailed discussion of quaternions and their connection to rotations, with applications to aerospace guidance systems and virtual reality. Appropriate for advanced undergraduates in mathematics, engineering, or the physical sciences.

## Additional Problems

**Problem 6.1** Revisit Example 2.2 from Chap. 2, which describes a sphere that rolls without slipping or pivoting on a horizontal surface.

(a) Using Euler angles $(\phi, \theta, \psi)$ to parametrize the angular degrees of freedom of the sphere, explicitly write down the constraints relating the coordinate differentials of the Euler angles and the center-of-mass coordinates $(x, y)$.
(b) Using techniques developed in Sect. 2.2.3, show that the system of constraints from part (a) is *non-holonomic*, consistent with the illustration shown in Fig. 2.1.

**Problem 6.2** Some authors, e.g., Goldstein et al. (2002), define the Euler angle rotation matrix $\mathsf{R}(\phi, \theta, \psi)$ using the $zxz$-convention—i.e.,

$$\mathsf{R}(\phi, \theta, \psi) \equiv \mathsf{R}_z(\psi)\mathsf{R}_x(\theta)\mathsf{R}_z(\phi)\,. \tag{6.98}$$

Show that, for this convention,

$$\mathsf{R}(\phi, \theta, \psi) = \begin{bmatrix} c\phi c\psi - c\theta s\phi s\psi & s\phi c\psi + c\theta c\phi s\psi & s\theta s\psi \\ -c\phi s\psi - c\theta s\phi c\psi & -s\phi s\psi + c\theta c\phi c\psi & s\theta c\psi \\ s\theta s\phi & -s\theta c\phi & c\theta \end{bmatrix}. \tag{6.99}$$

The above two expressions should be compared to (6.35) and (6.40), which are for the $zyz$-convention.

**Problem 6.3** Here you will rederive (6.55), which relates $\Psi$ to the Euler angles $(\phi, \theta, \psi)$, by using the fact that the trace of a matrix is invariant under a similarity transformation.

(a) First show that

$$1 + \operatorname{Tr} \mathsf{R}(\phi, \theta, \psi) = 4 \cos^2\left(\frac{\theta}{2}\right) \cos^2\left(\frac{\phi + \psi}{2}\right), \tag{6.100}$$

where $\mathsf{R}(\phi, \theta, \psi)$, given by (6.40), is the Euler angle representation of a rotation.

(b) Then show that

$$\mathsf{S}\mathsf{R}_{\hat{\mathbf{n}}}(\Psi)\mathsf{S}^{-1} = \begin{bmatrix} \cos\Psi & \sin\Psi & 0 \\ -\sin\Psi & \cos\Psi & 0 \\ 0 & 0 & 1 \end{bmatrix}, \tag{6.101}$$

where $\mathsf{S}$ is a similarity transformation to a new set of coordinates $(x', y', z')$ with the rotation axis $\hat{\mathbf{n}}$ aligned along $\hat{\mathbf{z}}'$.

(c) Using the previous result and the invariance of the trace of a matrix under a similarity transformation, show that

$$\cos\Psi = \frac{1}{2}\left(\operatorname{Tr} \mathsf{R}_{\hat{\mathbf{n}}}(\Psi) - 1\right). \tag{6.102}$$

(d) Finally, since $\mathsf{R}(\phi, \theta, \psi)$ and $\mathsf{R}_{\hat{\mathbf{n}}}(\Psi)$ both represent the *same* rotation, their traces are equal. Thus, combine (6.100) and (6.102) to obtain (6.55).

**Problem 6.4** Allowing *complex* solutions, find the eigenvectors and eigenvalues of the matrix

$$\begin{bmatrix} \cos\Psi & \sin\Psi & 0 \\ -\sin\Psi & \cos\Psi & 0 \\ 0 & 0 & 1 \end{bmatrix}. \tag{6.103}$$

You should find eigenvalues

$$1, \quad e^{i\Psi}, \quad e^{-i\Psi}, \tag{6.104}$$

and corresponding normalized eigenvectors

$$\begin{bmatrix} 0 \\ 0 \\ 1 \end{bmatrix}, \quad \frac{1}{\sqrt{2}}\begin{bmatrix} 1 \\ i \\ 0 \end{bmatrix}, \quad \frac{1}{\sqrt{2}}\begin{bmatrix} 1 \\ -i \\ 0 \end{bmatrix}. \tag{6.105}$$

For what values of $\Psi$ are the eigenvalues and eigenvectors all *real*. What are the corresponding eigenvectors for that case? (Thus, ignoring these two special cases, Euler's theorem for rigid body motion tells us that a general element of $SO(3)$ admits *one and only one* real eigenvector, and it has eigenvalue 1.)

**Problem 6.5** The **Pauli spin matrices** are defined by

$$\sigma_x \equiv \begin{bmatrix} 0 & 1 \\ 1 & 0 \end{bmatrix}, \quad \sigma_y \equiv \begin{bmatrix} 0 & -\mathrm{i} \\ \mathrm{i} & 0 \end{bmatrix}, \quad \sigma_z \equiv \begin{bmatrix} 1 & 0 \\ 0 & -1 \end{bmatrix}. \tag{6.106}$$

(a) Show that they obey the commutation relations

$$\left[\frac{-\mathrm{i}}{2}\sigma_i, \frac{-\mathrm{i}}{2}\sigma_j\right] = \sum_k \varepsilon_{ijk}\frac{-\mathrm{i}}{2}\sigma_k, \qquad i, j = 1, 2, 3, \tag{6.107}$$

which have the same form as (6.66), Exercise 6.8. Thus, like $\mathsf{L}_x$, $\mathsf{L}_y$, $\mathsf{L}_z$, the Pauli spin matrices (times $-\mathrm{i}/2$) can be thought of as generators of infinitesimal rotations in three-dimensions. They also arise when describing spin 1/2 particles in quantum mechanics, see e.g., Griffiths (2005).

(b) Show that together with the $2 \times 2$ unit matrix, $-\mathrm{i}\sigma_x$, $-\mathrm{i}\sigma_y$, and $-\mathrm{i}\sigma_z$ span the space (see Appendix D. 2) of **special unitary** $2 \times 2$ **matrices**, denoted $SU(2)$. *Hint:* The general element of $SU(2)$ can be written

$$\begin{bmatrix} a & b \\ -b^* & a^* \end{bmatrix}, \tag{6.108}$$

where $a$ and $b$ are complex numbers with $|a|^2 + |b|^2 = 1$.

(c) Show that $-\mathrm{i}\sigma_x$, $-\mathrm{i}\sigma_y$, and $-\mathrm{i}\sigma_z$ obey the same multiplicative relations as the quaternions **i**, **j**, and **k**.

**Problem 6.6** (a) Show that a continuous sequence of rotations about some fixed axis $\hat{\mathbf{n}}$, starting at $\Psi = 0$ (the identity) and ending at $\Psi = 2\pi$, defines a *closed curve* in the space of rotations $SO(3)$, and that this curve cannot be shrunk to a point. (b) Show that if you go around this closed curve *twice* (for a total rotation by $4\pi$), the resulting closed curve *can* be shrunk to a point. (*Hint*: Consider what this curve looks like in the space $S^3$ with antipodal points identified, see Fig. 6.16.) This result has important physical consequences related to the the *spin-statistics theorem* in quantum mechanics, see, e.g., Griffiths (2005).

# Chapter 7
# Rigid Body Dynamics

In the previous chapter, we developed the kinematical framework needed to describe rigid body motion. We discussed various representations of rotations, and how they can be used to describe the orientation of a rigid body relative to some fixed inertial frame. Here we extend our analysis of rigid body motion to include *dynamics*—i.e., the forces and torques that produce the complicated translational and rotational motion of a rigid body as it moves through space. In particular, we derive **Euler's equations for rigid body motion**, which extend the familiar freshman physics equation $\tau = I\alpha$ (for 2-dimensional rotational motion around a *fixed axis* in space) to general rotations in three dimensions around an axis that can change its orientation in response to external forces. We then apply Euler's equation to several examples: we analyze the motion of spinning tops and calculate the period of precession of the Earth's axis of rotation due to the gravitational forces exerted on Earth by the Sun and the Moon.

## 7.1 Angular Momentum and Kinetic Energy of a Rigid Body

In order to generalize the 2-dimensional rotation equation $\tau = I\alpha$ to rotational motion in three dimensions, we need to extend the definition of $I$ (the **rotational inertia** about a *fixed* axis) to a quantity that specifies the rotational inertia of a rigid body around an *arbitrary* direction in space. To do that, we will first write down an expression for the total angular momentum $\mathbf{L}$ of a rigid body and relate it to the instantaneous angular velocity vector $\boldsymbol{\omega}$. The result we find will allow us to generalize the relation $L = I\omega$ for 2-dimensional rotational motion to three dimensions.



M.J. Benacquista and J.D. Romano, *Classical Mechanics*, Undergraduate Lecture Notes in Physics, https://doi.org/10.1007/978-3-319-68780-3_7

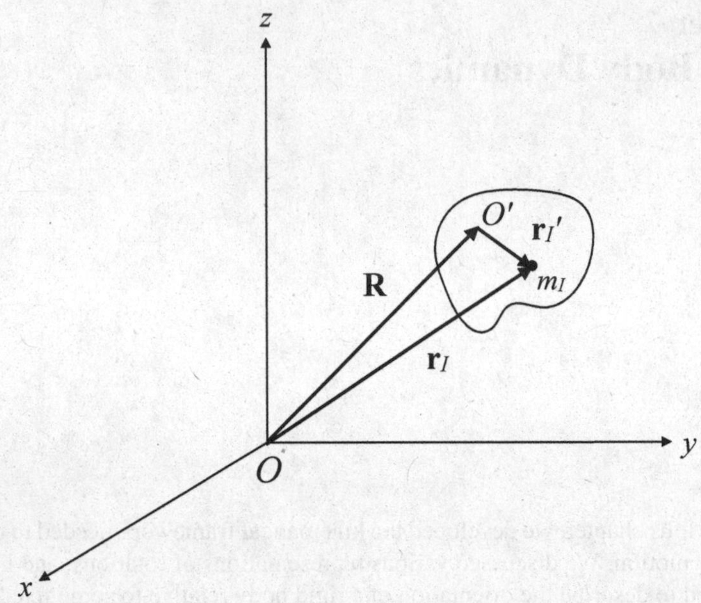


**Fig. 7.1** Definition of $\mathbf{r}'_I$, the position vector of mass point $m_I$ in a rigid body relative to a point $O'$ fixed in the body

So let $O$: $(x, y, z)$ denote an inertial reference frame, and let $\mathbf{R}$ be the position vector of some point $O'$ fixed in the rigid body.[1] Define $\mathbf{r}'_I$, the position vector of mass point $m_I$ relative to $O'$, via $\mathbf{r}_I = \mathbf{R} + \mathbf{r}'_I$ as shown in Fig. 7.1. Then using $\dot{\mathbf{r}}_I = \dot{\mathbf{R}} + \dot{\mathbf{r}}'_I$, it is fairly easy to show that the total kinetic energy $T \equiv \sum_I \frac{1}{2} m_I |\dot{\mathbf{r}}_I|^2$ can be written as

$$T = T' + \frac{1}{2} M |\dot{\mathbf{R}}|^2 + M\dot{\mathbf{R}} \cdot (\dot{\mathbf{R}}_{\text{COM}} - \dot{\mathbf{R}}) , \tag{7.1}$$

where $T'$ is the total kinetic energy relative to $O'$, and

$$M \equiv \sum_I m_I , \qquad \mathbf{R}_{\text{COM}} \equiv \frac{1}{M} \sum_I m_I \mathbf{r}_I , \tag{7.2}$$

are the *total mass* and the position vector of the center of mass of the rigid body, respectively. (In the above expressions, "dot" denotes time derivative with respect to the inertial frame; *not* with respect to the body frame.) Note that the second term on the right-hand side of (7.1) is the total kinetic energy of the rigid body treated as a

[1] Although we will often take $O'$ to be at the center of mass of the body, it doesn't have to be there, so we will keep things general at this stage of the calculation.

single mass point of total mass $M$ located at $O'$. The third term goes away if $O'$ is chosen to be at the center of mass.

Similarly, one can show that the total angular momentum $\mathbf{L} \equiv \sum_I m_I \mathbf{r}_I \times \dot{\mathbf{r}}_I$ of the rigid body can be written as

$$\mathbf{L} = \mathbf{L}' + M\mathbf{R} \times \dot{\mathbf{R}} + M(\mathbf{R}_{\text{COM}} - \mathbf{R}) \times \dot{\mathbf{R}} + M\mathbf{R} \times (\dot{\mathbf{R}}_{\text{COM}} - \dot{\mathbf{R}}), \quad (7.3)$$

where $\mathbf{L}'$ is the angular momentum of the mass points relative to $O'$. Similar to the expression for $T$, the second term on the right-hand side of (7.3) is the total angular momentum of the rigid body treated as a single mass point of total mass $M$ located at $O'$, and the third and fourth terms go away if $O'$ is chosen to be at the center of mass.

In what follows, we will consider only *two* different choices for the location of $O'$ in the rigid body: (i) the center of mass, for which the expressions for both $T$ and $\mathbf{L}$ simplify considerably, or (ii) a point in the rigid body which is *fixed with respect to an inertial frame*, like the fixed support of a gyroscope or a spinning top. When discussing rigid body motion, we will be primarily interested in $T'$ and $\mathbf{L}'$, which are the total kinetic energy and angular momentum relative to $O'$. So for the rest of this chapter we will drop the primes to simplify the notation. All position vectors, kinetic energies, angular momenta, etc. are defined respect to $O'$ unless stated otherwise.

**Exercise 7.1** Verify expressions (7.1) and (7.3) for $T$ and $\mathbf{L}$.

## 7.2 Rotational Inertia Tensor, Principal Axes

Consider the expression for the total angular momentum of the rigid body rotating with instantaneous angular velocity $\boldsymbol{\omega}$ about an origin $O$ fixed in the rigid body:

$$\mathbf{L} = \sum_I m_I \mathbf{r}_I \times \dot{\mathbf{r}}_I, \quad (7.4)$$

(remember that we have dropped the primes on $O$ and $\mathbf{r}_I$ to simplify the notation). Since $\dot{\mathbf{r}}_I$ denotes time derivative with respect to the inertial (i.e., space) frame, we have [2]

$$\dot{\mathbf{r}}_I \equiv \left(\frac{d\mathbf{r}_I}{dt}\right)_{\text{s}} = \left(\frac{d\mathbf{r}_I}{dt}\right)_{\text{b}} + \boldsymbol{\omega} \times \mathbf{r}_I = \boldsymbol{\omega} \times \mathbf{r}_I, \quad (7.5)$$

[2]See Sect. 1.5.2 and, in particular, (1.68). There we used the terminology *fixed* and *rotating* for the inertial and non-inertial (rotating) reference frames. Here we will use the terminology *space* and *body* for the these two reference frames, with corresponding subscripts 's' and 'b'.

where the last equality follows from the fact that the mass points are fixed with respect to the body. Using this result and the vector triple product identity (A.10):

$$\mathbf{A} \times (\mathbf{B} \times \mathbf{C}) = \mathbf{B}(\mathbf{A} \cdot \mathbf{C}) - \mathbf{C}(\mathbf{A} \cdot \mathbf{B}) , \tag{7.6}$$

one can show that

$$\mathbf{L} = \sum_I m_I \left( \boldsymbol{\omega} r_I^2 - \mathbf{r}_I \left( \mathbf{r}_I \cdot \boldsymbol{\omega} \right) \right) . \tag{7.7}$$

The above expression for $\mathbf{L}$ is linear in $\boldsymbol{\omega}$. As such, we can write this relation as a matrix equation $\mathsf{L} = \mathsf{I}\,\omega$ or, equivalently,

$$L_i = \sum_j I_{ij}\omega_j , \qquad I_{ij} \equiv \sum_I m_I (\delta_{ij}\, r_I^2 - r_{Ii} r_{Ij}) , \tag{7.8}$$

where the components $I_{ij}$ and $\omega_i$ are with respect to an arbitrary set of basis vectors in the body frame. Later on, we will take the basis vectors to be the **principal axes** of the rigid body, for which $I_{ij}$ becomes diagonal. For a continuous mass distribution,

$$I_{ij} = \int \mathrm{d}V \, \rho(\mathbf{r}) (\delta_{ij}\, r^2 - r_i r_j) , \tag{7.9}$$

where $\rho(\mathbf{r})$ is the mass density of the rigid body.

The $3 \times 3$ matrix $\mathsf{I}$ with components $I_{ij}$ is called the **rotational inertia tensor** (or, more simply, inertia tensor). This matrix should be distinguished from the **moment of inertia**

$$I(\hat{\mathbf{n}}) \equiv \sum_{i,j} n_i I_{ij} n_j , \tag{7.10}$$

which is a scalar quantity associated with a particular axis $\hat{\mathbf{n}}$. Note that the moment of inertia can also be written as

$$I(\hat{\mathbf{n}}) = \sum_I m_I \left( r_I^2 - (\mathbf{r}_I \cdot \hat{\mathbf{n}})^2 \right) = \sum_I m_I |\mathbf{r}_I \times \hat{\mathbf{n}}|^2 = \sum_I m_I r_I^2 \sin^2 \theta_I , \tag{7.11}$$

where $\theta_I$ is the angle between $\mathbf{r}_I$ and $\hat{\mathbf{n}}$. This is just the usual expression of the moment of inertia as the sum of the masses times their squared (perpendicular) distances from the axis of rotation—i.e.,

$$I(\hat{\mathbf{n}}) = \sum_I m_I d_I^2 , \tag{7.12}$$

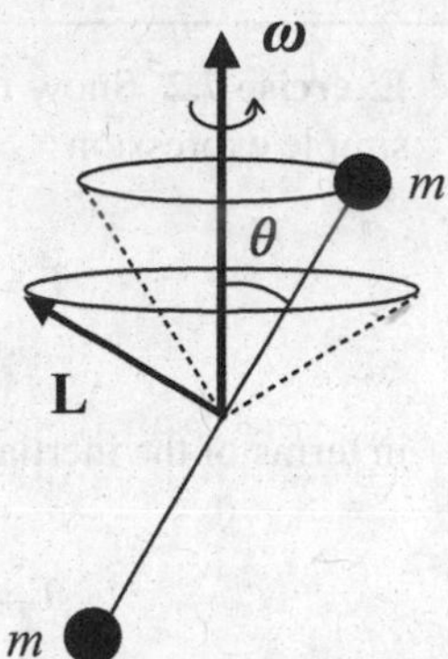


**Fig. 7.2** Rotating dumbbell consisting of two mass points, each of mass $m$. The angular momentum vector $\mathbf{L}$ is perpendicular to the line connecting the two masses. The angular velocity vector $\boldsymbol{\omega}$ points along the rotational axis, which for this case is not aligned with $\mathbf{L}$

where $d_I \equiv r_I \sin\theta_I$. This last expression for $I(\hat{\mathbf{n}})$ shows that the moment of inertia about an axis $\hat{\mathbf{n}}$ is *independent* of the choice of origin $O$ anywhere on the axis.

So we see that the matrix $I_{ij}$ and the scalar $I(\hat{\mathbf{n}})$ generalize the $I$ in $\tau = I\alpha$ and $L = I\omega$ to three dimensions. But note that since $I_{ij}$ is a matrix and not a single number, the angular momentum vector $L_i = \sum_j I_{ij}\omega_j$ need *not* point in the same direction as $\omega_i$, as illustrated in the following example.

***Example 7.1*** Consider a dumbbell that spins around an axis that makes an angle $\theta$ with respect to the symmetry axis of the dumbbell (i.e., the line connecting the two masses), as shown in Fig. 7.2. Then the angular velocity vector $\boldsymbol{\omega}$ is directed along the rotation axis, while $\mathbf{L}$ is directed perpendicular to the symmetry axis, lying in the plane spanned by $\boldsymbol{\omega}$ and the symmetry axis. This is most simply seen by using

$$\mathbf{L} = \sum_I m_I \mathbf{r}_I \times \dot{\mathbf{r}}_I = \sum_I m_I \mathbf{r}_I \times (\boldsymbol{\omega} \times \mathbf{r}_I)\,, \tag{7.13}$$

and the right-hand rule to determine the direction of the cross products $\boldsymbol{\omega} \times \mathbf{r}_I$ and then $\mathbf{r}_I \times (\boldsymbol{\omega} \times \mathbf{r}_I)$. Note that $\mathbf{L}$ precesses around the angular velocity vector $\boldsymbol{\omega}$ as the dumbbell rotates. Since $\mathbf{L}$ changes with time, a torque ($\boldsymbol{\tau} = d\mathbf{L}/dt$) is needed to sustain this rotational motion with $\boldsymbol{\omega}$ constant. We will talk more about torques later. □

**Exercise 7.2** Show that the total kinetic energy $T$ of a rigid body also has a simple expression

$$T = \frac{1}{2}\boldsymbol{\omega} \cdot \mathbf{L} = \frac{1}{2}\sum_{i,j} \omega_i I_{ij} \omega_j = \frac{1}{2} I(\hat{\mathbf{n}})\omega^2 , \tag{7.14}$$

in terms of the inertia tensor $I_{ij}$, where $\boldsymbol{\omega} = \omega\hat{\mathbf{n}}$. *Hint*: Write

$$T = \frac{1}{2}\sum_I m_I |\dot{\mathbf{r}}_I|^2 = \frac{1}{2}\sum_I m_I |\boldsymbol{\omega} \times \mathbf{r}_I|^2 , \tag{7.15}$$

and then use the scalar and vector triple product identities (A.9) and (A.10) to evaluate $|\boldsymbol{\omega} \times \mathbf{r}_I|^2 = (\boldsymbol{\omega} \times \mathbf{r}_I) \cdot (\boldsymbol{\omega} \times \mathbf{r}_I)$.

**Exercise 7.3** Extend the previous exercise to allow for *translational motion* of the rigid body. Assume that the origin of the body frame is located at the center of mass of the rigid body, specified by position vector $\mathbf{R}$ with respect to the fixed (inertial) frame. You should find

$$T = \frac{1}{2} M V^2 + \frac{1}{2}\sum_{i,j} \omega_i I_{ij} \omega_j , \tag{7.16}$$

where $M$ is the total mass of the rigid body and $\mathbf{V} \equiv \dot{\mathbf{R}}$ is the velocity of the center of mass. Thus, with respect to a body frame whose origin lies at the center of mass of the body, the kinetic energy splits into two parts: (i) a part for translational motion, which has the same form as if all of the mass of the body were located at its center of mass, and (ii) a part for rotational motion about the instantaneous angular velocity vector $\boldsymbol{\omega}$ passing through the center of mass.

### *7.2.1 Parallel-Axis Theorem*

From time to time, we will need to calculate the moment of inertia $I(\hat{\mathbf{n}})$ of a rigid body around an axis $\hat{\mathbf{n}}$ passing through some origin $O$. It turns out that if we already know the moment of inertia of the rigid body around a *parallel* axis passing through the *center of mass* $O'$, which we will denote by $I_{\mathrm{COM}}(\hat{\mathbf{n}})$, then there is a simple formula relating the two:

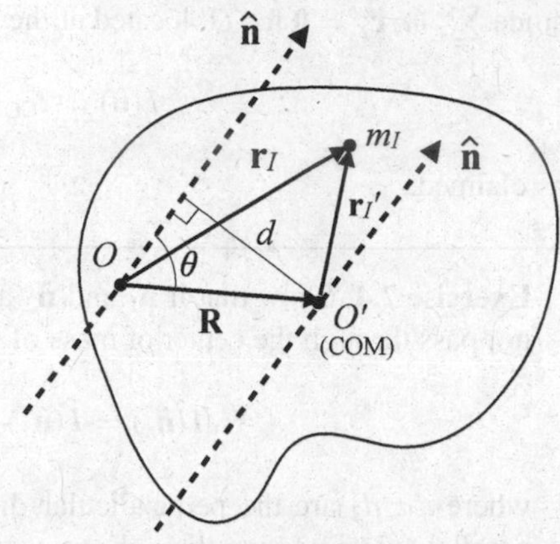


**Fig. 7.3** Parallel axes through $O$ and $O'$, the center of mass (COM) of the rigid body. Mass point $m_I$ is described by position vectors $\mathbf{r}_I$ and $\mathbf{r}'_I$ with respect to these two different origins. $d \equiv R\sin\theta$ is the perpendicular distance between the two axes

$$I(\hat{\mathbf{n}}) = I_{\rm COM}(\hat{\mathbf{n}}) + Md^2\,, \tag{7.17}$$

where $M$ is the total mass of the rigid body, and $d \equiv R\sin\theta$ is the perpendicular distance between the two axes as shown in Fig. 7.3. This result is called the **parallel-axis theorem**, which we prove below.

*Proof* Using the second expression for $I(\hat{\mathbf{n}})$ in (7.11), we can write

$$I(\hat{\mathbf{n}}) = \sum_I m_I |\mathbf{r}_I \times \hat{\mathbf{n}}|^2\,. \tag{7.18}$$

From Fig. 7.3 we see that $\mathbf{r}_I = \mathbf{R} + \mathbf{r}'_I$, which implies

$$|\mathbf{r}_I \times \hat{\mathbf{n}}|^2 = |(\mathbf{R} + \mathbf{r}'_I) \times \hat{\mathbf{n}}|^2 = |\mathbf{R} \times \hat{\mathbf{n}}|^2 + |\mathbf{r}'_I \times \hat{\mathbf{n}}|^2 + 2(\mathbf{R} \times \hat{\mathbf{n}}) \cdot (\mathbf{r}'_I \times \hat{\mathbf{n}}) \tag{7.19}$$

Substituting these terms into (7.18), the first term on the right-hand side gives

$$\sum_I m_I |\mathbf{R} \times \hat{\mathbf{n}}|^2 = \sum_I m_I\, R^2 \sin^2\theta = MR^2\sin^2\theta = Md^2\,, \tag{7.20}$$

while the second term gives

$$\sum_I m_I |\mathbf{r}'_I \times \hat{\mathbf{n}}|^2 = I_{\rm COM}(\hat{\mathbf{n}})\,. \tag{7.21}$$

The third term gives

$$\sum_I m_I\, 2(\mathbf{R} \times \hat{\mathbf{n}}) \cdot (\mathbf{r}'_I \times \hat{\mathbf{n}}) = 2(\mathbf{R} \times \hat{\mathbf{n}}) \cdot \left[\left(\sum_I m_I \mathbf{r}'_I\right) \times \hat{\mathbf{n}}\right] = \mathbf{0}\,, \tag{7.22}$$

since $\sum_I m_I \mathbf{r}'_I = \mathbf{0}$ for $O'$ located at the center of mass. Thus,

$$I(\hat{\mathbf{n}}) = I_{\rm COM}(\hat{\mathbf{n}}) + Md^2 \tag{7.23}$$

as claimed. □

**Exercise 7.4** Show that if $\hat{\mathbf{n}}_1$ and $\hat{\mathbf{n}}_2$ are *any* two parallel axes (i.e., they need not pass through the center of mass of the body), then

$$I(\hat{\mathbf{n}}_2) = I(\hat{\mathbf{n}}_1) + M(d_2^2 - d_1^2), \tag{7.24}$$

where $d_1$, $d_2$ are the perpendicular distances between the axes $\hat{\mathbf{n}}_1$, $\hat{\mathbf{n}}_2$ and a parallel axis $\hat{\mathbf{n}}$ passing through the center of mass.

**Exercise 7.5** Calculate the moments of inertia $I(\hat{\mathbf{n}})$ for the objects and axes shown in Fig. 7.4. (You should find: (a) $MR^2$, (b) $\frac{1}{2}MR^2$, (c) $\frac{2}{3}MR^2$, (d) $\frac{2}{5}MR^2$, (e) $\frac{1}{12}ML^2$, (f) $\frac{1}{3}ML^2$.)

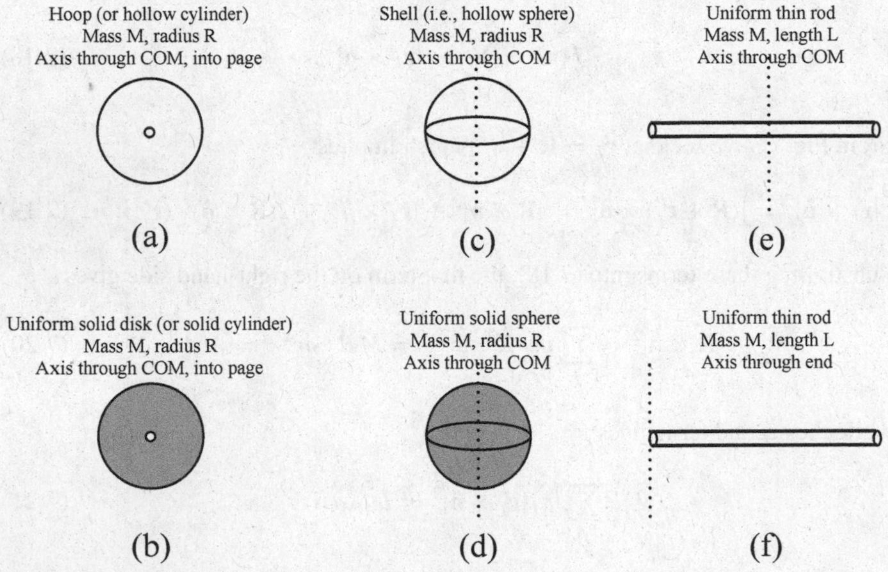


**Fig. 7.4** Several different objects and axes for calculating moments of inertia $I(\hat{\mathbf{n}})$

### 7.2.2 *Principal Axes*

Since the inertia tensor $\mathsf{I}$ is a real, symmetric matrix, it can be *diagonalized* by finding its eigenvalues and eigenvectors, and then switching to the basis of normalized eigenvectors (Appendix D.5.2). The normalized eigenvectors, which we will denote by $\hat{\mathbf{n}}_1$, $\hat{\mathbf{n}}_2$, $\hat{\mathbf{n}}_3$ are called the **principal axes** of the rigid body. The corresponding eigenvalues $I_1$, $I_2$, $I_3$ are called the **principal moments of inertia** of the rigid body. Note also that $I_{ij}$ is a non-negative matrix since $\sum_{i,j} I_{ij} v_i v_j \geq 0$ for all vectors $\mathbf{v}$. Thus, the principal moments of inertia $I_i$ satisfy $I_i \geq 0$.

Expressions for the inertia tensor $\mathsf{I}$, the angular momentum $\mathbf{L}$, and the kinetic energy $T$ of a rigid body greatly simplify when written in terms of components with respect to the principal axis basis:

$$I_i = \hat{\mathsf{n}}_i^T \, \mathsf{I} \, \hat{\mathsf{n}}_i \,, \qquad I_{ij} = I_i \, \delta_{ij} \,, \qquad L_i = I_i \omega_i \,, \qquad T = \frac{1}{2} \sum_i I_i \omega_i^2 \,. \tag{7.25}$$

In what follows, we will work in the principal axis basis unless stated otherwise. Note that for simple geometric objects, it is often possible to "guess" the directions of the principal axes using symmetry arguments, such as for a uniform ellipsoid shown in Fig. 7.5. But note that changing the location of the origin $O$ results in different components of the inertia tensor, and hence different principal axes and principal moments of inertia, in general.

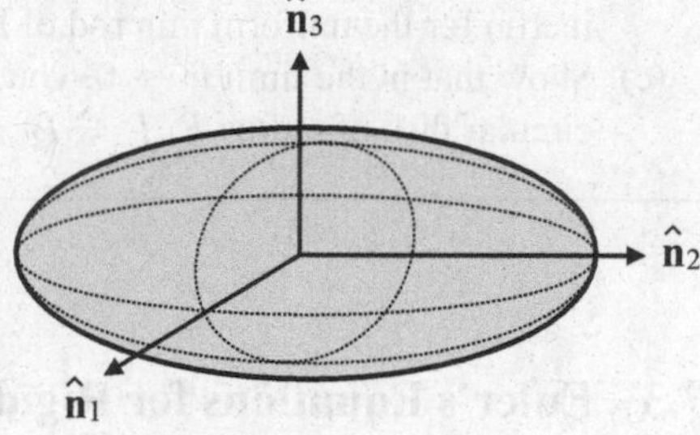


**Fig. 7.5** Principal axes for a uniform ellipsoid

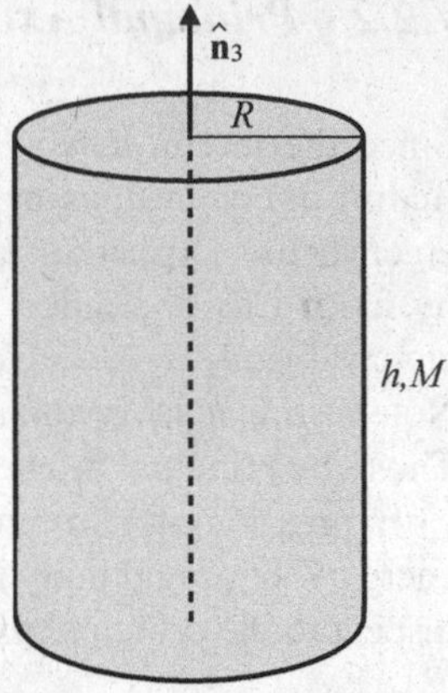


**Fig. 7.6** Uniform circular cylinder with radius $R$, height $h$, and mass $M$, with symmetry axis $\hat{\mathbf{n}}_3$

**Exercise 7.6** (a) Calculate the prinicipal moments of inertia (with respect to an origin passing through the center of mass) of a uniform circular cylinder with radius $R$, height $h$, and mass $M$ (See Fig. 7.6). You should find

$$I_1 = I_2 = \frac{1}{4}M\left(R^2 + \frac{1}{3}h^2\right), \quad I_3 = \frac{1}{2}MR^2, \tag{7.26}$$

where $I_3$ refers to the principal axis $\hat{\mathbf{n}}_3$, directed along the symmetry axis of the cylinder.

(b) Show that in the limit $R \to 0$, you recover from $I_1$ and $I_2$ the moment of inertia for the uniform thin rod of Exercise 7.5 part (e).

(c) Show that in the limit $h \to 0$, you obtain the moments of inertia for a thin circular disk of radius $R$, $I_1 = I_2 = \frac{1}{4}MR^2 = \frac{1}{2}I_3$.

## 7.3 Euler's Equations for Rigid Body Motion

We are now ready to write down **Euler's equations for rigid body motion**. These are simply the components of the torque equation

$$\boldsymbol{\tau} = \frac{\mathrm{d}\mathbf{L}}{\mathrm{d}t}, \tag{7.27}$$

with respect to the principal axes of the body. Here, the time derivative is with respect to the space frame, which we will take to be either: (i) an inertial frame attached to a *fixed* point in the rigid body (e.g., like the fixed support of a gyroscope or a spinning top), or (ii) moving with the center of mass of the body. In the latter case, the space frame can be accelerating, but it should not be rotating with respect to an inertial frame. So taking the time derivative of $\mathbf{L}$ using the general result (1.68), we have

$$\left(\frac{\mathrm{d}\mathbf{L}}{\mathrm{d}t}\right)_{\mathrm{s}} = \left(\frac{\mathrm{d}\mathbf{L}}{\mathrm{d}t}\right)_{\mathrm{b}} + \boldsymbol{\omega} \times \mathbf{L}\,. \tag{7.28}$$

Since $L_i = I_i \omega_i$ with respect to the principal axes, the $i$th component of $(\mathrm{d}\mathbf{L}/\mathrm{d}t)_{\mathrm{b}}$ is simply

$$\left(\frac{\mathrm{d}L_i}{\mathrm{d}t}\right)_{\mathrm{b}} = I_i \dot{\omega}_i\,, \tag{7.29}$$

where we used the fact that $I_i$ is fixed with respect to the body, and where we can write $\dot{\omega}_i$ without ambiguity since $(\mathrm{d}\boldsymbol{\omega}/\mathrm{d}t)_{\mathrm{s}} = (\mathrm{d}\boldsymbol{\omega}/\mathrm{d}t)_{\mathrm{b}}$. The $i$th component of the second term, $\boldsymbol{\omega} \times \mathbf{L}$, is given by

$$[\boldsymbol{\omega} \times \mathbf{L}]_i = \sum_{j,k} \varepsilon_{ijk} \omega_j L_k = \sum_{j,k} \varepsilon_{ijk} \omega_j \omega_k I_k\,. \tag{7.30}$$

Thus,

$$\tau_i = I_i \dot{\omega}_i + \sum_{j,k} \varepsilon_{ijk} \omega_j \omega_k I_k\,, \tag{7.31}$$

or, equivalently,

$$\begin{aligned} \tau_1 &= I_1 \dot{\omega}_1 - \omega_2 \omega_3 (I_2 - I_3)\,, \\ \tau_2 &= I_2 \dot{\omega}_2 - \omega_3 \omega_1 (I_3 - I_1)\,, \\ \tau_3 &= I_3 \dot{\omega}_3 - \omega_1 \omega_2 (I_1 - I_2)\,. \end{aligned} \tag{7.32}$$

Note that, via the time-dependence of the angular velocity vector $\boldsymbol{\omega}$, Euler's equations specify how the angular momentum vector $L_i = I_i \omega_i$ moves relative to the principal axes of the rigid body (which define the body frame). The dependence of the Euler angles $(\phi, \theta, \psi)$ on time can be found by integrating the expressions for $\boldsymbol{\omega}$ in terms of the Euler angles and their time derivatives, (6.71) or (7.110) below.

## 7.4 Solving Euler's Equations for Several Examples

We now solve Euler's equations in the context of several classic examples of rigid body motion, see e.g., Goldstein et al. (2002).

### 7.4.1 Torque-Free Motion with $\boldsymbol{\omega} = \text{const}$

For free-fall, or in the absencé of external forces, the motion of an isolated rigid body is *torque-free*. So setting $\tau_i = 0$ in (7.32), Euler's equations become

$$\begin{aligned} I_1\dot{\omega}_1 &= \omega_2\omega_3(I_2 - I_3)\,, \\ I_2\dot{\omega}_2 &= \omega_3\omega_1(I_3 - I_1)\,, \\ I_3\dot{\omega}_3 &= \omega_1\omega_2(I_1 - I_2)\,. \end{aligned} \tag{7.33}$$

These admit the solution $\boldsymbol{\omega} = \text{const}$ if and only if any two of the $\omega_i = 0$, e.g.,

$$\omega_1 = \text{const}\,, \quad \omega_2 = 0\,, \quad \omega_3 = 0\,, \qquad \text{(or cyclic permutation)}\,. \tag{7.34}$$

Although it is easy to see that (7.34) solve (7.33), it turns out that not all of these solutions are *stable*. In the most general case, all three principle moments of inertia will be different. Let's suppose initially that the rigid body has $I_1 < I_2 < I_3$, as shown in Fig. 7.7. In this case, the $\omega_1 = \text{const}$ and $\omega_3 = \text{const}$ solutions are *stable*, but the $\omega_2 = \text{const}$ solution is unstable, since small (but non-zero) perturbations to $\omega_1$, $\omega_3$ will grow exponentially and hence won't remain small (Exercise 7.7). If, instead, the rigid body has $I_1 = I_2 \neq I_3$, then only the $\omega_3 = \text{const}$ solution is stable. The $\omega_1 = \text{const}$ solution is unstable, as $\omega_3$ remains constant, but the $\omega_2$ perturbation grows linearly with time. Similarly for $\omega_2 = \text{const}$. This holds for either $I_3$ less than or greater than $I_1$.

*Proof* Here we prove the stability of the $\omega_1 = \text{const}$ solution for a rigid body with $I_1 < I_2 < I_3$. (The proof of stability of the $\omega_3 = \text{const}$ solution is similar. The proof that the $\omega_2 = \text{const}$ solution is unstable is left to Exercise 7.7.) Let's consider a perturbation to the solution given in (7.34), namely,

$$\omega_1 = \text{const}\,, \quad \omega_2 = \varepsilon_2\,, \quad \omega_3 = \varepsilon_3\,, \tag{7.35}$$

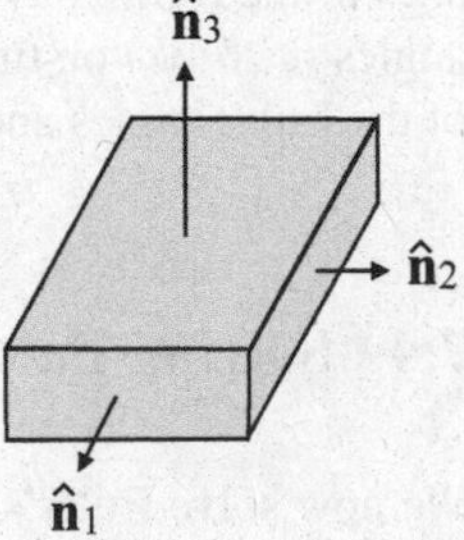


**Fig. 7.7** A rigid body (e.g., a textbook) with principal axes $\hat{\mathbf{n}}_1$, $\hat{\mathbf{n}}_2$, $\hat{\mathbf{n}}_3$ and moments of inertia $I_1 < I_2 < I_3$

where $\varepsilon_2$, $\varepsilon_3$ are small quantities, and see if it is an *approximate* solution to the equations (7.33). (By an approximate solution, we mean a solution to the equations ignoring terms that are 2nd-order or higher in $\varepsilon_2$, $\varepsilon_3$.) The first Euler equation in (7.33) becomes

$$\dot{\omega}_1 = \varepsilon_2\varepsilon_3\left(\frac{I_2 - I_3}{I_1}\right) = O(\varepsilon^2) = 0\,, \tag{7.36}$$

which is solved by $\omega_1 = \text{const}$. The other two equations become

$$\dot{\varepsilon}_2 = \varepsilon_3\omega_1\left(\frac{I_3 - I_1}{I_2}\right), \qquad \dot{\varepsilon}_3 = \omega_1\varepsilon_2\left(\frac{I_1 - I_2}{I_3}\right), \tag{7.37}$$

which are both 1st-order small, so we can't ignore any terms. Taking another time derivative of these equations yields

$$\begin{aligned}\ddot{\varepsilon}_2 &= (\dot{\varepsilon}_3\omega_1 + \varepsilon_3\dot{\omega}_1)\left(\frac{I_3 - I_1}{I_2}\right) = \varepsilon_2\omega_1^2\left(\frac{I_3 - I_1}{I_2}\right)\left(\frac{I_1 - I_2}{I_3}\right) = -\Omega^2\varepsilon_2\,,\\ \ddot{\varepsilon}_3 &= (\dot{\omega}_1\varepsilon_2 + \omega_1\dot{\varepsilon}_2)\left(\frac{I_1 - I_2}{I_3}\right) = \varepsilon_3\omega_1^2\left(\frac{I_1 - I_2}{I_3}\right)\left(\frac{I_3 - I_1}{I_2}\right) = -\Omega^2\varepsilon_3\,,\end{aligned} \tag{7.38}$$

where we used (7.36) and (7.37) to get the second equalities above, and where we defined

$$\Omega^2 \equiv \omega_1^2\left(\frac{I_3 - I_1}{I_2}\right)\left(\frac{I_2 - I_1}{I_3}\right) > 0\,. \tag{7.39}$$

Since these differential equations for $\varepsilon_2$, $\varepsilon_3$ are simple harmonic oscillator equations, the solutions to these equations will oscillate sinsoidally with angular frequency $\Omega$, never growing in size beyond their initial amplitudes. Thus, the $\omega_1 = \text{const}$, $\omega_2 = 0$, $\omega_3 = 0$ solution is stable against small perturbations. □

***Example 7.2*** The above statements can be demonstrated fairly easily by taking a typical textbook (with $I_1 < I_2 < I_3$), and tossing it vertically upward into the air as you simultaneously rotate it around one of the principal axes. (You will need to put a rubber band around the book to keep the pages from opening up as it is tossed in the air.) You should find that for rotations around the principal axis $\hat{\mathbf{n}}_1$ (or $\hat{\mathbf{n}}_3$), which corresponds to the smallest (or largest) moment of inertia $I_1$ (or $I_3$), the textbook continues to rotate smoothly about $\hat{\mathbf{n}}_1$ (or $\hat{\mathbf{n}}_3$) while it is in flight. But for rotations around $\hat{\mathbf{n}}_2$ the book will quickly start to "tumble" when it is tossed in the air, not being able to keep the rotation solely around $\hat{\mathbf{n}}_2$. The tumbling motion corresponds to perturbations away from $\omega_1 = 0$, $\omega_3 = 0$ growing exponentially with time. □

**Exercise 7.7** Prove that the $\omega_2 = \text{const}$, $\omega_1 = 0$, $\omega_3 = 0$ solution to (7.33) is unstable for a rigid body having $I_1 < I_2 < I_3$.
*Hint*: Proceed as in the proof above, but now take $\omega_1 = \varepsilon_1$, $\omega_3 = \varepsilon_3$ both small, and show that you obtain differential equations of the form

$$\ddot{\varepsilon}_1 = \kappa^2 \varepsilon_1 , \qquad \ddot{\varepsilon}_3 = \kappa^2 \varepsilon_3 . \tag{7.40}$$

Since these differential equations admit solutions which *grow* exponentially in time, the $\omega_2 = \text{const}$, $\omega_1 = 0$, $\omega_3 = 0$ solution is unstable against small perturbations.

### 7.4.2 Torque-Free Motion of a Symmetric Top

Euler's equations for torque-free motion for a *symmetric* top (which has $I_1 = I_2$) are

$$\dot{\omega}_1 = -\Omega\omega_2 , \qquad \dot{\omega}_2 = \Omega\omega_1 , \qquad \dot{\omega}_3 = 0 , \tag{7.41}$$

where

$$\Omega \equiv \omega_3 (I_3 - I_1)/I_1 . \tag{7.42}$$

The $\omega_3$ equation can be immediately solved, i.e., $\omega_3 = \text{const}$, implying that $\Omega$ is also a constant. The first two equations are *coupled* 1st-order equations that can be solved most simply by considering the complex combination

$$\zeta \equiv \omega_1 + \mathrm{i}\omega_2 . \tag{7.43}$$

Using (7.41), it follows that

$$\dot{\zeta} = \mathrm{i}\Omega\zeta , \tag{7.44}$$

which has solution

$$\zeta = C e^{\mathrm{i}\Omega t} , \tag{7.45}$$

where $C$ is a complex constant. This solution corresponds to uniform rotational motion of $\zeta$ in the complex $\hat{\mathbf{n}}_1$, $\hat{\mathbf{n}}_2$ plane. Thus,

$$\omega_1 = C \cos \Omega t , \qquad \omega_2 = C \sin \Omega t , \qquad \omega_3 = \text{const} . \tag{7.46}$$

The above results imply that $\boldsymbol{\omega}$ has constant magnitude $\omega \equiv |\boldsymbol{\omega}|$ and precesses about $\hat{\mathbf{n}}_3$ with constant angular velocity $\Omega$. The cone traced out by the precession of $\boldsymbol{\omega}$ about $\hat{\mathbf{n}}_3$ is called the **body cone**. Thus, we see that torque-free rigid body motion

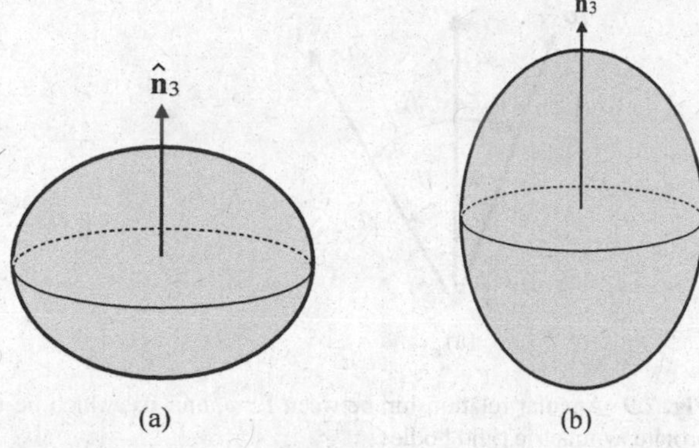


**Fig. 7.8** Oblate (panel (a)) and prolate (panel (b)) spheroids. For both cases, the horizontal cross sections (perpendicular to $\hat{\mathbf{n}}_3$) are circular disks with $I_1 = I_2$

that is *not* a simple rotation around one of the principal axes *cannot* have $\boldsymbol{\omega} = \text{const}$; only its *magnitude* can be constant.

The angular velocity $\Omega$ can be either positive or negative depending on the sign of $I_3 - I_1$. A positive value, corresponding to counter-clockwise rotation of $\boldsymbol{\omega}$ around $\hat{\mathbf{n}}_3$, occurs if $I_3 > I_1$; a negative value, corresponding to clockwise rotation, occurs if $I_3 < I_1$. These two cases correspond to an **oblate** rigid body ("fat" around the symmetry axis $\hat{\mathbf{n}}_3$) or a **prolate** rigid body ("skinny" around the symmetry axis), respectively. Examples of oblate and prolate spheroids are shown in Fig. 7.8.

Since there are no torques and no forces, the angular momentum $\mathbf{L}$ and total mechanical energy $E$ are conserved:

$$\mathbf{L} = \text{const}\,, \qquad E = T = \frac{1}{2}\boldsymbol{\omega}\cdot\mathbf{L} = \text{const}\,. \tag{7.47}$$

This means that we can take the $z$-axis of the (inertial) space frame to point along $\mathbf{L}$. In addition, as a consequence of $I_1 = I_2$, we find that $\mathbf{L}$, $\boldsymbol{\omega}$, and $\hat{\mathbf{n}}_3$ lie in a plane, since

$$\begin{aligned}\mathbf{L}\cdot(\boldsymbol{\omega}\times\hat{\mathbf{n}}_3) &= \mathbf{L}\cdot(\omega_2\hat{\mathbf{n}}_1 - \omega_1\hat{\mathbf{n}}_2) = L_1\omega_2 - L_2\omega_1\\ &= I_1\omega_1\omega_2 - I_2\omega_2\omega_1 = (I_1 - I_2)\omega_1\omega_2 = 0\,.\end{aligned} \tag{7.48}$$

Thus, both $\mathbf{L}$ and $\omega = |\boldsymbol{\omega}|$ are constant, and $\boldsymbol{\omega}$ and $\hat{\mathbf{n}}_3$ precess around $\mathbf{L}$ with constant angular velocity $\dot{\phi}$ (to be determined below). The cone traced out by the precession of $\boldsymbol{\omega}$ about $\mathbf{L}$ is called the **space cone**. Figures 7.9 and 7.10 illustrate the relationship between the vectors $\mathbf{L}$, $\boldsymbol{\omega}$, and $\hat{\mathbf{n}}_3$, and the corresponding space and body cones for both oblate and prolate rigid symmetric rigid bodies.

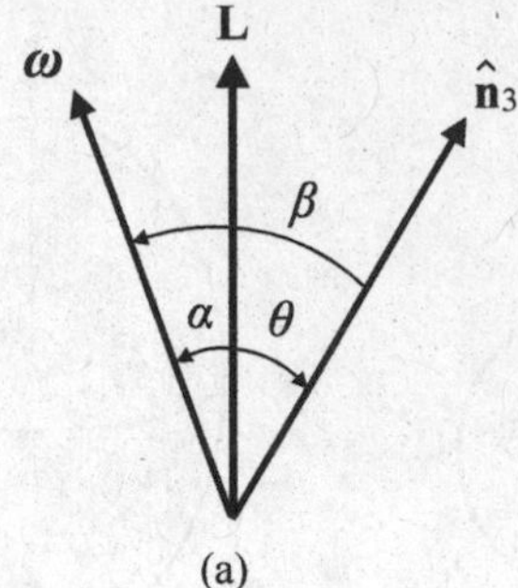


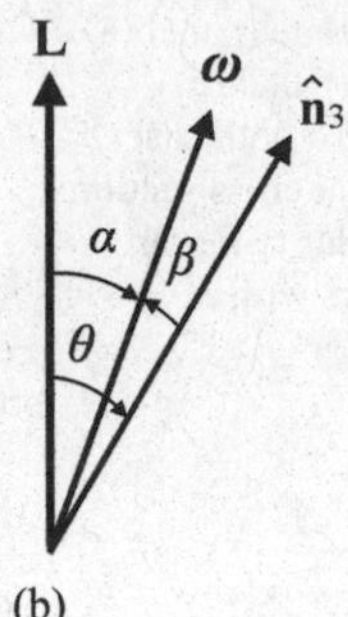


**Fig. 7.9** Angular relationship between $\mathbf{L}$, $\boldsymbol{\omega}$, and $\hat{\mathbf{n}}_3$, which lie in a plane, for (a) oblate and (b) prolate symmetric rigid bodies

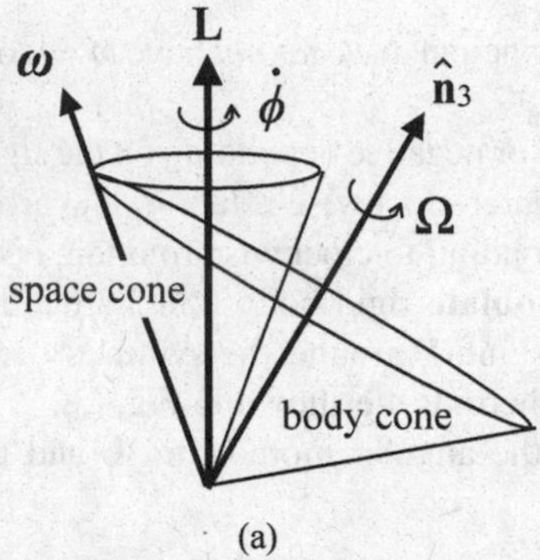


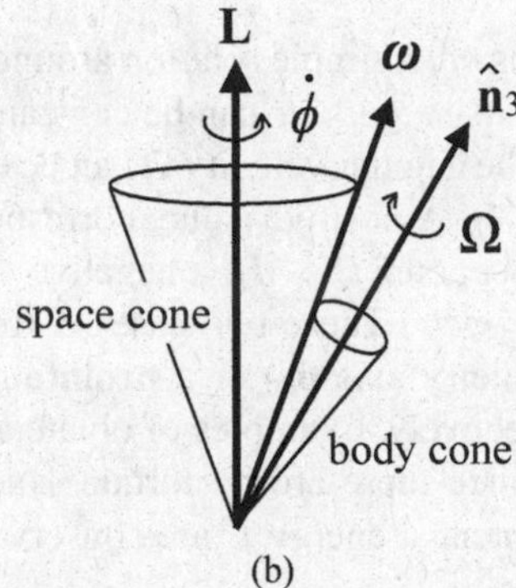


**Fig. 7.10** Space and body cones for (a) oblate and (b) prolate symmetric rigid bodies. The vectors $\mathbf{L}$, $\boldsymbol{\omega}$, and $\hat{\mathbf{n}}_3$ remain in a plane as $\boldsymbol{\omega}$ and $\hat{\mathbf{n}}_3$ precess around $\mathbf{L}$ with constant angular velocity $\dot{\phi}$

#### 7.4.2.1 Solving for the Euler Angles

From the above observations it follows immediately that the Euler angle

$$\theta = \text{const}\,. \tag{7.49}$$

This is the angle between $\hat{\mathbf{n}}_3$ and $\mathbf{L}$. The Euler angle $\psi$ and the angular velocity $\Omega$ are related by

$$\dot{\psi} = -\Omega\,, \tag{7.50}$$

which follows from the equivalence of an active counter-clockwise rotation of $\boldsymbol{\omega}$ about $\hat{\mathbf{n}}_3$ and a passive clockwise rotation of the principal axes $\hat{\mathbf{n}}_1$ and $\hat{\mathbf{n}}_2$ around $\hat{\mathbf{n}}_3$. Thus,

$$\psi(t) = -\Omega t + \psi_0 \,. \tag{7.51}$$

Since points on the body that lie along the rotation axis $\boldsymbol{\omega}$ are instantaneously at rest, the body cone rolls *without slipping* on the space cone. From this observation, it follows that

$$\Omega \sin\beta = \dot{\phi} \sin\alpha \tag{7.52}$$

where $\alpha$ is the angle between $\boldsymbol{\omega}$ and $\mathbf{L}$, and $\beta$ is the angle between $\boldsymbol{\omega}$ and $\hat{\mathbf{n}}_3$, as shown in Fig. 7.9.

From Fig. 7.10 we see that the space cone is inside the body cone for an oblate symmetric rigid body ($I_3 > I_1$), while the space cone is outside the body cone for a prolate symmetric rigid body ($I_3 < I_1$). We can demonstrate this algebraically by looking at the components of $\boldsymbol{\omega}$ and $\mathbf{L}$ with respect to the principal axes of the body at the instant of time when $\psi = 0$ (for which $\hat{\mathbf{n}}_1$ lies in the same plane as $\mathbf{L}$, $\boldsymbol{\omega}$, and $\hat{\mathbf{n}}_3$). Setting $\psi = 0$ in (6.71) we obtain:

$$\boldsymbol{\omega} = \begin{bmatrix} -\sin\theta\,\dot{\phi} \\ \dot{\theta} \\ \cos\theta\,\dot{\phi} + \dot{\psi} \end{bmatrix} = \begin{bmatrix} -\omega\sin\beta \\ 0 \\ \omega\cos\beta \end{bmatrix}, \tag{7.53}$$

and

$$\mathbf{L} = \begin{bmatrix} I_1\omega_1 \\ I_2\omega_2 \\ I_3\omega_3 \end{bmatrix} = \begin{bmatrix} -I_1\sin\theta\,\dot{\phi} \\ I_2\dot{\theta} \\ I_3(\cos\theta\,\dot{\phi} + \dot{\psi}) \end{bmatrix} = \begin{bmatrix} -I_1\omega\sin\beta \\ 0 \\ I_3\omega\cos\beta \end{bmatrix} = \begin{bmatrix} -L\sin\theta \\ 0 \\ L\cos\theta \end{bmatrix}. \tag{7.54}$$

Thus,

$$\frac{L_1}{L_3} \quad\Rightarrow\quad \tan\theta = \frac{I_1}{I_3}\tan\beta\,, \tag{7.55}$$

from which we can conclude that $\theta < \beta$ for $I_3 > I_1$, while $\theta > \beta$ for $I_3 < I_1$, consistent with Figs. 7.9 and 7.10.

Using the above expressions for the components of $\mathbf{L}$ and $\boldsymbol{\omega}$, we can also show that the angular velocity $\dot{\phi}$ of $\hat{\mathbf{n}}_3$ around $\mathbf{L}$ (the space frame $z$-axis) satisfies the following relations:

$$\dot{\phi} = \frac{L}{I_1} = \frac{I_3\omega_3}{I_1\cos\theta} = \frac{\Omega}{\cos\theta}\frac{I_3}{(I_3 - I_1)} \tag{7.56}$$

where the second equality follows from $L = L_3/\cos\theta = I_3\omega_3/\cos\theta$, and the third equality follows from the definition (7.42) of $\Omega$. Thus,

$$\phi(t) = \frac{\Omega}{\cos\theta}\frac{I_3}{(I_3 - I_1)}\,t + \phi_0\,. \tag{7.57}$$

Equations (7.49), (7.51) and (7.57) completely specify the Euler angles for the motion in terms of the constants $I_1$, $I_3$, and $\Omega$ (or $\omega_3$). The constant $C$, which appears in (7.46) for $\omega_1$ and $\omega_2$, is needed to determine the overall magnitude of the angular velocity vector $\boldsymbol{\omega}$ from $\omega^2 = |C|^2 + \omega_3^2$. All other constants, such as $L$ and $T$ can be determined from $I_1$, $I_3$, $\Omega$ (or $\omega_3$), and $C$.

**Exercise 7.8** Show that the relation $\dot{\psi} = -\Omega$, cf. (7.50), can be obtained directly using $\omega_3 = \cos\theta\,\dot{\phi} + \dot{\psi}$, (7.56), and (7.42).

***Example 7.3*** The angular velocities $\dot{\phi}$ and $\omega_3$ are often called the *wobble* and *spin* frequencies of the rigid body. The wobble frequency $\dot{\phi}$ is the precession rate of the symmetry axis $\hat{\mathbf{n}}_3$ around the angular momentum vector $\mathbf{L}$, which is fixed in the space frame. The spin frequency $\omega_3$ is the angular velocity of the rigid body about the symmetry axis $\hat{\mathbf{n}}_3$ of the body. From (7.56), we can write

$$\omega_3 = \frac{I_1 \cos\theta}{I_3}\,\dot{\phi}\,. \tag{7.58}$$

If we specialize to the case of nearly vertical motion (so $\theta \to 0$) and a thin, uniform circular disk (which has $I_3 = 2I_1$ as shown in part (c) of Exercise 7.6), then

$$\omega_3 \simeq \frac{1}{2}\dot{\phi}\,. \tag{7.59}$$

So the spin frequency is approximately *half* the wobble frequency for this case.

But this result is actually *inconsistent* with Feynman's anecdote about a wobbling dinner plate recounted in his book *"Surely You're Joking Mr. Feynman!"*, Feynman (1985), p. 173:

> I was in the cafeteria and some guy, fooling around, throws a plate in the air. As the plate went up in the air I saw it wobble, and I noticed the red medallion of Cornell on the plate going around. It was pretty obvious to me that the medallion went around faster than the wobbling. I had nothing to do, so I start figuring out the motion of the rotating plate. I discovered that when the angle is very slight, the medallion rotates twice as fast as the wobble rate—two to one.

Thus, according to Feynman, the spin frequency should be *twice* as large as the wobble frequency for this case. So have we made a mistake somewhere in our calculation?

As Feynman also said, "experiment is the sole judge of scientific truth." So you can check for yourself what the answer should be by doing the experiment with a (preferably *plastic*!) dinner plate with some black tape used to mark a point on its rim, and a high-speed video camera that allows you to record and playback the motion of the plate in slow motion. (There are also videos on-line that you can watch if you

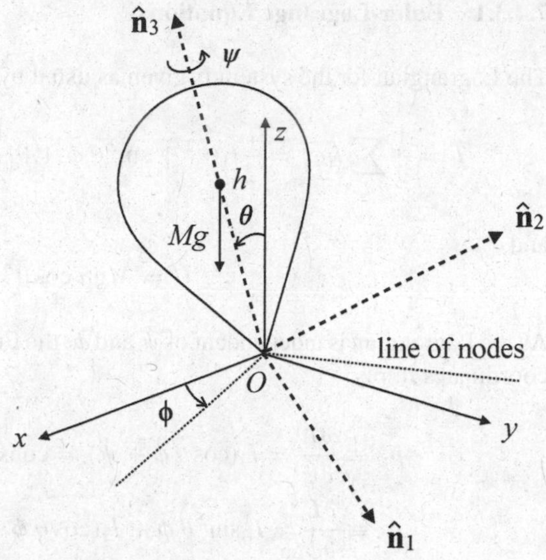


**Fig. 7.11** Symmetric top ($I_1 = I_2$) with fixed point $O$. The center of mass is located a distance $h$ from $O$, along the symmetry axis $\hat{\mathbf{n}}_3$. Gravity exerts a torque $Mgh \sin\theta$ about $O$, directed along the line of nodes

don't have access to a high-speed video camera.) By doing the experiment, you will find that we didn't make a mistake; (7.59) is okay afterall. Rather it was Feynman who made an error, which he failed to catch when proofreading his book. □

### 7.4.3 *Symmetric Top with One Point Fixed*

We now consider the motion of a symmetric rigid body (e.g., a top or gyroscope) about a fixed point $O$ located on the symmetry axis $\hat{\mathbf{n}}_3$ of the body ($I_1 = I_2$). We let $h$ denote the distance from $O$ to the center of mass, and $M$ denote the total mass of the body. We will use Euler angles $(\phi, \theta, \psi)$ to specify the orientation of the top, and we will decompose vectors with respect to the principal axes $\hat{\mathbf{n}}_1$, $\hat{\mathbf{n}}_2$, $\hat{\mathbf{n}}_3$.

Gravity is the only external force acting on the system,[3] exerting a torque directed along the line of nodes shown in Fig. 7.11. As $\hat{\mathbf{z}}$ and $\hat{\mathbf{n}}_3$ are perpendicular to the line of nodes, the angular momenta about these axes ($p_\phi$ and $p_\psi$) will be conserved. Since gravity is a *conservative* force, the total mechanical energy $E = T + U$ will also be conserved.

[3]There is also the support force that balances the weight of the top, but since it acts at $O$ it does not exert a torque on the system about this point.

#### 7.4.3.1 Euler-Lagrange Equations

The Lagrangian for the system is given as usual by $L = T - U$, where

$$T = \frac{1}{2}\sum_i I_i\omega_i^2 = \frac{1}{2}I_1(\dot{\theta}^2 + \sin^2\theta\,\dot{\phi}^2) + \frac{1}{2}I_3(\cos\theta\,\dot{\phi} + \dot{\psi})^2\,, \tag{7.60}$$

and

$$U = Mgh\cos\theta\,. \tag{7.61}$$

As the Lagrangian is independent of $\psi$ and $\phi$, the Euler-Lagrange equations for these coordinates imply

$$\begin{aligned} p_\psi &\equiv \frac{\partial L}{\partial\dot{\psi}} = I_3(\cos\theta\,\dot{\phi} + \dot{\psi}) = \text{const}\,, \\ p_\phi &\equiv \frac{\partial L}{\partial\dot{\phi}} = I_1\sin^2\theta\,\dot{\phi} + I_3(\cos\theta\,\dot{\phi} + \dot{\psi})\cos\theta = \text{const}\,. \end{aligned} \tag{7.62}$$

These equations can be inverted to yield

$$\dot{\phi} = \frac{p_\phi - p_\psi\cos\theta}{I_1\sin^2\theta}\,, \tag{7.63}$$

$$\dot{\psi} = \frac{p_\psi}{I_3} - \cos\theta\left(\frac{p_\phi - p_\psi\cos\theta}{I_1\sin^2\theta}\right). \tag{7.64}$$

The above equations can be integrated to yield $\phi(t)$ and $\psi(t)$ once we know $\theta(t)$.

> **Exercise 7.9** Show that $T$ has the above form in (7.60) using (6.71) for the components of $\boldsymbol{\omega}$ with respect to the prinicipal axes.

Rather than write down the Euler-Lagrange equation for $\theta$, which will be a 2nd-order ordinary differential equation, we can obtain a 1st-order equation by using the fact that $E = T + U$ is conserved (since the Lagrangian does not depend explicitly on $t$). After some straightforward algebra, we find

$$\begin{aligned} E &= T + U \\ &= \frac{1}{2}I_1(\dot{\theta}^2 + \sin^2\theta\,\dot{\phi}^2) + \frac{1}{2}I_3(\cos\theta\,\dot{\phi} + \dot{\psi})^2 + Mgh\cos\theta \\ &= \frac{1}{2}I_1\dot{\theta}^2 + \frac{1}{2}\frac{(p_\phi - p_\psi\cos\theta)^2}{I_1\sin^2\theta} + \frac{1}{2}\frac{p_\psi^2}{I_3} + Mgh\cos\theta\,. \end{aligned} \tag{7.65}$$

Since the second-to-last term above is constant, we can define a new conserved quantity

$$E' \equiv E - \frac{1}{2}\frac{p_\psi^2}{I_3}, \tag{7.66}$$

for which

$$\begin{aligned} E' &= \frac{1}{2} I_1 \dot{\theta}^2 + U_{\text{eff}}(\theta)\,, \\ U_{\text{eff}}(\theta) &\equiv \frac{1}{2}\frac{(p_\phi - p_\psi \cos\theta)^2}{I_1 \sin^2\theta} + Mgh\cos\theta\,. \end{aligned} \tag{7.67}$$

The one-dimensional effective potential equation for $\dot{\theta}$ is separable, leading to

$$\mathrm{d}t = \frac{\mathrm{d}\theta}{\sqrt{2(E' - U_{\text{eff}}(\theta))/I_1}}, \tag{7.68}$$

which can be integrated (numerically) to find $t = t(\theta)$. Inverting will give $\theta = \theta(t)$.

### 7.4.3.2 Qualitative Behavior of Motion

Qualitative behavior of the motion can be obtained from the shape of the *effective potential* $U_{\text{eff}}(\theta)$ for $0 \le \theta \le \pi$. The effective potential is generically U-shaped,

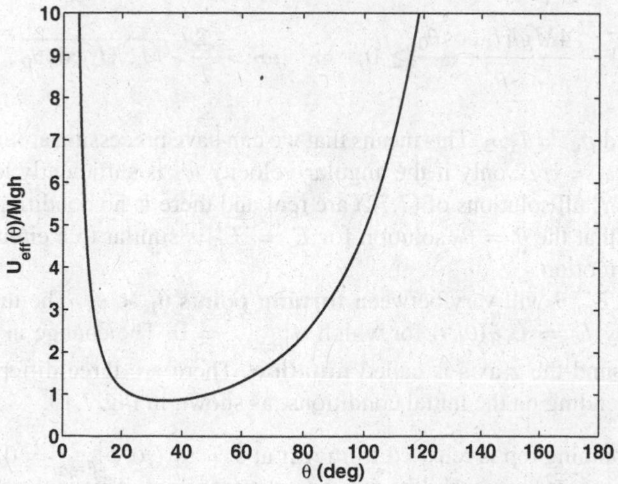


**Fig. 7.12** A typical effective potential $U_{\text{eff}}(\theta)$ normalized by $Mgh$

blowing up as $\theta \to 0$ and $\pi$, as shown in Fig. 7.12. There is a minimum in the potential ($E'_0 \equiv U_{\rm eff}(\theta_0)$) at $\theta_0$ determined by the equation

$$\left.\frac{dU_{\rm eff}(\theta)}{d\theta}\right|_{\theta=\theta_0} = 0\,. \tag{7.69}$$

This is a quadratic equation,

$$\cos\theta_0\,\beta^2 - p_\psi \sin^2\theta_0\,\beta + MghI_1\sin^4\theta_0 = 0\,, \tag{7.70}$$

in the quantity

$$\beta \equiv p_\phi - p_\psi\cos\theta_0 = I_1\sin^2\theta_0\,\dot\phi\,. \tag{7.71}$$

The solutions to the quadratic equation are

$$\beta_\pm = \frac{p_\psi\sin^2\theta_0}{2\cos\theta_0}\left(1 \pm \sqrt{1 - \frac{4MghI_1\cos\theta_0}{p_\psi^2}}\right), \tag{7.72}$$

with only one of these solutions corresponding to a minimum of the effective potential for $0 < \theta_0 < \pi$. The above equation for $\beta_\pm$ together with (7.71) give an implicit solution for $\theta_0$.

> **Exercise 7.10** Verify that the condition (7.69) for a minimum in the effective potential (7.67) leads to the quadratic equation (7.70) with solutions (7.72).

For $0 < \theta_0 < \pi/2$, the condition for a real solution is

$$1 - \frac{4MghI_1\cos\theta_0}{p_\psi^2} \geq 0 \quad \Leftrightarrow \quad \omega_3 > \frac{2}{I_3}\sqrt{MghI_1\cos\theta_0}\,, \tag{7.73}$$

where we used $p_\psi = I_3\omega_3$. This means that we can have precession around the $z$-axis for constant $\theta_0 < \pi/2$ only if the angular velocity $\omega_3$ is sufficiently large. (When $\pi/2 \leq \theta_0 < \pi$, all solutions of (7.72) are real and there is no condition on the size of $\omega_3$.) Note that the $\theta = \theta_0$ solution for $E' = E'_0$ is similar to a circular orbit for central force motion.

For $E' > E'_0$, $\theta$ will vary between **turning points** $\theta_1 < \theta_2$. The turning points are defined by $E' = U_{\rm eff}(\theta_{1,2})$ for which $\dot\theta\big|_{\theta=\theta_{1,2}} = 0$. The change in $\theta$ as the top precesses around the $\hat{\mathbf{z}}$-axis is called **nutation**. There are three different types of nutation, depending on the initial conditions, as shown in Fig. 7.13.

(i) If the spinning top is released from rest at $\theta = \theta_1$ (so $\dot\phi\big|_{\theta=\theta_1} = 0$), the axis of the top undergoes a cusp-like motion as it precesses. This is the standard way of releasing a spinning top.

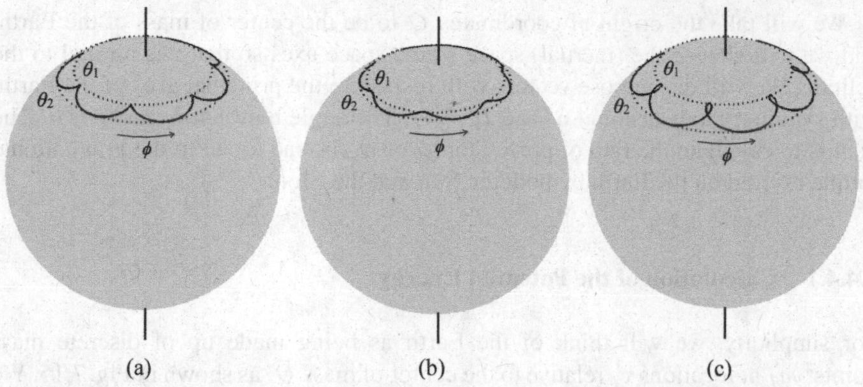


**Fig. 7.13** Three different types of nutation for a symmetric top corresponding to the initial conditions: $\dot{\phi}\big|_{\theta=\theta_1} = 0$, panel (a); $\dot{\phi}\big|_{\theta=\theta_1} > 0$, panel (b); and $\dot{\phi}\big|_{\theta=\theta_1} < 0$, panel (c)

(ii) If the spinning top is given an initial velocity in the direction of precession when it is released (so $\dot{\phi}\big|_{\theta=\theta_1} > 0$), the axis undergoes a sinusoidal-like motion as it precesses.

(iii) If the spinning top is given an initial velocity opposite the direction of precession when it is released (so $\dot{\phi}\big|_{\theta=\theta_1} < 0$), the axis undergoes a loop-the-loop-like motion as it precesses.

### 7.4.4 *Precession of the Equinoxes*

For our final example, we will calculate the rate of precession of Earth's rotational axis about the normal to the ecliptic (the plane of the Earth's orbit around the Sun). See Fig. 7.14. The precession is produced by the gravitational torques exerted on the (oblate) Earth by both the Sun and Moon. The main input for this calculation is the potential energy as function of the tilt $\theta$ of the Earth's rotational axis relative to the normal to the ecliptic.

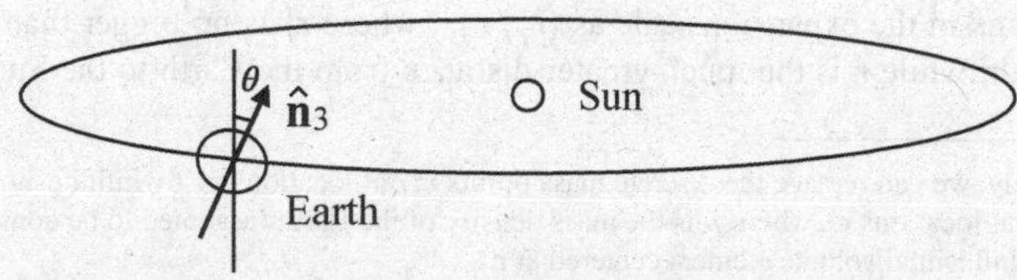


**Fig. 7.14** A perspective view of Earth's orbit around the Sun. The plane of the orbit is the *ecliptic*. The principal axis $\hat{\mathbf{n}}_3$, which defines the Earth's daily rotational motion, makes an angle $\theta$ with respect to the normal the ecliptic. (The Sun and Earth are not shown to scale)

We will take the origin of coordinates $O$ to be the center of mass of the Earth, and we will choose the (inertial) space frame space axes so that $\hat{\mathbf{z}}$ is normal to the ecliptic. We will decompose vectors with respect to the principal axes of the Earth, with symmetry axis denoted $\hat{\mathbf{n}}_3$ and $I_1 = I_2$. The angle between $\hat{\mathbf{n}}_3$ and $\hat{\mathbf{z}}$ is $\theta$. The goal is to calculate the rate of precession $\dot{\phi}$ of $\hat{\mathbf{n}}_3$ around $\hat{\mathbf{z}}$ due to the gravitational torque exerted on the Earth by both the Sun and the Moon.

#### 7.4.4.1 Calculation of the Potential Energy

For simplicity, we will think of the Earth as being made up of discrete mass points[4] $m_I$ at locations $\mathbf{r}'_I$ relative to the center of mass $O$, as shown in Fig. 7.15. We will let $\mathbf{r}$ denote the position vector of the Sun (or Moon) treated as a single mass point $M'$ relative to $O$, and we will let $\mathbf{r}_I$ denote the position vector of the Sun (or Moon) relative to $m_I$. The angle between $\mathbf{r}$ and $\mathbf{r}'_I$ will be denoted by $\gamma_I$. In terms of these quantities, the potential energy between the Sun (or Moon) and Earth is given by

$$U = \sum_I -\frac{GM'm_I}{r_I}. \tag{7.74}$$

Using the law of cosines we can write

$$r_I^2 = r^2 + r_I'^2 - 2rr'_I \cos\gamma_I\,, \tag{7.75}$$

which can then be substituted back into the expression for the potential energy:

$$\begin{aligned} U &= -\frac{GM'}{r}\sum_I \frac{m_I}{\sqrt{1 + (r'_I/r)^2 - 2(r'_I/r)\cos\gamma_I}} \\ &= -\frac{GM'}{r}\sum_I m_I \sum_{n=0}\left(\frac{r'_I}{r}\right)^n P_n(\cos\gamma_I)\,, \end{aligned} \tag{7.76}$$

where the last equality follows from the standard expansion for the generating function (E.27) of the Legendre polynomials $P_n(x)$. (For a review of Legendre polynomials,[5] see Appendix E.3.)

The terms in the expansion scale as $\left(r'_I/r\right)^n$, where $r'_I$ is no bigger than the size of the Earth, while $r$ is the much greater distance from the Earth to the Sun (or the

[4] Alternatively, we can replace the discrete mass points $m_I$ at locations $\mathbf{r}'_I$ by infinitesimal masses $dm = \rho dV$ at locations $\mathbf{r}'$, where $\rho$ is the mass density of the Earth (assumed to be constant), and $dV$ is an infinitesimal volume element centered at $\mathbf{r}'$.

[5] Recall that the first three Legendre polynomials are given by

$$P_0(x) = 1\,, \qquad P_1(x) = x\,, \qquad P_2(x) = \frac{1}{2}(3x^2 - 1)\,,$$

which are normalized so that $P_n(1) = 1$.

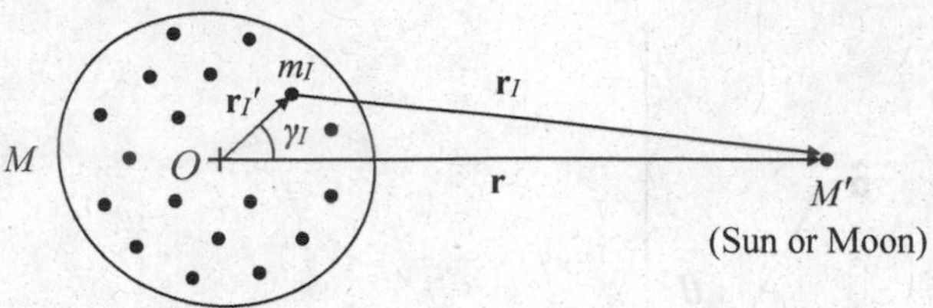


**Fig. 7.15** Position vector of mass point $m_I$ relative to the center of mass $O$ of the Earth, and relative to the Sun (or Moon). $\mathbf{r}$ is the position vector of the center of mass of the Sun (or Moon) relative to the center of mass $O$ of the Earth

Moon). As we are interested in the *largest* term that contributes to the precession, we will only need to evaluate the first *three* terms in the expansion, $U = U_0 + U_1 + U_2$. Substituting for $P_0(x)$, $P_1(x)$, and $P_2(x)$, it is fairly easy to show that

$$U_0 = -\frac{GM'M}{r}, \qquad U_1 = 0, \qquad U_2 = \frac{3}{2}\frac{GM'}{r^3}\sum_{i,j} Q_{ij}u_i u_j, \tag{7.77}$$

where $M \equiv \sum_I m_I$ is the total mass of the Earth, and

$$Q_{ij} \equiv I_{ij} - \frac{1}{3}\mathrm{Tr}(\mathsf{I})\delta_{ij} \tag{7.78}$$

is the **reduced** (or *trace-free*) **rotational inertia tensor**. Also, $u_i$ are the components of the unit vector $\hat{\mathbf{u}} \equiv \mathbf{r}/r$ with respect to an arbitrary basis. Note that $U_0$ is simply the potential for a point mass, and $U_1 = 0$ since $O$ is located at the center of mass of the Earth. Thus, the $U_2$ term is the first term in the potential that contains non-trivial information about the mass distribution of the Earth.

If we work in the basis defined by the principal axes $\hat{\mathbf{n}}_1$, $\hat{\mathbf{n}}_2$, $\hat{\mathbf{n}}_3$, then $I_{ij} = I_i\delta_{ij}$, for which

$$\sum_{i,j} Q_{ij}u_i u_j = \sum_i \left(I_i u_i^2 - \frac{1}{3}I_i\right) = I_1u_1^2 + I_2u_2^2 + I_3u_3^2 - \frac{1}{3}(I_1 + I_2 + I_3)\,. \tag{7.79}$$

But since $I_1 = I_2$ for a symmetric rigid body like the Earth, the above expression simplies to

$$\sum_{i,j} Q_{ij}u_i u_j = (I_3 - I_1)\left(u_3^2 - \frac{1}{3}\right) = \frac{2}{3}(I_3 - I_1)P_2(u_3)\,. \tag{7.80}$$

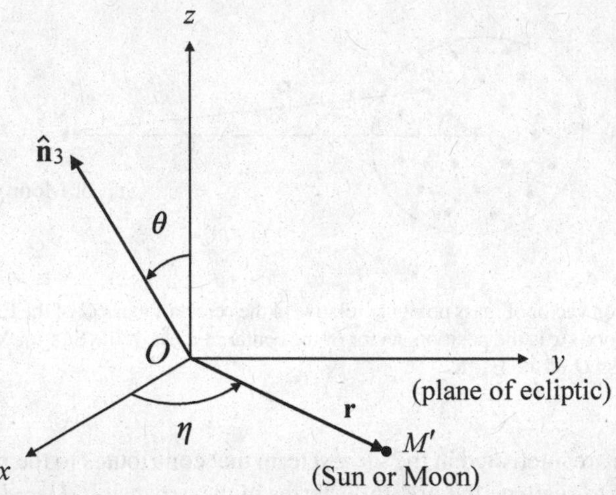


**Fig. 7.16** Definitions of the angles $\theta$ and $\eta$, which relate the principal axis $\hat{\mathbf{n}}_3$ with the position vector $\mathbf{r}$ of the Sun (or Moon) relative to the center of mass of the Earth

Using geometry, one can show that

$$u_3 \equiv \hat{\mathbf{n}}_3 \cdot \mathbf{r}/r = \sin\theta\,\cos\eta\,, \tag{7.81}$$

where $\eta$ is the angle that $\mathbf{r}$ makes with the $x$-axis in the plane of the ecliptic, as shown in Fig. 7.16. Thus,

$$U_2 = \frac{3}{2}\frac{GM'}{r^3}\sum_{i,j} Q_{ij}u_i u_j = \frac{1}{2}\frac{GM'}{r^3}(I_3 - I_1)(3\sin^2\theta\,\cos^2\eta - 1)\,. \tag{7.82}$$

Finally, if we average over one complete orbit, we get

$$\bar{U}_2 = -\frac{1}{2}\frac{GM'}{\bar{r}^3}(I_3 - I_1)P_2(\cos\theta)\,, \tag{7.83}$$

where $\bar{r}$ is the average radial distance for the orbit. This is the desired form of the potential.

**Exercise 7.11** Verify the calculations leading to the above expressions, (7.82) and (7.83), for $U_2$ and $\bar{U}_2$.

#### 7.4.4.2 Euler-Lagrange Equation

The Lagrangian is given as usual by $L = T - U$ where $U = \bar{U}_2$ (we can ignore $U_0$ since it is a constant when orbit-averaged) and $T$ is the same as for the symmetric top with one point fixed, i.e.,

$$T = \frac{1}{2}\sum_i I_i\omega_i^2 = \frac{1}{2}I_1(\dot{\theta}^2 + \sin^2\theta\,\dot{\phi}^2) + \frac{1}{2}I_3(\cos\theta\,\dot{\phi} + \dot{\psi})^2 . \tag{7.84}$$

Since the torque comes from the derivative of $U$ with respect to $\theta$, the relevant Euler-Lagrange equation is the $\theta$ equation:

$$\begin{aligned} 0 &= \frac{\mathrm{d}}{\mathrm{d}t}\left(\frac{\partial L}{\partial\dot{\theta}}\right) - \frac{\partial L}{\partial\theta} \\ &= I_1\ddot{\theta} - I_1\sin\theta\,\cos\theta\,\dot{\phi}^2 + I_3\omega_3\sin\theta\,\dot{\phi} + \frac{1}{2}\frac{GM'}{\bar{r}^3}(I_3 - I_1)\,3\cos\theta\,\sin\theta\,, \end{aligned} \tag{7.85}$$

where $\omega_3$ is shorthand in the above equation for $\omega_3 = \cos\theta\,\dot{\phi} + \dot{\psi}$ (See (6.71)). Since we are interested only in the rate of precession $\dot{\phi}$, we can ignore the $\ddot{\theta}$ term (in other words, for the calculation of the precession we can treat $\theta$ as a constant). In addition, the second term is much smaller than the third as $\dot{\phi} \ll \omega_3 \approx 2\pi$ rad/day (to leading order), and hence can be ignored. Thus, to a good approximation

$$\dot{\phi} \approx -\frac{3}{2}\frac{GM'}{\omega_3\bar{r}^3}\left(\frac{I_3 - I_1}{I_3}\right)\cos\theta\,. \tag{7.86}$$

Numerical values for $\dot{\phi}$ can be obtained for both the Sun and the Moon by substituting

$$\begin{aligned} &\theta \approx 23^\circ\,, \\ &\omega_3 \approx 2\pi \text{ rad/day}\,, \\ &(I_3 - I_1)/I_3 \approx (I_3 - I_1)/I_1 \approx 3.3\times 10^{-3}\,, \\ &G = 6.67\times 10^{-11}\ \mathrm{N\cdot m^2/kg^2}\,, \\ &M' = 2\times 10^{30}\ \text{kg (Sun)} \quad\text{or}\quad M' = 7.35\times 10^{22}\ \text{kg (Moon)}\,, \\ &\bar{r} = 1.5\times 10^{11}\ \text{m (Sun)} \quad\text{or}\quad \bar{r} = 3.84\times 10^{8}\ \text{m (Moon)}\,. \end{aligned} \tag{7.87}$$

The results are:

$$\dot{\phi}_{\text{Sun}} \approx \frac{16''}{\text{yr}}\,, \qquad \dot{\phi}_{\text{Moon}} \approx \frac{35''}{\text{yr}}\,, \tag{7.88}$$

where we've converted rad/s to ″/yr using $2\pi$ rad $= 360\cdot 60\cdot 60''$ and 1 yr $= 365\cdot 24\cdot 60\cdot 60$ s. As the orbit of the Moon around the Earth lies close to the ecliptic, and since the Moon and Sun both go around the Earth in the same direction, the above effects add, yielding

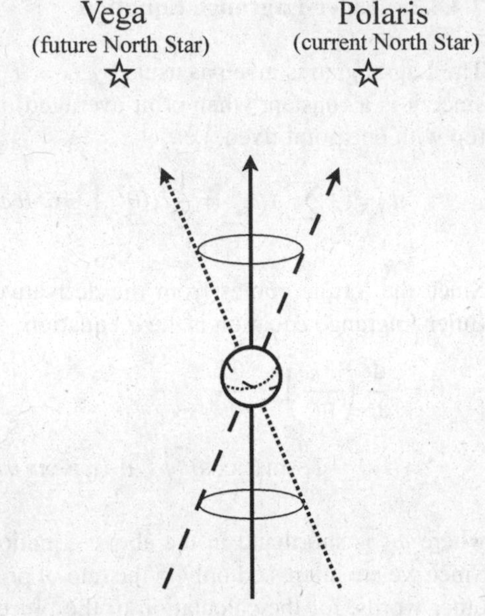


**Fig. 7.17** Precession of the Earth's rotational axis. The dashed arrow shows the current direction of the Earth's rotational axis; the dotted arrow shows its direction ≈13, 000 yr from now. The solid vertical arrow is the direction normal the Earth's orbital plane around the Sun, around which the Earth's rotational axis precesses with a period of ≈26, 000 yr.

$$\dot{\phi}_{\text{tot}} \approx \frac{51''}{\text{yr}} \quad \Leftrightarrow \quad \tau \approx \frac{26,000\text{yr}}{\text{cycle}} \,. \tag{7.89}$$

Currently, the Earth's rotational axis points in the direction of Polaris, the current *North Star*. But in 13,000 years, the axis will be pointing in the direction of Vega, our future North Star, as shown in Fig. 7.17.

## Suggested References

*Full references are given in the bibliography at the end of the book.*

Feynman (1985): A must read for every aspiring physicist. Pages 173–174 describe the wobbling plate anecdote.

Goldstein et al. (2002): One of the classic texts on classical mechanics. Our presentation of rigid body motion in this and the preceding chapter follows the basic structure of Chaps. 5 and 4 in Goldstein.

Landau and Lifshitz (1976): A classic graduate-level text on classical mechanics. Several of the additional problems in this chapter were adapted from problems in the relevant sections of this book.

Marion and Thornton (1995): Another classic text on classical mechanics, especially suited for undergraduate students. Chapter 11 is a detailed discussion of the dynamics of rigid body motion.

## Additional Problems

**Problem 7.1** Consider two reference frames which differ only in the choice of origin, so that the coordinate axes are parallel to one another. Assume that origin $O'$ is located at the center of mass of a rigid body, displaced from origin $O$ by the vector $\mathbf{R}$. Show that the components of the inertia tensor with respect to these two frames are given by

$$I_{ij} = I'_{ij} + M\left(R^2\delta_{ij} - R_iR_j\right), \tag{7.90}$$

where $I'_{ij}$ are the components of the inertia tensor with respect to the reference frame having origin $O'$ at the center of mass. The above relation can be thought of as the generalization of parallel-axis theorem (Sect. 7.2.1) to the full inertia tensor.

**Problem 7.2** A rigid body is composed of three equal mass points $m$ at $\mathbf{r}_1 = (a, 0, 0)$, $\mathbf{r}_2 = (0, a, 2a)$, and $\mathbf{r}_3 = (0, 2a, a)$.

(a) Show that the components $I_{ij} = \sum_I m_I(r_I^2\delta_{ij} - r_{Ii}r_{Ij})$ of the moment of inertia tensor are given by

$$I_{ij} = 2ma^2 \begin{bmatrix} 5 & 0 & 0 \\ 0 & 3 & -2 \\ 0 & -2 & 3 \end{bmatrix}. \tag{7.91}$$

(b) Find the principal moments of inertia and a set of principal axes for this body. (Note: Two of the principal axes are *not* uniquely determined for this case, but you can still choose them appropriately.)

**Problem 7.3** (*Adapted from several examples in Marion and Thornton (1995).*) Consider a uniform cube of mass $M$, side length $a$, and (constant) mass density $\rho$. Assume that it is described with respect to a coordinate system with origin at one corner of the cube, and with axes lying along three edges of the cube, as shown in Fig. 7.18.

(a) Calculate the components $I_{ij}$ of the inertia tensor with respect to this coordinate system.
(b) Find the prinicpal axes and corresponding principal moments of inertia of the cube. (Note: Two of the principal axes are *not* uniquely determined for this case, but you can still choose them appropriately.)
*Hint*: The right-hand side of the characteristic equation $0 = \det(\mathsf{I} - \lambda\mathsf{1})$ can be simplified by first performing elementary row and column operations on the matrix $\mathsf{I} - \lambda\mathsf{1}$ before taking its determinant (See part (c) of Exercise D.12).
(c) Now consider a new coordinate system with origin at the center of mass of the cube, and with axes parallel to the edges of the cube. Calculate the components of the inertia tensor with respect to this new coordinate system.
(d) What are the principal axes and corresponding principal moments of inertia with respect to this new coordinate system?

**Problem 7.4** (*Adapted from Landau and Lifshitz (1976), Sect. 32, Problem 2e.*) Calculate the principal moments of inertia (with respect to an origin located at the center of mass) of a uniform circular cone with base radius $R$, height $h$, and mass $M$ (See Fig. 7.19). You should find

$$I_1 = I_2 = \frac{3}{20}M\left[R^2 + \frac{1}{4}h^2\right], \qquad I_3 = \frac{3}{10}MR^2. \tag{7.92}$$

*Hint*: It is simplest to first calculate the prinicipal moments of inertia with respect to axes $\hat{\mathbf{n}}_{1'}$, $\hat{\mathbf{n}}_{2'}$, $\hat{\mathbf{n}}_{3'} \equiv \hat{\mathbf{n}}_3$, whose origin is at the vertex of the cone, and then use the pararallel axis theorem to get the principal moments of inertia about axes whose origin is at the center of mass.

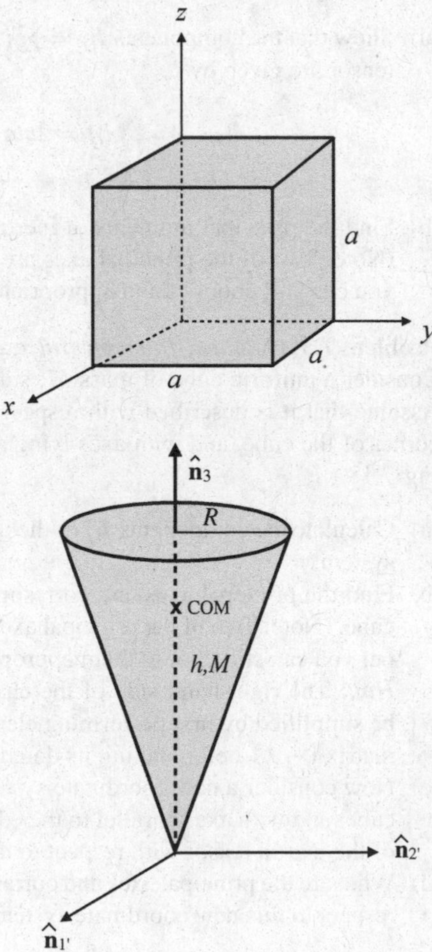


**Fig. 7.18** Uniform cube of mass $M$ and side length $a$, described in a coordinate system with origin at one corner of the cube, and with axes lying along three edges of the cube

**Fig. 7.19** Uniform circular cone having base radius $R$, height $h$, and mass $M$. The axes $\hat{\mathbf{n}}_{1'}$ and $\hat{\mathbf{n}}_{2'}$ pass through the vertex of the cone, perpendicular to the symmetry axis $\hat{\mathbf{n}}_3$. The center of mass of the cone is located at a distance of $3h/4$ above the vertex

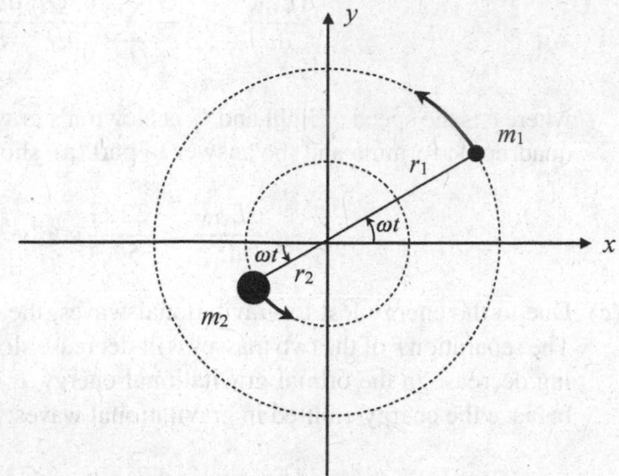


**Fig. 7.20** Two masses $m_1$ and $m_2$ in circular orbits (in the $xy$-plane) around their common center of mass; $r_1$ and $r_2$ denote the radii of these orbits; $r \equiv r_1 + r_2$ is their relative separation; and $\omega$ is the angular velocity of the orbits

**Problem 7.5** Calculate the principal moments of inertia for a uniform ellipsoid (Fig. 7.5) of total mass $M$, whose boundary is defined by

$$\left(\frac{x_1}{a}\right)^2 + \left(\frac{x_2}{b}\right)^2 + \left(\frac{x_3}{c}\right)^2 = 1, \tag{7.93}$$

where $x_1, x_2, x_3$ are coordinates along the principal axes $\hat{\mathbf{n}}_1, \hat{\mathbf{n}}_2, \hat{\mathbf{n}}_3$. You should find

$$I_1 = \frac{1}{5}M(b^2 + c^2), \quad I_2 = \frac{1}{5}M(c^2 + a^2), \quad I_3 = \frac{1}{5}M(a^2 + b^2). \tag{7.94}$$

**Problem 7.6** Consider two masses $m_1$ and $m_2$ in circular orbits of radii $r_1$ and $r_2$ around their common center of mass. Let $r \equiv r_1 + r_2$ denote their relative separation, and $\omega$ the angular velocity of the orbits. See Fig. 7.20.

(a) Treating the binary system initially as if it were a rigid body (i.e., a dumbell with unequal masses) calculate the components $Q_{ij}$ of the reduced rotational inertia tensor (7.78) where $x_i \equiv (x, y, z)$. You should find

$$Q_{ij} = -\frac{1}{2}\mu r^2 \begin{bmatrix} \frac{1}{3} + \cos(2\omega t) & \sin(2\omega t) & 0 \\ \sin(2\omega t) & \frac{1}{3} - \cos(2\omega t) & 0 \\ 0 & 0 & -\frac{2}{3} \end{bmatrix}, \tag{7.95}$$

where $\mu \equiv m_1 m_2/(m_1 + m_2)$ is the reduced mass of the system.

(b) In Einstein's theory of *general relativity* (the replacement for Newton's theory of gravity), such a system actually *loses energy* in the form of *gravitational waves.* To leading order in the velocity of the component masses, the rate of energy loss is given by the so-called **quadrupole formula**

$$\frac{\mathrm{d}E_{\mathrm{GW}}}{\mathrm{d}t} = \frac{1}{5}\frac{G}{c^5}\sum_{i,j}\frac{\mathrm{d}^3 Q_{ij}}{\mathrm{d}t^3}\frac{\mathrm{d}^3 Q_{ij}}{\mathrm{d}t^3}, \tag{7.96}$$

where $c$ is the speed of light and $G$ is Newton's gravitational constant. Using the quadrupole formula and the answer to part (a), show that

$$\frac{\mathrm{d}E_{\mathrm{GW}}}{\mathrm{d}t} = \frac{32}{5}\frac{G}{c^5}\mu^2 r^4 \omega^6. \tag{7.97}$$

(c) Due to the energy lost to gravitational waves, the orbits will begin to *inspiral*. The separation $r$ of the two masses will decrease slowly, leading to a corresponding decrease in the orbital gravitational energy $E_{\mathrm{orb}} = -GM\mu/2r$ in order to balance the energy emitted in gravitational waves:

$$\frac{\mathrm{d}E_{\mathrm{GW}}}{\mathrm{d}t} = -\frac{\mathrm{d}E_{\mathrm{orb}}}{\mathrm{d}t} = -\frac{GM\mu}{2r^2}\dot{r}, \tag{7.98}$$

where $M \equiv m_1 + m_2$ is the total mass of the system. Equating the above two expressions for $\mathrm{d}E_{\mathrm{GW}}/\mathrm{d}t$, and using Kepler's 3rd law ($\omega^2 r^3 = GM$) to relate $r$ and $\dot{r}$ in terms of $\omega$ and $\dot{\omega}$, show that

$$\dot{\omega}^3 = \left(\frac{96}{5}\right)^3 \left(\frac{G}{c^3}\right)^5 \omega^{11} \mathcal{M}_{\mathrm{c}}^5, \tag{7.99}$$

where

$$\mathcal{M}_{\mathrm{c}} \equiv \left(\mu^3 M^2\right)^{1/5} = \frac{(m_1 m_2)^{3/5}}{(m_1 + m_2)^{1/5}} \tag{7.100}$$

is the **chirp mass** of the binary system. The word "chirp" is used since both the emitted power and angular frequency of the motion *increase* as the masses spiral-in on one another.

(d) Invert (7.99) to find

$$\mathcal{M}_{\mathrm{c}} = \frac{c^3}{G}\left[\left(\frac{5}{96}\right)^3 \frac{\dot{\omega}^3}{\omega^{11}}\right]^{1/5}, \tag{7.101}$$

which expresses the chirp mass in terms of the instantaneous frequency and frequency derivative of the inspiraling binary.

Note: Such calculations applied to GW150914 (the first direct detection of gravitational waves from a binary black-hole merger) yield $\mathcal{M}_{\mathrm{c}} \approx 30\, M_{\odot}$, consistent with more careful calculations based on a numerical solution of Einstein's equations. See Abbott et al. (2016) and Abbott et al. (2017) for details.

**Problem 7.7** Consider a compound **physical pendulum** made up of a uniform rod (length $\ell$, mass $m_1$) attached to a pendulum bob (mass $m_2$), as shown in panel (a) of

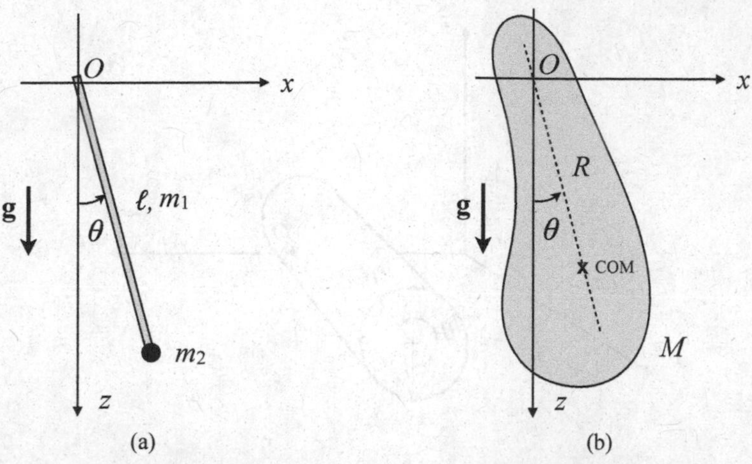


**Fig. 7.21** Panel (a) Compound physical pendulum made up of a uniform rod of length $\ell$, mass $m_1$, attached to a pendulum bob of mass $m_2$. The pendulum pivots about the $y$-axis, which points out of the page at $O$. There is a uniform gravitational field $\mathbf{g}$ pointing downward. Panel (b) A "generic" physical pendulum having total mass $M$, with its center of mass a distance $R$ from the axis of rotation

Fig. 7.21. The pendulum pivots about the $y$-axis (pointing out of the page at $O$) in response to a uniform gravitational field $\mathbf{g}$ (pointing downward).

(a) Write down the Lagrangian for the compound pendulum assuming small angular deviations of the pendulum away from vertical, i.e., $\theta \ll 1$.
(b) Solve the Euler-Lagrange equation in this *small-angle* limit, showing that the pendulum undergoes simple harmonic motion with angular frequency

$$\omega = \sqrt{\frac{3g}{2\ell}\frac{(m_1 + 2m_2)}{(m_1 + 3m_2)}}\,. \tag{7.102}$$

(c) Check that $\omega$ has the correct limiting behavior for $m_1 \ll m_2$ and $m_2 \ll m_1$.
(d) Consider a "generic" physical pendulum having total mass $M$, with its center of mass a distance $R$ from the axis of rotation, as shown in panel (b) of Fig. 7.21. Show that for this more general case,

$$\omega = \sqrt{\frac{gR}{\kappa^2}}\,, \tag{7.103}$$

where $\kappa$ is the **radius of gyration** defined by $I \equiv M\kappa^2$, where $I$ is the moment of inertia of the object about the $y$-axis (pointing out of the page at $O$).
(e) Show that (7.103) reduces to (7.102) by explicitly calculating $R$ and $\kappa$ for the specific physical pendulum shown in panel (a).

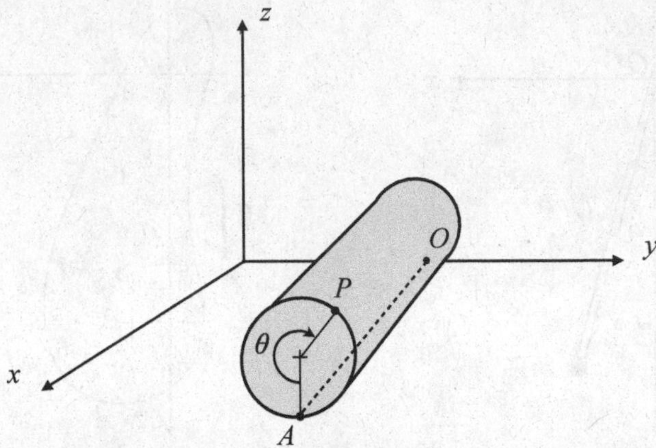


**Fig. 7.22** Uniform circular cylinder (radius $R$, height $h$, mass $M$) that rolls without slipping on a horizontal surface. The position of the cylinder is specified by the angle $\theta$ through which a point $P$ on the edge of the cylinder has turned while rolling on the surface. Line segment $OA$ is the instantaneous axis of rotation. See also Fig. 7.6 for more details

**Problem 7.8** Calculate the kinetic energy $T$ of a uniform cylinder of radius $R$, height $h$, and mass $M$ that rolls without slipping on a horizontal surace, as shown in Fig. 7.22. You should find

$$T = \frac{3}{4} M R^2 \dot{\theta}^2 , \tag{7.104}$$

where $\theta$ is the angle through which the cylinder has turned while rolling. *Hint*: Use the result of Exercise 7.3, noting that the line $OA$ is instantaneously at rest and points in the direction of the instantaneous angular velocity vector $\boldsymbol{\omega}$.

**Problem 7.9** (*Adapted from Landau and Lifshitz (1976), Sect. 32, Problem 7.*) Calculate the kinetic energy of a uniform circular cone (base radius $R$, height $h$, mass $M$) that rolls without slipping on a horizontal surface, as shown in Fig. 7.23. You should find

$$T = \frac{1}{2} M V^2 + \frac{1}{2} \left( I_1 \sin^2 \alpha + I_3 \cos^2 \alpha \right) \omega^2 , \tag{7.105}$$

where

$$V = a \cos\alpha\, \dot{\theta} , \qquad \omega = \frac{V}{a \sin\alpha} = \cot\alpha\, \dot{\theta} , \tag{7.106}$$

are the magnitudes of the velocity of the center of the mass and the angular velocity of the cone, respectively; $2\alpha$ is the opening angle of the cone (i.e., $\tan\alpha = R/h$); and $\theta$ specifies the location of the cone, as defined in the figure. Then, using the results of Problem 7.4, show that the above expression for $T$ reduces to

$$T = \frac{3}{40} M h^2 \dot{\theta}^2 \left( 1 + 5 \cos^2 \alpha \right) . \tag{7.107}$$

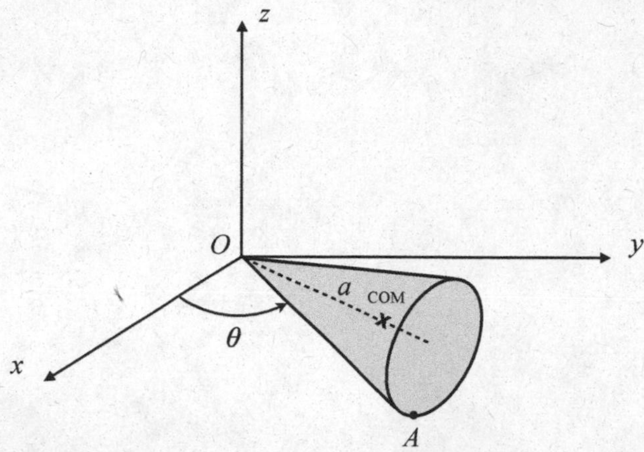


**Fig. 7.23** Circular cone (base radius $R$, height $h$, mass $M$) that rolls without slipping on a horizontal surface. The position of the cone is specified by the angle $\theta$ that the instantaneous axis of rotation $OA$ makes with the $x$-axis of the fixed (inertial) frame. The center of mass, which lies at a distance $a = 3h/4$ from the vertex of the cone, is indicated by an $\times$. See also Fig. 7.19 for more details

*Hint*: Same as for the previous problem.

**Problem 7.10** Show that the third Euler equation in (7.32) is just the Euler-Lagrange equation for the Euler angle $\psi$—i.e.,

$$\frac{\mathrm{d}}{\mathrm{d}t}\left(\frac{\partial T}{\partial \dot{\psi}}\right) - \frac{\partial T}{\partial \psi} - F_\psi = 0\,, \tag{7.108}$$

where

$$F_\psi \equiv \sum_I \frac{\partial \mathbf{r}_I}{\partial \psi} \cdot \mathbf{F}_I = \sum_I (\hat{\mathbf{n}}_3 \times \mathbf{r}_I) \cdot \mathbf{F}_I\,. \tag{7.109}$$

The other equations follow from cyclic permutation of 1, 2, 3. *Hint*: Use the expressions for the components of the angular velocity vector with respect to the principal axes, (6.71):

$$\begin{aligned}
\omega_1 &= -\sin\theta\cos\psi\,\dot{\phi} + \sin\psi\,\dot{\theta}\,,\\
\omega_2 &= \sin\theta\sin\psi\,\dot{\phi} + \cos\psi\,\dot{\theta}\,,\\
\omega_3 &= \cos\theta\,\dot{\phi} + \dot{\psi}\,,
\end{aligned} \tag{7.110}$$

to write the kinetic $T = \frac{1}{2}\sum_i I_i\omega_i^2$ in terms of the Euler angles $(\phi, \theta, \psi)$.

# Chapter 8
# Small Oscillations

One of the more common applications of classical mechanics is the study of small oscillations about an equilibrium state. In this chapter, we will apply the Lagrangian formalism to the general case of coupled $N$-body systems perturbed from equilibrium. The material introduced in this chapter will also provide the framework for the extension to continuous systems and fields in the limit that $N \to \infty$. We will first look at the simple one-dimensional oscillator as a refresher, and then go on to develop the general formulation of the problem, applying it to solve a few simple example problems.

## 8.1 One-Dimensional Oscillator

Consider a particle of mass $m$ that is constrained to move along a 1-dimensional curve under the influence of a time-independent potential. In this case, there is a generalized coordinate $q$ in terms of which the Lagrangian can be written as

$$L = \frac{1}{2}M\dot{q}^2 - U(q), \tag{8.1}$$

where

$$M \equiv m\,\frac{\partial \mathbf{r}}{\partial q} \cdot \frac{\partial \mathbf{r}}{\partial q} \tag{8.2}$$

is independent of $q$.[1] Note that since $q$ may not be a Cartesian coordinate, the $M$ defined by (8.2) need not have the dimensions of mass (Exercise 8.1). The equations

[1]If $M$ is not already independent of $q$, we can always change variables to a new generalized coordinate $q' \equiv \int \mathrm{d}q\,\sqrt{M(q)/M_0}$, in terms of which $M(q)\dot{q}^2 = M_0\dot{q}'^2$, with $M_0$ independent of $q$.



M.J. Benacquista and J.D. Romano, *Classical Mechanics*, Undergraduate Lecture Notes in Physics, https://doi.org/10.1007/978-3-319-68780-3_8

of motion for this system are quickly found from the Euler-Lagrange equations for $L$:

$$M\ddot{q} + \frac{\partial U}{\partial q} = 0\,. \tag{8.3}$$

**Exercise 8.1** (a) Show that for a particle of mass $m$ constrained to move in a plane on the end of a massless rod of length $\ell$ (e.g., a simple pendulum), the $M$ in (8.2) is just

$$M = m\ell^2\,. \tag{8.4}$$

(b) Suppose that instead of a single particle, we have a rigid body that is constrained to move in a plane around an axis $\hat{\mathbf{n}}$—i.e., a *physical pendulum* as discussed in Problem 7.7. Then for this more general case, (8.2) should be replaced by

$$M \equiv \sum_I m_I \frac{\partial \mathbf{r}_I}{\partial q} \cdot \frac{\partial \mathbf{r}_I}{\partial q}\,, \tag{8.5}$$

where the sum is over the individual mass points in the body. Show that with this replacement,

$$M = \sum_I m_I r_I^2 \sin^2\theta_I = \sum_I m_I d_I^2\,, \tag{8.6}$$

where $d_I \equiv r_I \sin\theta_I$ is the perpendicular distance of mass point $m_I$ from $\hat{\mathbf{n}}$. Thus, $M$ has the interpretation of the moment of inertia $I(\hat{\mathbf{n}})$ of the rigid body around the axis, (7.12).

At this point, the potential is still arbitrary (other than the time-independence mentioned above). If the potential has local maxima or minima, then there will exist values of $q$ for which $\partial U/\partial q = 0$. These are points of **equilibrium**. If a particle is placed at an equilibrium point $q_0$ with the initial condition that $\dot{q}(0) = 0$, then the particle will simply remain at this point.

Points of equilibrium can be either **stable** or **unstable**. If small perturbations of the particle about $q_0$ drive the particle away from $q_0$, then the equilibrium is unstable; if the particle is driven back toward $q_0$, then the equilibrium is stable. To see what requirements this places on the behavior of the potential around an equilibrium point, let's consider a small perturbation about equilibrium. The generalized coordinate can then be written as

$$q = q_0 + \eta\,, \tag{8.7}$$

where $\eta$ is assumed to be small relative to $q_0$ or to other relevant scales in the problem. It is clear that $\dot{q} = \dot{\eta}$, but we also need to express $T$ and $U$ in terms of $\eta$. We can use a Taylor expansion for each of these, keeping only the lowest order non-zero terms in $\eta$ and $\dot{\eta}$. Since the kinetic energy is already second-order in $\dot{\eta}$, we can just use the equilibrium value for the mass $M_0 \equiv M(q_0)$, if $M$ is not already independent of $q$. The potential is expanded as

$$U(q) = U(q_0) + \eta \left.\frac{\mathrm{d}U}{\mathrm{d}q}\right|_0 + \frac{1}{2}\eta^2 \left.\frac{\mathrm{d}^2U}{\mathrm{d}q^2}\right|_0 , \tag{8.8}$$

where $|_0$ indicates evaluation at $q = q_0$. Since $q_0$ is an equilibrium point, the term that is linear in $\eta$ is zero. Furthermore, the leading term is simply a constant, so it can be eliminated by resetting the zero of the potential.

The perturbed Lagrangian can now be written entirely in terms of the perturbation $\eta$, which can be taken to be the generalized coordinate for the problem. Thus,

$$L = \frac{1}{2}M_0\dot{\eta}^2 - \frac{1}{2}K\eta^2 , \tag{8.9}$$

where $K \equiv \mathrm{d}^2U/\mathrm{d}q^2|_0$. The equation of motion obtained from this Lagrangian is

$$M_0\ddot{\eta} + K\eta = 0 , \tag{8.10}$$

which can be easily solved to give the general solution

$$\eta(t) = A\mathrm{e}^{\sqrt{-K/M_0}t} + B\mathrm{e}^{-\sqrt{-K/M_0}t} , \tag{8.11}$$

where $A$ and $B$ are integration constants to be determined by the initial conditions. Clearly, if $K < 0$, this solution will exponentially run away from equilibrium, and thus the equilibrium is unstable. If $K > 0$, the solution will oscillate about the equilibrium, indicating a stable equilibrium. If $K = 0$, then the type of equilibrium is indeterminate (explained in more detail below). Since $K$ is the value of the second derivative of the potential at equilibrium, $K < 0$ corresponds to a local maximum of the potential, while $K > 0$ corresponds to a local minimum. This makes sense in that a perturbation about a local maximum will drive the system away from equilibrium, while a perturbation about a local minimum will drive the system back toward the minimum. If the equilibrium point is an *inflection point* (e.g., if the potential is *cubic* about equilibrium), then $K = 0$ and the system is unstable. Potentials that are *quartic* about equilibrium also have $K = 0$, but the system can be either in stable (or unstable) equilibrium depending on whether the equilbrium point is a local minimum (or maximum) of the potential. A graphical representation of stable and unstable equilibrium points are shown in Fig. 8.1.

Stable equilibria where $K = 0$ (e.g., the quartic potential discussed above) do not produce linear equations of motion such as (8.10), since the first non-zero term in the potential would be proportional, e.g., to $\eta^4$. Since we are interested here in *small*

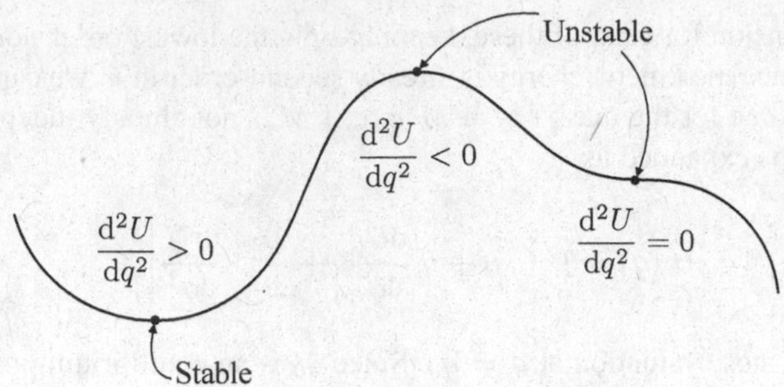


**Fig. 8.1** Stable and unstable equilibrium points for an arbitrary potential. Stable equilibria occur at local minima, while unstable equilibria occur at local maxima and inflection points

oscillations (for which terms quadratic or higher in the amplitude of the oscillation are negligible), we will ignore these quartic-like potentials, and restrict our attention to the stable solutions of *linear* equations. In this case, the solution is given by (8.11) with $K > 0$. But since this solution is *complex*, we will ultimately need to take its real part. So let's return to (8.10), and go through the solution in a little more detail. This will prove to be illustrative for the general $N$-body solution to be discussed later.

### 8.1.1 Free Oscillations

For this more general analysis, we allow $\eta$ to initially be complex and have the periodic form

$$\eta = Z e^{i\omega t}, \tag{8.12}$$

where $Z$ is a complex constant amplitude and $\omega$ is a real constant. (To get the final *real-valued* solution we will need to take its real part at the end of the calculation.) Inserting this trial solution into (8.10), we find that these constants must obey

$$\left(\omega^2 - \frac{K}{M_0}\right) Z = 0. \tag{8.13}$$

This can be thought of as a one-dimensional eigenvalue equation, similar to what we have seen in Appendix D.5. In this case, the non-trivial solutions (which have $Z \neq 0$), are those for which

$$\omega^2 - \frac{K}{M_0} = 0. \tag{8.14}$$

Thus, the eigenvalues are the two allowed frequencies $\omega = \pm\omega_0$, with

$$\omega_0 \equiv \sqrt{\frac{K}{M_0}}. \tag{8.15}$$

The general solution is a linear combination of the two eigenfunctions, giving

$$\eta = \mathrm{Re}\left(Z_+ \mathrm{e}^{\mathrm{i}\omega_0 t} + Z_- \mathrm{e}^{-\mathrm{i}\omega_0 t}\right) = \mathrm{Re}\left(C\mathrm{e}^{\mathrm{i}\omega_0 t}\right), \tag{8.16}$$

where $C \equiv Z_+ + Z_-^*$. Writing $C = A\mathrm{e}^{\mathrm{i}\phi}$, where $A$ and $\phi$ are real, we have

$$\eta = A\cos(\omega_0 t + \phi). \tag{8.17}$$

The constants $A$ and $\phi$ are determined by the initial conditions. The **natural frequency** $\omega_0$ is solely determined by the second derivative (or curvature) of the potential evaluated at $q_0$.

### 8.1.2 Damped Oscillations

All real oscillators have some dissipation mechanism which removes energy from the system. Such forces are often (but not always) associated with some sort of friction, and so they cannot be described by a potential. But we can insert, by hand, a frictional force $f$ into the equation of motion, so that

$$M_0\ddot{\eta} + K\eta = f. \tag{8.18}$$

Although frictional forces can take on a variety of forms, we will consider only *velocity-dependent* frictional forces of the form $f = -f_0\dot{\eta}$. For a damped free oscillator, the equation of motion can then be written as

$$\ddot{\eta} + 2\varphi\dot{\eta} + \omega_0^2\eta = 0, \tag{8.19}$$

where $\varphi \equiv f_0/2M_0$. Since (8.19) is a linear differential equation with constant coefficients, it can be solved as before, assuming a solution of the form

$$\eta = Z\mathrm{e}^{\mathrm{i}\omega_\mathrm{d} t}, \tag{8.20}$$

and then taking its real part. Substituting this function into (8.19) yields an algebraic equation for $\omega_\mathrm{d}$:

$$-\omega_\mathrm{d}^2 + 2\mathrm{i}\varphi\omega_\mathrm{d} + \omega_0^2 = 0. \tag{8.21}$$

There are two complex solutions to this equation,

$$\omega_{\rm d} = \pm\sqrt{\omega_0^2 - \varphi^2} + {\rm i}\varphi\,. \tag{8.22}$$

Substituting these expressions for $\omega_{\rm d}$ back into (8.20), and taking the real part of the complex solution, we find

$$\eta = A{\rm e}^{-\varphi t}\cos\left(\omega_0\, t\sqrt{1 - \frac{\varphi^2}{\omega_0^2}} + \phi\right), \tag{8.23}$$

where $A$ and $\phi$ are the two constants needed to match the initial conditions. Note that the solution exponentially decays over time with a decay time constant $\tau \equiv 1/\varphi = 2M_0/f_0$.

**Exercise 8.2** Obtain (8.23) from the complex solution for the damped oscillator.

Note that the oscillation frequency of the damped oscillator is *not* the natural frequency of the free oscillator. It is lower by the multiplicative factor $\sqrt{1-(\varphi/\omega_0)^2}$. Thus, increasing the strength of the damping coefficient will decrease the oscillation frequency. Furthermore, the decay time scales as $1/\varphi$, so an increase in the damping coefficient will also decrease the decay time. For a sufficiently small damping coefficient, the decay time will be much longer than the oscillation period and the system is said to be **underdamped**; the system will undergo many cycles before the amplitude changes substantially. If the damping coefficient is large, then the oscillation frequency will be imaginary and the system will exponentially decay to zero. In this case the system is said to be **overdamped**. These situations are shown in Fig. 8.2. When $\varphi = \omega_0$, then the system is **critically damped**. In this case, the oscillation frequency is 0. In a critically damped system, the system returns to equilibrium as quickly as possible without oscillating, as can be seen in Fig. 8.3. The ratio $\varphi/\omega_0$ is sometimes called the **damping ratio**.

### *8.1.3 Damped and Driven Oscillations*

The equation of motion for the combined damped, driven oscillator is[2]

$$\ddot{\eta} + 2\varphi\dot{\eta} + \omega_0^2\eta = F_0\cos(\omega t + \delta)\,, \tag{8.24}$$

[2]There is no loss of generality in assuming a sinusoidal driving force as we have done here, since most real driving forces can be expressed as a Fourier series or a Fourier transform, which involve sums of such oscillatory terms.

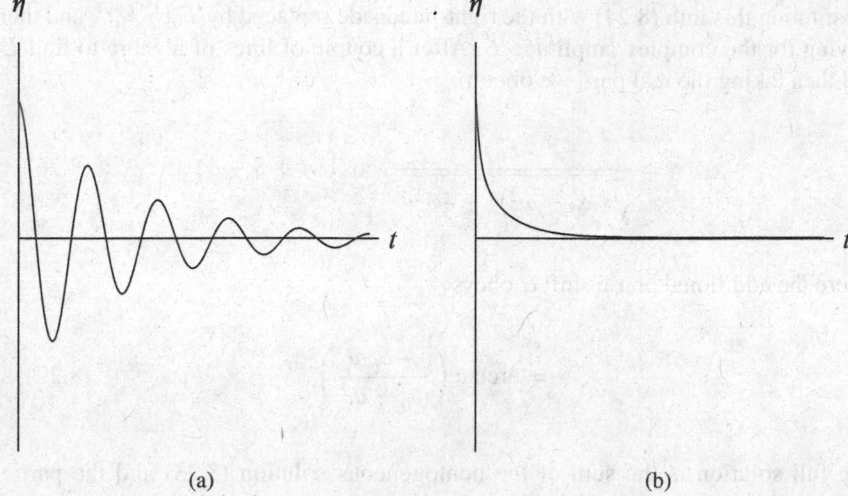


**Fig. 8.2** Time evolution of (a) an underdamped oscillator, and (b) an overdamped oscillator

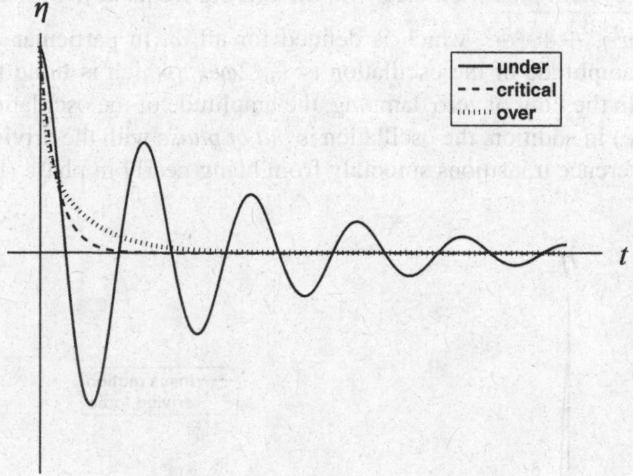


**Fig. 8.3** Underdamped, overdamped, and critically damped motion. Note that the critically damped motion arrives at equilibrium before the overdamped motion

where $\varphi$ is as before, and $F_0$, $\omega$, and $\delta$ are the amplitude, frequency, and phase of the driving force $F_0 \cos(\omega t + \delta)$. This equation should look familiar to anyone who has studied $RLC$-circuits in an introductory physics class. The general homogeneous solution is simply the damped free oscillator solution given by (8.23), while the particular solution can be obtained by assuming a complex solution of the form

$$\eta_P = E e^{i\omega t} , \tag{8.25}$$

substituting this into (8.24) with the right-hand side replaced by $F_0 e^{i(\omega t+\delta)}$, and then solving for the complex amplitude $E$. After a couple of lines of algebra to find $E$, and then taking the real part, we obtain

$$\eta_p = \frac{F_0}{\sqrt{\left(\omega_0^2 - \omega^2\right)^2 + 4\varphi^2\omega^2}} \cos\left(\omega t + \delta + \alpha\right), \tag{8.26}$$

where the additional phase shift $\alpha$ obeys

$$\alpha = \arctan\left(\frac{-2\varphi\omega}{\omega_0^2 - \omega^2}\right). \tag{8.27}$$

The full solution is the sum of the homogeneous solution (8.23) and the particular solution $\eta_p$. As shown in Fig. 8.4, the homogeneous solution decays away with a time scale given by $\tau$, leaving the particular solution, which is long-term solution. The system then oscillates with the driving frequency, with an amplitude $F_0/\sqrt{\left(\omega_0^2 - \omega^2\right)^2 + 4\varphi^2\omega^2}$, which is defined for all $\omega$. In particular, note that at $\omega = \omega_0$ the amplitude of the oscillation is $F_0/2\varphi\omega_0$, which is finite for non-zero damping $\varphi$. (In the limit of zero damping, the amplitude of the oscillation is infinite when $\omega = \omega_0$.) In addition, the oscillation is *out of phase* with the driving force, but the phase difference transitions smoothly from being nearly in phase (for $\omega \ll \omega_0$)

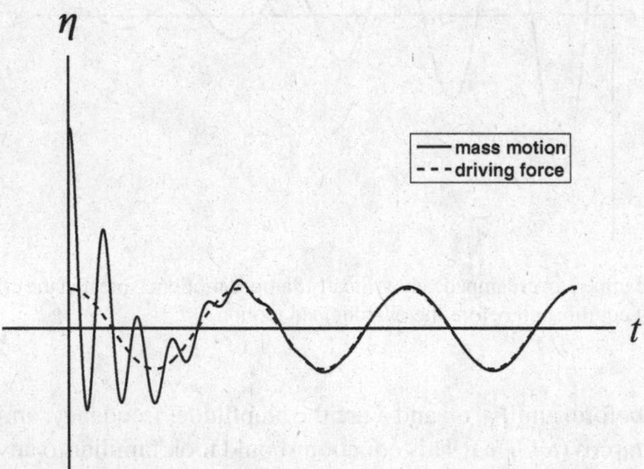


**Fig. 8.4** A damped driven oscillator, showing the initial transient behavior of the homogeneous solution, which eventually decays away, leaving the system oscillating with the driving frequency. In this example, the driving frequency is less than the natural frequency

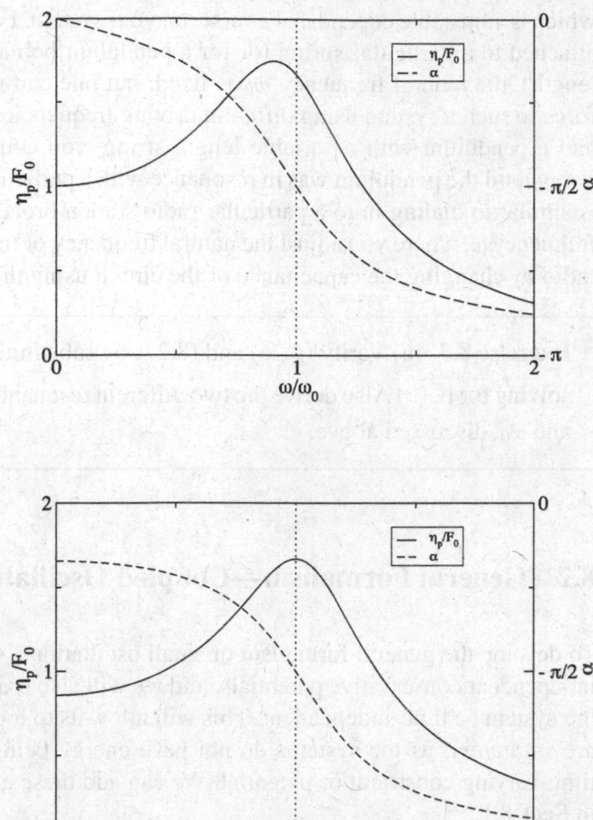


**Fig. 8.5** The amplitude (solid curve) and phase (dashed curve) of the steady-state damped, driven oscillator with $\varphi = 0.3\omega_0$. (The amplitude scale is on the left-hand side of the figures; the phase scale is on the right-hand side the figures.) Note that the phase passes smoothly through $-\pi/2$ at $\omega = \omega_0$ and that the amplitude reaches a maximum in the vicinity of $\omega = \omega_0$ (Exercise 8.3). The top panel corresponds to the case where the natural frequency $\omega_0$ is fixed, but the driving frequency $\omega$ is variable. The bottom panel corresponds to the opposite case where the driving frequency $\omega$ is fixed, but the natural frequency $\omega_0$ is variable

to being nearly 180° out of phase (for $\omega \gg \omega_0$). When $\omega = \omega_0$, the phase difference is exactly 90°. The amplitude and phase behavior can be seen in Fig. 8.5.

### *8.1.4 Resonance*

Note that there are *two* possible behaviors for the amplitude and the phase of the damped, driven oscillator, depending on which frequency (natural or driving) is held fixed and which is variable. For fixed natural frequency $\omega_0$, the amplitude of the steady-state oscillations is maximum when $\omega = \sqrt{\omega_0^2 - 2\varphi^2}$, assuming weak damping, $\varphi \ll \omega_0$. For the case where the driving frequency $\omega$ is fixed, the amplitude of the oscillations is maximum when $\omega = \omega_0$. (You can see this behavior in the two panels of Fig. 8.5.) These two different values for $\omega$ are the **resonant frequencies** for the two different scenarios. The difference between which frequency is fixed and which is adjustable depends, of course, on your system. For example, for a fixed mass attached to a particular spring (or for a pendulum bob attached to a string of fixed length), the natural frequency $\omega_0$ is fixed, but one can imagine applying a driving force to such a system using different driving frequencies. On the other hand, if you had a pendulum with a variable length string, you could adjust the length of the string until the pendulum was in resonance with a particular fixed driving force. This is similar to dialing in to a particular radio station broadcasting at a fixed (driving) frequency $\omega$, where you adjust the natural frequency of the $LC$-tuning circuit in your radio by changing the capacitance of the circuit using the tuning knob.

**Exercise 8.3** (a) Verify (8.26) and (8.27) by substituting (8.25) into (8.24) and solving for E. (b) Also derive the two different resonant frequencies, $\sqrt{\omega_0^2 - 2\varphi^2}$ and $\omega_0$, discussed above.

## 8.2 General Formalism—Coupled Oscillations

To develop the general formalism of small oscillations, we will consider only time-independent conservative potentials, and we will also require that any constraints on the system be time-independent. This will allow us to look at equilibrium states that are *stationary*, as the systems do not have energy being added to them through a time-varying constraint or potential. We can add these effects later on, as was done in Sect. 8.1.

The general system of oscillators is described by the $N$ position vectors $\mathbf{r}_I$, which are functions of $n \leq 3N$ independent generalized coordinates $q^a$, so $\mathbf{r}_I(q^1, q^2, \ldots, q^n)$. The generalized potential is then $U = U(q^1, q^2, \ldots, q^n)$, and the Lagrangian is $L = T - U$. From (2.76), we have

$$\frac{\mathrm{d}}{\mathrm{d}t}\left(\frac{\partial T}{\partial \dot{q}^a}\right) - \frac{\partial T}{\partial q^a} - F_a = 0\,, \qquad a = 1, 2, \ldots, n\,, \tag{8.28}$$

where $F_a$ is the generalized force. The generalized force is expressible as the gradient of the potential with respect to the generalized coordinates, $F_a = -\partial U/\partial q^a$. Since the constraints are time-independent, the kinetic energy $T$ can be written in terms of the generalized coordinates and their time-derivatives as

$$T = \frac{1}{2}\sum_{a,b} T_{ab}\dot{q}^a\dot{q}^b\,, \tag{8.29}$$

with

$$T_{ab} = \sum_I m_I \frac{\partial \mathbf{r}_I}{\partial q^a}\cdot\frac{\partial \mathbf{r}_I}{\partial q^b}\,. \tag{8.30}$$

Additional details can be found in Sect. 2.7.

Let $q_0^a$ denote the values of the generalized coordinates at equilibrium, and let's assume that the system is initially at rest, so that $\dot{q}^a|_0 = 0$. Then the condition $\ddot{q}^a = 0$ guarantees that if the system is *initially* in equilibrium, then it will *remain* in equilibrium. Using (8.28), these conditions lead to

$$F_a|_0 = -\left.\frac{\partial U}{\partial q^a}\right|_0 = 0, \quad a = 1, 2, \ldots, n, \tag{8.31}$$

where $|_0$ indicates that the derivatives are evaluated at the equilibrium values $q_0^a$. Note that this is the statement that the components of the generalized force vanish at equilibrium.

Now let's look at the effects of small perturbations about these equilibrium positions. We define the perturbations as $\eta^a$, where

$$q^a = q_0^a + \eta^a, \tag{8.32}$$

with the $\eta^a$ assumed to be small. Remembering that the $q_0^a$ are fixed quantities, the time derivatives of $q^a$ are

$$\dot{q}^a = \dot{\eta}^a, \quad \ddot{q}^a = \ddot{\eta}^a. \tag{8.33}$$

The kinetic energy is then

$$T = \frac{1}{2}\sum_{a,b} T_{ab}^0 \dot{\eta}^a \dot{\eta}^b, \tag{8.34}$$

where

$$T_{ab}^0 = \sum_I m_I \left.\frac{\partial \mathbf{r}_I}{\partial q^a}\right|_0 \left.\frac{\partial \mathbf{r}_I}{\partial q^b}\right|_0. \tag{8.35}$$

Note that this is an expansion in terms of powers of the $\eta^a$ and their derivatives $\dot{\eta}^a$. Since the leading term in the expansion of $T$ is already quadratic in $\dot{\eta}^a$, we keep only the 0th-order terms in $T_{ab}$.

The expansion of the potential is the standard $n$-dimensional Taylor series expansion about $q_0^a$:

$$U = U(q_0^a) + \sum_a \left.\frac{\partial U}{\partial q^a}\right|_0 \eta_a + \frac{1}{2}\sum_{a,b} \left.\frac{\partial^2 U}{\partial q^a \partial q^b}\right|_0 \eta^a \eta^b + \cdots. \tag{8.36}$$

The condition for equilibrium ensures that the 1st-order term in the expansion vanishes, and we can always add an arbitrary constant to the potential to set $U(q_0^a) = 0$. Thus, the leading non-zero term in the expansion of the potential is

$$U = \frac{1}{2} \sum_{a,b} U_{ab} \eta^a \eta^b \,, \quad \text{with} \quad U_{ab} \equiv \left. \frac{\partial^2 U}{\partial q^a \partial q^b} \right|_0 . \tag{8.37}$$

The Lagrangian can now be written entirely in terms of the perturbations $\eta^a$, which can be taken as new generalized coordinates. Thus, for small $\eta$, we have

$$L = \frac{1}{2} \sum_{a,b} \left( T^0_{ab} \dot{\eta}^a \dot{\eta}^b - U_{ab} \eta^a \eta^b \right) = \frac{1}{2} \left( \dot{\eta}^T \mathsf{T} \dot{\eta} - \eta^T \mathsf{U} \eta \right) , \tag{8.38}$$

where we have dropped the superscript 0 from the matrix representation $\mathsf{T}$ of $T^0_{ab}$ to simplify the notation in what follows. We can now obtain the equations of motion for the system in the standard way, finding

$$\sum_b \left( T^0_{ab} \ddot{\eta}^b + U_{ab} \eta^b \right) = 0 \,, \qquad a = 1, 2, \ldots, n \,. \tag{8.39}$$

In general, each of these equations of motion will involve all of the new generalized coordinates $\eta^a$, so we must solve this *coupled* set of differential equations simultaneously to determine the motion near the equilibrium point.

Following the procedure used in the one-dimensional case, we assume a set of oscillatory solutions of the form

$$\eta^a = A^a e^{i\omega t} \,, \qquad a = 1, 2, \ldots, n \,, \tag{8.40}$$

where the $A^a$ are complex amplitudes. At the end of the calculation we take the real part of the solution. Plugging this form of the solution into (8.39), we find the following equation governing the amplitudes:

$$\sum_b \left[ U_{ab} - \omega^2 T^0_{ab} \right] A^b = 0 \,, \qquad a = 1, 2, \ldots, n \,. \tag{8.41}$$

In matrix notation,

$$\left[ \mathsf{U} - \omega^2 \mathsf{T} \right] \mathsf{A} = 0 \,, \tag{8.42}$$

which resembles the standard eigenvalue/eigenvector equation (D.100) with $\omega^2$ playing the role of $\lambda$. Although $\mathsf{T}$ is not the unit matrix, it is still the case that non-zero solutions to (8.42) for $\mathsf{A}$ require that

$$\det[\mathsf{U} - \omega^2 \mathsf{T}] = 0 \,. \tag{8.43}$$

So we will still call the values of $\omega^2$ which solve (8.43) the *eigenvalues* $\omega_\alpha^2$, and the vectors $\mathsf{A}_\alpha$ for which

$$[\mathsf{U} - \omega_\alpha^2 \mathsf{T}]\mathsf{A}_\alpha = 0\,, \qquad \alpha = 1, 2, \ldots, n\,, \tag{8.44}$$

the corresponding *eigenvectors* of this equation. Note that we will be using Greek indices like $\alpha$ and $\beta$ to label a particular eigenvector $\mathsf{A}_\alpha$ and its corresponding eigenvalue $\omega_\alpha$, while components of vectors and generalized coordinates will be labeled (as usual) with Latin indices like $a$ and $b$. Thus, for example, $A_\alpha^a$ will denote the $a$th component of the $\alpha$th eigenvector $\mathsf{A}_\alpha$.

## 8.3 Solving the Eigenvalue/Eigenvector Equation

To solve the eigenvalue/eigenvector equation (8.42), we begin by considering the $\alpha$th eigenvalue $\omega_\alpha$ and its associated eigenvector $\mathsf{A}_\alpha$:

$$\left[\mathsf{U} - \omega_\alpha^2 \mathsf{T}\right]\mathsf{A}_\alpha = 0\,. \tag{8.45}$$

The adjoint (complex conjugate transpose) equation is

$$\mathsf{A}_\beta^\dagger \left[\mathsf{U} - \omega_\beta^{*2} \mathsf{T}\right] = 0\,, \tag{8.46}$$

where we have changed the label $\alpha$ to $\beta$ for later convenience. So far, we have made no assumptions about the nature of the eigenvectors, so the adjoint $\mathsf{A}_\beta^\dagger$ is left as a complex conjugate row vector. On the other hand, we used the fact that both $\mathsf{U}$ and $\mathsf{T}$ are real and symmetric, so they are Hermitian (i.e., $\mathsf{U}^\dagger = \mathsf{U}$ and similarly for $\mathsf{T}$). Now, let's multiply (8.45) on the left by $\mathsf{A}_\beta^\dagger$, and (8.46) on the right by $\mathsf{A}_\alpha$. Taking the difference between these two expressions gives

$$\left(\omega_\alpha^2 - \omega_\beta^{*2}\right) \mathsf{A}_\beta^\dagger \mathsf{T} \mathsf{A}_\alpha = 0\,. \tag{8.47}$$

Let's look first at the case $\alpha = \beta$. Clearly, $\mathsf{A}_\beta^\dagger \mathsf{T} \mathsf{A}_\beta$ is real, because $\mathsf{T}$ is Hermitian. Since $\omega_\beta^2 - \omega_\beta^{*2}$ is twice the imaginary part of $\omega_\beta^2$, if $\mathsf{A}_\beta^\dagger \mathsf{T} \mathsf{A}_\beta$ is non-zero, then the eigenvalues $\omega_\beta^2$ must be real. To show that this indeed the case, begin by splitting the eigenvector $\mathsf{A}_\beta$ into its real and imaginary parts:

$$\mathsf{A}_\beta = \mathsf{a}_\beta + \mathrm{i}\mathsf{b}_\beta\,. \tag{8.48}$$

It then follows that

$$\mathsf{A}_\beta^\dagger \mathsf{T} \mathsf{A}_\beta = \mathsf{a}_\beta^T \mathsf{T} \mathsf{a}_\beta + \mathsf{b}_\beta^T \mathsf{T} \mathsf{b}_\beta + \mathrm{i}\left(\mathsf{a}_\beta^T \mathsf{T} \mathsf{b}_\beta - \mathsf{b}_\beta^T \mathsf{T} \mathsf{a}_\beta\right)\,. \tag{8.49}$$

Again, because $\mathsf{T}$ is Hermitian, $\mathsf{b}_\beta^T\mathsf{T}\mathsf{a}_\beta = \mathsf{a}_\beta^T\mathsf{T}\mathsf{b}_\beta$, so the imaginary part is identically zero, confirming that the product is real. The remaining part consists of $\mathsf{T}$ "sandwiched" between two real vectors ($\mathsf{a}$ and $\mathsf{b}$). Since we already know that the kinetic energy term in the Lagrangian satisfies

$$\frac{1}{2}\dot{\eta}^T\mathsf{T}\dot{\eta} > 0 \tag{8.50}$$

for a real, non-zero vector $\dot{\eta}$, it follows that

$$\mathsf{A}_\beta^\dagger\mathsf{T}\mathsf{A}_\beta = \mathsf{a}_\beta^T\mathsf{T}\mathsf{a}_\beta + \mathsf{b}_\beta^T\mathsf{T}\mathsf{b}_\beta > 0\,. \tag{8.51}$$

*Thus, the eigenvalues $\omega_\beta^2$ are real.*

Let's return to (8.45), and multiply it on the left by $\mathsf{A}_\alpha^\dagger$. The resulting equation can then be solved for the eigenvalue:

$$\omega_\alpha^2 = \frac{\mathsf{A}_\alpha^\dagger\mathsf{U}\mathsf{A}_\alpha}{\mathsf{A}_\alpha^\dagger\mathsf{T}\mathsf{A}_\alpha}\,, \qquad \alpha = 1, 2, \ldots, n\,. \tag{8.52}$$

Although we don't yet know the eigenvectors, we do know that the denominator is positive. If we are going to require that the equilibrium point be stable, then we must have real frequencies $\omega_\alpha$. This requires $\eta^\dagger\mathsf{U}\eta > 0$ for all $\eta$. In the language of components,

$$\sum_{a,b} \left.\frac{\partial^2 U}{\partial q^a \partial q^b}\right|_0 \eta^a\eta^b > 0\,, \tag{8.53}$$

which is the $n$-dimensional generalization of the requirement that the one-dimensional stable equilibria occur at local minima of the potential.

It can be shown (Problem 8.1) that if all the components in the matrix $\mathsf{U} - \omega_\alpha^2\mathsf{T}$ are real, then all the components in the vector $\mathsf{A}_\alpha$ are real up to a common complex phase factor. Thus, we can write the eigenvectors as

$$\mathsf{A}_\alpha = \mathrm{e}^{\mathrm{i}\phi_\alpha}\mathsf{z}_\alpha\,, \qquad \alpha = 1, 2, \ldots, n\,, \tag{8.54}$$

where all of the components of $\mathsf{z}_\alpha$ are real and are determined up to one free component, which we will take to be $z_\alpha^n$. Thus, without loss of generality, we can remove the phase factor from $\mathsf{A}_\alpha$ and work with the *real* eigenvector $\mathsf{z}_\alpha$.

Returning to (8.47) and working with the knowledge that everything is real, we have

$$\left(\omega_\alpha^2 - \omega_\beta^2\right)\mathsf{z}_\beta^T\mathsf{T}\mathsf{z}_\alpha = 0\,. \tag{8.55}$$

If all the eigenvalues are distinct, then we are left with something similar to an orthogonality condition on the eigenvectors. In principle, this condition only requires that $\mathsf{z}_\beta^T \mathsf{T} \mathsf{z}_\alpha$ be diagonal. But recalling that the definition of $\mathsf{z}_\alpha$ retained the freedom in choosing the value of one component $z_\alpha^n$, we can also normalize the eigenvectors so that

$$\mathsf{z}_\beta^T \mathsf{T} \mathsf{z}_\alpha = \delta_{\alpha\beta} \,. \tag{8.56}$$

This choice is made to simplify the construction of the normal modes of oscillation in Sect. 8.4. Note that with this choice of normalization, $\mathsf{z}^2$ has dimensions equal to the inverse of the dimensions of $\mathsf{T}$.

Equation (8.56) is not a pure orthonormality condition since it involves the matrix $\mathsf{T}$. In addition, if the eigenvalues are not all distinct, then there will be more freedom than simply choosing $z_\alpha^n$, and the additional eigenvectors for these degenerate eigenvalues may not necessarily be orthogonal. However, we can always use the Gram-Schmidt procedure to make this set of eigenvectors orthonormal (See Appendix D.3.1 for details). Thus, combining all the eigenvectors $\mathsf{z}_\alpha$ into a single matrix $\mathsf{Z}$, as is done in Appendix D.5.2, we have

$$\mathsf{Z}^T \mathsf{T} \mathsf{Z} = \mathsf{1} \,. \tag{8.57}$$

The matrix $\mathsf{Z}$ is also known as the **modal matrix** and has the property that it diagonalizes $\mathsf{T}$.

Now, let's consider writing the equations of motion as a matrix equation. First we introduce the matrix $\Omega$ which is diagonal in the eigenvalues $\omega_\alpha^2$. In component notation, we have $\Omega_{\alpha\beta} = \omega_\alpha^2 \delta_{\alpha\beta}$. Remembering that the modal matrix is simply the matrix built from the eigenvectors, (8.45) can be written as

$$\mathsf{U}\mathsf{Z} - \mathsf{T}\mathsf{Z}\Omega = 0 \,. \tag{8.58}$$

Multiplying this equation on the left by $\mathsf{Z}^T$ gives

$$\mathsf{Z}^T \mathsf{U} \mathsf{Z} = \mathsf{Z}^T \mathsf{T} \mathsf{Z} \Omega = \Omega \,, \tag{8.59}$$

so the modal matrix manages to diagonalize *both* $\mathsf{T}$ *and* $\mathsf{U}$.

We now have enough information to construct the general solution to the equations of motion (8.45) for arbitrary initial conditions. We will do this in the next section and introduce new coordinates (called *normal coordinates*) that will allow us to decouple the eigenvector solutions.

## 8.4 Normal Modes, Normal Coordinates, and General Solution

From the previous section, we have seen that the equations of motion for small oscillations about equilibrium can be cast in the form of an eigenvalue/eigenvector problem. The solutions were a set of possibly complex eigenvectors $\mathsf{A}_\alpha$ that were associated with the eigenvalues $\omega_\alpha^2$. The eigenvectors could be expressed in terms of real eigenvectors $\mathsf{z}_\alpha$ times an overall complex phase factor $e^{i\phi_\alpha}$, so the generalized coordinates could be written as

$$\eta_\alpha^a = z_\alpha^a e^{i(\omega_\alpha t + \phi_\alpha)} . \tag{8.60}$$

The general solution is then a linear combination of these eigenvalue solutions:

$$\eta^a = \sum_\alpha C_\alpha z_\alpha^a e^{i\omega_\alpha t} , \qquad a = 1, 2, \ldots, n , \tag{8.61}$$

where we have absorbed the phase factors $e^{i\phi_\alpha}$ into the complex constants $C_\alpha$, which are determined by the initial conditions. The eigenvectors $\mathsf{z}_\alpha$ and the (positive) square root of the eigenvalues $\omega_\alpha$ are called the **normal modes** and **normal mode** (or resonant) **frequencies** of the oscillating system.

Recalling that the components of the modal matrix $\mathsf{Z}$ are

$$Z_{a\alpha} = z_\alpha^a , \tag{8.62}$$

we can rewrite (8.61) as

$$\eta^a = \sum_\alpha Z_{a\alpha} C_\alpha e^{i\omega_\alpha t} , \tag{8.63}$$

which can be viewed as the components of a matrix equation for the vector $\eta$. If we specify the initial conditions as $\eta(0) = \eta_0$ and $\dot{\eta}(0) = \dot{\eta}_0$, and recall that the final solution is just the real part of (8.63), then the initial conditions constrain the real and imaginary parts of the vector $\mathsf{C}$ through

$$\eta_0 = \mathsf{Z}\,(\mathrm{Re}\,\mathsf{C}) , \qquad \dot{\eta}_0 = -\mathsf{Z}\Omega^{1/2}\,(\mathrm{Im}\,\mathsf{C}) , \tag{8.64}$$

where $\Omega^{1/2}$ is the square root of the diagonal eigenvalue matrix defined in (8.59). If we then multiply the equations in (8.64) on the left by $\mathsf{Z}^T\mathsf{T}$, we obtain the solution for $\mathsf{C}$ as

$$\mathsf{C} = \mathsf{Z}^T \mathsf{T} \eta_0 - \mathrm{i}\Omega^{-1/2} \mathsf{Z}^T \mathsf{T} \dot{\eta}_0 \,. \tag{8.65}$$

**Exercise 8.4** Verify (8.64) and (8.65).

The solution for $\eta$ given by either (8.61) or (8.63) shows that each generalized coordinate $\eta^a$ moves (in general) in a non-periodic fashion being a sum over the normal modes $\mathsf{Z}_\alpha$ corresponding to different normal mode frequencies $\omega_\alpha$. There is a way, however, to use the modal matrix to obtain a *new* set of coordinates, which individually oscillate with a single normal mode frequency of the system. These new coordinates $Q_\alpha$ are called **normal coordinates**, and they are related to the generalized coordinates $\eta^a$ via

$$\eta \equiv \mathsf{Z}\mathsf{Q} \quad \Leftrightarrow \quad \mathsf{Q} \equiv \mathsf{Z}^T \mathsf{T} \eta \,. \tag{8.66}$$

Note that the normal coordinates are just the coefficients multiplying the eigenvectors in an eigenvector expansion of $\eta$. Using the above definition, we can rexpress the Lagrangian (8.38) in term of $\mathsf{Q}$ as follows. For the kinetic energy, we have

$$T = \frac{1}{2}\dot{\eta}^T \mathsf{T} \dot{\eta} = \frac{1}{2}\dot{\mathsf{Q}}^T \mathsf{Z}^T \mathsf{T} \mathsf{Z} \dot{\mathsf{Q}} = \frac{1}{2}\dot{\mathsf{Q}}^T \mathsf{1} \dot{\mathsf{Q}} = \frac{1}{2}\sum_\alpha \dot{Q}_\alpha^2 \,. \tag{8.67}$$

For the potential energy,

$$U = \frac{1}{2}\eta^T \mathsf{U} \eta = \frac{1}{2}\mathsf{Q}^T \mathsf{Z}^T \mathsf{U} \mathsf{Z} \mathsf{Q} = \frac{1}{2}\mathsf{Q}^T \Omega \mathsf{Q} = \frac{1}{2}\sum_\alpha \omega_\alpha^2 Q_\alpha^2 \tag{8.68}$$

Thus, in terms of the normal coordinates, the Lagrangian is simply

$$L = \frac{1}{2}\sum_\alpha \left(\dot{Q}_\alpha^2 - \omega_\alpha^2 Q_\alpha^2\right) \,. \tag{8.69}$$

The equations of motion are then

$$\ddot{Q}_\alpha + \omega_\alpha^2 Q_\alpha = 0 \,, \qquad \alpha = 1, 2, \ldots, n \,, \tag{8.70}$$

which are easily solved by

$$Q_\alpha = A_\alpha \cos(\omega_\alpha t + \phi_\alpha)\,, \qquad \alpha = 1, 2, \ldots, n\,, \tag{8.71}$$

for real constants $A_\alpha$ and $\phi_\alpha$. Thus, each normal coordinate $Q_\alpha$ oscillates sinusoidally with a single normal frequency $\omega_\alpha$. Note that the definition given in (8.66) together with the above solution for $Q_\alpha$ are consistent with the real part of (8.63), as one would expect.

## 8.5 Examples

After the above lengthy digression into the mathematical formalism underlying small oscillations, we turn now to several simple (and standard) examples to illustrate how this formalism can be used to solve specific problems. In particular, we shall consider the double pendulum (Sect. 8.5.1), the linear triatomic molecule (Sect. 8.5.2), and the loaded string (Sect. 8.5.3). We outline the relevant calculations, leaving exercises for the reader to fill in the details.

### 8.5.1 Double Pendulum

We discussed the double pendulum in Problem 1.4 in Chap. 1. Here, we will simplify the problem by requiring that the two masses and the lengths of the two rods be equal to one another ($m_1 = m_2 \equiv m$ and $\ell_1 = \ell_2 \equiv \ell$); see Fig. 8.6. Again, we choose Cartesian coordinates with the $x$-axis pointing down and the $y$-axis pointing to the right. The generalized coordinates for this problem are the two angles $\phi_1$ and $\phi_2$. These are related to the Cartesian coordinates through

$$\begin{aligned}
x_1 &= \ell \cos\phi_1\,,\\
y_1 &= \ell \sin\phi_1\,,\\
x_2 &= \ell \cos\phi_1 + \ell\cos\phi_2\,,\\
y_2 &= \ell \sin\phi_1 + \ell\sin\phi_2\,.
\end{aligned} \tag{8.72}$$

With these, we can compute the kinetic energy terms using

$$T_{ab} = m\frac{\partial \mathbf{r}_1}{\partial \phi_a}\cdot\frac{\partial \mathbf{r}_1}{\partial \phi_b} + m\frac{\partial \mathbf{r}_2}{\partial \phi_a}\cdot\frac{\partial \mathbf{r}_2}{\partial \phi_b}\,. \tag{8.73}$$

The kinetic energy is then computed from

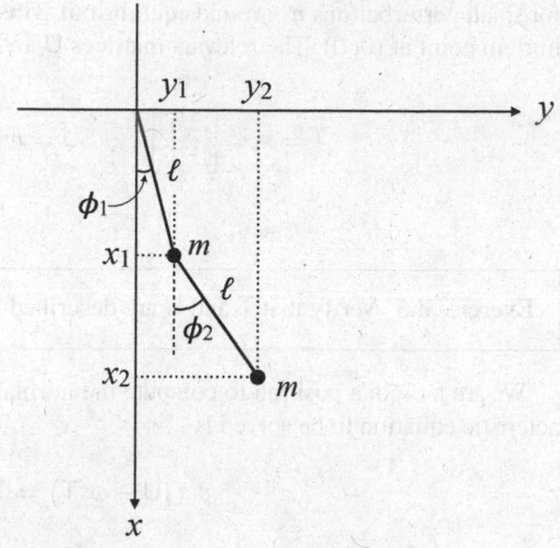


**Fig. 8.6** The double pendulum with equal masses and equal length rods. The position of the first mass is given by $(x_1, y_1)$, and the position of the second mass is given by $(x_2, y_2)$. The angles $\phi_1$ and $\phi_2$ are measured with respect to the vertical in the counterclockwise direction

$$\begin{aligned} T &= \frac{1}{2}m\ell^2 \begin{bmatrix} \dot{\phi}_1 & \dot{\phi}_2 \end{bmatrix} \begin{bmatrix} 2 & \cos(\phi_1 - \phi_2) \\ \cos(\phi_1 - \phi_2) & 1 \end{bmatrix} \begin{bmatrix} \dot{\phi}_1 \\ \dot{\phi}_2 \end{bmatrix} \\ &= \frac{1}{2}m\ell^2 \left(2\dot{\phi}_1^2 + \dot{\phi}_2^2 + 2\dot{\phi}_1\dot{\phi}_2 \cos(\phi_1 - \phi_2)\right) . \end{aligned} \tag{8.74}$$

The potential energy for this system is simply

$$U = -mg\,(x_1 + x_2) = -mg\ell\,(2\cos\phi_1 + \cos\phi_2) . \tag{8.75}$$

Since we want to find small oscillations about stable equilibria, we need to first identify the equilibria. These are found by setting

$$\begin{aligned} \left.\frac{\partial U}{\partial \phi_1}\right|_0 &= 2mg\ell \sin\phi_{10} = 0 , \\ \left.\frac{\partial U}{\partial \phi_2}\right|_0 &= mg\ell \sin\phi_{20} = 0 , \end{aligned} \tag{8.76}$$

where the extra subscript 0 denotes the value at equilibrium. These equations yield four equilibrium points: $(\phi_{10}, \phi_{20}) = (0, 0)$, $(0, \pi)$, $(\pi, 0)$, and $(\pi, \pi)$. In order to determine the stable equilibria, we must further require

$$\sum_{a,b} \left.\frac{\partial^2 U}{\partial \phi_a \partial \phi_b}\right|_0 \eta^a \eta^b > 0 , \tag{8.77}$$

for small perturbations $\eta^a$ around equilibrium. This identifies the single stable equilibrium point at (0, 0). The relevant matrices $\mathsf{U}$ and $\mathsf{T}$ that appear in (8.42) are then

$$\mathsf{T} = m\ell^2 \begin{bmatrix} 2 & 1 \\ 1 & 1 \end{bmatrix}, \qquad \mathsf{U} = mg\ell \begin{bmatrix} 2 & 0 \\ 0 & 1 \end{bmatrix}. \tag{8.78}$$

**Exercise 8.5** Verify that $\mathsf{T}$ and $\mathsf{U}$ are described by (8.78).

We are now in a position to compute the normal modes of the system. The characteristic equation to be solved is

$$\det\left(\mathsf{U} - \omega^2 \mathsf{T}\right) = 0. \tag{8.79}$$

This yields a quadratic equation in $\omega^2$ that is solved by

$$\omega_\pm^2 \equiv \frac{g}{\ell}\left(2 \pm \sqrt{2}\right). \tag{8.80}$$

The associated eigenvectors (normal modes) are found to be

$$z_\pm = N_\pm \begin{bmatrix} 1 \\ \mp\sqrt{2} \end{bmatrix}, \tag{8.81}$$

where $N_\pm$ are normalization constants that can be adjusted so that these vectors satisfy the orthonormality condition (8.57). (See Exercise 8.7 below.)

**Exercise 8.6** Solve (8.79) to obtain the eigenvalues in (8.80), and then the corresponding eigenvectors in (8.81).

**Exercise 8.7** Use the orthonormality condition of (8.57) to show that

$$N_\pm = \frac{1}{\sqrt{2m\ell^2\left(2 \mp \sqrt{2}\right)}}. \tag{8.82}$$

Note that the dimensionality of the normalized eigenvectors has dimensions of $\left[\text{length}\right]^{-1} [\text{mass}]^{-1/2}$.

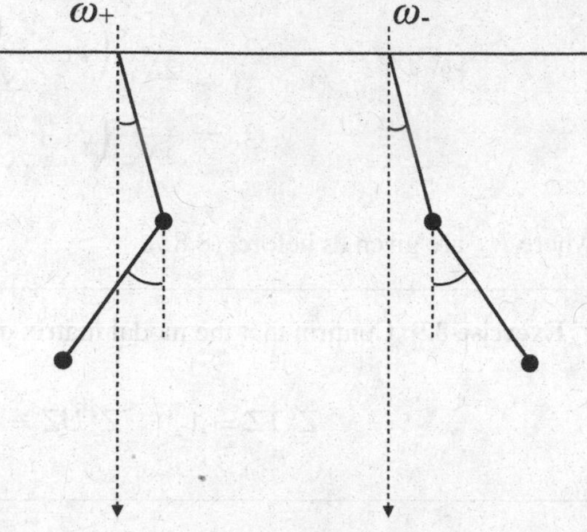


**Fig. 8.7** The two normal modes of oscillation, $\mathsf{z}_+$ and $\mathsf{z}_-$, for the double pendulum. The high-frequency solution $\mathsf{z}_+$ is on the left and the low-frequency solution $\mathsf{z}_-$ is on the right

From (8.81) we see that the normal mode $\mathsf{z}_+$, which has the larger of the two normal mode frequencies, corresponds to the two pendula oscillating *out of phase* with one another, and with the amplitude of the bottom pendulum oscillation being $\sqrt{2}$ times larger than that for the top pendulum. Similarly, the normal mode $\mathsf{z}_-$, which has the smaller of the two normal mode frequencies, corresponds to the two pendula oscillating *in phase* with one another, and with the relative amplitudes of the two oscillating pendula as before. These normal mode oscillations are illustrated graphically in Fig. 8.7.

**Exercise 8.8** Show that the normal mode $\mathsf{z}_+$ for the double pendulum has a "node" on the $x$-axis given by $x_{\rm node} = \ell(1 + 1/\sqrt{2}) \approx 1.707\ell$, assuming (as usual) small angular displacements $\phi_1, \phi_2$ from equilibrium.

The modal matrix can be constructed from the normal modes to give

$$\mathsf{Z} = \frac{1}{\sqrt{2m\ell^2}} \begin{bmatrix} \frac{1}{\sqrt{2-\sqrt{2}}} & \frac{1}{\sqrt{2+\sqrt{2}}} \\ \frac{-\sqrt{2}}{\sqrt{2-\sqrt{2}}} & \frac{\sqrt{2}}{\sqrt{2+\sqrt{2}}} \end{bmatrix} . \tag{8.83}$$

The normal coordinates $\mathsf{Q}$ can then be calculated from (8.66) with $\mathsf{Z}$ given as above and

$$\eta = \begin{bmatrix} \phi_1 \\ \phi_2 \end{bmatrix} . \tag{8.84}$$

The result of the calculation is

$$Q_+ = \frac{1}{2N_+}\left(\phi_1 - \frac{1}{\sqrt{2}}\phi_2\right),$$
$$Q_- = \frac{1}{2N_-}\left(\phi_1 + \frac{1}{\sqrt{2}}\phi_2\right), \tag{8.85}$$

where $N_\pm$ are given as before, (8.82).

**Exercise 8.9** Confirm that the modal matrix diagonalizes both $\mathsf{T}$ and $\mathsf{U}$, with

$$\mathsf{Z}^T\mathsf{T}\mathsf{Z} = 1\,, \qquad \mathsf{Z}^T\mathsf{U}\mathsf{Z} = \begin{bmatrix} \omega_+^2 & 0 \\ 0 & \omega_-^2 \end{bmatrix}. \tag{8.86}$$

**Exercise 8.10** Verify (8.85) for the normal coordinates $Q_+$, $Q_-$.

### 8.5.2 *Linear Triatomic Molecule*

In this example, we will consider the classical vibrational modes of oscillation for a linear triatomic molecule such as carbon dioxide. Carbon dioxide ($CO_2$) consists of a carbon atom with two oxygen atoms on either side of it. In this case, the true potential energy of the interactions between the atoms is a complicated result of the quantum interactions between the outer-shell electrons that are shared by the atoms in the molecule. Instead of determining the equilibrium separation by finding local minima of this potential, we will simply assume an equilibrium separation of $b$ (which is 116 pm for $CO_2$) and take the approximate potential to be quadratic in small oscillations about this equilibrium. In other words, we will assume that the atoms are separated by two little springs with spring constant $k$. We'll let the two oxygen atoms each have mass $M$, while the carbon atom will have mass $m$. The configuration is shown in Fig. 8.8.

Since this configuration is linear, the positions of all molecules can be simply described by the $x$-coordinates $x_a$, $a = 1, 2, 3$. The equilibrium positions are defined in terms of the equilibrium separation $b$:

$$x_{20} - x_{10} = x_{30} - x_{20} = b\,. \tag{8.87}$$

The perturbations are deviations from equilibrium, so $\eta_a \equiv x_a - x_{a0}$. The kinetic energy of the small oscillations is then given by

$$T = \frac{1}{2}\left[M\left(\dot{\eta}_1^2 + \dot{\eta}_3^2\right) + m\dot{\eta}_2^2\right]. \tag{8.88}$$

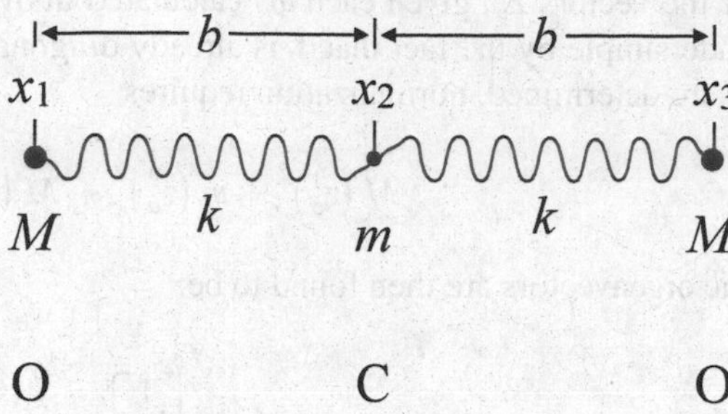


**Fig. 8.8** The configuration for the linear triatomic molecule $CO_2$ with separation $b = 116$ pm

We have already assumed that the perturbed potential is adequately described by the potential of a spring, so the potential energy for small oscillations is

$$U = \frac{1}{2}k\left[(\eta_2 - \eta_1)^2 + (\eta_3 - \eta_2)^2\right]. \tag{8.89}$$

From these expressions for $T$ and $U$, we can easily compute the relevant *matrices* $\mathsf{T}$ and $\mathsf{U}$, which are simply

$$\mathsf{T} = \begin{bmatrix} M & 0 & 0 \\ 0 & m & 0 \\ 0 & 0 & M \end{bmatrix}, \qquad \mathsf{U} = \begin{bmatrix} k & -k & 0 \\ -k & 2k & -k \\ 0 & -k & k \end{bmatrix}. \tag{8.90}$$

Next, we solve for the eigenvalues (i.e., the normal mode frequencies) using

$$\det\left(\mathsf{U} - \omega^2\mathsf{T}\right) = \det\begin{bmatrix} k - \omega^2 M & -k & 0 \\ -k & 2k - \omega^2 m & -k \\ 0 & -k & k - \omega^2 M \end{bmatrix} = 0, \tag{8.91}$$

which yields the characteristic equation

$$\omega^2\left(k - \omega^2 M\right)\left(\omega^2 M m - k\left(2M + m\right)\right) = 0. \tag{8.92}$$

The solutions are

$$\omega_1 = 0, \qquad \omega_2 = \sqrt{\frac{k}{M}}, \qquad \omega_3 = \sqrt{\frac{k\left(2M + m\right)}{Mm}}. \tag{8.93}$$

With the eigenvalues in hand, we can now calculate the eigenvectors. This is done by solving

$$\left(\mathsf{U} - \omega_\alpha^2\mathsf{T}\right)\mathsf{z}_\alpha = 0, \qquad \alpha = 1, 2, 3, \tag{8.94}$$

for the vectors $\mathsf{z}_\alpha$, given each $\omega_\alpha$ calculated above. The normalization condition is made simple by the fact that $\mathsf{T}$ is already diagonalized, so once the components of $\mathsf{z}_\alpha$ are determined, normalization requires

$$M\left(z_\alpha^1\right)^2 + m\left(z_\alpha^2\right)^2 + M\left(z_\alpha^3\right)^2 = 1\,. \tag{8.95}$$

The eigenvectors are then found to be:

$$\begin{aligned}
\mathsf{z}_1 &= \tfrac{1}{\sqrt{2M+m}}\begin{bmatrix}1\\1\\1\end{bmatrix},\\
\mathsf{z}_2 &= \tfrac{1}{\sqrt{2M}}\begin{bmatrix}1\\0\\-1\end{bmatrix},\\
\mathsf{z}_3 &= \tfrac{1}{\sqrt{2(M/m)(2M+m)}}\begin{bmatrix}1\\-2M/m\\1\end{bmatrix}.
\end{aligned} \tag{8.96}$$

These are the normal modes corresponding to the normal mode frequencies given in (8.93).

**Exercise 8.11** Use the results of (8.93) to compute the eigenvectors given in (8.96).

For the first normal mode, $\omega_1 = 0$. There is no oscillation and the motion of all three atoms is identical. This is simply a *linear translation* of the entire molecule. In retrospect, we could have eliminated an extra degree of freedom by working in the center-of-mass frame. In the center-of-mass frame, an additional constraint would relate $\eta_1$, $\eta_2$, and $\eta_3$ with each other, leaving just two generalized coordinates.

For the second normal mode, the central atom (carbon) is at rest (i.e., $\eta_2 = 0$), while the two oxygen atoms oscillate out of phase with one another about the carbon atom with angular frequency $\omega_2 = \sqrt{k/M}$. The motion is such that the carbon atom plays no role in the oscillation, and it is as if the oxygen atoms were separated by a distance $2b$ and connected by a spring with spring constant $2k$. This motion is shown in Fig. 8.9, panel (b). Note that the center of mass of the molecule remains at rest for this mode.

For the third normal mode, all of the atoms oscillate with angular frequency $\omega_3 = \sqrt{k\,(2M+m)\,/Mm}$. The two oxygen atoms move in phase with one another, while the carbon atom moves out of phase with the oxygen atoms. The ratios of the amplitudes of the oscillations are such that the center of mass of the molecule remains at rest. This motion is also shown in Fig. 8.9, panel (c).

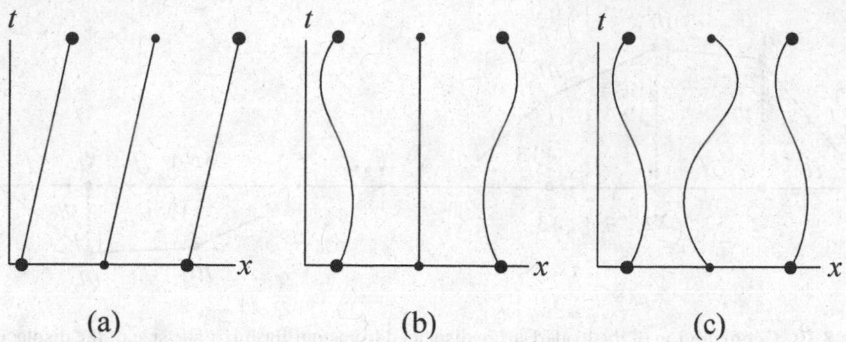


**Fig. 8.9** The three normal modes $\mathbf{z}_1, \mathbf{z}_2, \mathbf{z}_3$ for the motion of the linear triatomic molecule $CO_2$. Panel (a) shows the first normal mode, where the whole molecule simply moves to the right, without oscillations. Panel (b) shows the second normal mode, where the two oxygen atoms vibrate out of phase with one another about a stationary carbon atom. Panel (c) shows the third normal mode, where the two oxygen atoms oscillate in phase with one another, while the carbon atom oscillates out of phase with the oxygen atoms, with an amplitude that keeps the center of mass of the molecule at rest

**Exercise 8.12** Using (8.96), show that the center of mass remains at rest for the motions of the two oscillatory normal modes in the linear $CO_2$ molecule.

### 8.5.3 *Loaded String*

The loaded string is a discrete model of a real, massive string of length $\ell$ and total mass $M$, with a linear mass density $\mu \equiv M/\ell$. In this model, we consider the string to be massless, with $N$ point masses of mass $m$ evenly spaced along the string to provide an equivalent linear mass density. We will consider only small vertical displacements of the masses, and the equilibrium configuration will consist of the straight string. The massless string will have a constant tension $\tau$ (we use $\tau$ instead of $T$ to avoid the obvious confusion with the kinetic energy). The string can stretch, but does not have any restoring force, so it doesn't act like a spring. The general configuration is shown in Fig. 8.10.

The restoring force that tends to bring each mass back toward equilibrium comes from the tension in the string. For a string of length $\ell$ with $N$ masses equally separated by a distance $d \equiv \ell/(N+1)$, the restoring force on the $a$th mass is related to the displacement of the mass itself, $y_a$, and the displacement of its two neighbors, $y_{a-1}$ and $y_{a+1}$. If we look at the forces acting on $a$th mass as shown in Fig. 8.11, we see that the net force can be written in terms of the tension as

$$F_a = \tau (\sin \beta - \sin \alpha) , \tag{8.97}$$

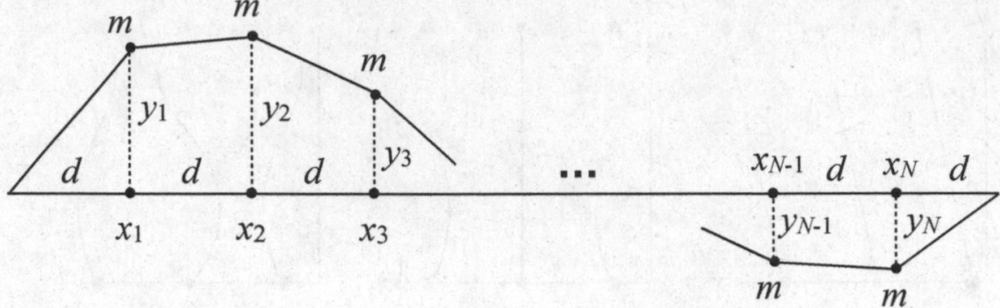


**Fig. 8.10** Configuration of the loaded string displaced from equilibrium. (The size of the displacements have been greatly exaggerated in this figure.) There are $N$ masses on a massless string of length $\ell$. They are separated by a distance $d \equiv \ell/(N+1)$, and each is displaced vertically by $y_a$, where $a = 1, 2, \ldots, N$ labels the masses

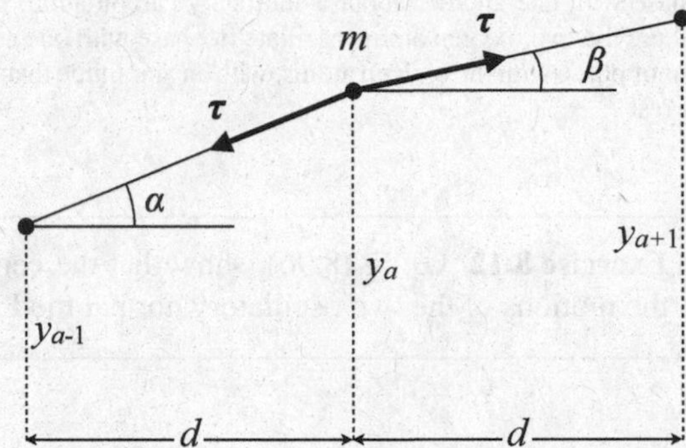


**Fig. 8.11** The forces acting on the $a$th mass due to the tension $\tau$ in the string. Since we are assuming that all displacements are in the vertical direction, we ignore any residual force in the horizontal direction. The sum of the vertical components of the tension is then simply $\tau \sin\beta - \tau \sin\alpha$, where vertically upward is taken to be positive

where the angles $\alpha$ and $\beta$ are as shown in the figure. It is at this point that we can impose the condition that the deviations from equilibrium $y_a$ be small. We have not computed the potential energy yet, but if we require that the angles $\alpha$ and $\beta$ be small so that we can use the small-angle approximation for the sines, then the above expression for the restoring force becomes

$$F_a \simeq \tau(\beta - \alpha) \simeq \frac{\tau}{d}\left[(y_{a+1} - y_a) - (y_a - y_{a-1})\right]. \tag{8.98}$$

We will also need to impose an additional constraint to deal with the endpoints ($y_0$ and $y_{N+1}$) which are not associated with a mass. The conventional constraint is to require the ends of the string to be fixed so that $y_0 = y_{N+1} = 0$.

We have so far avoided the issue of the Lagrangian for this problem, but we have nonetheless obtained the equations of motion

$$m\ddot{y}_a - \frac{\tau}{d}(y_{a-1} - 2y_a + y_{a+1}) = 0, \qquad a = 1, 2, \ldots, N. \tag{8.99}$$

It is simple to show that these equations of motion can be obtained from a Lagrangian of the form

$$L = \frac{1}{2} m \sum_{a=1}^{N} (\dot{y}_a)^2 - \frac{\tau}{2d} \sum_{a=0}^{N} (y_{a+1} - y_a)^2. \tag{8.100}$$

From here, we can easily read off the two $(N \times N)$-matrices $\mathsf{T}$ and $\mathsf{U}$ that will be used in the characteristic equation to determine the normal mode frequencies:

$$\mathsf{T} = m\mathbb{1}, \qquad \mathsf{U} = \frac{\tau}{d} \begin{bmatrix} 2 & -1 & 0 & 0 & \cdots & 0 & 0 \\ -1 & 2 & -1 & 0 & \cdots & 0 & 0 \\ 0 & -1 & 2 & -1 & \cdots & 0 & 0 \\ 0 & 0 & -1 & 2 & \cdots & 0 & 0 \\ \vdots & \vdots & \vdots & \vdots & \ddots & \vdots & \vdots \\ 0 & 0 & 0 & 0 & \cdots & 2 & -1 \\ 0 & 0 & 0 & 0 & \cdots & -1 & 2 \end{bmatrix}. \tag{8.101}$$

The characteristic equation is then $\det\left(\mathsf{U} - \omega^2 \mathsf{T}\right) = 0$, but we will find it convenient to make the substitution

$$\lambda \equiv \frac{md\omega^2}{\tau} - 2, \tag{8.102}$$

so that solving the characteristic equation for $\omega^2$ is equivalent to solving

$$\left(\frac{\tau}{d}\right)^N \det \mathsf{A} = 0 \tag{8.103}$$

for $\lambda$, where

$$\mathsf{A} = \begin{bmatrix} -\lambda & -1 & 0 & 0 & \cdots & 0 & 0 \\ -1 & -\lambda & -1 & 0 & \cdots & 0 & 0 \\ 0 & -1 & -\lambda & -1 & \cdots & 0 & 0 \\ 0 & 0 & -1 & -\lambda & \cdots & 0 & 0 \\ \vdots & \vdots & \vdots & \vdots & \ddots & \vdots & \vdots \\ 0 & 0 & 0 & 0 & \cdots & -\lambda & -1 \\ 0 & 0 & 0 & 0 & \cdots & -1 & -\lambda \end{bmatrix}. \tag{8.104}$$

We are now left with the problem of finding the determinant of an $N \times N$ matrix, which for arbitrary $N$ could be rather complicated. But, as we shall see below, the particular form of $\mathsf{A}$ allows for a relatively simple expression. Using the method of cofactors (Appendix D.4.3.2), we note that the determinant of an $N \times N$ matrix can be expressed as a sum of $N$ determinants of $(N-1) \times (N-1)$ matrices. For a matrix with the specific form of $\mathsf{A}$, this leads to

$$\det \mathsf{A}^{(N)} = -\lambda \det \mathsf{A}^{(N-1)} - \det \mathsf{A}^{(N-2)}\,, \tag{8.105}$$

where $\mathsf{A}^{(M)}$ is the $M \times M$ matrix with the same structural form as $\mathsf{A}$. This recursion relation will allow us to find the determinant for any value of $N$, since the determinants for $N = 1$ and $N = 2$ are trivial.

**Exercise 8.13** Show that (8.105) follows from the definition of $\mathsf{A}$ in (8.104).

The coefficients in the recursion relation in (8.105) are independent of $N$, which suggests proposing an $N$-dependence to $\det \mathsf{A}^{(N)}$ of the form

$$\det \mathsf{A}^{(N)} = C\mathrm{e}^{\mathrm{i}N\gamma}\,, \tag{8.106}$$

where $C$ and $\gamma$ are complex numbers that are independent of $N$. Substituting this expression into (8.105) and solving for $\gamma$ gives

$$\gamma = \pm \arccos\left(\frac{-\lambda}{2}\right) \quad \Leftrightarrow \quad \lambda = -2\cos\gamma\,. \tag{8.107}$$

At this point, we will take $\gamma$ to be positive, and express the general solution for $\det \mathsf{A}^{(N)}$ as the linear combination

$$\det \mathsf{A}^{(N)} = C_+\mathrm{e}^{\mathrm{i}N\gamma} + C_-\mathrm{e}^{-\mathrm{i}N\gamma}\,. \tag{8.108}$$

For the trivial cases of $N = 1$ and $N = 2$, we have

$$\begin{aligned} \det \mathsf{A}^{(1)} &= -\lambda = 2\cos\gamma\,, \\ \det \mathsf{A}^{(2)} &= \lambda^2 - 1 = 4\cos^2\gamma - 1\,. \end{aligned} \tag{8.109}$$

Substituting (8.108) in for the determinants in (8.109) allows us to solve for $C_+$ and $C_-$.

**Exercise 8.14** Use (8.108) and (8.109) to solve for the constants $C_+$ and $C_-$. You should find

$$C_+ = \frac{\mathrm{e}^{\mathrm{i}\gamma}}{2\mathrm{i}\sin\gamma}\,, \qquad C_- = C_+^*\,. \tag{8.110}$$

(This might be a good opportunity to use the tools of Appendix D.)

We are now in a position to find the general expression for $\det \mathsf{A}$ for arbitrary $N$. Substitution of (8.110) into (8.108) gives

$$\det \mathsf{A} = \frac{\sin\left[(N+1)\,\gamma\right]}{\sin\gamma}. \tag{8.111}$$

Using this general solution, we can now solve the characteristic equation $\det \mathsf{A} = 0$, which has solutions $\gamma = \gamma_n$, where

$$\gamma_n \equiv \frac{n\pi}{N+1}, \tag{8.112}$$

for integer $n$. Note, however, that not all integer values of $n$ will work. If $n = 0$ or $n = N+1$, then $\det \mathsf{A}$ is undefined as the denominator is 0. With the above solution for $\gamma$, we can use (8.107) to find $\lambda$, and then (8.102) to finally arrive at the eigenfrequencies $\omega$. After clever use of a trig identity, we find

$$\omega_\alpha^2 = \frac{4\tau}{md}\sin^2\left(\frac{\alpha\pi}{2\,(N+1)}\right), \qquad \alpha = 1, 2, \ldots, N\,. \tag{8.113}$$

It is the square of the sine function in (8.113) that restricts the number of unique solutions for the eigenfrequencies to $\alpha = 1, 2, \ldots, N$.

The components of the eigenvectors $\mathsf{z}_\alpha$ for this problem give the displacement of each of the masses for the normal mode oscillation of the system having angular frequency $\omega_\alpha$. The eigenvector equation $\left(\mathsf{U} - \omega_\alpha^2\mathsf{T}\right)\mathsf{z}_\alpha = 0$ yields a recursion relation between the components of $\mathsf{z}_\alpha$:

$$z_\alpha^{a+1} = -\lambda_\alpha z_\alpha^a - z_\alpha^{a-1}, \tag{8.114}$$

where $\lambda_\alpha$ is related to $\omega_\alpha^2$ via (8.102). Comparing the above equation with (8.105), we see that the two recursion relations are identical with the dimension $N$ of the matrix replaced by the eigenvector component index $a$. Thus, we can propose the same type of solution

$$z_\alpha^a = C_\alpha \mathrm{e}^{\mathrm{i}a\kappa_\alpha}, \tag{8.115}$$

which when substituted into (8.114) yields the same answer for the quantity in the exponential,

$$\kappa_\alpha = \pm\gamma_\alpha = \pm\frac{\alpha\pi}{N+1}. \tag{8.116}$$

(Recall that we are taking $\gamma_\alpha$ to be positive.) Thus, the general solution for $\mathsf{z}_\alpha$ is

$$z_\alpha^a = C_\alpha^+ \mathrm{e}^{\mathrm{i}a\gamma_\alpha} + C_\alpha^- \mathrm{e}^{-\mathrm{i}a\gamma_\alpha}. \tag{8.117}$$

Imposing the conditions that the loaded string be fixed at the boundaries (i.e., $z_\alpha^0 = 0$ and $z_\alpha^{N+1} = 0$) implies $z_\alpha^a = N_\alpha \sin(a\gamma_\alpha)$, where $N_\alpha$ is a normalization constant, which we can choose to be real (Problem 8.1). Normalizing the normal modes so that $\mathsf{Z}^T\mathsf{T}\mathsf{Z} = \mathsf{1}$ yields (Problem 8.6):

$$N_\alpha = \sqrt{\frac{2}{m(N+1)}}\,. \tag{8.118}$$

Thus,

$$z_\alpha^a = \sqrt{\frac{2}{m(N+1)}}\sin\left(\frac{a\alpha\pi}{N+1}\right), \qquad \alpha = 1, 2, \ldots, N\,. \tag{8.119}$$

**Exercise 8.15** Sketch the shape of the normal modes $\mathsf{z}_\alpha$ for a loaded string consisting of $N = 5$ mass points, fixed at both ends.

Note that the index $a$ gives the location of the $a$th particle, which is at $x$-coordinate $x_a = ad$. Thus, we can also write the argument of the sine function above as $k_\alpha x_a$ where

$$k_\alpha \equiv \frac{\gamma_\alpha}{d} = \frac{\alpha\pi}{d(N+1)}\,, \qquad \alpha = 1, 2, \ldots, N\,, \tag{8.120}$$

is the **wave number** of the $\alpha$th normal mode. Putting all of these results together, the general solution for the motion of the $a$th mass point is the linear combination

$$\begin{aligned} y_a &= \sum_\alpha A_\alpha \sin\left(k_\alpha x_a\right)\cos(\omega_\alpha t + \phi_\alpha) \\ &= \sum_\alpha \frac{1}{2}A_\alpha\left[\cos\left(k_\alpha x_a + \omega_\alpha t + \phi_\alpha\right) + \cos\left(k_\alpha x_a - \omega_\alpha t - \phi_\alpha\right)\right]\,, \end{aligned} \tag{8.121}$$

where $A_\alpha$ and $\phi_\alpha$ are the amplitudes and phases of the individual normal modes, to be determined by the initial conditions of the string. The above solution for $y_a$ is a sum of **standing waves**, which we have represented on the second line as a superposition of left and right-moving waves. The relationship between $\omega_\alpha$ and $k_\alpha$ can be simply computed as

$$\omega_\alpha^2 = \frac{4\tau}{md}\sin^2\left(\frac{k_\alpha d}{2}\right), \qquad \alpha = 1, 2, \ldots, N\,. \tag{8.122}$$

The functional dependence of $\omega_\alpha$ on $k_\alpha$ is known as a **dispersion relation**.

Models such as the loaded string are useful starting points when going over to the continuum limit. For example, we can allow $N \to \infty$ while simultaneously letting $m \to 0$ and $d \to 0$ such that $m/d \equiv \mu$, where $\mu$ is a constant linear mass density. This will result in a description of small oscillations for a real, massive string under tension. We will discuss the transition to the continuous limit in Chap. 9.

## 8.6 Damped and Driven Coupled Oscillations

The formalism developed in Sect. 8.2 does not include dissipative or driving forces. As was done for the one-dimensional oscillator in Sects. 8.1.2 and 8.1.3, we will introduce these additional forces at the point where the formal discussion has led us to the equations of motion for the normal coordinates.

### 8.6.1 Damped Systems

We can introduce frictional forces to the coupled oscillator problem in the same way that we did in Sect. 8.1.2, but the simple frictional forces that we will look at depend upon the velocities of the particles, not the velocities of the normal coordinates. Thus, the equations of motion should be of the form

$$\mathsf{T}\ddot{\eta} + 2\Phi\dot{\eta} + \mathsf{U}\eta = 0\,, \tag{8.123}$$

where $\Phi$ is a matrix representing the coupling of the frictional forces between generalized coordinates.

The techniques developed in Sect. 8.2 allowed us to find a modal matrix $\mathsf{Z}$ that simultaneously diagonalized $\mathsf{T}$ and $\mathsf{U}$, so that we could obtain the simple, decoupled equations of motion in terms of the normal coordinates. Unfortunately, it is not possible to also diagonalize $\Phi$ for general velocity-dependent damping forces. Thus, in general, we cannot expect to arrive at $N$ copies of the simple one-dimensional damped oscillator equation of motion in normal coordinates. But there are some special cases where we can simultaneously diagonalize $\mathsf{U}$, $\mathsf{T}$, and $\Phi$. In these cases, we find the equations of motion to be

$$\ddot{Q}_\alpha + 2\varphi_\alpha \dot{Q}_\alpha + \omega_\alpha^2 Q_\alpha = 0\,. \tag{8.124}$$

These result in the standard solution

$$Q_\alpha = A_\alpha e^{-\varphi_\alpha t} \cos\left(\omega_\alpha t \sqrt{1 - \frac{\varphi_\alpha^2}{\omega_\alpha^2}} + \phi_\alpha\right) . \tag{8.125}$$

If the frictional forces do not allow for this simple formulation, then the problem becomes substantially more complicated. (See, e.g., Goldstein et al. (2002) for a discussion of solution techniques.)

### 8.6.2 Damped and Driven Systems

Remarkably enough, the solution for the damped, driven coupled oscillation problem is actually simpler than the undriven case described above. As we did in Sect. 8.1.3, let's assume a sinusoidal driving force $F^a(t) = F_0^a e^{i\omega t}$ (allowing $F_0^a$ to be complex to allow for arbitrary phase shifts). The equations of motion then read

$$\mathsf{T}\ddot{\eta} + 2\Phi\dot{\eta} + \mathsf{U}\eta = \mathsf{F}_0 e^{i\omega t} . \tag{8.126}$$

If we consider a steady-state particular solution of the form

$$\eta_\mathrm{p} = \mathsf{A}e^{i\omega t} , \tag{8.127}$$

then (8.126) becomes

$$\mathsf{M}(\omega)\mathsf{A} = \mathsf{F}_0 , \tag{8.128}$$

where

$$\mathsf{M}(\omega) = -\omega^2\mathsf{T} + 2i\omega\Phi + \mathsf{U} . \tag{8.129}$$

This is solved directly using the inverse of $\mathsf{M}(\omega)$, giving

$$\mathsf{A} = \mathsf{M}(\omega)^{-1}\mathsf{F}_0 = \frac{\mathsf{C}^T\mathsf{F}_0}{\det \mathsf{M}(\omega)} , \tag{8.130}$$

where $\mathsf{C}$ is the matrix of cofactors described Appendix D.4.3.3.

The numerator $\mathsf{C}^T\mathsf{F}_0$ relates the amplitude of the particular solution to the amplitude of the driving force. The denominator, however, gives resonance peaks at values of $\omega$ that minimize $\det \mathsf{M}(\omega)$. This is similar to finding the minimum of the denominator in (8.26) for the one-dimensional damped, driven oscillator.

We can determine the minima of $\det \mathsf{M}(\omega)$ by considering the homogeneous solution to (8.126). Let's assume a homogeneous solution of the form

$$\eta = \mathsf{B}e^{i\gamma t} , \qquad \gamma \equiv \omega + i\kappa . \tag{8.131}$$

Plugging this into (8.123), we find

$$\left[-\gamma^2 \mathsf{T} + 2i\gamma \Phi + \mathsf{U}\right] \mathsf{B} = \mathsf{M}(\gamma)\mathsf{B} = 0 \,. \tag{8.132}$$

The non-trivial solutions to this equation occur when $\det \mathsf{M}(\gamma) = 0$. The roots of this equation will give the $2n$ eigenvalues $\gamma_\alpha$ for $\mathsf{M}(\gamma)$. These eigenvalues consist of $n$ solutions $\gamma_\alpha$ of the characteristic equation, and their complex conjugates $-\gamma_\alpha^*$. Thus, using (D.128), we can write

$$\det \mathsf{M}(\omega) = D \prod_{\alpha=1}^{n} (\omega - \gamma_\alpha)\left(\omega + \gamma_\alpha^*\right). \tag{8.133}$$

Writing

$$\mathsf{A} = \frac{(\det \mathsf{M}(\omega))^* \, \mathsf{C}^T \mathsf{F}_0}{(\det \mathsf{M}(\omega))^* \det \mathsf{M}(\omega)}, \tag{8.134}$$

we find that the denominator becomes

$$D^* D \prod_{\alpha=1}^{n} \left((\omega - \omega_\alpha)^2 + \kappa_\alpha^2\right)\left((\omega + \omega_\alpha)^2 + \kappa_\alpha^2\right). \tag{8.135}$$

Although we haven't provided values for $\omega_\alpha$ and $\kappa_\alpha$ in terms of the relevant damping terms from $\Phi$, the procedure is now straightforward for solving any real problem. What is important here is that the system doesn't acquire arbitrarily large amplitudes at resonance, and that for small damping the system can oscillate with frequencies close to the normal mode frequencies of the system.

At this point we have developed the tools for describing the behavior of systems that are perturbed slightly from equilibrium. In addition, we have laid down the foundation for describing continuous systems and fields based on allowing the number of particles to go to infinity while keeping the total mass of the system constant. We will explore this in more detail in Chap. 9.

## Suggested References

*Full references are given in the bibliography at the end of the book.*

Fetter and Walecka (1980): Chapter 4 has very good coverage of small oscillations leading to the continuous limit.

Goldstein et al. (2002): A similar treatment of small oscillations is given in Chap. 6.

## Additional Problems

**Problem 8.1** Consider the matrix equation

$$\mathsf{C}\mathsf{u} = 0\,, \tag{8.136}$$

where all the components of C are real. Following the steps below, show that the components of a solution u to this equation are also real, up to an overall multiplicative complex phase factor.

(a) Show that if $\det \mathsf{C} \neq 0$, then the only solution for (8.136) is $\mathsf{u} = 0$.

(b) If $\det \mathsf{C} = 0$, then at least one of the linear equations described by (8.136) is redundant and carries no additional information. Assume that this is the case and that the $n$th equation is the only redundant equation. The other $n-1$ equations are

$$\begin{aligned} &C_{11}u_1 + C_{12}u_2 + \cdots + C_{1,n-1}u_{n-1} + C_{1n}u_n = 0\,,\\ &C_{21}u_1 + C_{22}u_2 + \cdots + C_{2,n-1}u_{n-1} + C_{2n}u_n = 0\,,\\ &\cdots\\ &C_{n-1,1}u_1 + C_{n-1,2}u_2 + \cdots + C_{n-1,n-1}u_{n-1} + C_{n-1,n}u_n = 0\,. \end{aligned} \tag{8.137}$$

Use these $n-1$ equations to write a second matrix equation

$$\mathsf{D}\mathsf{v} = \mathsf{w}\,, \tag{8.138}$$

where D is the $(n-1) \times (n-1)$-dimensional matrix found by removing the $n$th row and $n$th column from C; v is the $(n-1)$-dimensional vector with components $v_a \equiv u_a/u_n$; and w is the $(n-1)$-dimensional vector formed from the $n$th column of C ($w_a \equiv -C_{an}$). [Note: $a = 1, 2, \ldots, n-1$ in the definitions of D, v, and w.]

(c) Since D is invertible by construction,[3] we have $\det \mathsf{D} \neq 0$ and $\mathsf{v} = \mathsf{D}^{-1}\mathsf{w}$. Use this last equation for v and the definitions of D and w given above to argue that all of the components of v must be real. Hence the components of u are real up to an overall multiplicative complex phase factor $u_n$.

**Problem 8.2** A simple pendulum of length $\ell$ and mass $m$ hangs from the ceiling of a railroad car, which is accelerating with constant acceleration $a$, as shown in Fig. 8.12.

(a) Find the equilibrium angle $\theta_0$ that the pendulum makes with the vertical.

(b) Calculate the frequency for small oscillations of the pendulum bob away from equilibrium.

---

[3]If there were more than one redundant equation, then we can simply repeat this process until we arrive at an invertible matrix with a non-zero determinant.

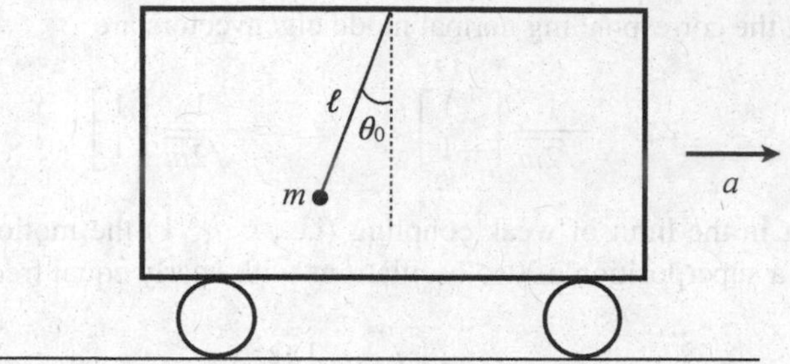


**Fig. 8.12** Simple pendulum attached to the ceiling of an accelerating railroad car

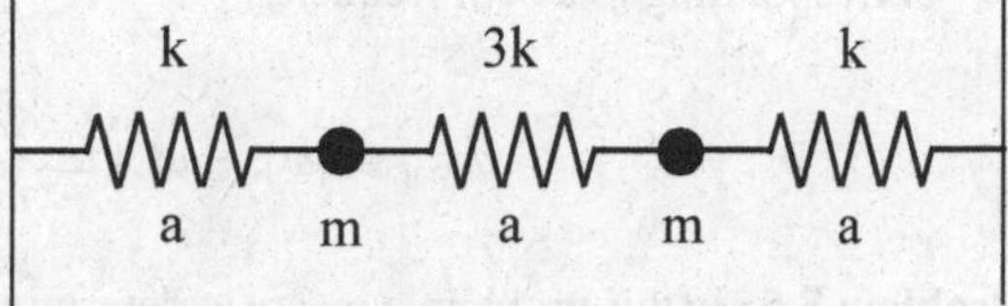


**Fig. 8.13** Two masses $m_1 = m_2 \equiv m$ attached to three springs, with spring constants $k$, $3k$, and $k$, respectively

**Problem 8.3** Two particles of mass $m$ move in one dimension at the junction of three springs, as shown in Fig. 8.13. The springs all have unstretched lengths $a$ and force constants $k$, $3k$, and $k$.

(a) Find the components of the kinetic and potential energy matrices T and U.
(b) Find the normal mode frequencies of the system.
(c) Find the corresponding normalized eigenvectors.
(d) Verify that the eigenvectors are orthogonal.
(e) Discuss the physical nature of the normal mode solutions.

**Problem 8.4** Consider a slight generalization of Problem 8.3 by replacing the middle spring of that problem with a spring having spring constant $\kappa$.

(a) Show that the Lagrangian for this mass-plus-spring system can be written as

$$L = \frac{1}{2}m(\dot{x}^2 + \dot{y}^2) - \frac{1}{2}m\omega_0^2(x^2 + y^2) + m\Omega^2 xy\,, \tag{8.139}$$

where

$$\omega_0^2 \equiv \frac{k+\kappa}{m}\,, \qquad \Omega^2 \equiv \frac{\kappa}{m}\,. \tag{8.140}$$

(b) Show that the normal mode frequencies of oscillation are given by

$$\omega_\pm^2 = \omega_0^2 \pm \Omega^2\,. \tag{8.141}$$

(c) Show that the corresponding normal mode eigenvectors are

$$\mathsf{z}_+ = \frac{1}{\sqrt{2m}}\begin{bmatrix}1\\-1\end{bmatrix}, \quad \mathsf{z}_- = \frac{1}{\sqrt{2m}}\begin{bmatrix}1\\1\end{bmatrix}. \tag{8.142}$$

(d) Show that in the limit of weak coupling (i.e., $\kappa \ll k$) the motion of the two masses is a superposition of two oscillations with nearly equal frequencies

$$\omega_\pm \simeq \omega_0\left(1 \pm \frac{1}{2}\frac{\Omega^2}{\omega_0^2}\right), \tag{8.143}$$

corresponding to a *beat* frequency

$$\omega_{\text{beat}} \equiv |\omega_+ - \omega_-| \simeq \frac{\Omega^2}{\omega_0} \simeq \sqrt{\frac{\kappa^2}{km}}. \tag{8.144}$$

**Problem 8.5** In this problem, you are to determine the small oscillations of a coplanar double pendulum (see Fig. 8.6), with equal lengths $\ell_1 = \ell_2 \equiv \ell$, but different masses $m_1 \neq m_2$.

(a) Show that to leading order, the kinetic and potential energy matrices are given by

$$\mathsf{T} = \ell^2\begin{bmatrix}m_1+m_2 & m_2\\ m_2 & m_2\end{bmatrix}, \quad \mathsf{U} = g\ell\begin{bmatrix}m_1+m_2 & 0\\ 0 & m_2\end{bmatrix}. \tag{8.145}$$

(b) Find the two normal mode frequencies for this problem.

(c) Show that if $m_1 \gg m_2$, then the normal mode frequencies have the approximate form

$$\omega_\pm^2 \approx \frac{g}{\ell}(1 \pm \varepsilon), \quad \varepsilon \equiv \sqrt{\frac{m_2}{m_1}}, \tag{8.146}$$

ignoring all higher-order terms in $\varepsilon$.

(d) Calculate the *beat* frequency $\omega_{\text{beat}} \equiv |\omega_+ - \omega_-|$ for this motion.

(e) Find the eigenvectors in this approximation.

(f) Show that if the pendula are set in motion by pulling the upper mass $m_1$ slightly away from the vertical ($\phi_1(0) = \phi_0$) while keeping the lower mass $m_2$ located on the $x$-axis ($\phi_2(0) = -\phi_0$), and then releasing both from rest, then the motion for the two masses in this approximation has the form

$$\begin{aligned}\phi_1(t) &\approx +\frac{1}{2}\phi_0\left[(\cos\omega_+t + \cos\omega_-t) + \varepsilon(\cos\omega_+t - \cos\omega_-t)\right],\\ \phi_2(t) &\approx -\frac{1}{2}\frac{\phi_0}{\varepsilon}\left[(\cos\omega_+t - \cos\omega_-t) + \varepsilon(\cos\omega_+t + \cos\omega_-t)\right].\end{aligned} \tag{8.147}$$

**Problem 8.6** Verify the orthonormality of the normal modes of the loaded string by showing that

$$\sum_{a=1}^{N} \sin\left(\frac{a\alpha\pi}{N+1}\right) \sin\left(\frac{a\beta\pi}{N+1}\right) = \frac{N+1}{2}\,\delta_{\alpha\beta}\,. \tag{8.148}$$

*Hint*: Expand the left-hand side of the above summation in terms of complex exponentials, and then use the identity

$$\sum_{n=1}^{N} \mathrm{e}^{\mathrm{i}nx} = \frac{\mathrm{e}^{\mathrm{i}x} - \mathrm{e}^{\mathrm{i}(N+1)x}}{1-\mathrm{e}^{\mathrm{i}x}} = \mathrm{e}^{\mathrm{i}(N+1)x/2}\,\frac{\sin(Nx/2)}{\sin(x/2)}\,. \tag{8.149}$$

# Chapter 9
# Wave Equation

Here, we extend the analysis of the previous chapter from many-particle coupled oscillators to continuous systems. In so doing, we will be able to describe the oscillatory motion of strings, membranes, and solid objects. We discuss both the eigenfunction and normal form solutions of the wave equation, paying particular attention to various boundary conditions and initial conditions. The Lagrangian and Hamiltonian formalism for continuous systems and fields will be described in Chap. 10.

## 9.1 Transition from Discrete to Continuous Systems

In Chap. 8, we developed the tools necessary for analyzing small oscillations of systems of $N$ coupled oscillators. These tools could be applied to systems with large values of $N$ to produce discrete models for what were effectively continuous systems. Here we consider the transition from these discrete models to *truly* continuous systems. Recall that discrete models generically consist of $N$ particles of mass $m$ separated by a characteristic distance $d$, which are coupled by some kind of idealized restoring force such as springs or a string under tension. Small displacements of these masses are described by a set of individual displacements $\eta^a$, with $a = 1, 2, \ldots, N$, subject to a coupling potential between adjacent masses that is quadratic in the displacements. Applying standard tools of Lagrangian mechanics to the problem then yields a set of eigenfrequencies of the motion, representing normal modes. The eigenvectors associated with these eigenfrequencies describe the motion of all the masses of the $N$-body system for each normal mode. The components of the eigenvectors are the displacements of each mass point.

When going to the continuous limit, we let the number of particles go to infinity ($N \to \infty$), while at the same time keeping the total mass $M$ of the system constant. This implies that the mass of the individual particles goes to zero ($m \to 0$) in such a way that

 
M.J. Benacquista and J.D. Romano, *Classical Mechanics*, Undergraduate Lecture Notes in Physics, https://doi.org/10.1007/978-3-319-68780-3_9

$$M \equiv \sum_a m_a = Nm \qquad \longrightarrow \qquad M \equiv \lim_{\substack{N\to\infty \\ m\to 0}} Nm\,. \tag{9.1}$$

Of course, as the number of particles goes to infinity, their spacing must also shrink to zero in such a way that the basic size and shape of the object is preserved. The exact form of this limit will depend on the dimensionality of the object, but for a one-dimensional object of length $\ell$, it can be simply expressed as

$$\ell \equiv \sum_a d_a = Nd \qquad \longrightarrow \qquad \ell \equiv \lim_{\substack{N\to\infty \\ d\to 0}} Nd\,. \tag{9.2}$$

The displacement of each particle about its equilibrium position can be expressed in terms of the equilibrium position vector $\mathbf{r}_a$ of each particle, so that

$$\eta^a(t) \equiv \eta(\mathbf{r}_a, t)\,, \qquad a = 1, 2, \ldots, N\,. \tag{9.3}$$

Thus, in the continuous limit, the discrete components $\eta^a$ of the displacement vector are replaced by a single displacement *function* $\eta(\mathbf{r}, t)$.

When going over to the continuous limit, the results obtained for the discrete case can be used to directly obtain the appropriate values for the continuous case. This is often simpler than attacking the continuous problem from first principles. We shall discuss this procedure in the context of a vibrating string in the next section, using our results from the loaded string analysis from Sect. 8.5.3.

## 9.2 Vibrating String

A vibrating string of mass $M$ and length $\ell$ can be approximated by a massless string with regularly spaced discrete mass points located along its length. This is the loaded string of Sect. 8.5.3. In this case, the eigenfrequencies $\omega_\alpha$ of the loaded string are given by (8.113) for integer values of $\alpha$ ranging from 1 to $N$. In the limit $N \to \infty$, it is clear that $\alpha \ll N$ for any reasonable normal mode. Thus, we can use the small angle approximation to obtain

$$\omega_\alpha^2 = \frac{4\tau}{md}\sin^2\left(\frac{\alpha\pi}{2\,(N+1)}\right) \simeq \frac{4\tau}{md}\left(\frac{\alpha\pi}{2\,(N+1)}\right)^2\,. \tag{9.4}$$

In order to go to the continuous limit, we note that the total mass of the string is indeed given by $M = Nm$, but that the total length of the string is given by $\ell = (N+1)\,d$, since there are no masses on either endpoint of the string (See Fig. 8.10). Thus, in the continuous limit, we require that

$$\lim_{\substack{N\to\infty \\ d\to 0}} (N+1)\,d = \ell\,, \tag{9.5}$$

and so the eigenfrequencies become

$$\omega_\alpha^2 = \lim_{\text{cont}} \frac{\tau}{m/d}\left(\frac{\alpha\pi}{d\,(N+1)}\right)^2 = \frac{\tau}{\mu}\left(\frac{\alpha\pi}{\ell}\right)^2, \tag{9.6}$$

where $\lim_{\text{cont}}$ simply means the appropriate continuous limit for $N$, $m$, and $d$. The linear mass density is defined as $\mu \equiv M/\ell$.

The wave number for a loaded string with fixed endpoints is given in (8.120). Under the transition to the continuous limit, it becomes

$$k_\alpha = \frac{\alpha\pi}{d\,(N+1)} \quad\longrightarrow\quad k_\alpha = \frac{\alpha\pi}{\ell} = \frac{2\pi}{\lambda_\alpha}\,, \tag{9.7}$$

where

$$\lambda_\alpha \equiv \frac{2\ell}{\alpha}\,, \qquad \alpha = 1, 2, \ldots\,, \tag{9.8}$$

is the expected wavelength for standing waves on a string. In terms of the wave number, $\omega_\alpha$ can be written as $\omega_\alpha = \pm k_\alpha v$, where

$$v \equiv \sqrt{\frac{\tau}{\mu}} \tag{9.9}$$

is the standard expression for the velocity of transverse waves on an ideal string of mass density $\mu$ under tension $\tau$.

The eigenvectors for the loaded string are (modulo the normalization constant) given by (8.119), with

$$z_\alpha^a = A_\alpha \sin\left(\frac{a\alpha\pi}{N+1}\right), \qquad \alpha = 1, 2, \ldots, N\,. \tag{9.10}$$

When we go to the continuous limit, the eigenvectors become continuous functions of the variable $x$, where $x$ is the continuous limit of $x_a \equiv ad$ and $z_\alpha^a \equiv z_\alpha(x_a)$:

$$z_\alpha^a = A_\alpha \sin\left(\frac{ad\alpha\pi}{d\,(N+1)}\right) \quad\longrightarrow\quad z_\alpha(x) = A_\alpha \sin(k_\alpha x)\,. \tag{9.11}$$

The constant $A_\alpha$ can be complex; if we write it in the form $|A_\alpha|\,\mathrm{e}^{\mathrm{i}\phi_\alpha}$, then

$$y_\alpha(t) = \mathrm{Re}\left[z_\alpha \mathrm{e}^{\mathrm{i}\omega_\alpha t}\right] = |A_\alpha| \sin(k_\alpha x) \cos(k_\alpha v t + \phi_\alpha) \,, \tag{9.12}$$

which are simply the standing waves on a string.

So we see that we can recover many of the basic properties of standing waves on a string by analyzing the problem in the discrete approximation and then taking the continuous limit of these results. In the next section, we show how we can apply the continuous limit earlier in the process to obtain the equations of motion for a continuous system.

## 9.3 One-Dimensional Wave Equation

In the last section, we showed how we can reproduce the basic eigenfrequencies and other properties of continuous systems by taking the continuous limit of results derived from discrete coupled oscillators. Here, we obtain the equations of motion for a continuous system by taking the continuous limit of the discrete equations of motion. We will focus on the vibrating string as an example, but the approach applies broadly to many systems.

The equations of motion for the discrete case of the loaded string are given by (8.99), which we repeat here for clarity in the upcoming discussion:

$$m\ddot{y}_a - \frac{\tau}{d}(y_{a-1} - 2y_a + y_{a+1}) = 0 \,, \tag{9.13}$$

where $a = 1, 2, \ldots, N$. The first issue we face is how to treat $y_a$ which is defined at discrete values $x_a \equiv ad$. In the continuous limit, $x_a \to x$, where $x$ is a continuous variable. Thus, we should treat $y_a$ as a function of both $x$ and $t$, so $y_a \to y(x, t)$. In anticipation of the constraints of the continuous limit, we now rewrite (9.13) as

$$\ddot{y}_a - \frac{\tau}{m/d}\left[\frac{1}{d}\left(\frac{y_{a+1} - y_a}{d} - \frac{y_a - y_{a-1}}{d}\right)\right] = 0 \,. \tag{9.14}$$

Since $d \to 0$ in the continuous limit,

$$\lim_{d\to 0} \frac{y_{a+1} - y_a}{d} = \lim_{d\to 0} \frac{y((a+1)d, t) - y(ad, t)}{d} = \frac{\partial y}{\partial x} \,. \tag{9.15}$$

Thus, the term in the square brackets in (9.14) is the *second* derivative of $y$ with respect to $x$. In the continuous limit, the equation of motion then becomes the one-dimensional **wave equation**:

$$\frac{1}{v^2}\frac{\partial^2 y}{\partial t^2} - \frac{\partial^2 y}{\partial x^2} = 0\,, \tag{9.16}$$

where $v \equiv \sqrt{\tau/\mu}$ is the speed of the wave on the string.

Although we derived the wave equation from the continuous limit of the loaded string, note that the only reference to the string is hidden in $v$. Thus, the wave equation is valid for any one-dimensional continuous system with a well-defined constant wave speed. In the following subsections, we will explore different methods of solving the wave equation.

### 9.3.1 Eigenfunction Solution (Separation of Variables)

As was done for the discrete case, we postulate a normal mode solution of the form $y(x,t) = z(x)e^{-i\omega t}$, noting that we are ultimately interested in just its real part. Substituting this proposed solution back into (9.16) yields the *ordinary* differential equation for $z(x)$:

$$\frac{d^2 z}{dx^2} + k^2 z = 0\,, \tag{9.17}$$

where $k^2 \equiv \omega^2/v^2$. This is known as the **Helmholtz equation**. It has the standard solutions

$$z(x) = Ae^{ikx} + Be^{-ikx}\,. \tag{9.18}$$

Thus, for a given value of $\omega$, the solution for $y(x,t)$ can be written as

$$y(x,t) = \mathrm{Re}\left[\left(Ae^{ikx} + Be^{-ikx}\right)e^{-i\omega t}\right], \tag{9.19}$$

or, equivalently,

$$y(x,t) = \mathrm{Re}\left[Ae^{ik(x-vt)} + Be^{-ik(x+vt)}\right], \tag{9.20}$$

where we used $\omega = kv$. Note that the first term represents a wave with wavevector $k$ moving to the *right* with speed $v$, while the second term represents a wave moving to the *left* with the same speed. In the above expressions, $A$ and $B$ are complex constants whose values are to be determined by the boundary conditions and initial conditions of the problem.

At this point, we usually obtain the *general* solution of the differential equation by forming a linear combination of the eigenfunction solutions over the *allowed* eigenvalues for the problem. These eigenvalues are determined by requiring that the eigenfunctions satisfy the appropriate *boundary conditions*. The *particular* solution for a given problem is then obtained by matching the general solution to the *initial conditions*, which usually specify the displacement function and its time derivative at $t = 0$:

$$y(x,0) = f(x), \qquad \dot{y}(x,0) = g(x). \tag{9.21}$$

## 9.3.2 Normal Form Solution (Characteristic Coordinates)

An alternative approach to solving the wave equation that side-steps the separation of variables procedure is hinted at by (9.20), which is a sum of right-moving and left-moving waves. The form of this solution suggests making a change of variables from $x$ and $t$ to

$$\xi \equiv x - vt, \qquad \eta \equiv x + vt, \tag{9.22}$$

which are called **characteristic coordinates** for the wave equation. When expressed in terms of $\xi$ and $\eta$, the wave equation takes the very simple form

$$\frac{\partial^2 y}{\partial \xi \, \partial \eta} = 0, \tag{9.23}$$

which is immediately solved by

$$y(\xi, \eta) = \Psi(\xi) + \Phi(\eta), \tag{9.24}$$

where $\Psi$ and $\Phi$ are *arbitrary* functions to be determined by the boundary values and initial conditions for the particular problem. Equation (9.23) is the so-called **normal form** for the wave equation. Partial differential equations that can be expressed in the form of (9.23) are known as **hyperbolic equations**.

In this form, the initial conditions ($t = 0$) correspond to $\xi = x$ and $\eta = x$, with

$$f(x) = \Psi(x) + \Phi(x), \tag{9.25}$$
$$g(x) = -v\Psi'(x) + v\Phi'(x). \tag{9.26}$$

By integrating (9.26) and adding or subtracting it from (9.25), we find

$$\Psi(x) = \frac{1}{2} f(x) - \frac{1}{2v} \int g(x)\mathrm{d}x\,, \tag{9.27}$$

$$\Phi(x) = \frac{1}{2} f(x) + \frac{1}{2v} \int g(x)\mathrm{d}x\,. \tag{9.28}$$

The full solution is then $y(x,t) = \Psi(x - vt) + \Phi(x + vt)$.

**Exercise 9.1** Consider a wave with an initial displacement of a Gaussian,

$$y(x,0) = f(x) \equiv A\mathrm{e}^{-x^2/2\sigma_x^2}\,, \tag{9.29}$$

and no initial displacement velocity. Determine the displacement for any later time $t$.

**Exercise 9.2** Consider a wave with an initial displacement of zero everywhere, but an initial displacement velocity given by a Gaussian,

$$\dot{y}(x,0) = g(x) \equiv A\mathrm{e}^{-x^2/2\sigma_v^2}\,. \tag{9.30}$$

Determine the displacement for any later time $t$.

**Exercise 9.3** Determine the displacement function $y(x,t)$ for a wave with the initial conditions

$$y(x,0) = A\cos kx\,, \qquad \dot{y}(x,0) = \omega A \sin kx\,, \tag{9.31}$$

where $\omega = kv$.

We now have two approaches to solving the wave equation, but so far we have only discussed how to handle the *initial conditions*; we have not yet discussed what happens at the *boundaries* of the string. In the next three sections, we will look at imposing boundary conditions to describe some of the more common applications of the one-dimensional wave equation. We will consider a string with (i) fixed endpoints, (ii) periodic boundary conditions, or (iii) infinite boundary conditions.

## 9.4 String with Fixed Endpoints

As our first case, consider a string of length $\ell$ with fixed endpoints, so that the boundary conditions are $y(0,t) = y(\ell,t) = 0$ for all times $t$. If we approach this problem using the solutions to the normal form of the wave equation, we find a difficulty in imposing the initial conditions. Since the string is of finite length, the initial condition functions $f(x)$ and $g(x)$ from (9.21) are defined only on the interval $0 \le x \le \ell$. If we then apply the solutions for the functions $\Psi(\xi)$ and $\Phi(\eta)$ from (9.27) and (9.28), we find that the solution for the displacement is only valid for values of $x$ and $t$ that simultaneously satisfy $0 \le \xi \le \ell$ and $0 \le \eta \le \ell$. This is the shaded region shown in Fig. 9.1. In order to use this solution, we need to extend the functions $f$ and $g$ beyond the length of the finite string. But before doing this, let us look at the eigenvalue/eigenfunction solution of (9.20) for guidance.

Imposing the boundary conditions on the eigenfunctions given by (9.18) results in the following constraints for each eigenfrequency $\omega$:

$$z(0) = A + B = 0\,, \qquad z(\ell) = A\mathrm{e}^{\mathrm{i}k\ell} + B\mathrm{e}^{-\mathrm{i}k\ell} = 0\,. \tag{9.32}$$

The first condition requires that $A = -B$, so the second equation now reads

$$z(\ell) = 2\mathrm{i}A\sin(k\ell) = 0\,, \tag{9.33}$$

which is solved by requiring $\sin(k\ell) = 0$. Thus, $k\ell$ must be an integer multiple of $\pi$:

$$k_n = n\pi/\ell\,, \qquad n = 1, 2, \ldots\,. \tag{9.34}$$

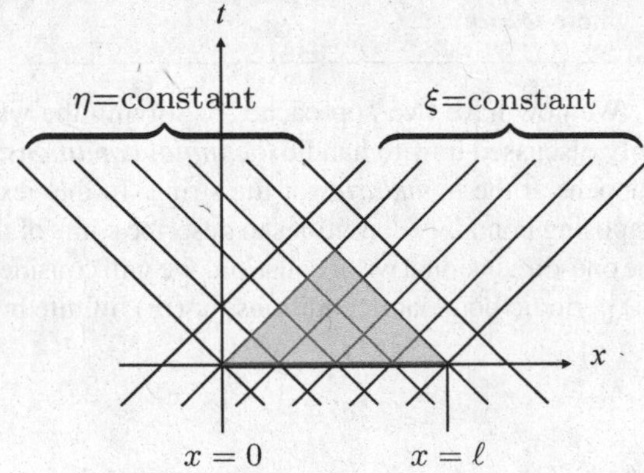


**Fig. 9.1** The characteristic coordinates $\xi$ and $\eta$ with the usual $x$ and $t$ coordinates. The initial conditions of the string are defined on the bold line between $x = 0$ and $x = \ell$ at $t = 0$. The characteristic solution is then valid only in the shaded region bounded by the lines $t = 0$, $\xi = 0$ and $\eta = \ell$

Note that we can restrict attention to positive values of $n$, since $n = 0$ gives the trivial (zero) solution, and negative integers simply change the sign of the sine functions, which can be absorbed in the multiplicative constant $A$. Thus, the boundary conditions have imposed a *discrete* structure to the eigenvalues, $\omega_n = k_n v = n\pi v/\ell$.

With the discrete set of eigenfunctions, we can now build the general solution as a sum over $n$, so

$$y(x,t) = \frac{1}{2}\sum_{n=1}^{\infty} \sin(n\pi x/\ell)\left[C_n \mathrm{e}^{-\mathrm{i}\omega_n t} + C_n^* \mathrm{e}^{\mathrm{i}\omega_n t}\right], \tag{9.35}$$

where $C_n \equiv a_n + ib_n$ are complex coefficients. (Note that we have made $y(x,t)$ real by including the complex-conguate terms.) The initial conditions that must be satisfied by this general solution are the same as those given by (9.21). They can be written in terms of real expansion coefficients $a_n$ and $b_n$ as follows:

$$f(x) = \sum_{n=1}^{\infty} a_n \sin(n\pi x/\ell), \qquad g(x) = \sum_{n=1}^{\infty} \omega_n b_n \sin(n\pi x/\ell). \tag{9.36}$$

These summations are reminiscent of a *Fourier sine series* over the interval $-\ell < x < \ell$. Using the orthonormality property of the sine functions, these equations can be inverted to give expressions for the $a_n$ and $b_n$:

$$\begin{aligned} a_n &= \frac{1}{\ell}\int_{-\ell}^{\ell} \mathrm{d}x\, f(x)\sin(n\pi x/\ell) = \frac{2}{\ell}\int_0^{\ell} \mathrm{d}x\, f(x)\sin(n\pi x/\ell), \\ \omega_n b_n &= \frac{1}{\ell}\int_{-\ell}^{\ell} \mathrm{d}x\, g(x)\sin(n\pi x/\ell) = \frac{2}{\ell}\int_0^{\ell} \mathrm{d}x\, g(x)\sin(n\pi x/\ell), \end{aligned} \tag{9.37}$$

where the second equality assumes that $f(x)$ and $g(x)$ were extended as *odd* functions for $-\ell \le x \le 0$ (See a more detailed discussion of this right after Exercise 9.4). The general solution for $y(x,t)$ is then

$$y(x,t) = \sum_{n=1}^{\infty} |C_n| \sin\left(\frac{n\pi x}{\ell}\right)\cos\left(\frac{n\pi v t}{\ell} - \phi_n\right), \tag{9.38}$$

where $|C_n| \equiv \sqrt{a_n^2 + b_n^2}$, $\phi_n \equiv \tan^{-1}(b_n/a_n)$, with $a_n$, $b_n$ determined by (9.37).

**Exercise 9.4** Consider a guitar string of length 65 cm and mass 2 gm, which is under a tension of 100 lb. Assume that you pluck it by pulling it up 1 cm at a point that is 1/4 of its length from one end, and then release it from rest. (The initial displacement thus has a triangle shape.) (a) What is the fundamental frequency? (b) What are the relative amplitudes of the harmonics? (c) Are any harmonics completely missing from the spectrum?

In our discussion above, we noted that the general solution suggested a Fourier sine series over an interval that was twice the length of the vibrating string. This information provides us with a suggestion on how to extend the initial conditions beyond the endpoints of the string. If we continue the functions $f(x)$ and $g(x)$ over all $x$ while keeping them both continuous in $x$, then the boundary conditions can be satisfied for all $f$ and $g$ provided they are both *odd* functions about $x = 0$ and $x = \ell$:

$$\begin{aligned} f(-x) &= -f(x)\,, \\ g(-x) &= -g(x)\,, \\ f(\ell - x) &= -f(\ell + x)\,, \\ g(\ell - x) &= -g(\ell + x)\,. \end{aligned} \tag{9.39}$$

A little manipulation of these requirements shows that

$$f(x + 2\ell) = f(x)\,, \qquad g(x + 2\ell) = g(x)\,, \tag{9.40}$$

which is equivalent to saying that the initial conditions must be odd functions (about $x = 0$) and have a periodicity equal to $2\ell$. Thus, the Fourier sine series exhibits the appropriate periodicity of *twice* the length of the string. This extension of the initial conditions is exactly what is needed to have the solutions for the normal form of the wave equation be defined for all $t > 0$.

**Exercise 9.5** Verify that (9.40) follows from (9.39).

## 9.5 Periodic Boundary Conditions

We can relax the requirement that $f(x)$ and $g(x)$ be odd functions, but keep the requirements of (9.40) to obtain **periodic boundary conditions**. To determine the eigenfunctions for periodic boundary conditions, we return to the solutions of the wave equation (9.20) and require $y(x, t) = y(x + 2\ell, t)$ for all times $t$. Thus,

$$A\mathrm{e}^{ikx} + B\mathrm{e}^{-ikx} = A\mathrm{e}^{ikx}\mathrm{e}^{ik2\ell} + B\mathrm{e}^{-ikx}\mathrm{e}^{-ik2\ell}\,, \tag{9.41}$$

which implies

$$e^{ik2\ell} = e^{-ik2\ell} = 1\,, \tag{9.42}$$

or $k_n = n\pi/\ell$. We see that periodic boundary conditions with a periodicity of $2\ell$ reproduce the discrete spectrum of $k_n$ that were obtained from the fixed endpoints of a string of length $\ell$, but do not require that the eigenfunctions be sines. The general solution for an arbitrary wave on a string with periodic boundary conditions then becomes

$$y(x,t) = \frac{1}{2}\sum_{n=-\infty}^{\infty}\left[C_n e^{-i\omega_n t} + C_{-n}^* e^{i\omega_n t}\right] e^{ik_n x}\,, \tag{9.43}$$

where $C_n$ are complex coefficients to be determined by the initial conditions $y(x,0) = f(x)$ and $\dot{y}(x,0) = g(x)$. Note that for this case, the summation is over both positive and negative values of $n$. Since we will want the eigenfrequencies $\omega_n$ to have non-negative values, we define

$$\omega_n \equiv \frac{|n|\pi v}{\ell} = |k_n| v\,, \qquad n = 0, \pm 1, \pm 2, \ldots\,. \tag{9.44}$$

By including the $C_{-n}^* e^{i\omega_n t} e^{ik_n x}$ terms in the summation, we make $y(x,t)$ real.

In terms of the complex coefficients $C_n$, the initial conditions become

$$\begin{aligned} f(x) &= \sum_{n=-\infty}^{\infty} \frac{1}{2}\left(C_n + C_{-n}^*\right) e^{ik_n x}\,, \\ g(x) &= \sum_{n=-\infty}^{\infty} -\frac{i}{2}\omega_n \left(C_n - C_{-n}^*\right) e^{ik_n x}\,. \end{aligned} \tag{9.45}$$

These equations recall the complex form of the Fourier series to describe an arbitrary function with periodicity $2\ell$. Using the orthogonality property of the complex exponentials $e^{ik_n x} = e^{in\pi x/\ell}$, we can invert these equations to find

$$\begin{aligned} \frac{1}{2}(C_n + C_{-n}^*) &= \frac{1}{2\ell}\int_{-\ell}^{\ell} dx\, f(x) e^{-ik_n x}\,, \\ -\frac{i}{2}\omega_n (C_n - C_{-n}^*) &= \frac{1}{2\ell}\int_{-\ell}^{\ell} dx\, g(x) e^{-ik_n x}\,. \end{aligned} \tag{9.46}$$

For $n = 0$ we have,

$$\frac{1}{2}\left(C_0 + C_0^*\right) = \frac{1}{2\ell}\int_{-\ell}^{\ell} \mathrm{d}x \; f(x)\,, \tag{9.47}$$

and for $n \neq 0$:

$$C_n = \frac{1}{2\ell}\int_{-\ell}^{+\ell} \mathrm{d}x \, \left( f(x) + \frac{\mathrm{i}}{\omega_n} g(x) \right) \mathrm{e}^{-\mathrm{i}k_n x}\,. \tag{9.48}$$

### *9.5.1 Equivalence of Eigenfunction and Normal Form Solutions*

We now have the two solutions for waves with periodic boundary conditions. In this subsection, we show the equivalence between the two solutions.

Starting with the eigenfunction solution, we note that the general solution (9.43) can be written as

$$y(x,t) = \frac{1}{2}\sum_{n=-\infty}^{\infty}\left[C_n \mathrm{e}^{\mathrm{i}(k_n x - \omega_n t)} + C_{-n}^* \mathrm{e}^{\mathrm{i}(k_n x + \omega_n t)}\right]. \tag{9.49}$$

Since we eventually want to write the solution in terms of the characteristic coordinates $\xi \equiv x - vt$ and $\eta \equiv x + vt$, we break up the summations into terms involving only positive and negative values of $n$ so that we can easily extract the $x \pm vt$ contributions:

$$\begin{aligned} y(x,t) = \frac{1}{2}\left(C_0 + C_0^*\right) &+ \frac{1}{2}\sum_{n=1}^{\infty}\left[C_n \mathrm{e}^{\mathrm{i}k_n(x-vt)} + C_{-n}^* \mathrm{e}^{\mathrm{i}k_n(x+vt)}\right] \\ &+ \frac{1}{2}\sum_{n=-1}^{-\infty}\left[C_n \mathrm{e}^{\mathrm{i}k_n(x+vt)} + C_{-n}^* \mathrm{e}^{\mathrm{i}k_n(x-vt)}\right], \end{aligned} \tag{9.50}$$

where we used $\omega_n = |k_n| v$ with $k_n = n\pi/\ell$. Using (9.47) and (9.48) for the coefficients $C_n$, and doing a couple of lines of algebra to combine the summations so that they run again over both positive and negative values of $n$, we find

$$y(x,t) = \frac{1}{2}\left\{\int_{-\ell}^{+\ell} \mathrm{d}u\, f(u)\frac{1}{2\ell}\sum_{n=-\infty}^{+\infty}\left(\mathrm{e}^{\mathrm{i}k_n(\xi-u)} + \mathrm{e}^{\mathrm{i}k_n(\eta-u)}\right)\right.$$
$$\left. + \int_{-\ell}^{+\ell} \mathrm{d}u\, g(u)\frac{1}{2\ell}\sum_{n=-\infty,\neq 0}^{+\infty}\frac{i}{k_n v}\left(\mathrm{e}^{\mathrm{i}k_n(\xi-u)} - \mathrm{e}^{\mathrm{i}k_n(\eta-u)}\right)\right\}, \quad (9.51)$$

where we substituted $\xi$ and $\eta$ for $x - vt$ and $x + vt$, and where the summation inside the integral for $g(u)$ does not include the $n = 0$ term. These formulas can be simplified by noting that the summations inside the integrals can be written in terms of the **Dirac delta function** $\delta(x)$ on the interval $-\ell \le x \le \ell$,

$$\delta(x - x') = \frac{1}{2\ell}\sum_{n=-\infty}^{+\infty} \mathrm{e}^{\pm\mathrm{i}n\pi(x-x')/\ell}, \quad (9.52)$$

and the **Heaviside theta function** $\Theta(x)$, which is defined by

$$\int_{-\infty}^{x} \mathrm{d}u\, \delta(u) = \Theta(x) = \begin{cases} 1 & x > 0 \\ 0 & x < 0 \end{cases} \quad (9.53)$$

and can be written as

$$\Theta(x - x') = \mp\frac{\mathrm{i}}{2\ell}\sum_{n=-\infty,\neq 0}^{+\infty} \frac{\ell}{n\pi}\mathrm{e}^{\pm\mathrm{i}n\pi(x-x')/\ell}. \quad (9.54)$$

Thus, making these substitutions, it follows that

$$\begin{aligned} y(x,t) &= \frac{1}{2}\left\{\int_{-\ell}^{+\ell} \mathrm{d}u\, f(u)\left[\delta(\xi - u) + \delta(\eta - u)\right]\right. \\ &\qquad \left. - \frac{1}{v}\int_{-\ell}^{+\ell} \mathrm{d}u\, g(u)\left[\Theta(\xi - u) - \Theta(\eta - u)\right]\right\} \\ &= \frac{1}{2}\left(f(\xi) + f(\eta)\right) - \frac{1}{2v}\left(\int_{-\ell}^{\xi} \mathrm{d}u\, g(u) - \int_{-\ell}^{\eta} \mathrm{d}u\, g(u)\right) \\ &= \Psi(\xi) + \Phi(\eta), \end{aligned} \quad (9.55)$$

which is the normal form solution. So for periodic boundary conditions, the solution written as an expansion in eigenfunctions is equivalent to the solution obtained through the normal form of the wave equation, as it should be.

**Exercise 9.6** Verify that (9.54) follows from (9.52) and (9.53).

**Exercise 9.7** Fill in the details to go from (9.49) to (9.55).

## 9.6 Infinite Boundary Conditions

If both $f(x) = y(x, 0)$ and $g(x) = \dot{y}(x, 0)$ are specified, the initial conditions are known as **Cauchy conditions**. When Cauchy conditions are specified, the normal form of the solution is usually the most convenient way to solve the wave equation. If, however, the initial conditions are restricted to have zero initial displacement velocity (i.e., $g(x) = 0$), then the eigenfunction solution very easily admits of a Fourier series solution. In the following discussion, we will restrict ourselves to this simpler class of initial conditions. In Exercise 9.9, you are asked to extend the analysis given here to the general case where $g(x) \neq 0$.

In order to understand the behavior of an infinite string, we will start with a finite string of length $2\ell$ and extend the solutions in the limit that $\ell \to \infty$. For periodic boundary conditions with $g(x) = 0$, it follows that

$$C_{-n}^* = C_n\,, \quad \text{with} \quad C_n = \frac{1}{2\ell}\int_{-\ell}^{+\ell} \mathrm{d}x\; f(x)\mathrm{e}^{-\mathrm{i}k_n x}\,. \tag{9.56}$$

In addition, the Fourier expansion for $y(x, t)$ reduces to

$$y(x,t) = \frac{1}{2}\sum_{n=-\infty}^{\infty} C_n\left(\mathrm{e}^{-\mathrm{i}\omega_n t} + \mathrm{e}^{\mathrm{i}\omega_n t}\right)\mathrm{e}^{\mathrm{i}k_n x} = \sum_{n=-\infty}^{\infty} C_n\cos(\omega_n t)\mathrm{e}^{\mathrm{i}k_n x}\,. \tag{9.57}$$

We can now obtain an expression in the limit where $k$ becomes continuous, which arises when we have no boundary conditions (or when we move the boundaries out to infinity). What follows is not a rigorous mathematical proof of the **Fourier transform** integrals (9.61) and (9.62) but rather a plausibility argument for them.

For the initial conditions that we are now considering, periodic boundary conditions give

$$f(x) = \sum_{n=-\infty}^{+\infty} C_n \mathrm{e}^{\mathrm{i}k_n x}\,, \qquad C_n = \frac{1}{2\ell}\int_{-\ell}^{+\ell} \mathrm{d}u\; f(u)\mathrm{e}^{-\mathrm{i}k_n u}\,, \tag{9.58}$$

where $k_n = n\pi/\ell$, and we have replaced the dummy variable $x$ in the integral with $u$ in order to avoid some confusion later in this argument. To prepare for the eventual transition to continuous $k$, let's imagine that the $k_n$ are a discrete sampling of an underlying continuous real variable $k$. In this case, we can define the discrete steps in $k$ with

$$\Delta k \equiv k_{n+1} - k_n = \frac{\pi}{\ell} . \tag{9.59}$$

Now, combining the series expression for $f(x)$ with the integral expression for the coefficients $C_n$, we find

$$f(x) = \sum_{n=-\infty}^{+\infty} \frac{1}{2\ell} \int_{-\ell}^{+\ell} du\ f(u) e^{-ik_n(u-x)} = \frac{1}{2\pi} \sum_{n=-\infty}^{+\infty} \Delta k \left[ \int_{-\ell}^{+\ell} du\ f(u) e^{-ik_n u} \right] e^{ik_n x} . \tag{9.60}$$

In the limit $\ell \to \infty$, $\Delta k$ will go to the infinitesimal d$k$, the sum over $n$ will go to an integral, and the term in square brackets will transition from discrete coefficients to a continuous function of $k$. We will define this continuous function as

$$F(k) \equiv \frac{1}{\sqrt{2\pi}} \int_{-\infty}^{+\infty} du\ f(u) e^{-iku} , \tag{9.61}$$

so that in the limit $\ell \to \infty$, the discrete summation over $k_n$ transitions to an integral over the continuous variable $k$, yielding

$$f(x) = \frac{1}{\sqrt{2\pi}} \int_{-\infty}^{+\infty} dk\ F(k) e^{ikx} . \tag{9.62}$$

The function $F(k)$ in (9.61) is the **Fourier transform** of the function $f(x)$. The two functions $f(x)$ and $F(k)$ form a **Fourier transform pair**.[1]

The solutions to the Helmholtz equation (9.17) are orthogonal. The orthogonality condition for functions that are parameterized by the continuous variable $k$ (as opposed to a discrete parameterization $k_n$), is

$$\int_{-\infty}^{+\infty} dx\ f^*(k; x) f(k'; x) = \mathcal{N} \delta(k - k') , \tag{9.63}$$

where $\mathcal{N}$ is some constant that may depend upon $k$. If the orthogonal functions are normalized, then $\mathcal{N} = 1$. For the Helmholtz equation, we can determine the value of $\mathcal{N}$ from the definition of the Dirac delta function given in (A.106), which gives

[1] There is no consensus on the placement of the factors of $2\pi$ in the Fourier transform. Some authors use a factor of $1/2\pi$ in front of the definition of $f(x)$ (with nothing in front of the definition of $F(k)$), while others put the factor of $1/2\pi$ in front of the definition of $F(k)$. We have chosen to share it evenly across both definitions, but you should check to see what convention is being used when referring to software or tables of Fourier transform pairs.

$$\int_{-\infty}^{+\infty} \mathrm{d}x\, \mathrm{e}^{-\mathrm{i}kx}\mathrm{e}^{\mathrm{i}k'x} = 2\pi\delta(k-k')\,, \tag{9.64}$$

and so $\mathcal{N} = 2\pi$ in this case.

**Exercise 9.8** Show that $F(k)$ satisfies the continuous equivalent of $C_{-n} = C_n^*$ for real $f(x)$—i.e., $F(-k) = F^*(k)$.

### *9.6.1 Equivalence of Fourier Transform and Normal Form Solutions*

Using the above results, we can now show the equivalence between the Fourier transform solution of the wave equation and the normal form solution $y(x,t) = \Psi(\xi) + \Phi(\eta)$. We do this here for the case where the initial conditions are such that $g(x) = 0$. In Exercise 9.9, you are asked to extend this analysis to the case where $g(x) \neq 0$.

We will start then with the Fourier transform solution

$$y(x,t) = \frac{1}{\sqrt{2\pi}}\int_{-\infty}^{+\infty} \mathrm{d}k\, \frac{1}{2}C(k)\left[\mathrm{e}^{\mathrm{i}(kx-\omega t)} + \mathrm{e}^{\mathrm{i}(kx+\omega t)}\right]\,, \tag{9.65}$$

where

$$C(k) = \frac{1}{\sqrt{2\pi}}\int_{-\infty}^{+\infty} \mathrm{d}x\, f(x)\mathrm{e}^{-\mathrm{i}kx} \equiv F(k) \tag{9.66}$$

is the Fourier transform of the function $f(x) \equiv y(x,0)$. Remembering that $\omega$ is always positive and related to $k$ by $\omega = |k|v$, where $v$ is the wave speed, we can split the two integrals into contributions involving only positive and negative values of $k$:

$$\begin{aligned} y(x,t) = \frac{1}{\sqrt{2\pi}}\frac{1}{2}\Bigg[\int_0^{\infty} \mathrm{d}k\, C(k)\mathrm{e}^{\mathrm{i}k(x-vt)} + \int_{-\infty}^{0} \mathrm{d}k\, C(k)\mathrm{e}^{\mathrm{i}k(x+vt)} \\ + \int_0^{\infty} \mathrm{d}k\, C(k)\mathrm{e}^{\mathrm{i}k(x+vt)} + \int_{-\infty}^{0} \mathrm{d}k\, C(k)\mathrm{e}^{\mathrm{i}k(x-vt)}\Bigg]\,. \end{aligned} \tag{9.67}$$

But notice that we can combine the 1st and 4th integrals, and also the 2nd and 3rd integrals, to get

$$
\begin{aligned}
y(x,t) &= \frac{1}{\sqrt{2\pi}}\frac{1}{2}\left[\int_{-\infty}^{\infty} \mathrm{d}k\; C(k)\mathrm{e}^{\mathrm{i}k(x-vt)} + \int_{-\infty}^{\infty} \mathrm{d}k\; C(k)\mathrm{e}^{\mathrm{i}k(x+vt)}\right] \\
&= \frac{1}{\sqrt{2\pi}}\frac{1}{2}\int_{-\infty}^{\infty} \mathrm{d}k\; C(k)\left[\mathrm{e}^{\mathrm{i}k\xi} + \mathrm{e}^{\mathrm{i}k\eta}\right],
\end{aligned}
\tag{9.68}
$$

where we have made the substitution $\xi \equiv x - vt, \eta \equiv x + vt$ to obtain the last equality. Then we can substitute for $C(k)$ using (9.66), and use the intergral representation of the Dirac delta function (9.64) to simplify the right-hand side:

$$
\begin{aligned}
y(x,t) &= \frac{1}{\sqrt{2\pi}}\frac{1}{2}\int_{-\infty}^{\infty} \mathrm{d}k\; \frac{1}{\sqrt{2\pi}}\int_{-\infty}^{\infty} \mathrm{d}u\; f(u)\mathrm{e}^{-\mathrm{i}ku}\left[\mathrm{e}^{\mathrm{i}k\xi} + \mathrm{e}^{\mathrm{i}k\eta}\right] \\
&= \frac{1}{2}\int_{-\infty}^{\infty} \mathrm{d}u\; f(u)\frac{1}{2\pi}\int_{-\infty}^{\infty} \mathrm{d}k\left[\mathrm{e}^{\mathrm{i}k(\xi-u)} + \mathrm{e}^{\mathrm{i}k(\eta-u)}\right] \\
&= \frac{1}{2}\int_{-\infty}^{\infty} \mathrm{d}u\; f(u)\left[\delta(\xi-u) + \delta(\eta-u)\right] \\
&= \frac{1}{2}f(\xi) + \frac{1}{2}f(\eta) = \Psi(\xi) + \Phi(\eta),
\end{aligned}
\tag{9.69}
$$

which is the desired result for the case $g(x) \equiv \dot{y}(x,0) = 0$.

**Exercise 9.9** Extend the analysis of the previous two subsections to allow non-zero initial velocity—i.e., $g(x) \neq 0$. *Hint*: You should start with the Fourier transform solution

$$
y(x,t) = \frac{1}{\sqrt{2\pi}}\int_{-\infty}^{+\infty} \mathrm{d}k\; \frac{1}{2}\left[C(k)\mathrm{e}^{-\mathrm{i}\omega t} + C^*(-k)\mathrm{e}^{\mathrm{i}\omega t}\right]\mathrm{e}^{\mathrm{i}kx}, \tag{9.70}
$$

where

$$
C(k) = \frac{1}{\sqrt{2\pi}}\int_{-\infty}^{+\infty} \mathrm{d}x\; \left(f(x) + \frac{i}{\omega}g(x)\right)\mathrm{e}^{-\mathrm{i}kx} \equiv F(k) + \frac{i}{\omega}G(k), \tag{9.71}
$$

with $F(k)$ and $G(k)$ being the Fourier transforms of $f(x) \equiv y(x,0)$ and $g(x) \equiv \dot{y}(x,0)$. You will also need to use the following intergral representation of the Heaviside theta function, cf. (9.54):

$$
\Theta(x-x') = \mp\frac{i}{2\pi}\int_{-\infty}^{\infty} \mathrm{d}k\; \frac{1}{k}\mathrm{e}^{\pm\mathrm{i}k(x-x')}. \tag{9.72}
$$

## 9.7 Three-Dimensional Wave Equation

So far, we have discussed the properties of the wave equation in the context of the continuous limit of the discrete equations of motion for the loaded string. As such, we have restricted ourselves to *one-dimensional* waves. But the wave equation can be generalized to more dimensions by simply replacing the partial derivative with respect to $x$, i.e., $\partial^2/\partial x^2$, with the Laplacian $\nabla^2$. The three-dimensional wave equation is thus

$$\Box^2 y(\mathbf{r}, t) \equiv \left(\nabla^2 - \frac{1}{v^2}\frac{\partial^2}{\partial t^2}\right) y(\mathbf{r}, t) = 0\,, \tag{9.73}$$

where $\Box^2$ is the **D'Alembertian** operator.

Solutions to (9.73) can be found as before using separation of variables. Let's first separate the spatial and temporal variables by postulating a solution of the form

$$y(\mathbf{r}, t) = \zeta(\mathbf{r})\mathrm{e}^{-\mathrm{i}\omega t}\,. \tag{9.74}$$

This results in the three-dimensional Helmholtz equation

$$\nabla^2\zeta + k^2\zeta = 0\,, \tag{9.75}$$

where $k^2 \equiv \omega^2/v^2$. The solution of the Helmholtz equation can also be accomplished through separation of variables, with the choice of coordinates depending upon the symmetries of the boundary conditions and the initial conditions. In order to gain a better understanding of the meaning of $k^2$, let us consider the specific case of Cartesian coordinates. In this case, we let

$$\zeta(\mathbf{r}) = X(x)Y(y)Z(z)\,, \tag{9.76}$$

for which separation of variables results in the three ordinary differential equations

$$\begin{aligned} X'' + k_x^2 X &= 0\,, \\ Y'' + k_y^2 Y &= 0\,, \\ Z'' + k_z^2 Z &= 0\,, \end{aligned} \tag{9.77}$$

where

$$k_x^2 + k_y^2 + k_z^2 = k^2\,. \tag{9.78}$$

The solutions to these three equations result in a solution for $y$ of the form

$$y(\mathbf{r}, t) = C e^{i(k_x x + k_y y + k_z z - \omega t)} = C e^{i(\mathbf{k}\cdot\mathbf{r} - \omega t)}, \tag{9.79}$$

where the **wave vector k** is defined by

$$\mathbf{k} = k_x \hat{\mathbf{x}} + k_y \hat{\mathbf{y}} + k_z \hat{\mathbf{z}}, \quad \text{with} \quad |\mathbf{k}\cdot\mathbf{k}| = k^2 \equiv \omega^2/v^2. \tag{9.80}$$

Thus, the wavelength $\lambda$ of the waves is related to the magnitude of the wave vector **k** via $k = 2\pi/\lambda$. In addition, the wave vector also has the property that it points in the direction of propagation of a plane wave in three dimensions. In Cartesian coordinates, we can build up the general solution by summing such solutions over **k**. For infinite boundary conditions, the general solution becomes

$$y(\mathbf{r}, t) = \mathrm{Re}\left[\frac{1}{(2\pi)^{3/2}} \int \mathrm{d}V_{\mathbf{k}}\, C(\mathbf{k}) e^{i(\mathbf{k}\cdot\mathbf{r} - \omega t)}\right], \tag{9.81}$$

or, explicitly,

$$y(\mathbf{r}, t) = \frac{1}{(2\pi)^{3/2}} \int \mathrm{d}V_{\mathbf{k}}\, \frac{1}{2}\left[C(\mathbf{k}) e^{-i\omega t} + C^*(-\mathbf{k}) e^{i\omega t}\right] e^{i\mathbf{k}\cdot\mathbf{r}}, \tag{9.82}$$

where $\mathrm{d}V_{\mathbf{k}}$ is the 3-dimensional volume element in **k**-space. The expansion coefficients $C(\mathbf{k})$ are determined by the initial conditions via

$$C(\mathbf{k}) = F(\mathbf{k}) + \frac{i}{\omega} G(\mathbf{k}), \tag{9.83}$$

where $F(\mathbf{k})$ and $G(\mathbf{k})$ are the 3-dimensional Fourier transforms of the functions $f(\mathbf{r}) \equiv y(\mathbf{r}, 0)$ and $g(\mathbf{r}) \equiv \dot{y}(\mathbf{r}, 0)$:

$$\begin{aligned} F(\mathbf{k}) &= \frac{1}{(2\pi)^{3/2}} \int \mathrm{d}V\, f(\mathbf{r}) e^{-i\mathbf{k}\cdot\mathbf{r}}, \\ G(\mathbf{k}) &= \frac{1}{(2\pi)^{3/2}} \int \mathrm{d}V\, g(\mathbf{r}) e^{-i\mathbf{k}\cdot\mathbf{r}}. \end{aligned} \tag{9.84}$$

Note that the above equations are the generalizations of the solution of the 1-dimensional wave equation for the case of infinite boundary conditions, which is discussed in Exercise 9.9.

***Example 9.1*** (Solar oscillations) The Sun undergoes normal mode oscillations in pressure about the value of the equilibrium pressure. We can approximate these oscillations as sound waves within the Sun. In practice, the equilibrium pressure in the Sun varies with depth, but we will use a simplifying assumption that the speed of sound is constant throughout the Sun, so the motion can be described using the wave equation. Denoting the pressure variations by $u(\mathbf{r}, t)$, we have

$$\Box^2 u \equiv \nabla^2 u - \frac{1}{v^2}\frac{\partial^2 u}{\partial t^2} = 0\,. \tag{9.85}$$

As we interested in normal mode oscillations, we can immediately separate out the time dependence assuming a solution of the form

$$u(\mathbf{r}, t) = \zeta(\mathbf{r})\mathrm{e}^{-\mathrm{i}\omega t}\,. \tag{9.86}$$

This leads to the three-dimensional Helmholtz equation

$$\nabla^2\zeta + k^2\zeta = 0\,, \tag{9.87}$$

where $k^2 \equiv \omega^2/v^2$.

Since the solar oscillations exhibits spherically symmetric boundary conditions with the pressure oscillations going to zero at the surface of the Sun, it is best to use spherical coordinates $(r, \theta, \phi)$ for the spatial part of the analysis. Separating variables in the spatial function $\zeta(\mathbf{r})$ as

$$\zeta(r, \theta, \phi) \equiv R(r)\Theta(\theta)\Phi(\phi)\,, \tag{9.88}$$

the Helmholtz equation becomes

$$\frac{\sin^2\theta}{R}\left[\frac{\mathrm{d}}{\mathrm{d}r}\left(r^2\frac{\mathrm{d}R}{\mathrm{d}r}\right)\right] + \frac{\sin\theta}{\Theta}\left[\frac{\mathrm{d}}{\mathrm{d}\theta}\left(\sin\theta\frac{\mathrm{d}\Theta}{\mathrm{d}\theta}\right)\right] + \frac{1}{\Phi}\left[\frac{\mathrm{d}^2\Phi}{\mathrm{d}\phi^2}\right] + k^2r^2\sin^2\theta = 0\,. \tag{9.89}$$

Since the $\mathrm{d}^2\Phi/\mathrm{d}\phi^2$ term is the only term that depends on $\phi$ (the others depend on both $r$ and $\theta$), it must equal a constant:

$$\frac{\mathrm{d}^2\Phi}{\mathrm{d}\phi} = -m^2\Phi\,. \tag{9.90}$$

This equation is immediately solved by

$$\Phi(\phi) = A\mathrm{e}^{\mathrm{i}m\phi} + B\mathrm{e}^{-\mathrm{i}m\phi}\,. \tag{9.91}$$

But since $\phi$ is a cyclic coordinate, we must impose *periodic* boundary conditions, $\Phi(\phi + 2\pi) = \Phi(\phi)$, which imply that $m$ be an integer. Allowing both positive and negative values of $m$, we can write the solution as

$$\Phi(\phi) \propto e^{im\phi}, \qquad m = 0, \pm 1, \pm 2, \ldots . \tag{9.92}$$

In terms of $m$, the Helmholtz equation can be further separated between $\theta$ and $r$ by dividing through by $\sin^2\theta$, leaving

$$\frac{1}{R}\left[\frac{d}{dr}\left(r^2\frac{dR}{dr}\right)\right] + \frac{1}{\Theta}\frac{1}{\sin\theta}\left[\frac{d}{d\theta}\left(\sin\theta\frac{d\Theta}{d\theta}\right)\right] - \frac{m^2}{\sin^2\theta} + r^2k^2 = 0. \tag{9.93}$$

As before, the terms involving only $\theta$ must equal a constant, which we choose (with some hindsight) to be $-l(l+1)$:

$$\frac{1}{\sin\theta}\left[\frac{d}{d\theta}\left(\sin\theta\frac{d\Theta}{d\theta}\right)\right] - \frac{m^2}{\sin^2\theta}\Theta = -l\,(l+1)\,\Theta. \tag{9.94}$$

This is the differential equation for the **associated Legendre functions** $P_l^m(x)$ with argument $x = \cos\theta$. Therefore, we have the solutions

$$\Theta(\theta) \propto P_l^m(\cos\theta). \tag{9.95}$$

The Properties of the associated Legendre functions can be found in Appendix E.3.3. Since the associated Legendre functions are not finite at the poles ($\cos\theta = \pm 1$) unless $l = 0, 1, \ldots$ and $m = -l, -l+1, \ldots, l$, we must restrict the values of $l$ and $m$ accordingly. (Allowing negative integer values of $l$ does not introduce any new solutions.) The product of the two angular functions form a set of orthogonal functions called **spherical harmonics**:

$$\Theta(\theta)\Phi(\phi) \propto Y_{lm}(\theta,\phi) = N_l^m P_l^m(\cos\theta)e^{im\phi}, \tag{9.96}$$

where the $N_l^m$ are chosen to normalize the $Y_{lm}$ when integrated over the unit two-dimensional sphere. (See Appendix E.4 for more details.)

Finally, the radial equation can be obtained by multiplying the Helmholtz equation by $R/r^2$, with the $\theta$ and $\phi$ dependence replaced by the separation constants $l$ and $m$:

$$\frac{d^2R}{dr^2} + \frac{2}{r}\frac{dR}{dr} + \left(k^2 - \frac{l\,(l+1)}{r^2}\right)R = 0. \tag{9.97}$$

This equation is solved by the **spherical Bessel functions** of the 1st and 2nd kind, $j_l$ and $n_l$:

$$R(r) = A_l j_l(kr) + B_l n_l(kr), \tag{9.98}$$

the properties of which you can find in Appendix E.5.5. Of particular importance for the solar oscillation problem is the property that all of the $n_l$ blow up at the origin

$(r = 0)$, so all of the $B_l$ must equal zero. In addition, since the radial solution must vanish at the radius of the Sun (which we will denote by $r = a$), we need

$$j_l(ka) = 0\,, \tag{9.99}$$

which can be satisfied by having $ka$ equal the zeros of the spherical Bessel function $j_l(x)$. (Recall from Appendix E.5.5 that the spherical Bessel functions behave much like damped sinusoids.) Thus, this boundary condition forces the separation constant $k$ to take on only discrete values

$$k_{ln}a = x_{ln}\,, \qquad n = 1, 2, \ldots\,, \tag{9.100}$$

where $x_{ln}$ denotes the $n$th zero of the $l$th spherical Bessel function $j_l(x)$. Thus,

$$R(r) \propto j_l(x_{ln}r/a)\,. \tag{9.101}$$

Putting together all of the above results, the normal mode oscillations of the Sun have the form

$$u_{lmn}(r, \theta, \phi, t) = j_l\left(\frac{x_{ln}r}{a}\right) Y_{lm}(\theta, \phi)\mathrm{e}^{-\mathrm{i}\omega_{ln}t}\,, \tag{9.102}$$

with normal mode frequencies

$$\omega_{ln} = \frac{x_{ln}v}{a}\,, \qquad l = 0, 1, \ldots\,, \qquad n = 1, 2, \ldots\,. \tag{9.103}$$

□

**Exercise 9.10** Verify that the differential equation (9.94) for $\Theta(\theta)$ reduces to the standard form given in (E.32) when you make the change of variables $\Theta(\theta) = y(x)|_{x=\cos\theta}$.

## Suggested References

*Full references are given in the bibliography at the end of the book.*

Fetter and Walecka (1980): An excellent treatment of continua and the wave equation can be found in Chaps. 4, 7, and 8.

Mathews and Walker (1970): For a more thorough treatment of hyperbolic equations and the application of boundary and initial conditions, see Chap. 8, Sect. 2.

## Additional Problems

**Problem 9.1** A guitar string of length $\ell$ is plucked a fractional distance $\alpha$ from the bridge. (The initial conditions are such that the string is released from rest.) Show that the Fourier coefficients for a Fourier series expansion of the displacement $y(x, t)$ of the string are proportional to $\sin(n\pi\alpha)/n^2$, where $n = 1, 2, \ldots$. Note that if $\alpha = 1/N$, then there is no contribution from the $N$th harmonic and its multiples.

**Problem 9.2** Sound waves in air are local deviations in atmospheric pressure about equilibrium, $p(\mathbf{r}, t) \equiv P(\mathbf{r}, t) - P_0$, which satisfy the 3-dimensional wave equation:

$$\left(\nabla^2 - \frac{1}{v^2}\frac{\partial^2}{\partial t^2}\right) p(\mathbf{r}, t) = 0\,, \tag{9.104}$$

where $v$ is the velocity of sound in air ($v \approx 340$ m/s at room temperature). The 1-dimensional problem (analogous to waves on a stretched string) consists of sound waves in a narrow column of air (a *tube*), like that in an organ pipe, a flute, or a clarinet. Show that the general solution for sound waves in a tube of length $\ell$ can be written as a superposition of standing wave pressure deviations, with standing wave frequencies

$$f_n = \frac{nv}{2\ell}\,, \qquad n = 1, 2, 3, \ldots\,, \tag{9.105}$$

for a tube which is open or closed at both ends, and

$$f_n = \frac{nv}{4\ell}\,, \qquad n = 1, 3, 5, \ldots\,, \tag{9.106}$$

for a tube which is closed at one end only. Note, in particular, that a tube which is closed at only one end admits only *odd* harmonics of the fundamental frequency. (*Hint*: The boundary conditions for standing waves are such that the pressure deviation is zero (a *node*) at the open end of a tube and a maximum (an *anti-node*) at a closed end.)

**Problem 9.3** Consider an ideal rectangular membrane of dimensions $0 \leq x \leq a$, $0 \leq y \leq b$, rigidly attached to supports along its sides. By solving the 2-dimensional wave equation

$$\frac{\partial^2 u}{\partial x^2} + \frac{\partial^2 u}{\partial y^2} - \frac{1}{v^2}\frac{\partial^2 u}{\partial t^2} = 0\,, \tag{9.107}$$

show that the characteristic frequencies of vibration of the membrane are:

$$f_{mn} = \frac{v}{2}\sqrt{\left(\frac{m}{a}\right)^2 + \left(\frac{n}{b}\right)^2}\,, \qquad m, n = 1, 2, \ldots\,, \tag{9.108}$$

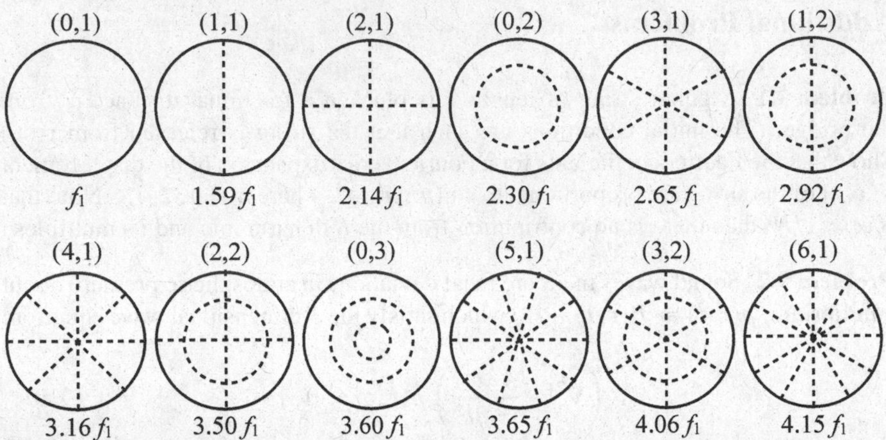


**Fig. 9.2** Several vibrational modes of an ideal circular membrane, fixed at the circumference. The modes are labeled by two integers $(m, n)$, where $m$ is the number of nodal diameters and $n$ is the number of nodal circles. Adjacent segments of the drum head move in opposite directions. (There is always one nodal circle, located at the circumference of the drum head.) Also given are the vibrational frequencies of each mode in terms of the fundamental frequency $f_1 \equiv \omega_{01} v/(2\pi a)$. (This figure is modeled after a similar figure in Berg and Stork (2005).)

where $v$ is the velocity of the waves on the membrane. Sketch the normal modes of vibration for the first few integers, indicating the nodal lines and positive and negative displacements of the membrane.

**Problem 9.4** Solve for the vibrational modes of an ideal circular drum head of radius $a$, fixed at the circumference.
You should find

$$u_{mn}(\rho, \phi, t) = J_m\left(\frac{x_{mn}\rho}{a}\right) e^{i\phi} e^{-i\omega_{mn} t}\,, \tag{9.109}$$

where

$$\omega_{mn} = \frac{x_{mn} v}{a}\,, \qquad m = 0, 1, \ldots\,, \qquad n = 1, 2, \ldots\,, \tag{9.110}$$

are the normal mode frequencies. (Here $x_{mn}$ is the $n$th zero of the $m$th Bessel function $J_m(x)$.) See Figs. 9.2 and 9.3 for a graphical illustration of the first several vibrational modes.

**Problem 9.5** Consider the wave equation with a source term:

$$\nabla^2 u - \frac{1}{v^2}\frac{\partial^2 u}{\partial t^2} = g(\mathbf{r}, t)\,. \tag{9.111}$$

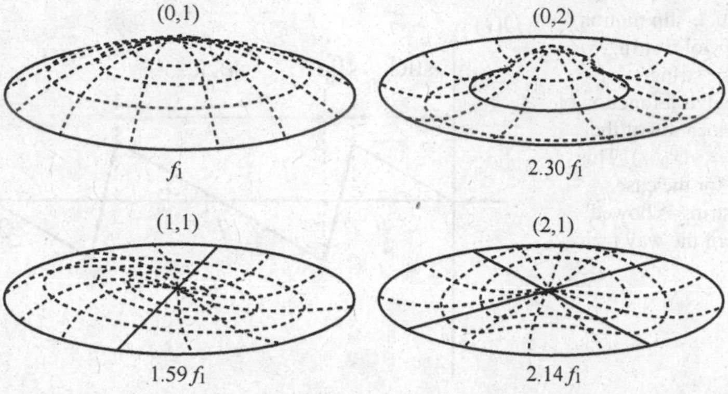


**Fig. 9.3** A perspective view of the first four vibrational modes of an ideal circular membrane, fixed at the circumference. (See also Fig. 9.2.) The nodal lines and circles are indicated by solid curves on these figures. (This figure is modeled after a similar figure in Rossing et al. (2002).)

Show that

$$u(\mathbf{r}, t) = -\frac{1}{4\pi}\int \mathrm{d}V' \, \frac{g(\mathbf{r}', t - |\mathbf{r} - \mathbf{r}'|/v)}{|\mathbf{r} - \mathbf{r}'|}. \tag{9.112}$$

is a solution to the above equation. This is called the *retarded* solution, since the value of $u$ at $\mathbf{r}$ and $t$ depends on what the source at $\mathbf{r}'$ was doing at an earlier time $t_{\mathrm{ret}} \equiv t - |\mathbf{r} - \mathbf{r}'|/v$. *Hint*: Substitute the assumed form of the solution into (9.111), and then do the differentiations using (A.115).

**Problem 9.6** Consider a rectangular box-shaped room with dimensions $L \times W \times H$. Show that the resonant frequencies for sound waves propagating in the room are:

$$f_{lmn} = \frac{v}{2}\sqrt{\left(\frac{l}{L}\right)^2 + \left(\frac{m}{W}\right)^2 + \left(\frac{n}{H}\right)^2}, \qquad l, m, n = 0, 1, 2, \ldots, \tag{9.113}$$

where $v$ is the speed of sound in air. (This is just a generalization of standing waves in a 1-dimensional tube closed at both ends to three dimensions; see Problem 9.2.) Using the above result, show that the first few resonant frequencies for a rectangular-shaped shower stall having dimensions 1 m × 1 m × 2m are:

$$\begin{aligned} &f_{001} = 85\text{ Hz}, \quad f_{010} = 170\text{ Hz}, \quad f_{100} = 170\text{ Hz},\\ &f_{002} = 170\text{ Hz}, \quad f_{110} = 240\text{ Hz}, \quad f_{111} = 255\text{ Hz},\\ &\ldots \end{aligned} \tag{9.114}$$

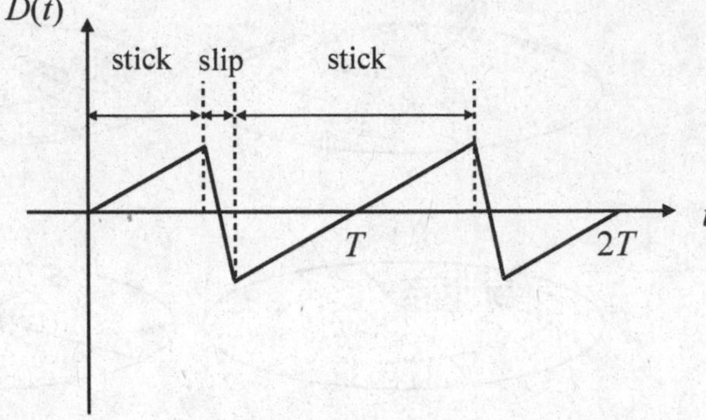


**Fig. 9.4** Stick-slip motion of a bowed violin string. Plotted is the string displacement as a function of time at the location of the bow, $D(t) = y(x_0, t)$. This example is for the case where the string is bowed one-eighth of the way from the bridge

**Problem 9.7** Find the Fourier series solution for the displacement of a bowed violin string of length $\ell$, fixed at both ends. Assume that at $t = 0$ the string is in its equilibrium configuration $y(x, 0) = 0$ for $0 \le x \le \ell$. Bowing imposes a boundary condition on the displacement of the string at the location of the bow where the string sticks to the bow and is dragged along until it slips. We describe this condition at the bow location ($x = x_0 \equiv \alpha\ell$) for all $t$ as

$$y(x_0, t) = D(t) \equiv \begin{cases} \frac{2y_0}{(1-\alpha)} \frac{t}{T} & \frac{t}{T} \in \left[0, \frac{1}{2}(1-\alpha)\right], \\ -\frac{2y_0}{\alpha}\left(\frac{t}{T} - \frac{1}{2}\right) & \frac{t}{T} \in \left[\frac{1}{2}(1-\alpha), \frac{1}{2}(1+\alpha)\right], \\ \frac{2y_0}{(1-\alpha)}\left(\frac{t}{T} - 1\right) & \frac{t}{T} \in \left[\frac{1}{2}(1+\alpha), 1\right], \end{cases} \tag{9.115}$$

where $y_0$ is the maximum displacement of the string at the location of bowing, and $T$ is the period of the "stick-slip" motion of the string on the bow. Note that $\alpha$ specifies the location of the bowing as a fraction of length of the string as measured from the bridge of the violin; it is also equal to the fraction of the period $T$ during which the string slips on the bow, as shown in Fig. 9.4.
You should find

$$y(x, t) = \frac{2}{\pi^2} \frac{y_0}{\alpha(1-\alpha)} \sum_{n=1}^{\infty} \frac{(-1)^{(n+1)}}{n^2} \sin\left(\frac{n\pi x}{\ell}\right) \sin\left(\frac{n\pi v t}{\ell}\right), \tag{9.116}$$

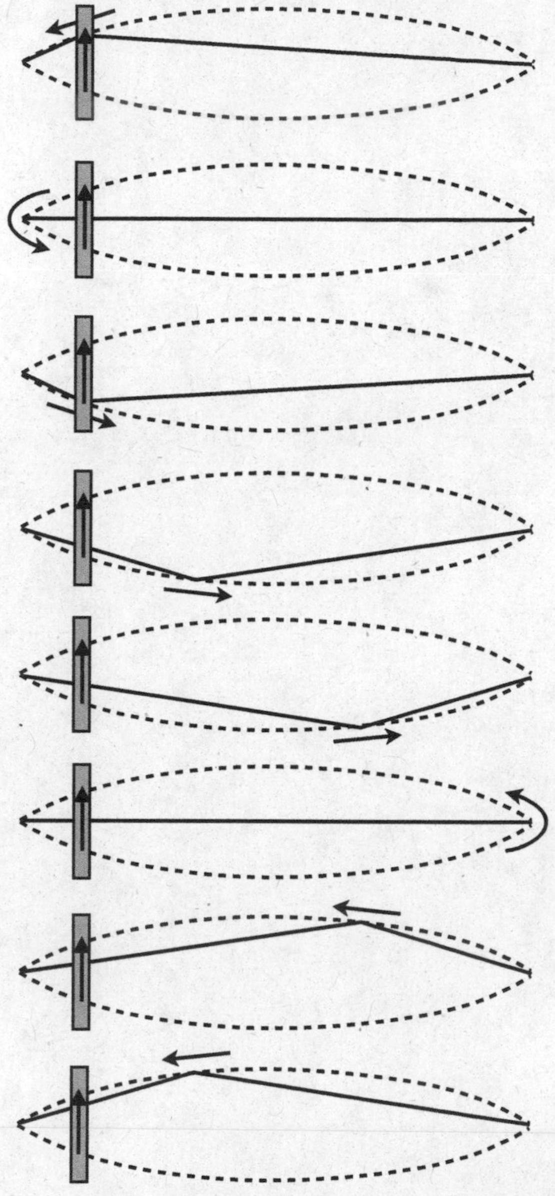

**Fig. 9.5** Schematic diagram showing the motion of a bowed string. In the top three panels, the string slips from underneath the bow, with the kink traveling counter-clockwise from bow to bridge to bow again. In the bottom five panels, the string sticks to the bow and is dragged along with it, with the kink traveling counter-clockwise from bow to nut (the support on the far right-hand side), and eventually back to the bow again. For this illustration, $\alpha = 1/8$. (This figure is modeled after a similar figure from http://www.colorado.edu/physics/phys1240/.)

where $v \equiv 2\ell/T$. By plotting the above displacement as a function of time, one sees that the vibrational motion of a bowed string is basically a triangular wave that runs around the curved profile of the vibrating string, as shown in Fig. 9.5.

# Chapter 10
# Lagrangian and Hamiltonian Formulations for Continuous Systems and Fields

In this chapter, we develop the Lagrangian and Hamiltonian formulations of continuous systems, which serves as an introduction to the analysis of fields. This chapter is optional in the sense that it is dependent on the previous chapters, but will not be necessary for the following chapter on relativistic mechanics.

## 10.1 Lagrangian Formulation for a Continuous System

In Chap. 9, we obtained the wave equation by taking the continuous limit of the equations of motion for the loaded string from Chap. 8. The full three-dimensional version of the wave equation was then arrived at by a natural extension of the one-dimensional case. In this section, we will use the Lagrangian for the loaded string to develop the Lagrangian formulation for continuous systems, and then generalize to fields that go beyond the simple one-dimensional string.

### *10.1.1 Lagrangian Density*

The Lagrangian for a loaded string of length $\ell$ is (See (8.100)):

$$L = \frac{1}{2}m\sum_{a=1}^{N}(\dot{y}_a)^2 - \frac{\tau}{2d}\sum_{a=0}^{N}(y_{a+1} - y_a)^2\,, \tag{10.1}$$

where $m$ is the mass of the individual particles, $d$ is the interparticle separation, and $\tau$ is the constant tension in the string. We can rewrite the Lagrangian in a more suggestive form by multiplying and dividing each term by $d$:



M.J. Benacquista and J.D. Romano, *Classical Mechanics*, Undergraduate Lecture Notes in Physics, https://doi.org/10.1007/978-3-319-68780-3_10

$$L = \left[ \frac{1}{2} \frac{m}{d} \sum_{a=1}^{N} (\dot{y}_a)^2 - \frac{1}{2} \tau \sum_{a=0}^{N} \left( \frac{y_{a+1} - y_a}{d} \right)^2 \right] d \,, \tag{10.2}$$

and recall that

$$\lim_{d \to 0} \frac{y((a+1)d, t) - y(ad, t)}{d} = \frac{\partial y}{\partial x} \,. \tag{10.3}$$

In the continuous limit ($d \to 0$ and $N \to \infty$ such that $(N+1)d = \ell$), the summations transition to integrals. Thus, with the definition of the linear mass density $\mu \equiv M/\ell$, we have

$$L = \int_0^\ell \mathrm{d}x \, \frac{1}{2} \left[ \mu \left( \frac{\partial y}{\partial t} \right)^2 - \tau \left( \frac{\partial y}{\partial x} \right)^2 \right] , \tag{10.4}$$

or, equivalently,

$$L \equiv \int_0^\ell \mathrm{d}x \, \mathcal{L} \,, \tag{10.5}$$

where the **Lagrangian density** is defined as

$$\mathcal{L} \equiv \frac{1}{2} \left[ \mu \left( \frac{\partial y}{\partial t} \right)^2 - \tau \left( \frac{\partial y}{\partial x} \right)^2 \right] . \tag{10.6}$$

Note that the units of the Lagrangian density here are [energy/length].

For the specific case of the loaded string, the Lagrangian density depends only upon the partial derivatives of $y(x, t)$. More generally, the Lagrangian density could also depend on $y(x, t)$ in addition to its partial derivatives $\partial y/\partial x$, $\partial y/\partial t$, as well as the coordinates $x$ and $t$. Thus, the most general form of the Lagrangian density for the field $y(x, t)$ in one dimension is

$$\mathcal{L} = \mathcal{L} \left( y, \frac{\partial y}{\partial x}, \frac{\partial y}{\partial t}, x, t \right) . \tag{10.7}$$

### *10.1.2 Equations of Motion*

The equations of motion for $y(x, t)$ follow from Hamilton's principle, Sect. 3.1. The action is defined in terms of the Lagrangian density as

$$S[y] = \int_{t_1}^{t_2} \mathrm{d}t \int_{x_1}^{x_2} \mathrm{d}x\, \mathcal{L}(y, y_{,x}, y_{,t}, x, t)\,, \tag{10.8}$$

where we have introduced the shorthand notation

$$y_{,x} \equiv \frac{\partial y}{\partial x}\,, \qquad y_{,t} \equiv \frac{\partial y}{\partial t}\,, \tag{10.9}$$

for partial derivatives. When evaluating the integral on the right-hand side of (10.8), we need to express $y$ $y_{,x}$, $y_{,t}$ as functions of $x$ and $t$ via the field $y(x,t)$. But for the variational calculations that follow, we simply treat $\mathscr{L}$ as a function of five *independent* variables $y$, $y_{,x}$, $y_{,t}$, $x$, and $t$. The fact that $y$, $y_{,x}$, and $y_{,t}$ are related to one another only shows up later on when we need to relate the variations $\delta y_{,x}$ and $\delta y_{,t}$ to $\delta y$ (See (10.13)).

When finding the stationary points of the action, we will consider infinitesimal variations of the field

$$y(x,t) \to y(x,t) + \delta y(x,t)\,, \tag{10.10}$$

which are fixed at the spatial and temporal boundaries—i.e.,

$$\delta y|_{x_1,x_2} = 0 \quad \forall t\,, \qquad \delta y|_{t_1,t_2} = 0 \quad \forall x\,. \tag{10.11}$$

We then demand that the corresponding variation in the action[1]

$$\delta S = \int_{t_1}^{t_2} \mathrm{d}t \int_{x_1}^{x_2} \mathrm{d}x \left( \frac{\partial \mathcal{L}}{\partial y}\delta y + \frac{\partial \mathcal{L}}{\partial y_{,x}}\delta y_{,x} + \frac{\partial \mathcal{L}}{\partial y_{,t}}\delta y_{,t} \right)\,, \tag{10.12}$$

vanish to 1st order in the variation $\delta y$. In the above equation,

$$\delta y_{,x} \equiv \delta\left(\frac{\partial y}{\partial x}\right) = \frac{\partial}{\partial x}\delta y\,, \qquad \delta y_{,t} \equiv \delta\left(\frac{\partial y}{\partial t}\right) = \frac{\partial}{\partial t}\delta y\,, \tag{10.13}$$

which means that variations of the partial derivatives of the field are the same as the partial derivatives of the variation of the field. Making these substitutions into (10.12) and integrating by parts, we find

$$\delta S = \int_{t_1}^{t_2} \mathrm{d}t \int_{x_1}^{x_2} \mathrm{d}x \left( \frac{\partial \mathcal{L}}{\partial y} - \frac{\mathrm{d}}{\mathrm{d}x}\left(\frac{\partial \mathcal{L}}{\partial y_{,x}}\right) - \frac{\mathrm{d}}{\mathrm{d}t}\left(\frac{\partial \mathcal{L}}{\partial y_{,t}}\right)\right)\delta y\,, \tag{10.14}$$

[1] For this and all subsequent variations, we ignore all 2nd-order and higher terms in $\delta y$ and its partial derivatives.

where we used (10.11) to eliminate the boundary terms. (Note that we write the last two terms using *total* derivatives $\mathrm{d}/\mathrm{d}x$ and $\mathrm{d}/\mathrm{d}t$, since we are allowing the Lagrangian density $\mathcal{L}$ to have, in general, both *explicit* and *implicit* dependence on $x$ and $t$.) Thus, by requiring that $\delta S = 0$ for all $\delta y$ leads to the equation of motion

$$\frac{\partial \mathcal{L}}{\partial y} - \frac{\mathrm{d}}{\mathrm{d}x}\left(\frac{\partial \mathcal{L}}{\partial y_{,x}}\right) - \frac{\mathrm{d}}{\mathrm{d}t}\left(\frac{\partial \mathcal{L}}{\partial y_{,t}}\right) = 0 . \tag{10.15}$$

This is the generalization of the Euler-Lagrange equation

$$\frac{\partial L}{\partial y} - \frac{\mathrm{d}}{\mathrm{d}t}\left(\frac{\partial L}{\partial \dot{y}}\right) = 0 \tag{10.16}$$

for a *function* $y(t)$ to a *field* $y(x, t)$.

**Exercise 10.1** Fill in the missing steps in going from (10.8) to (10.15).

***Example 10.1*** Here we show that the Euler-Lagrange equation for the Lagrangian density

$$\mathcal{L} = \frac{1}{2}\left[\mu y_{,t}^2 - \tau y_{,x}^2\right] \tag{10.17}$$

is the one-dimensional wave equation (9.16).

From (10.15), we see that we first need to evaluate the partial derivatives of $\mathcal{L}$ with respect to $y$, $y_{,x}$, and $y_{,t}$. These are simply

$$\frac{\partial \mathcal{L}}{\partial y} = 0 , \qquad \frac{\partial \mathcal{L}}{\partial y_{,x}} = -\tau y_{,x} , \qquad \frac{\partial \mathcal{L}}{\partial y_{,t}} = \mu y_{,t} . \tag{10.18}$$

The relevant total derivatives with respect to $x$ and $t$ are then

$$\frac{\mathrm{d}}{\mathrm{d}x}\left(\frac{\partial \mathcal{L}}{\partial y_{,x}}\right) = -\tau y_{,xx} , \qquad \frac{\mathrm{d}}{\mathrm{d}t}\left(\frac{\partial \mathcal{L}}{\partial y_{,t}}\right) = \mu y_{,tt} . \tag{10.19}$$

Thus, the Euler-Lagrange equation for this Lagrangian density is

$$\tau y_{,xx} - \mu y_{,tt} = 0 , \tag{10.20}$$

or, equivalently,

$$y_{,xx} - \frac{1}{v^2} y_{,tt} = 0 . \tag{10.21}$$

This is the one-dimensional wave equation (9.16) with wave speed $v \equiv \sqrt{\tau/\mu}$.

It is interesting to note that in the discrete case of the weighted string, there were $N$ equations of motion for the $N$ degrees of freedom in the problem, while here we have only one equation of motion. This is because we have gone from $N$ ordinary differential equations in one variable $t$ to one partial differential equation in two variables $t$ and $x$ in the limit $N \to \infty$. The expected infinite number of equations of motion have been replaced by a single equation in the continuous variable $x$. □

**Exercise 10.2** Show that the equation of motion obtained from the Lagrangian density

$$\mathcal{L} = \frac{1}{2}\left[\frac{1}{c^2} y_{,t}^2 - y_{,x}^2 - \mu^2 y^2\right] \tag{10.22}$$

is the one-dimensional **Klein-Gordon equation**

$$y_{,xx} - \frac{1}{c^2} y_{,tt} = \mu^2 y\,, \tag{10.23}$$

where $c$ is the speed of light and $\mu$ is a constant.

### 10.1.3 Functional Derivative Notation

In deriving the equation of motion (10.15) for a field $y(x, t)$, we found that the variation of the action $S[y]$ with respect to the field was equal to

$$\delta S = \int_{t_1}^{t_2} \mathrm{d}t \int_{x_1}^{x_2} \mathrm{d}x \left[\frac{\partial \mathcal{L}}{\partial y} - \frac{\mathrm{d}}{\mathrm{d}x}\left(\frac{\partial \mathcal{L}}{\partial y_{,x}}\right) - \frac{\mathrm{d}}{\mathrm{d}t}\left(\frac{\partial \mathcal{L}}{\partial y_{,t}}\right)\right] \delta y\,, \tag{10.24}$$

where we assumed that the variation $\delta y$ vanished on the spatial and temporal boundaries, (10.11). As discussed in Appendix C.3, it is common to denote the expression in square brackets in the above integral as the **functional derivative** of the action with respect to the field[2]:

$$\frac{\delta S}{\delta y} \equiv \frac{\partial \mathcal{L}}{\partial y} - \frac{\mathrm{d}}{\mathrm{d}x}\left(\frac{\partial \mathcal{L}}{\partial y_{,x}}\right) - \frac{\mathrm{d}}{\mathrm{d}t}\left(\frac{\partial \mathcal{L}}{\partial y_{,t}}\right). \tag{10.25}$$

[2]To explicitly denote the $x$ and $t$ dependence of the functional derivative, one should write $\delta S/\delta y(x, t)$ instead of the shorthand notation $\delta S/\delta y$.

Thus, we can write the equation of motion (10.15) as simply $\delta S/\delta y = 0$. Since $\delta S/\delta y$ is defined *inside* an integral, the physical dimensions of $\mathrm{d}t\, \mathrm{d}x\, \delta S/\delta y$ are equal to the dimensions of $S$ divided by the dimensions of $y$.

More generally, given a functional $I[\varphi_1, \varphi_2, \ldots, \varphi_N]$ of $N$ fields $\varphi_I(x^1, x^2, \ldots, x^n)$, $I = 1, 2, \ldots, N$, which are defined on some region $U$ of an $n$-dimensional space, the functional derivative of $I$ with respect to $\varphi_I$ is defined by

$$\delta I \equiv \int_U \mathrm{d}^n x \sum_I \frac{\delta I}{\delta \varphi_I} \delta \varphi_I , \tag{10.26}$$

where $\delta I$ is the total change in the functional for variations $\delta\varphi_I$ in all of the fields. If the functional $I$ can be written as an integral of a (local) function $F$ of the fields and its derivatives at point:

$$I[\varphi_1, \ldots, \varphi_N] \equiv \int_U \mathrm{d}^n x\, F(\varphi_I, \varphi_{I,i}, x^i) , \tag{10.27}$$

then for variations in the fields which are fixed on the boundary of $U$, one finds

$$\delta I = \int_U \mathrm{d}^n x \sum_I \left[ \frac{\partial F}{\partial \varphi_I} - \sum_i \frac{\mathrm{d}}{\mathrm{d}x^i} \left( \frac{\partial F}{\partial \varphi_{I,i}} \right) \right] \delta \varphi_I . \tag{10.28}$$

Thus,

$$\frac{\delta I}{\delta \varphi_I} = \frac{\partial F}{\partial \varphi_I} - \sum_i \frac{\mathrm{d}}{\mathrm{d}x^i} \left( \frac{\partial F}{\partial \varphi_{I,i}} \right) . \tag{10.29}$$

The use of functional derivative notation will allow us to simplify some of the expressions in the following sections.

### *10.1.4 Generalization to Multiple Fields and Dimensions*

Given the discussion of the previous subsection, it is easy to extend the Lagrangian formulation for continuous systems to multiple fields and dimensions. In general, we will consider $N$ interacting fields $\varphi_I(\mathbf{r}, t)$, $I = 1, 2, \ldots, N$, with $\mathbf{r}$ labeling a position in 3-dimensional space. The Lagrangian density will then take the form

$$\mathcal{L} = \mathcal{L}\left(\varphi_I, \varphi_{I,i}, \varphi_{I,t}, x^i, t\right) , \tag{10.30}$$

where

$$\varphi_{I,i} \equiv \frac{\partial \varphi_I}{\partial x^i}, \qquad i = 1, 2, 3 \tag{10.31}$$

denotes the partial derivatives of $\varphi_I$ with respect to the spatial coordinates $x^i$, where $i = 1, 2, 3$. The action $S$ will be a functional of the fields $S = S[\varphi_1, \varphi_2, \ldots, \varphi_N]$, and can be written as the integral

$$S = \int_{t_1}^{t_2} \mathrm{d}t \int_V \mathrm{d}^3x \, \mathcal{L}\left(\varphi_I, \varphi_{I,i}, \varphi_{I,t}, x^i, t\right), \tag{10.32}$$

where $V$ is the spatial volume of interest on which the fields are defined.[3] The equations of motion are obtained by finding the stationary points of the action in the usual way, leading to

$$\frac{\partial \mathcal{L}}{\partial \varphi_I} - \frac{\mathrm{d}}{\mathrm{d}t}\left(\frac{\partial \mathcal{L}}{\partial \varphi_{I,t}}\right) - \sum_i \frac{\mathrm{d}}{\mathrm{d}x^i}\left(\frac{\partial \mathcal{L}}{\partial \varphi_{I,i}}\right) = 0, \tag{10.33}$$

for $I = 1, 2, \ldots, N$. In general, these partial differential equations will be coupled across the different fields, and thus must be solved as a *system* of such equations.

Note that the time derivatives appear in the same form as the spatial derivatives in the Euler-Lagrange equation. So we can introduce a simplification by using Greek indices to label both time and spatial coordinates. Our convention will be to use $\alpha = 0$ to indicate the time coordinate and $\alpha = 1, 2, 3 \equiv i$ to indicate the spatial coordinates, so sums over $\alpha$ will run from 0 to 3.[4] We then have

$$\mathcal{L} = \mathcal{L}\left(\varphi_I, \varphi_{I,\alpha}, x^\alpha\right), \tag{10.34}$$

and the corresponding Euler-Lagrange equations

$$\mathscr{E}_I \equiv \frac{\partial \mathcal{L}}{\partial \varphi_I} - \sum_\alpha \frac{\mathrm{d}}{\mathrm{d}x^\alpha}\left(\frac{\partial \mathcal{L}}{\partial \varphi_{I,\alpha}}\right) = 0, \tag{10.35}$$

for $I = 1, 2, \ldots, N$. For convenience, we have introduced the notation $\mathscr{E}_I$ for the particular combination of derivatives that enter in the equations of motion (10.35).

[3] Note that the action integral is defined with respect to the 3-dimensional *coordinate* volume element $\mathrm{d}^3x$ and not with respect to the 3-dimensional *invariant* volume element $\mathrm{d}V$, which differs from $\mathrm{d}^3x$ by a factor of $\sqrt{\det g}$ (See (A.61) in Appendix A.5). The above definition is consistent with the interpretation of $\mathscr{L}$ being a *density* with respect to coordinate transformation. If you don't want to worry about such a distinction, just always work in Cartesian coordinates $x^i = (x, y, z)$ for which $\mathrm{d}^3x$ and $\mathrm{d}V$ are numerically equal to one another.

[4] In Chap. 11, we will introduce similar notation to describe the components of four-dimensional vectors in the context of special relativity. In that context, there is an implied geometric structure to the spacetime, which is Minkowskian; but we can get by without introducing that formalism here.

Not surprisingly, they are called the *Euler-Lagrange derivatives* of the Lagrangian density $\mathscr{L}$, and are equal to the functional derivatives $\delta S/\delta\varphi_I$ for field variations that vanish on the boundary of the integration region defining the action $S$. The $\mathscr{E}_I$ are just part of the *full* expression for the variation of the Lagrangian density, as we will discuss in the next subsection.

It is important to note that the generalization to multiple fields and dimensions may not be derivable by taking the continuous limit of a discrete system. But from a practical perspective, the only requirement on the Lagrangian density is that it produce the correct equations of motion. This means that it is not necessarily the case that the Lagrangian density will be written in the form of a kinetic energy density $\mathcal{T}$ minus a potential energy density $\mathcal{U}$.

### 10.1.5 Variational Derivative of the Lagrangian Density

It is also convenient at this stage to write down the *full* expression for the variation $\delta\mathscr{L}$ of the Lagrangian density which led to the Euler-Lagrange equations (10.35) Piecing together the calculations discussed above, we have

$$\delta\mathscr{L} = \sum_I \mathscr{E}_I \delta\varphi_I + \sum_\alpha \frac{\mathrm{d}}{\mathrm{d}x^\alpha}\left(\sum_I \frac{\partial\mathscr{L}}{\partial\varphi_{I,\alpha}}\delta\varphi_I\right) . \tag{10.36}$$

Since all of the terms in (10.36) are infinitesimal, we can effectively factor out a common small parameter (which we'll denote by $\lambda$), and write the equation in terms of finite quantities (i.e., derivatives) defined by

$$\begin{aligned} \varphi_I &\to \varphi_I + \delta\varphi_I = \varphi_I + \lambda \left.\frac{\mathrm{d}\varphi_I}{\mathrm{d}\lambda}\right|_{\lambda=0} , \\ \mathscr{L} &\to \mathscr{L} + \delta\mathscr{L} = \mathscr{L} + \lambda \left.\frac{\mathrm{d}\mathscr{L}}{\mathrm{d}\lambda}\right|_{\lambda=0} . \end{aligned} \tag{10.37}$$

In this approach, we are basically thinking of $(\mathrm{d}\varphi_I/\mathrm{d}\lambda)|_{\lambda=0}$ as a tangent vector to a 1-parameter family of fields $\varphi_I(\lambda)$ passing through the "point" $\varphi_I \equiv \varphi_I(0)$. (This is similar to what we did in Appendix C.3, when describing a more formal treatment of the variational procedure, and also what we did in Sect. 3.6.1 for infinitesimal canonical transformations, e.g., (3.105).) In terms of the derivatives, (10.36) can then be written as

$$\left.\frac{\mathrm{d}\mathscr{L}}{\mathrm{d}\lambda}\right|_{\lambda=0} = \sum_I \mathscr{E}_I \left.\frac{\mathrm{d}\varphi_I}{\mathrm{d}\lambda}\right|_{\lambda=0} + \sum_\alpha \frac{\mathrm{d}}{\mathrm{d}x^\alpha}\left(\sum_I \frac{\partial\mathscr{L}}{\partial\varphi_{I,\alpha}} \left.\frac{\mathrm{d}\varphi_I}{\mathrm{d}\lambda}\right|_{\lambda=0}\right) . \tag{10.38}$$

Equation (10.38) will come in handy when discussing the connection between symmetries and conservation laws in the context of Noether's theorem, Sect. 10.4.

**Exercise 10.3** Verify (10.36) for the variation of $\mathscr{L}$.

**Exercise 10.4** Consider two Lagrangian densities $\mathscr{L} = \mathscr{L}(\varphi, \varphi_{,\alpha}, x^\alpha)$ and $\mathscr{L}' = \mathscr{L}'(\varphi, \varphi_{,\alpha}, x^\alpha)$ that differ only by a divergence term—i.e.,

$$\mathscr{L}' = \mathscr{L} + \sum_\alpha \frac{\mathrm{d}V^\alpha}{\mathrm{d}x^\alpha}, \tag{10.39}$$

where $V^\alpha \equiv V^\alpha(\varphi)$. Show that the Euler-Lagrange equations for these two Lagrangians are identical.

## 10.2 Hamiltonian Formulation for a Continuous System

We have seen that in the Lagrangian formulation for a continuous system, the Euler-Lagrange equations for the fields could be expressed in a way that treated both space and time equally, (10.35). This feature makes the Lagrangian density formulation ideal for studying relativistic fields which exist in spacetime. The Hamiltonian formulation preferentially selects out a time coordinate as a means of defining the canonical momentum. Thus, the Hamiltonian density that we will describe in this section will be harder to adapt to a relativistic field. Nonetheless, since the Hamiltonian operator is essential in quantum theory, the Hamiltonian formulation is ideal for studying the classical version of quantum fields.

### *10.2.1 Hamiltonian Density*

To develop a continuous description of the Hamiltonian, we will return now to the discrete model of the loaded string. Recalling from (10.2), the Lagrangian for the loaded string is

$$L = \left[\frac{1}{2}\frac{m}{d}\sum_a (\dot{y}_a)^2 - \frac{1}{2}\tau\sum_a\left(\frac{y_{a+1} - y_a}{d}\right)^2\right] d = d\sum_a L_a\,, \tag{10.40}$$

where

$$L_a \equiv \frac{1}{2}\frac{m}{d}\dot{y}_a^2 - \frac{1}{2}\tau\left(\frac{y_{a+1} - y_a}{d}\right)^2. \tag{10.41}$$

Recall also that in the Hamiltonian formulation, the conjugate momentum is defined as

$$p_a \equiv \frac{\partial L}{\partial \dot{y}_a} . \tag{10.42}$$

It is also assumed that one can invert this relationship to solve for the velocities $\dot{y}_a$ in terms of the displacements, momenta, and time—i.e., $\dot{y}_a = \dot{y}_a(y, p, t)$. One needs to be able to do this to pass from the Lagrangian to the Hamiltonian via a Legendre transform, as described in Sects. 3.4.1 and 3.4.2.

From the relationship between $L$ and $L_a$, we see that the derivative of $L$ with respect to $\dot{y}_a$ enters only through $L_a$, so

$$p_a = d \frac{\partial L_a}{\partial \dot{y}_a} . \tag{10.43}$$

In the continuous limit $p_a \to 0$, since $d \to 0$ and $L_a \to \mathcal{L}$, but we can define a **conjugate momentum density**[5]

$$\pi(x, t) = \lim_{\text{cont}} \frac{p_a}{d} = \frac{\partial \mathcal{L}}{\partial \dot{y}} , \tag{10.44}$$

which remains finite in the continuous limit. Given our assumption regarding $p_a$, this relation can be inverted to give $\dot{y}$ in terms of the $y$, $y_{,x}$, $\pi$, and the coordinates $x$ and $t$.

The discrete Hamiltonian is defined as

$$H = \left( \sum_a p_a \dot{y}_a - L \right) \Big|_{\dot{y}=\dot{y}(y,p,t)} = \sum_a d \left[ \frac{p_a}{d} \dot{y}_a - L_a \right] \Big|_{\dot{y}=\dot{y}(y,p,t)} ; \tag{10.45}$$

and so when we go to the continuous limit, the sum turns into an integral and we have

$$H = \int_0^\ell \mathrm{d}x \, (\pi \dot{y} - \mathcal{L}) \big|_{\dot{y}=\dot{y}(y,y_{,x},\pi,x,t)} . \tag{10.46}$$

In the spirit of the Lagrangian formalism, we can now define a **Hamiltonian density** as

$$\mathcal{H} \equiv (\pi \dot{y} - \mathcal{L}) \big|_{\dot{y}=\dot{y}(y,y_{,x},\pi,x,t)} . \tag{10.47}$$

Note that $\mathcal{H}$ is a local function of the field $y$, its spatial derivative $y_{,x}$, the momentum density $\pi$, and the coordinates $x$ and $t$, i.e., $\mathcal{H} = \mathcal{H}(y, y_{,x}, \pi, x, t)$.

[5] For the field $y(x, t)$, we will often use $\dot{y}$ to denote the partial derivative $y_{,t} \equiv \partial y / \partial t$.

For a set of fields $\varphi_I(\mathbf{r}, t)$, $I = 1, 2, \ldots, N$, defined over 3-dimensional space, the above formulae generalize to

$$H = \int_V \mathrm{d}^3x\, \mathcal{H}\,, \tag{10.48}$$

where

$$\mathcal{H}(\varphi_I, \varphi_{I,i}, \pi_I, x^i, t) = \left( \sum_I \pi_I \dot{\varphi}_I - \mathcal{L} \right) \Bigg|_{\dot{\varphi}=\dot{\varphi}(\varphi,\varphi_{,i},\pi,x^i,t)}\,, \tag{10.49}$$

with

$$\pi_I \equiv \frac{\partial \mathcal{L}}{\partial \dot{\varphi}_I}\,. \tag{10.50}$$

***Example 10.2*** As a simple example, let us construct the Hamiltonian density for 1-dimensional waves starting from the Lagrangian density

$$\mathcal{L} = \frac{1}{2}\left[\mu \dot{y}^2 - \tau y_{,x}^2\right]. \tag{10.51}$$

The momentum density conjugate to $y$ is given by

$$\pi \equiv \frac{\partial \mathcal{L}}{\partial \dot{y}} = \mu \dot{y}\,, \tag{10.52}$$

which can be trivially inverted for $\dot{y}$ in terms of $\pi$. Then

$$\mathcal{H} = (\pi \dot{y} - \mathcal{L})|_{\dot{y}=\pi/\mu} = \pi\,\frac{\pi}{\mu} - \frac{1}{2}\left[\mu\left(\frac{\pi}{\mu}\right)^2 - \tau y_{,x}^2\right], \tag{10.53}$$

which simplifies to

$$\mathcal{H} = \frac{1}{2}\left[\frac{1}{\mu}\pi^2 + \tau y_{,x}^2\right]. \tag{10.54}$$

This has the form of a kinetic energy density plus a potential energy density. Note also that, for this example, $\mathcal{H}$ depends only on $y_{,x}$ and $\pi$, while in general it could have depended on $y$, $y_{,x}$, $\pi$, $x$, and $t$. □

**Exercise 10.5** Construct the Hamiltonian density $\mathcal{H}$ associated with the Lagrangian density $\mathcal{L}$ for the one-dimensional Klein-Gordon field (10.22).

### 10.2.2 Equations of Motion

With the Hamiltonian density in hand, we can now turn to obtaining the continuous analog of Hamilton's equations. For the discrete case, Hamilton's equations were found in Chap. 3 as

$$\dot{q}^a = \frac{\partial H}{\partial p_a}, \qquad \dot{p}_a = -\frac{\partial H}{\partial q^a}, \tag{10.55}$$

where $a$ indicates any of the discrete canonical variables in the system. For the continuous case, we will simplify the derivation of Hamilton's equations by initially considering a *single* field $\varphi(\mathbf{r}, t)$. The extension to multiple fields $\varphi_I(\mathbf{r}, t)$, $I = 1, 2, \ldots, N$, is straightforward.

So let's begin with the action written in 1st-order form,

$$S[\varphi, \pi] = \int_{t_1}^{t_2} \mathrm{d}t \int_V \mathrm{d}^3x \, \left[\pi\dot{\varphi} - \mathcal{H}(\varphi, \varphi_{,i}, \pi, x^i, t)\right] . \tag{10.56}$$

(See Sect. 3.4.3 for the discrete case.) Hamilton's equations arise by varying $S$ with respect to both $\varphi(\mathbf{r}, t)$ and $\pi(\mathbf{r}, t)$, for variations which vanish at both the spatial and temporal boundaries. To obtain the $\dot{\varphi}$ equation, we first vary $S$ with respect to the canonical momentum density $\pi$, finding

$$\delta S = \int_{t_1}^{t_2} \mathrm{d}t \int_V \mathrm{d}^3x \, \left[\dot{\varphi} - \frac{\partial \mathcal{H}}{\partial \pi}\right] \delta\pi \, . \tag{10.57}$$

Requiring that $\delta S = 0$ for all $\delta\pi$ thus gives

$$\dot{\varphi} = \frac{\partial \mathcal{H}}{\partial \pi} . \tag{10.58}$$

The above equation can also be written in terms of the Hamiltonian $H$, since for a variation of $\pi$,

$$\delta H = \int_V \mathrm{d}^3x \, \frac{\partial \mathcal{H}}{\partial \pi} \, \delta\pi \quad \Rightarrow \quad \frac{\delta H}{\delta \pi} = \frac{\partial \mathcal{H}}{\partial \pi} . \tag{10.59}$$

Thus, in functional derivative notation,

$$\dot{\varphi} = \frac{\delta H}{\delta \pi} . \tag{10.60}$$

To obtain Hamilton's equation for $\dot{\pi}$, we now vary $S$ respect to $\varphi$:

$$\delta S = \int_{t_1}^{t_2} \mathrm{d}t \int_V \mathrm{d}^3x \left[ \pi\,\delta\dot{\varphi} - \frac{\partial \mathcal{H}}{\partial \varphi}\delta\varphi - \sum_i \frac{\partial \mathcal{H}}{\partial \varphi_{,i}}\delta\varphi_{,i} \right]. \tag{10.61}$$

But since

$$\delta\dot{\varphi} = \delta\left(\frac{\partial \varphi}{\partial t}\right) = \frac{\partial}{\partial t}\delta\varphi\,, \qquad \delta\varphi_{,i} = \delta\left(\frac{\partial \varphi}{\partial x^i}\right) = \frac{\partial}{\partial x^i}\delta\varphi\,, \tag{10.62}$$

we can integrate the first and third terms by parts, throwing away the boundary terms since the variation $\delta\varphi$ is assumed to vanish on both the spatial and temporal boundaries. Thus, we can rewrite $\delta S$ as

$$\delta S = \int_{t_1}^{t_2} \mathrm{d}t \int_V \mathrm{d}^3x \left[ -\dot{\pi} - \frac{\partial \mathcal{H}}{\partial \varphi} + \sum_i \frac{\mathrm{d}}{\mathrm{d}x^i}\left(\frac{\partial \mathcal{H}}{\partial \varphi_{,i}}\right) \right] \delta\varphi\,, \tag{10.63}$$

where we factored out the common $\delta\varphi$. (Just as we did for the Lagrangian formulation, we use a total derivative $\mathrm{d}/\mathrm{d}x^i$ in the above expression, since $\mathcal{H}$ may depend both explicitly and implictly on $x^i$ and $t$.) Finally, by requiring that $\delta S = 0$ for all $\delta\varphi$ gives

$$\dot{\pi} = -\frac{\partial \mathcal{H}}{\partial \varphi} + \sum_i \frac{\mathrm{d}}{\mathrm{d}x^i}\left(\frac{\partial \mathcal{H}}{\partial \varphi_{,i}}\right), \tag{10.64}$$

which is Hamilton's equation for $\dot{\pi}$.

As before, this last equation can be written in terms of the Hamiltonian $H$ using the fact that for variations in $\varphi$,

$$\delta H = \int_V \mathrm{d}^3x \left( \frac{\partial \mathcal{H}}{\partial \varphi}\,\delta\varphi + \sum_i \frac{\partial \mathcal{H}}{\partial \varphi_{,i}}\,\delta\varphi_{,i} \right). \tag{10.65}$$

Proceeding as above, we can interchange the variations and partial derivatives in the last term, and integrate by parts, throwing away the boundary terms. The final results is

$$\delta H = \int_V \mathrm{d}^3x \left[ \frac{\partial \mathcal{H}}{\partial \varphi} - \sum_i \frac{\mathrm{d}}{\mathrm{d}x^i}\left(\frac{\partial \mathcal{H}}{\partial \varphi_{,i}}\right) \right] \delta\varphi\,. \tag{10.66}$$

Comparing (10.64) and (10.66) we see that

$$\dot{\pi} = -\frac{\delta H}{\delta \varphi}. \tag{10.67}$$

**Exercise 10.6** Extend the above analysis to obtain Hamilton's equations of motion for multiple fields $\varphi_I, \pi_I$.

**Exercise 10.7** Using Hamilton's equations of motion and (10.49), show that

$$\frac{\partial \mathcal{H}}{\partial \varphi_I} = -\frac{\partial \mathcal{L}}{\partial \varphi_I}, \quad \frac{\partial \mathcal{H}}{\partial \varphi_{I,i}} = -\frac{\partial \mathcal{L}}{\partial \varphi_{I,i}}, \quad \frac{\partial \mathcal{H}}{\partial t} = -\frac{\partial \mathcal{L}}{\partial t}. \tag{10.68}$$

***Example 10.3*** For the case of one-dimensional waves, we showed in Example 10.2 that the Hamiltonian density is given by

$$\mathcal{H} = \frac{1}{2}\left[\frac{1}{\mu}\pi^2 + \tau y_{,x}^2\right]. \tag{10.69}$$

The corresponding Hamilton's equations, (10.58) and (10.64), are then

$$\dot{y} = \frac{\partial \mathcal{H}}{\partial \pi} = \frac{\pi}{\mu}, \tag{10.70}$$

and

$$\dot{\pi} = -\frac{\partial \mathcal{H}}{\partial y} + \frac{\mathrm{d}}{\mathrm{d}x}\left(\frac{\partial \mathcal{H}}{\partial y_{,x}}\right) = -0 + \frac{\mathrm{d}}{\mathrm{d}x}\left(\tau y_{,x}\right) = \tau y_{,xx}, \tag{10.71}$$

which are two 1st-order differential equations for $y$ and $\pi$. Note that we can recover the single 2nd-order one-dimensional wave equation for $y$, (10.21), by taking a time derivative of the $\dot{y}$ equation and then substituting for $\dot{\pi}$ using the $\dot{\pi}$ equation:

$$\ddot{y} = \frac{\dot{\pi}}{\mu} = \frac{\tau}{\mu}y_{,xx} \quad \Rightarrow \quad y_{,xx} - \frac{1}{v^2}\ddot{y} = 0, \tag{10.72}$$

where $v \equiv \sqrt{\tau/\mu}$ is the wave speed. □

***Example 10.4*** We can also obtain Hamilton's equations for one-dimensional waves by taking the functional derivatives of the Hamiltonian

$$H \equiv \int_{x_1}^{x_2} \mathrm{d}x\, \mathcal{H} = \int_{x_1}^{x_2} \mathrm{d}x\, \frac{1}{2}\left[\frac{1}{\mu}\pi^2 + \tau y_{,x}^2\right], \tag{10.73}$$

with respect to $\pi$ and $y$, according to (10.60) and (10.67). To show this explicitly, let us first vary $\pi$:

$$\delta H = \int_{x_1}^{x_2} \mathrm{d}x \, \frac{1}{\mu}\pi \, \delta\pi \quad \Rightarrow \quad \frac{\delta H}{\delta \pi} = \frac{\pi}{\mu} \,. \tag{10.74}$$

Thus,

$$\dot{y} = \frac{\delta H}{\delta \pi} = \frac{\pi}{\mu} , \tag{10.75}$$

recovering (10.70). Similarly, by varying $y$:

$$\delta H = \int_{x_1}^{x_2} \mathrm{d}x \, \tau y_{,x} \, \delta y_{,x} = \int_{x_1}^{x_2} \mathrm{d}x \, \tau y_{,x} \frac{\partial}{\partial x} \delta y \,, \tag{10.76}$$

where we interchanged variation and partial derivative to get the last equality. Requiring that the variation $\delta y$ vanishes at $x_1$ and $x_2$, we can integrate this last integral by parts, and throw away the boundary terms, yielding

$$\delta H = -\int_{x_1}^{x_2} \mathrm{d}x \, \tau y_{,xx} \, \delta y \quad \Rightarrow \quad \frac{\delta H}{\delta y} = -\tau y_{,xx} \,. \tag{10.77}$$

Thus,

$$\dot{\pi} = -\frac{\delta H}{\delta y} = \tau y_{,xx} \,, \tag{10.78}$$

recovering (10.71). □

**Exercise 10.8** Repeat the analyses of the previous two examples for the case of the Klein-Gordon Hamiltonian density found in Exercise 10.5.

### *10.2.3 Conserved Quantities*

The discrete form of the Hamiltonian is a conserved quantity if and only if it (or the Lagrangian) does not depend explicitly on time. Thus,

$$\frac{\mathrm{d}H}{\mathrm{d}t} = 0 \quad \Leftrightarrow \quad \frac{\partial H}{\partial t} = -\frac{\partial L}{\partial t} = 0 \,. \tag{10.79}$$

When we go to the continuous limit, the Hamiltonian is the integral of the Hamiltonian density over the relevant volume of space, (10.48). Here, we want to determine the conditions that are placed on the Hamiltonian density in order for $H$ to be a conserved quantity.

To calculate the total time derivative of $\mathcal{H}$ we simply use the chain rule

$$\frac{\mathrm{d}\mathcal{H}}{\mathrm{d}t} = \sum_I \left[ \frac{\partial \mathcal{H}}{\partial \varphi_I}\dot{\varphi}_I + \sum_i \frac{\partial \mathcal{H}}{\partial \varphi_{I,i}}\dot{\varphi}_{I,i} + \frac{\partial \mathcal{H}}{\partial \pi_I}\dot{\pi}_I \right] + \frac{\partial \mathcal{H}}{\partial t}, \tag{10.80}$$

where the dots on the right-hand side are partial derivatives with respect to $t$, e.g., $\dot{\varphi}_{I,i} \equiv \partial \varphi_{I,i}/\partial t = \varphi_{I,it}$. Using Hamilton's equations (10.58) and (10.64), we can substitute for $\dot{\varphi}_I$ and $\dot{\pi}_I$, obtaining:

$$\begin{aligned} \frac{\mathrm{d}\mathcal{H}}{\mathrm{d}t} = \sum_I \Bigg[ & \frac{\partial \mathcal{H}}{\partial \varphi_I}\frac{\partial \mathcal{H}}{\partial \pi_I} + \sum_i \frac{\partial \mathcal{H}}{\partial \varphi_{I,i}}\dot{\varphi}_{I,i} \\ & + \frac{\partial \mathcal{H}}{\partial \pi_I}\left( -\frac{\partial \mathcal{H}}{\partial \varphi_I} + \sum_i \frac{\mathrm{d}}{\mathrm{d}x^i}\left( \frac{\partial \mathcal{H}}{\partial \varphi_{I,i}} \right) \right) \Bigg] + \frac{\partial \mathcal{H}}{\partial t}. \end{aligned} \tag{10.81}$$

Note that the first and third terms in the square brackets cancel, while the second and fourth terms can be rewritten as

$$\begin{aligned} \sum_i \left[ \frac{\partial \mathcal{H}}{\partial \varphi_{I,i}}\dot{\varphi}_{I,i} + \frac{\partial \mathcal{H}}{\partial \pi_I}\frac{\mathrm{d}}{\mathrm{d}x^i}\left( \frac{\partial \mathcal{H}}{\partial \varphi_{I,i}} \right) \right] &= \sum_i \left[ \frac{\partial \mathcal{H}}{\partial \varphi_{I,i}}\frac{\partial \dot{\varphi}_I}{\partial x^i} + \dot{\varphi}_I \frac{\mathrm{d}}{\mathrm{d}x^i}\left( \frac{\partial \mathcal{H}}{\partial \varphi_{I,i}} \right) \right] \\ &= \sum_i \frac{\mathrm{d}}{\mathrm{d}x^i}\left( \frac{\partial \mathcal{H}}{\partial \varphi_{I,i}}\dot{\varphi}_I \right), \end{aligned} \tag{10.82}$$

where we used (10.58) again and the product rule for derivatives. Thus,

$$\frac{\mathrm{d}\mathcal{H}}{\mathrm{d}t} = \sum_I \sum_i \frac{\mathrm{d}}{\mathrm{d}x^i}\left( \frac{\partial \mathcal{H}}{\partial \varphi_{I,i}}\dot{\varphi}_I \right) + \frac{\partial \mathcal{H}}{\partial t}. \tag{10.83}$$

If we now assume that $\partial\mathcal{H}/\partial t = 0$ and use $\partial\mathcal{H}/\partial\varphi_{I,i} = -\partial\mathcal{L}/\partial\varphi_{I,i}$ (from Exercise 10.7), we have

$$\frac{\partial \mathcal{H}}{\partial t} = 0 \quad \Leftrightarrow \quad \frac{\mathrm{d}\mathcal{H}}{\mathrm{d}t} + \sum_i \frac{\mathrm{d}}{\mathrm{d}x^i}\left( \sum_I \frac{\partial \mathcal{L}}{\partial \varphi_{I,i}}\dot{\varphi}_I \right) = 0. \tag{10.84}$$

This last equation has the form of a **continuity equation**

$$\frac{\mathrm{d}J^0}{\mathrm{d}t} + \sum_i \frac{\mathrm{d}J^i}{\mathrm{d}x^i} = 0, \tag{10.85}$$

where $J^0$ and $J^i$ are thought of here as functions of the fields $\varphi_I$, their partial derivatives $\varphi_{I,i}$, $\dot{\varphi}_I$, and the coordinates $x^i$ and $t$, i.e.,

$$J^0 \equiv J^0(\varphi_I, \varphi_{I,i}, \dot{\varphi}_I, x^i, t), \qquad J^i \equiv J^i(\varphi_I, \varphi_{I,i}, \dot{\varphi}_I, x^i, t). \tag{10.86}$$

If we substitute for $\varphi_I$, $\varphi_{I,i}$, $\dot{\varphi}_I$ their explicit expressions in terms of $x^i$ and $t$, then $J^0$ and $J^i$ can be interpreted as *fields*, $J^0 \equiv J^0(x^i, t)$, $J^i \equiv J^i(x^i, t)$. The total derivatives then become *partial* derivatives, and the continuity equation, (10.85), takes the more familiar form

$$\frac{\partial J^0}{\partial t} + \nabla \cdot \mathbf{J} = 0. \tag{10.87}$$

(We will try to make it clear in the text which interpretation we are using for the functions $J^\alpha$ and their derivatives.) The continuity equation describes how a field quantity that is conserved *globally* can change *locally*.

***Example 10.5*** (Conservation of mass in a fluid) In this example, we will consider a fluid with mass density $\rho(\mathbf{r}, t)$ and velocity distribution $\mathbf{v}(\mathbf{r}, t)$ defined with respect to the invariant volume element $\mathrm{d}V$. Let's look at the rate of change of mass contained in a volume $V$ over a time interval $\Delta t$. At any time, the amount of mass contained in this volume is

$$m = \int_V \mathrm{d}V \, \rho(\mathbf{r}, t). \tag{10.88}$$

Since mass is conserved, the amount of matter flowing into (or out of) $V$ through the boundary surface $S$ in time $\Delta t$ is equal to the change in $m$ during that time. Since the vector surface area element $\hat{\mathbf{n}}\,\mathrm{d}a$ has its unit normal $\hat{\mathbf{n}}$ directed *outward* from the volume, the amount of matter flowing into the volume in time $\Delta t$ is

$$\Delta m = -\Delta t \oint_S \rho \mathbf{v} \cdot \hat{\mathbf{n}}\,\mathrm{d}a, \tag{10.89}$$

so, in the infinitesimal limit, we have

$$\frac{\mathrm{d}m}{\mathrm{d}t} = \frac{\mathrm{d}}{\mathrm{d}t}\left(\int_V \mathrm{d}V \, \rho\right) = -\oint_S \rho \mathbf{v} \cdot \hat{\mathbf{n}}\,\mathrm{d}a. \tag{10.90}$$

We can bring the total time derivative inside the volume integral (converting it to a partial derivative of $\rho$), and we can use the divergence theorem (A.87) to convert the surface integral into a volume integral of a divergence. Thus,

$$\int_V \mathrm{d}V \left[\frac{\partial \rho}{\partial t} + \nabla \cdot (\rho \mathbf{v})\right] = 0. \tag{10.91}$$

Since the volume we chose was arbitrary, this equation can be satisfied if and only if the integrand is zero:

$$\frac{\partial \rho}{\partial t} + \nabla \cdot \mathbf{j} = 0 \,, \qquad \mathbf{j} \equiv \rho \mathbf{v} \,. \tag{10.92}$$

This is a continuity equation, which relates the time derivative of a density $\rho$ to the divergence of a corresponding current density $\mathbf{j} \equiv \rho\mathbf{v}$. Although the mass density is not locally conserved at every point, if we integrate $\rho$ over all space (or at least beyond the size of the system), then the total mass $M$ is conserved:

$$\frac{\mathrm{d}M}{\mathrm{d}t} = 0 \,, \qquad M \equiv \int_{\text{all space}} \mathrm{d}V \, \rho(\mathbf{r}, t) \,. \tag{10.93}$$

This is because there is no mass flowing through the boundary of the volume once the volume exceeds the extent of the fluid. □

Returning to the continuity equation for the Hamiltonian density, (10.84), we can now conclude that the total Hamiltonian $H = \int \mathrm{d}^3x\, \mathcal{H}$ is a conserved quantity if and only if $\mathcal{H}$ does not explicitly depend on time:

$$\frac{\mathrm{d}H}{\mathrm{d}t} = 0 \quad \Leftrightarrow \quad \frac{\partial \mathcal{H}}{\partial t} = 0 \,. \tag{10.94}$$

We also see that the quantity

$$\sum_I \frac{\partial \mathcal{L}}{\partial \varphi_{I,i}} \dot{\varphi}_I \tag{10.95}$$

plays the role of a current density corresponding to $\mathcal{H}$. In the next section, we will see how $\mathcal{H}$ and its corresponding current density are part of a larger, more comprehensive object that describes the momentum and energy densities of the continuous system.

**Exercise 10.9** Using the definition (10.49) of the Hamiltonian density $\mathcal{H}$ in terms of $\mathcal{L}$, show that the continuity equation (10.84) can be written as

$$\sum_\beta \frac{\mathrm{d}}{\mathrm{d}x^\beta} \left( \sum_I \frac{\partial \mathcal{L}}{\partial \varphi_{I,\beta}} \dot{\varphi}_I - \delta_{t\beta} \mathcal{L} \right) = 0 \,, \tag{10.96}$$

where the sum is over both the space and time coordinates $x^\beta \equiv (t, x^i)$.

## 10.3 Stress-Energy Tensor

If we are given a Lagrangian density $\mathcal{L}$ for a set of fields $\varphi_I$, then we can take the total derivative of $\mathcal{L}$ with respect to both the temporal and spatial coordinates $x^\alpha$:

$$\frac{\mathrm{d}\mathcal{L}}{\mathrm{d}x^\alpha} = \sum_I \frac{\partial \mathcal{L}}{\partial \varphi_I}\varphi_{I,\alpha} + \sum_I\sum_\beta \frac{\partial \mathcal{L}}{\partial \varphi_{I,\beta}}\varphi_{I,\beta\alpha} + \frac{\partial \mathcal{L}}{\partial x^\alpha}. \tag{10.97}$$

Using (10.35), we can rewrite the first term on the right-hand side as

$$\sum_I \frac{\partial \mathcal{L}}{\partial \varphi_I}\varphi_{I,\alpha} = \sum_I\sum_\beta \frac{\mathrm{d}}{\mathrm{d}x^\beta}\left(\frac{\partial \mathcal{L}}{\partial \varphi_{I,\beta}}\right)\varphi_{I,\alpha}, \tag{10.98}$$

and then combine it with the second term using $\varphi_{I,\beta\alpha} = \varphi_{I,\alpha\beta}$ and the product rule for derivatives. This leads to

$$\frac{\mathrm{d}\mathcal{L}}{\mathrm{d}x^\alpha} = \sum_\beta \frac{\mathrm{d}}{\mathrm{d}x^\beta}\left(\sum_I \frac{\partial \mathcal{L}}{\partial \varphi_{I,\beta}}\varphi_{I,\alpha}\right) + \frac{\partial \mathcal{L}}{\partial x^\alpha}, \tag{10.99}$$

which can be rearranged to give

$$-\frac{\partial \mathcal{L}}{\partial x^\alpha} = \sum_\beta \frac{\mathrm{d}}{\mathrm{d}x^\beta}\left(\sum_I \frac{\partial \mathcal{L}}{\partial \varphi_{I,\beta}}\varphi_{I,\alpha} - \delta_{\alpha\beta}\mathcal{L}\right). \tag{10.100}$$

Thus, for solutions to the field equations,

$$\frac{\partial \mathcal{L}}{\partial x^\alpha} = 0 \quad \Leftrightarrow \quad \sum_\beta \frac{\mathrm{d}T_{\alpha\beta}}{\mathrm{d}x^\beta} = 0, \tag{10.101}$$

where

$$T_{\alpha\beta} \equiv \sum_I \frac{\partial \mathcal{L}}{\partial \varphi_{I,\beta}}\varphi_{I,\alpha} - \delta_{\alpha\beta}\mathcal{L} \tag{10.102}$$

is the so-called **stress-energy tensor**. Equation (10.101) is a compact form of writing the continuity equation, which holds for solutions to the field equations when the Lagrangian density $\mathcal{L}$ does not depend explicitly on the spatial and temporal coordinates $x^\alpha$.

From its definition, we see that $T_{00}$ is just the Hamiltonian density $\mathcal{H}$, and $T_{0i}$ is its associated current (See also Exercise 10.9). In those cases where the Lagrangian density can be written as $\mathcal{L} = \mathcal{T} - \mathcal{U}$, the kinetic energy density has the form

$$\mathcal{T} = \sum_I \frac{1}{2}\mu\dot{\varphi}_I^2 , \tag{10.103}$$

and the interpretation of

$$T_{00} = \mathcal{T} + \mathcal{U} , \tag{10.104}$$

as the *total energy density* in the fields $\varphi_I$ is obvious. Consequently, we can interpret $T_{0i}$ as the components of the *energy density current* vector $\mathbf{T}_{(0)}$, specifying the flow of energy density in the fields. In this form we can write (10.101) as

$$\frac{\partial T_{00}}{\partial t} + \nabla \cdot \mathbf{T}_{(0)} = 0 , \tag{10.105}$$

where we have replaced the total derivatives in (10.101) by partial derivatives $\partial/\partial x^\alpha$, interpreting the components $T_{\alpha\beta}$ of the stress-energy tensor here as functions of just the coordinates $x^\alpha$ (See (10.87) and the surrounding discussion). The conserved quantity is $H = \int d^3x\, T_{00} = \int d^3x\, \mathcal{H}$.

Continuing in this spirit, we separate out the spatial and temporal parts of the continuity equation (10.101), so that

$$\frac{\partial T_{\alpha 0}}{\partial t} + \sum_j \frac{\partial T_{\alpha j}}{\partial x^j} = 0 . \tag{10.106}$$

The time component is just (10.105), while the spatial components become:

$$\frac{\partial T_{i0}}{\partial t} + \nabla \cdot \mathbf{T}_{(i)} = 0 , \tag{10.107}$$

where the vector $\mathbf{T}_{(i)}$ has components $T_{ij}$. Thus, we see that we have three conserved quantities $\int d^3x\, T_{i0}$ corresponding to the three densities

$$T_{i0} \equiv \sum_I \frac{\partial \mathcal{L}}{\partial \dot{\varphi}_I}\varphi_{I,i} = \sum_I \pi_I \varphi_{I,i} . \tag{10.108}$$

These three densities correspond to the three components of the mechanical momentum density of the system. In particular, $-T_{i0} =$ $i$th component of the mechanical momentum density associated with the fields. In this sense, the spatial components $T_{ij}$ correspond to the flux of momentum density. In Cartesian coordinates, $-T_{ij} =$ flow of the $i$th component of mechanical momentum density in the $j$th direction.

The components of the stress-energy tensor can now be described as

$$\mathsf{T} = \left[\begin{array}{c|c} \text{energy density} & \left(\begin{array}{c}\text{components of the}\\ \text{flux of energy density}\end{array}\right) \\ \hline -\left(\begin{array}{c}\text{components of}\\ \text{momentum density}\end{array}\right) & -\left(\begin{array}{c}\text{components of the flux}\\ \text{of the components of}\\ \text{momentum density}\end{array}\right) \end{array}\right] \tag{10.109}$$

Conservation of mechanical energy and mechanical momentum are described by the continuity equation governing the stress-energy tensor. This equation holds for Lagrangian densities that do not explicitly depend upon time or the spatial coordinates.

Note that we have taken pains to specify that these components of the stress-energy tensor are related to the *mechanical momentum density*. The astute reader may have noticed that we have already defined the *canonical momentum density* as

$$\pi_I \equiv \frac{\partial \mathcal{L}}{\partial \dot{\varphi}_I}, \tag{10.110}$$

which is *not* equal to any component of the stress-energy tensor—in fact, it doesn't even have a reference to the coordinates $x^i$. The meaning of any conservation of canonical momentum will become clear when we discuss Noether's theorem in the next section.

***Example 10.6*** To make things a bit more concrete, let's calculate the components $T_{\alpha\beta}$ of the stress-energy tensor for the Lagrangian density for one-dimensional waves,

$$\mathcal{L} = \frac{1}{2}\left[\mu y_{,t}^2 - \tau y_{,x}^2\right]. \tag{10.111}$$

For this example, the general expression (10.102) for $T_{\alpha\beta}$ reduces to

$$T_{\alpha\beta} = \frac{\partial \mathcal{L}}{\partial y_{,\beta}} y_{,\alpha} - \delta_{\alpha\beta}\mathcal{L}, \tag{10.112}$$

with $x^\alpha \equiv (t, x)$. The relevant partial derivatives of the Lagrangian density are simply

$$\frac{\partial \mathcal{L}}{\partial y_{,t}} = \mu y_{,t}, \qquad \frac{\partial \mathcal{L}}{\partial y_{,x}} = -\tau y_{,x}. \tag{10.113}$$

Thus,

$$\begin{aligned}
T_{tt} &= \frac{\partial \mathcal{L}}{\partial y_{,t}} y_{,t} - \mathcal{L} = \mu y_{,t} y_{,t} - \frac{1}{2}\left[\mu y_{,t}^2 - \tau y_{,x}^2\right] = \frac{1}{2}\left[\mu y_{,t}^2 + \tau y_{,x}^2\right], \\
T_{tx} &= \frac{\partial \mathcal{L}}{\partial y_{,x}} y_{,t} = -\tau y_{,x} y_{,t}, \qquad T_{xt} = \frac{\partial \mathcal{L}}{\partial y_{,t}} y_{,x} = \mu y_{,t} y_{,x}, \\
T_{xx} &= \frac{\partial \mathcal{L}}{\partial y_{,x}} y_{,x} - \mathcal{L} = -\tau y_{,x} y_{,x} - \frac{1}{2}\left[\mu y_{,t}^2 - \tau y_{,x}^2\right] = -\frac{1}{2}\left[\mu y_{,t}^2 + \tau y_{,x}^2\right].
\end{aligned} \tag{10.114}$$

Note that if we replace $y_{,t}$ with $\pi/\mu$, then $T_{tt}$ is equal to the Hamiltonian density given in (10.54), as one would expect in general. In addition, for this example, $T_{xx} = -T_{tt}$. We also see that the stress-energy tensor is not necessarily symmetric, as $T_{tx} \neq T_{xt}$. □

## 10.4 Noether's Theorem

Throughout this book, we have seen many instances where a conserved quantity is associated with a cyclic variable in the Lagrangian. For example, in Chap. 4, we found that conservation of angular momentum in the two-body problem is directly related to the Lagrangian being independent of the azimuthal coordinate $\phi$. Similarly, for a system where the Lagrangian is independent of a particular Cartesian coordinate, e.g., $x$, then the $x$-component of the total linear momentum is conserved. (More generally, if $\partial L/\partial q^{\underline{a}} = 0$, then $p_{\underline{a}} \equiv \partial L/\partial \dot{q}^{\underline{a}} = \text{const.}$) And, of course, if the Lagrangian is independent of the time variable $t$, then the energy function $h \equiv \sum_a p_a \dot{q}^a - L$ is conserved.

For the case of fields, the Lagrangian $L$ is replaced by the Lagrangian density $\mathscr{L}$, which is a function of $\varphi_I$, $\varphi_{I,\alpha}$, and the coordinates $x^\alpha \equiv (t, x^i)$. But there is still a correspondence between symmetries of the system and conserved quantities, which is given by **Noether's theorem**:

> For every continuous symmetry $\varphi_I \to \varphi_I + \delta\varphi_I$ of the Lagrangian density $\mathscr{L}$, there is a conserved current $J^\alpha$ constructed, in general, from the fields, their derivatives, and the coordinates $x^\alpha$.

By continuous *symmetry*, we mean an infinitesimal change in the fields

$$\varphi_I \to \varphi_I + \delta\varphi_I \equiv \varphi_I + \lambda \left.\frac{\mathrm{d}\varphi_I}{\mathrm{d}\lambda}\right|_{\lambda=0}, \tag{10.115}$$

which depends locally on the coordinates $x^\alpha$, the fields $\varphi_I$, and their derivatives $\varphi_{I,\alpha}, \varphi_{I,\alpha\beta}, \cdots$ to some finite order, for which

$$\mathscr{L} \to \mathscr{L} + \delta\mathscr{L} \equiv \mathscr{L} + \lambda \left.\frac{\mathrm{d}\mathscr{L}}{\mathrm{d}\lambda}\right|_{\lambda=0}, \qquad \left.\frac{\mathrm{d}\mathscr{L}}{\mathrm{d}\lambda}\right|_{\lambda=0} = \sum_\alpha \frac{\mathrm{d}W^\alpha}{\mathrm{d}x^\alpha}, \tag{10.116}$$

for some $W^\alpha$. Note that we are only requiring that $(\mathrm{d}\mathscr{L}/\mathrm{d}\lambda)_{\lambda=0}$ equal a divergence, since two Lagrangians that differ by a divergence yield the *same* equations of motion (Exercise 10.4).[6] By conserved current we mean

$$\sum_\alpha \frac{\mathrm{d}J^\alpha}{\mathrm{d}x^\alpha} = 0 \tag{10.117}$$

whenever $\varphi_I$ satisfy the field equations (i.e., $\mathscr{E}_I = 0$, for $I = 1, 2, \ldots, N$). Assuming that the fields fall-off sufficiently fast as $r \to \infty$, the associated conserved quantity (sometimes called a **conserved charge**) is given by the integral

$$Q \equiv \int_{\text{all space}} \mathrm{d}^3x\, J^0(\varphi_I(\mathbf{r}, t), \varphi_{I,i}(\mathbf{r}, t), \dot{\varphi}_I(\mathbf{r}, t), \mathbf{r}, t)\,. \tag{10.118}$$

In the following subsections we give a proof of Noether's theorem, followed by several examples showing how conservation of energy, momentum, and angular momentum arise in the context of a field theory. We also discuss a conserved quantity for an internal (field) symmetry (Example 10.9). For many more details, we recommend Torre (2016).

### 10.4.1 Proof of Noether's Theorem

In the form stated above, Noether's theorem is a simple consequence of the variational identity (10.38) for the Lagrangian density $\mathscr{L}$, together with the assumed form (10.116) for a divergence symmetry. Equating those two expressions for $(\mathrm{d}\mathscr{L}/\mathrm{d}\lambda)|_{\lambda=0}$ yields

$$\sum_I \mathscr{E}_I \left.\frac{\mathrm{d}\varphi_I}{\mathrm{d}\lambda}\right|_{\lambda=0} + \sum_\alpha \frac{\mathrm{d}}{\mathrm{d}x^\alpha}\left(\sum_I \frac{\partial\mathscr{L}}{\partial\varphi_{I,\alpha}} \left.\frac{\mathrm{d}\varphi_I}{\mathrm{d}\lambda}\right|_{\lambda=0}\right) = \sum_\alpha \frac{\mathrm{d}W^\alpha}{\mathrm{d}x^\alpha}\,. \tag{10.119}$$

But since $\mathscr{E}_I = 0$ for $I = 1, 2, \ldots, N$ on a solution to the field equations, the above expression reduces to

[6] Such a symmetry is sometimes called a *divergence symmetry*, as opposed to a *variational symmetry*, which would have $\delta\mathscr{L} = 0$.

$$\sum_\alpha \frac{\mathrm{d}J^\alpha}{\mathrm{d}x^\alpha} = 0\,, \quad \text{where} \quad J^\alpha \equiv \sum_I \frac{\partial \mathscr{L}}{\partial \varphi_{I,\alpha}} \left.\frac{\mathrm{d}\varphi_I}{\mathrm{d}\lambda}\right|_{\lambda=0} - W^\alpha\,. \tag{10.120}$$

This is the conserved current of Noether's theorem.

### 10.4.2 Some Simple Examples

Probably the best way to appreciate this theorem and its consequences is to see it in action in the context of a few simple examples.[7]

***Example 10.7*** (Time translation symmetry) Consider the Lagrangian density for the Klein-Gordon field in three dimensions:

$$\mathscr{L} = \frac{1}{2}\left[\frac{1}{c^2}\varphi_{,t}^2 - \nabla\varphi\cdot\nabla\varphi - \mu^2\varphi^2\right]. \tag{10.121}$$

A time translation

$$t \to t + \lambda \tag{10.122}$$

induces the following change in the field

$$\varphi(\mathbf{r}, t) \to \varphi(\mathbf{r}, t+\lambda) = \varphi(\mathbf{r}, t) + \lambda\varphi_{,t}(\mathbf{r}, t) + O(\lambda^2)\,, \tag{10.123}$$

so that

$$\left.\frac{\mathrm{d}\varphi}{\mathrm{d}\lambda}\right|_{\lambda=0} = \varphi_{,t}\,. \tag{10.124}$$

This in turn leads to the following change in the Lagrangian density

$$\begin{aligned}\left.\frac{\mathrm{d}\mathscr{L}}{\mathrm{d}\lambda}\right|_{\lambda=0} &= \frac{\partial\mathscr{L}}{\partial\varphi}\left.\frac{\mathrm{d}\varphi}{\mathrm{d}\lambda}\right|_{\lambda=0} + \frac{\partial\mathscr{L}}{\partial\varphi_{,t}}\left.\frac{\mathrm{d}\varphi_{,t}}{\mathrm{d}\lambda}\right|_{\lambda=0} + \sum_i \frac{\partial\mathscr{L}}{\partial\varphi_{,i}}\left.\frac{\mathrm{d}\varphi_{,i}}{\mathrm{d}\lambda}\right|_{\lambda=0}\\ &= -\mu^2\varphi\varphi_{,t} + \frac{1}{c^2}\varphi_{,t}\varphi_{,tt} - \sum_i \varphi_{,i}\varphi_{,it}\\ &= \frac{\mathrm{d}\mathscr{L}}{\mathrm{d}t} = \sum_\alpha \frac{\mathrm{d}}{\mathrm{d}x^\alpha}\left(\delta_t^\alpha\mathscr{L}\right)\,,\end{aligned} \tag{10.125}$$

[7] For all these examples, we will work in Cartesian coordinates $x^i = (x, y, z)$ for which $\sqrt{\det g} = 1$ and the Lagrangian densities for the 3-dimensional Klein-Gordon field have the explicit forms given in (10.121) and (10.137). If, instead, you would like to do a particular calculation in e.g., spherical coordinates $(r, \theta, \phi)$, then you should first multiply the Lagrangian densities given in (10.121) and (10.137) by $\sqrt{\det g} = r^2\sin\theta$ *before* assessing the form of the symmetry and calculating the conserved current $J^\alpha$, etc.

where we've have written the last equality in such away that it is manifestly a divergence with $W^\alpha \equiv \delta_t^\alpha \mathscr{L}$. This means that the conserved current associated with the time translation symmetry is

$$J^\alpha = \frac{\partial \mathscr{L}}{\partial \varphi_{,\alpha}} \varphi_{,t} - \delta_t^\alpha \mathscr{L} . \tag{10.126}$$

Note that $J^\alpha$ are the $t\alpha$ components of the stress-energy tensor, $T_{t\alpha}$, given by (10.102). Evaluating the individual components of $J^\alpha$ for the above form of the Lagrangian density yields

$$J^0 = \frac{1}{2}\left[\frac{1}{c^2}\varphi_{,t}^2 + \nabla\varphi \cdot \nabla\varphi + \mu^2\varphi^2\right], \qquad J^i = -\varphi_{,i}\varphi_{,t} . \tag{10.127}$$

Note that $J^0$ is just the energy density of the Klein-Gordon field, so the conserved "charge"

$$Q \equiv \int_{\text{all space}} \mathrm{d}^3x \, J^0 \tag{10.128}$$

is the total energy in the field. Hence, we have recovered conservation of total energy from time translation symmetry as we might have expected. □

***Example 10.8*** (Spatial translation symmetry) A similar analysis can be applied to the case of a spatial translation

$$x^i \to x^i + \lambda n^i , \tag{10.129}$$

where $n^i$ are the components of a constant unit vector $\hat{\mathbf{n}}$. Following the same procedure as before, it is fairly easy to show that

$$\left.\frac{\mathrm{d}\varphi}{\mathrm{d}\lambda}\right|_{\lambda=0} = \hat{\mathbf{n}} \cdot \nabla\varphi , \tag{10.130}$$

leading to

$$\left.\frac{\mathrm{d}\mathscr{L}}{\mathrm{d}\lambda}\right|_{\lambda=0} = \sum_\alpha \frac{\mathrm{d}W^\alpha}{\mathrm{d}x^\alpha}, \qquad W^\alpha = (0, \hat{\mathbf{n}}\,\mathscr{L}), \tag{10.131}$$

and

$$J^\alpha = \sum_i n^i \left(\frac{\partial \mathscr{L}}{\partial \varphi_{,\alpha}} \varphi_{,i} - \delta_i^\alpha \mathscr{L}\right) = \sum_i n^i T_{i\alpha} . \tag{10.132}$$

Now, since $T_{i0}$ has the interpretation of minus the momentum density in the field $\varphi$, it follows that the conserved "charge"

$$Q \equiv \int_{\text{all space}} \mathrm{d}^3x \, J^0 = \sum_i n^i \left( \int_{\text{all space}} \mathrm{d}^3x \, T_{i0} \right) = \hat{\mathbf{n}} \cdot \left( \frac{1}{c^2} \int_{\text{all space}} \mathrm{d}^3x \, \varphi_{,t} \nabla \varphi \right), \tag{10.133}$$

is minus the total momentum of the field projected along $\hat{\mathbf{n}}$. □

**Exercise 10.10** (Rotational symmetry) Repeat the calculation of the previous two examples, but this time for an infinitesimal rotation

$$x^i \to x^i + \lambda \sum_{j,k} \varepsilon^{ijk} n^j x^k \,, \tag{10.134}$$

where $n^i$ are the components of a constant unit vector $\hat{\mathbf{n}}$. You should find:

$$\left.\frac{\mathrm{d}\varphi}{\mathrm{d}\lambda}\right|_{\lambda=0} = (\hat{\mathbf{n}} \times \mathbf{r}) \cdot \nabla \varphi \,, \qquad W^\alpha = (0, (\hat{\mathbf{n}} \times \mathbf{r}) \mathscr{L}) \,,$$
$$J^\alpha = \sum_i (\hat{\mathbf{n}} \times \mathbf{r})^i \left( \frac{\partial \mathscr{L}}{\partial \varphi_{,\alpha}} \varphi_{,i} - \delta^\alpha_i \mathscr{L} \right) = \sum_i (\hat{\mathbf{n}} \times \mathbf{r})^i T_{i\alpha} \,. \tag{10.135}$$

The conserved charge is thus

$$Q = \hat{\mathbf{n}} \cdot \left( \frac{1}{c^2} \int_{\text{all space}} \mathrm{d}^3x \, \varphi_{,t} \mathbf{r} \times \nabla \varphi \right), \tag{10.136}$$

which is minus the total angular momentum of the field projected along $\hat{\mathbf{n}}$.

***Example 10.9*** **(Internal symmetries)** In addition to the *coordinate symmetries* discussed above, we can also consider the case of *internal symmetries* of the Lagrangian density, which are associated with infinitesimal field variations $\varphi_I \to \varphi_I + \delta\varphi_I$ that aren't induced by coordinate transformations. In this example, we consider the so-called *charged* Klein-Gordon field, which is described by a *complex-valued* field $\varphi = \varphi_1 + \mathrm{i}\varphi_2$, where $\varphi_1$ and $\varphi_2$ are real. But rather than work directly with $\varphi_1$ and $\varphi_2$, it turns out to be simpler to work with $\varphi$ and $\varphi^*$ (its complex conjugate), which are treated as *independent* variables in the variational process.

The Lagrangian density for the complex-valued Klein-Gordon field in three dimensions can be written in terms of $\varphi$ and $\varphi^*$ as

$$\mathscr{L} = \frac{1}{2} \left[ \frac{1}{c^2} \varphi_{,t} \varphi^*_{,t} - \nabla \varphi \cdot \nabla \varphi^* - \mu^2 \varphi \varphi^* \right], \tag{10.137}$$

which is manifestly real. The equations of motion that you obtain by varying the Lagrangian density with respect to $\varphi^*$ and $\varphi$ are the standard equation

$$-\frac{1}{c^2}\varphi_{,tt} + \nabla^2\varphi = \mu^2\varphi\,, \tag{10.138}$$

(See Exercise 10.2 for the 1-dimensional version of this equation) and its complex conjugate

$$-\frac{1}{c^2}\varphi^*_{,tt} + \nabla^2\varphi^* = \mu^2\varphi^*\,. \tag{10.139}$$

It is also easy to see that that the Lagrangian density is invariant under the continuous field transformation

$$\varphi \to \varphi' = \mathrm{e}^{\mathrm{i}\lambda}\varphi\,, \qquad \varphi^* \to \varphi^{*\prime} = \mathrm{e}^{-\mathrm{i}\lambda}\varphi^*\,, \qquad \lambda \in \mathbb{R}\,, \tag{10.140}$$

which can be thought of as a complex rotation in field space. Expanding the exponential, these transformations become

$$\varphi \to \varphi + i\lambda\varphi + O(\lambda^2)\,, \qquad \varphi^* \to \varphi^* - i\lambda\varphi^* + O(\lambda^2)\,, \tag{10.141}$$

for which

$$\left.\frac{\mathrm{d}\varphi}{\mathrm{d}\lambda}\right|_{\lambda=0} = i\varphi\,, \qquad \left.\frac{\mathrm{d}\varphi^*}{\mathrm{d}\lambda}\right|_{\lambda=0} = -i\varphi^*\,. \tag{10.142}$$

Since the Lagrangian density is *invariant* under such a change (i.e., $\mathrm{d}\mathscr{L}/\mathrm{d}\lambda|_{\lambda=0} = 0$), it follows that the conserved current $J^\alpha$ is just

$$J^\alpha = \frac{\partial\mathscr{L}}{\partial\varphi_{,\alpha}}i\varphi + \frac{\partial\mathscr{L}}{\partial\varphi^*_{,\alpha}}(-i\varphi^*)\,, \tag{10.143}$$

which includes contributions from both $\varphi$ and $\varphi^*$. Explicitly evaluating the components, we find

$$J^0 = \frac{i}{2c^2}\left(\varphi^*_{,t}\varphi - \varphi_{,t}\varphi^*\right)\,, \qquad J^i = -\frac{i}{2}\left(\varphi^*_{,i}\varphi - \varphi_{,i}\varphi^*\right)\,, \tag{10.144}$$

and the conserved charge

$$Q = \frac{i}{2c^2}\int_{\text{all space}} \mathrm{d}^3x\ \left(\varphi^*_{,t}\varphi - \varphi_{,t}\varphi^*\right)\,. \tag{10.145}$$

□

**Exercise 10.11** Explicitly verify that $\mathrm{d}J^\alpha/\mathrm{d}x^\alpha = 0$ for $J^\alpha$ given in the above example. (*Hint*: You will need to use the field equations for $\varphi$ and $\varphi^*$.)

As the above examples and exercises illustrate, Noether's theorem is quite powerful for identifying conserved quantities for systems of particles and fields that admit continuous symmetries. But its true value lies in the creation of Lagrangian densities to describe fields with certain *prescribed* conservation laws. In other words, there is actually a *one-to-one* correspondence between symmetries and conservation laws (which Noether also proved), which is stronger than the statement of Noether's theorem that we gave above. (We only proved that symmetries *imply* the existence of conservation laws; we didn't prove the converse.) Thus, when one is developing a field theory to describe a given phenomenon or interaction, the functional form of the Lagrangian density must exhibit the necessary symmetries to reflect the known conserved quantities in the interaction.

## Suggested References

*Full references are given in the bibliography at the end of the book.*

Goldstein et al. (2002): Chap. 13 introduces classical field theory, but this is after covering relativity, so there is a good discussion of relativistic field theory as well.

Torre (2016): An introduction to classical field theory from a more mathematical perspective, appropriate for graduate students. Our presentation of Noether's theorem in Sect. 10.4 is based on that from Chaps. 2 and 3 in this reference. In addition, Problems 10.3 and 10.4 below are adapted from similar examples and exercises in Torre (2016).

## Additional Problems

**Problem 10.1** For the Hamiltonian description of a discrete system of particles, we defined in Sect. 3.5.1 the Poisson bracket of two functions $f = f(q, p)$, $g = g(q, p)$ to be

$$\{f, g\} \equiv \sum_a \left( \frac{\partial f}{\partial q^a}\frac{\partial g}{\partial p_a} - \frac{\partial f}{\partial p_a}\frac{\partial g}{\partial q^a} \right). \tag{10.146}$$

From its definition, we could easily show that

$$\{q^a, q^b\} = 0\,, \qquad \{p_a, p_b\} = 0\,, \qquad \{q^a, p_b\} = \delta^a_b\,, \tag{10.147}$$

and rewrite Hamilton's equations of motion as

$$\dot{q}^a = \{q^a, H\}, \qquad \dot{p}_a = \{p_a, H\}. \tag{10.148}$$

For a field theory described by $\varphi(\mathbf{r}, t)$ and its conjugate momentum $\pi(\mathbf{r}, t)$, we can define a similar Poisson bracket structure via[8]

$$\{F, G\} \equiv \int_V \mathrm{d}^3x \left( \frac{\delta F}{\delta\varphi(\mathbf{r}, t)} \frac{\delta G}{\delta\pi(\mathbf{r}, t)} - \frac{\delta F}{\delta\pi(\mathbf{r}, t)} \frac{\delta G}{\delta\varphi(\mathbf{r}, t)} \right). \tag{10.149}$$

Compared to (10.146), we see that the functions $f$ and $g$ have been replaced by *functionals* $F = F[\varphi, \pi]$ and $G = G[\varphi, \pi]$; the partial derivatives have been replaced by *functional derivatives* with respect to $\varphi(\mathbf{r}, t)$ and $\pi(\mathbf{r}, t)$; and the summation over $a$ has been replaced by *integration* over $\mathbf{r}$.

(a) Show that the fundamental Poisson brackets are

$$\begin{aligned} \{\varphi(\mathbf{r}, t), \varphi(\mathbf{r}', t)\} &= 0, \\ \{\pi(\mathbf{r}, t), \pi(\mathbf{r}', t)\} &= 0, \\ \{\varphi(\mathbf{r}, t), \pi(\mathbf{r}', t)\} &= \delta(\mathbf{r} - \mathbf{r}'). \end{aligned} \tag{10.150}$$

*Hint*: Recall Exercise C.4.

(b) Show that Hamilton's equations (10.60) and (10.67) can be rewritten as

$$\dot{\varphi}(\mathbf{r}, t) = \{\varphi(\mathbf{r}, t), H\}, \qquad \dot{\pi}(\mathbf{r}, t) = \{\pi(\mathbf{r}, t), H\}, \tag{10.151}$$

where $H = \int_V \mathrm{d}^3x\, \mathscr{H}$.

**Problem 10.2** The Lagrangian density for the one-dimensional Klein-Gordon field $\varphi(x, t)$ is

$$\mathscr{L} = \frac{1}{2}\left[\frac{1}{c^2}\varphi_{,t}^2 - \varphi_{,x}^2 - \mu^2\varphi^2\right], \tag{10.152}$$

where $\mu$ and $c$ are constants.

(a) Calculate the components $T_{tt}$, $T_{xx}$, $T_{tx}$, and $T_{xt}$ of the stress-energy tensor for the Klein-Gordon field.
(b) Calculate the conserved current $J_t^\alpha$ associated with the time-translation symmetry of the Lagrangian density, i.e., $\partial\mathscr{L}/\partial t = 0$. (Here $\alpha = t, x$.)
(c) Calculate the conserved current $J_x^\alpha$ associated with the space-translation symmetry of the Lagrangian density, i.e., $\partial\mathscr{L}/\partial x = 0$. (Again, $\alpha = t, x$.)

**Problem 10.3** The *self-interacting* one-dimensional Klein-Gordon field is defined by the Lagrangian density

[8] It turns out that this definition of Poisson brackets is *independent* of the choice of time $t$. See e.g., Torre (2016) for a proof.

$$\mathscr{L} = \frac{1}{2}\left[\frac{1}{c^2}\varphi_{,t}^2 - \varphi_{,x}^2\right] - V(\varphi)\,, \tag{10.153}$$

where

$$V(\varphi) \equiv -\frac{1}{2}a^2\varphi^2 + \frac{1}{4}b^2\varphi^4\,. \tag{10.154}$$

with $a$, $b$, and $c$ constants. Consider the fields to be defined over a large but finite space.

(a) Sketch the potential $V(\varphi)$. (It should look like a double well.)
(b) Write down the field equation for $\varphi(x,t)$.
(c) Find the subset of solutions to the field equation having $\varphi = \text{const}$.
(d) Calculate the Hamiltonian density $\mathscr{H}$ and Hamiltonian $H \equiv \int \mathrm{d}x\, \mathscr{H}$.
(e) Show that the $\varphi = \text{const}$ solutions found above are stationary values of the energy function $E \equiv H$.

**Problem 10.4** Extend the self-interacting Klein-Gordon field discussed in Problem 10.3 to allow for *complex-valued* $\varphi(x,t)$, similar to what we discussed in Example 10.9. That is, consider the Lagrangian density

$$\mathscr{L} = \frac{1}{2}\left[\frac{1}{c^2}\varphi_{,t}\varphi_{,t}^* - \varphi_{,x}\varphi_{,x}^*\right] - V(\varphi)\,, \tag{10.155}$$

where

$$V(\varphi) \equiv -\frac{1}{2}a^2\varphi\varphi^* + \frac{1}{4}b^2\varphi^2\varphi^{*2}\,. \tag{10.156}$$

with $a$, $b$, and $c$ constants.

(a) Sketch the potential $V(\varphi)$, this time as a function of the complex variable $\varphi$. (*Hint*: It should look like a familiar type of hat.)
(b) Show that the Lagrangian density is again invariant under a complex rotation of the fields

$$\varphi \to \varphi' = \mathrm{e}^{\mathrm{i}\lambda}\varphi\,, \qquad \varphi^* \to \varphi^{*\prime} = \mathrm{e}^{-\mathrm{i}\lambda}\varphi^*\,, \qquad \lambda \in \mathbb{R}\,. \tag{10.157}$$

(c) Write down the field equations for $\varphi$ and $\varphi^*$, and then find the subset of solutions having $\varphi = \text{const}$. How does this subset of solutions differ from what you found in part c of Problem 10.3?

**Problem 10.5** (*Adapted from an example in Goldstein et al. (2002).*) Sound waves in a gas can be described by the Lagrangian density

$$\mathscr{L} \equiv \frac{1}{2}\left[\mu_0|\dot{\boldsymbol{\eta}}|^2 - \gamma P_0(\nabla\cdot\boldsymbol{\eta})^2\right]\,, \tag{10.158}$$

where $\boldsymbol{\eta} = \boldsymbol{\eta}(\mathbf{r}, t)$ is a vector describing the displacement of the gas molecules away from their nominal positions $\mathbf{r}$ at time $t$; $\mu_0$ is the mean mass density of the gas; $P_0$ is the mean gas pressure; and $\gamma$ is the ratio of specific heats for the gas.

(a) Show that the Euler-Lagrange equations are

$$-\frac{1}{v^2}\ddot{\boldsymbol{\eta}} + \boldsymbol{\nabla}(\boldsymbol{\nabla} \cdot \boldsymbol{\eta}) = 0\,, \qquad v \equiv \sqrt{\frac{\gamma P_0}{\mu_0}}\,. \tag{10.159}$$

(b) The above equation can be written in a more familiar form if we replace the vector field $\boldsymbol{\eta}$ by the fractional change $\sigma$ in the mass density $\mu$, which is a scalar field defined by

$$\mu(\mathbf{r}, t) = \mu_0\,(1 + \sigma(\mathbf{r}, t))\,. \tag{10.160}$$

To do this, first show that conservation of mass in the fluid

$$\frac{\partial \mu}{\partial t} + \boldsymbol{\nabla} \cdot (\mu \mathbf{u}) = 0 \quad \Leftrightarrow \quad \sigma = -\boldsymbol{\nabla} \cdot \boldsymbol{\eta}\,, \tag{10.161}$$

where $\mathbf{u} \equiv \Delta\mathbf{s}/\Delta t = \boldsymbol{\eta}/\Delta t$ denotes the velocity of the gas molecules away from their nominal positions.

(c) Then by taking the divergence of (10.159), show that

$$-\frac{1}{v^2}\frac{\partial^2 \sigma}{\partial t^2} + \nabla^2 \sigma = 0\,, \tag{10.162}$$

which is the ordinary 3-dimensional wave equation for $\sigma$ (See also Problem 9.2).

**Problem 10.6** Consider the Lagrangian density

$$\mathscr{L} \equiv \frac{i\hbar}{2}(\Psi^*\dot{\Psi} - \Psi\dot{\Psi}^*) - \frac{\hbar^2}{2m}\boldsymbol{\nabla}\Psi^* \cdot \boldsymbol{\nabla}\Psi - U(\mathbf{r}, t)\Psi^*\Psi\,, \tag{10.163}$$

where $\Psi = \Psi(\mathbf{r}, t)$ is a complex field, $m$ is a constant with dimensions of mass, and $\hbar$ is Planck's constant $h$ divided by $2\pi$. Since $\Psi$ is complex, we can obtain its equation of motion by treating $\Psi$ and $\Psi^*$ as independent fields in the variational process, similar to what we did in Example 10.9.

(a) Show that the Euler-Lagrange equations obtained by varying $\Psi^*$ and $\Psi$ are

$$-\frac{\hbar^2}{2m}\nabla^2\Psi + U\Psi = i\hbar\frac{\partial \Psi}{\partial t}\,, \tag{10.164}$$

and

$$-\frac{\hbar^2}{2m}\nabla^2\Psi^* + U\Psi^* = -i\hbar\frac{\partial \Psi^*}{\partial t}\,, \tag{10.165}$$

respectively, which are complex conjugates of one another (as you would expect). Note that the equation for $\Psi$ is just the *Schrödinger equation* for a particle of mass $m$ in quantum mechanics (See (3.133)).

(b) Show that the Hamiltonian density for the above Lagrangian is

$$\mathscr{H} = \frac{\hbar^2}{2m}\nabla\Psi^* \cdot \nabla\Psi + U(\mathbf{r}, t)\Psi^*\Psi\,. \tag{10.166}$$

(c) Show that the Hamiltonian is

$$H \equiv \int_{\text{all space}} \mathrm{d}^3x\ \Psi^* \left(-\frac{\hbar^2}{2m}\nabla^2 + U\right)\Psi\,, \tag{10.167}$$

assuming that the $\Psi$ falls off sufficiently fast at infinity that we can ignore any boundary term that arises in calculating $H$. Note that $H$ has the interpretation of the *expectation value* of the Hamiltonian operator $\hat{H}$ in the state $|\Psi\rangle$ in the configuration representation (See (3.134)).

(d) Similar to what we found for the charged Klein-Gordon field in Example 10.9, show that the above Lagrangian density is invariant under the continuous field transformation

$$\Psi \to \Psi' = \mathrm{e}^{\mathrm{i}\lambda}\Psi\,, \qquad \Psi^* \to \Psi^{*\prime} = \mathrm{e}^{-\mathrm{i}\lambda}\Psi^*\,, \qquad \lambda \in \mathbb{R}\,, \tag{10.168}$$

and hence there exists a conserved current $J^\alpha$ and conserved charge $Q$. Show that

$$J^0 = \hbar|\Psi|^2\,, \qquad \mathbf{J} = \frac{i\hbar^2}{2m}\left(\Psi\nabla\Psi^* - \Psi^*\nabla\Psi\right)\,, \tag{10.169}$$

and

$$Q = \hbar\int_{\text{all space}} \mathrm{d}^3x\ |\Psi|^2\,. \tag{10.170}$$

Comment on $Q$ in light of the standard quantum-mechanical interpretation of $|\Psi(\mathbf{r}, t)|^2$ as the probability density of finding the particle in a small volume around $\mathbf{r}$.

# Chapter 11
# Special Relativity

Einstein's theory of **special relativity** extends Newtonian mechanics to the realm of objects moving with speeds close to or at the speed of light. Special relativity was developed in part by Lorentz, Poincaré, and Einstein to reconcile Newton's laws of mechanics with Maxwell's equations of electrodynamics with respect to measurements made by different inertial observers. Since its publication in 1905, special relativity has passed all observational tests; it is a cornerstone on which all other fundamental physical theories must be based. Special relativity can be thought of as a theory of the properties of *space and time* (**spacetime**) in the *absence* of gravitational fields. To properly incorporate gravity in a relativistic fashion, one needs to use Einstein's theory of **general relativity**, which extends the notion of an inertial frame to one that is *freely falling* in a gravitational field.

In this chapter we introduce the basic concepts of special relativity—i.e., spacetime, 4-vectors, the relativistic form of Newton's 2nd law, and the relativistic Lagrangian formulation, which could serve as the starting point for a course devoted to relativity or relativistic dynamics. For more details, readers should consult Mermin (2005), Taylor and Wheeler (1992), and the relevant chapters in books on general relativity, e.g., Schutz (2009) and Hartle (2003).

## 11.1 Why Do We Need Special Relativity?

As mentioned early on in Chap. 1, an inertial frame is *any* frame of reference in which Newton's laws of motion are valid. The emphasis on the word "any" means that there is no *preferred* inertial frame; all such frames are equally valid for Newton's laws. This seemingly simple observation—that Newton's laws of motion are valid in any inertial frame—can actually be promoted to a fundamental principle of physics if we simply replace "Newton's laws of motion" with "laws of physics", which includes not only mechanics, but also electricity and magnetism, optics, etc..



M.J. Benacquista and J.D. Romano, *Classical Mechanics*, Undergraduate Lecture Notes in Physics, https://doi.org/10.1007/978-3-319-68780-3_11

**Fig. 11.1** Two inertial reference frames for observers $O$ and $O'$, moving with respect to one another with constant speed $u$ in the common $x$ (or $x'$) direction

**Principle of relativity**: The laws of physics should be the same for all inertial observers.

To check whether or not this more general principle is valid, we need to know how measurements made in one inertial frame are related to those made in another inertial frame.

So let $O$ and $O'$ be two inertial frames that move with respect to one another with constant speed $u$ in the $x$-direction, as shown in Fig. 11.1, with the origins coinciding at $t = 0$. Then a **Galilean transformation** relates the time and space coordinates in the two frames $(t, x, y, z)$ and $(t', x', y', z')$ according to

$$\begin{aligned} t' &= t\,, \\ x' &= x - ut\,, \\ y' &= y\,, \\ z' &= z\,. \end{aligned} \tag{11.1}$$

The Galilean transformation is simply the coordinate transformation in (1.54), where we have now made *explicit* the *implicit* assumption that all clocks run at the same rate (i.e., there is a universal time that is valid in all inertial reference frames). Application of the Galilean transformation leads to the familiar "addition of velocities" formula

$$v = u + v' \qquad \text{(for Galilean transformation)}\,, \tag{11.2}$$

where $v$ and $v'$ are the speeds of a particle moving in the common $x$ (or $x'$) direction as measured by $O$ and $O'$, respectively. These were the transformation equations used since the time of Galileo, and in terms of these transformation equations, Newton's 2nd law $\mathbf{F} = m\mathbf{a}$ (or, equivalently, $\mathbf{F} = d\mathbf{p}/dt$) is invariant, as we discussed in Chap. 1.

Shortly after Maxwell formulated the final laws of electricity and magnetism (See (11.6)), it was found that they are not invariant under a Galilean transformation. In particular, the values of physical constants $\mu_0$ and $\varepsilon_0$ would depend upon the velocity $u$ of an inertial frame with respect to a *preferred* frame, defined by the

**ether** (a hypothetical substance through which electromagnetic waves were thought to travel). So Lorentz proposed instead an alternative set of transformation equations under which Maxwell's equations are invariant. A **Lorentz transformation** relates the two sets of inertial coordinates $(t, x, y, z)$ and $(t', x', y', z')$ via

$$\begin{aligned} t' &= \gamma(t - ux/c^2)\,, \\ x' &= \gamma(x - ut)\,, \\ y' &= y\,, \\ z' &= z\,, \end{aligned} \tag{11.3}$$

where

$$\gamma \equiv \frac{1}{\sqrt{1 - u^2/c^2}}\,, \tag{11.4}$$

and $c$ is the speed of light in vacuum. The major difference between Galilean and Lorentz transformations is that the time of an event as measured by two different inertial observers can differ according to (11.3), but not according to (11.1).

With respect to the Lorentz transformation equations, the "addition of velocites" formula (11.2) is replaced by

$$v = \frac{u + v'}{1 + uv'/c^2} \quad \text{(for Lorentz transformation)}\,, \tag{11.5}$$

which has a denominator that involves the speed of light $c$. Note that for speeds $u, v' \ll c$, the right-hand side of (11.5) is approximately $u + v'$, so (11.2) and (11.5) agree quite well for speeds that are small compared to the speed of light. But if $v' = c$, then for a Galilean transformation $v = u + c \neq c$ for any (non-zero) value of $u$, while $v = c$ (*always*) for a Lorentz transformation. Thus, the speed of light $c$ is a constant for all inertial observers according to Lorentz transformations. This is in agreement with the predictions of Maxwell's equations, which have $c = 1/\sqrt{\varepsilon_0 \mu_0}$, and the experimental confirmation of this by Michelson and Morley in 1887, making the existence of the ether an unnecessary hypothesis. As we shall see shortly, this result can also be promoted to a principle of physics.

**Principle of the constancy of the speed of light**: The speed of light in vacuum has the same value $c$ independent of the motion of the source or the observer.

**Exercise 11.1** (a) Derive (11.5) for the "addition of velocities" formula using the Lorentz transformation equations (11.3). (b) Suppose $u$ and $v'$ are both equal to 200 km/h. By how much does $v$ given by (11.5) differ from 400 km/h?

### 11.1.1 Conflict Between Newtonian Mechanics and Electrodynamics

As mentioned above, the equations of Newtonian mechanics (in particular Newton's 2nd law $\mathbf{F} = \mathrm{d}\mathbf{p}/\mathrm{d}t$) are invariant under a Galilean transformation (11.1), but not under a Lorentz transformation (11.3). On the other hand, Maxwell's equations of electrodynamics:[1]

$$\begin{aligned} \nabla \cdot \mathbf{E} &= \rho/\varepsilon_0 \,, \\ \nabla \times \mathbf{E} &= -\frac{\partial \mathbf{B}}{\partial t} \,, \\ \nabla \cdot \mathbf{B} &= 0 \,, \\ \nabla \times \mathbf{B} &= \mu_0 \varepsilon_0 \frac{\partial \mathbf{E}}{\partial t} + \mu_0 \mathbf{J} \,, \end{aligned} \tag{11.6}$$

are invariant under a Lorentz transformation, but not under a Galilean transformation. Hence, in order that the principle of relativity hold for *all* of physics (e.g., electrodynamics in addition to mechanics) we can either (i) keep the Galilean transformations and Newtonian mechanics as they are, and look for a modification of Maxwell's equations, or (ii) keep the Lorentz transformations and Maxwell's equations as they are, but change Newtonian mechanics. Einstein decided to do the latter, which is equivalent to *postulating* the constancy of the speed of light in vacuum for all inertial observers in addition to the principle of relativity. Taking this point of view, Einstein *derived* the Lorentz transformation equations as the necessary relationship between the coordinates of two inertial observers, in order that both the principle of the constancy of the speed of light and the principle of relativity hold.

In this sense, then, you can think of special relativity as simply a modification of Newtonian mechanics for which the equations are invariant with respect to Lorentz transformations. We shall see later on that this modification amounts to using a different expression for the momentum, $\mathbf{p} \equiv \gamma m \mathbf{v}$, in Newton's 2nd law $\mathbf{F} = \mathrm{d}\mathbf{p}/\mathrm{d}t$. The "special" part of special relativity corresponds to the fact that "all of physics" does not include gravitation. That addition comes later, in 1915, with Einstein's theory of general relativity.

[1]These equations for the electric and magnetic fields **E** and **B** are written in MKS units. The quantities $\rho$ and **J**, which appear on the right-hand side, are the electric charge density and electric current density, respectively, which act as sources for the fields. $\varepsilon_0$ is a physical constant called the *permitivity of free space* ($\varepsilon_0 = 8.85 \times 10^{-12}\ \mathrm{C}^2/\mathrm{N}\cdot\mathrm{m}^2$), and $\mu_0$ is the *permeability of free space* ($\mu_0 = 4\pi \times 10^{-7}\ \mathrm{N/A}^2$). See e.g., Griffiths (1999) for more details.

Since the speed of light $c$ is constant for all observers, we can treat it as a simple conversion factor, expressing time measurements in units of distance or vice-versa—i.e., 1 m of time is defined to be the time that it takes light to travel 1 m, or 1 s of distance is the distance that light travels in 1 s. In these units $c = 1$. Although it is often convenient to do calculations in these units, we will continue to treat time and space measurements as having different units, and hence will explicitly display the necessary factors of $c$ in all our equations.

## 11.2 *k* Calculus

There are many methods available for exploring the consequences of special relativity. In this text, we will use a method called the *k* **calculus**, developed by Hermann Bondi, see e.g., Bondi (1962). This is an appealing approach because it uses *radar methods* (i.e., sending and receiving pulses of electromagnetic radiation—e.g., light) to measure distances and time intervals between events in spacetime, allowing a single observer to map out the position of objects (and thus their trajectories) in spacetime. It is often convenient to describe spacetime by drawing a simplified 2-dimensional **spacetime diagram** as shown in Fig. 11.2, where time increases vertically upward and the spatial coordinate increases to the right. If we adopt standard MKS units such as seconds and meters, then the paths of light rays become almost indistinguishable from horizontal lines (this is why we perceive light to move nearly instantaneously in our everyday experience). However, if we plot either $ct$ versus $x$ (or $t$ versus $x/c$), then light rays travel on 45° degree lines in the spacetime diagram. *All* of our spacetime diagrams will be such that light travels on 45° lines.

The *k* calculus is based on the fact that distances can be measured using only a clock and light, which is a consequence of the constancy of the speed of light,

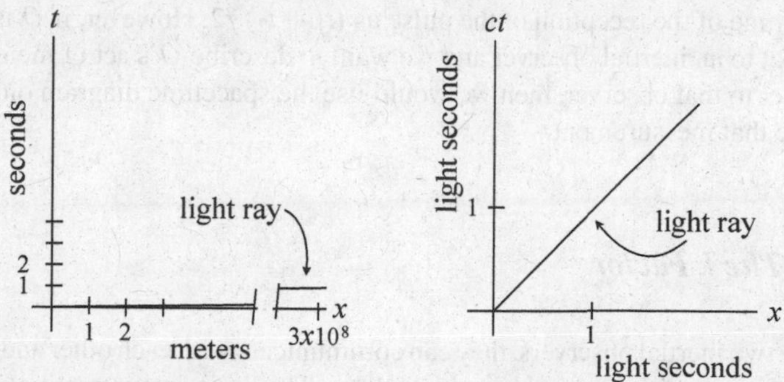


**Fig. 11.2** Spacetime diagrams using different units. Left-hand panel: In MKS units of meters and seconds, the path of a light ray rises by only one unit in time after $3 \times 10^8$ units in space. Right-hand panel: A light ray travels along a 45° path when *light-seconds* are used for both the spatial and temporal dimensions

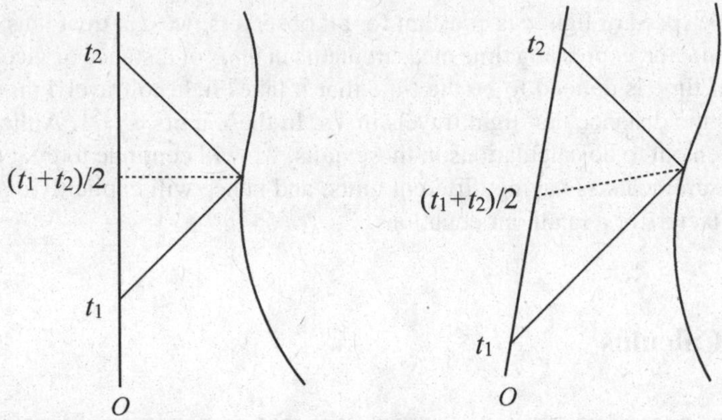


**Fig. 11.3** Determining the distance to an object using radar methods. Time $t$ increases upward, while the spatial coordinate $x$ increases to the right. $t_1$ and $t_2$ are the times of emission and reception of a pulse of light bounced off an object. The spacetime diagram on the left assumes that $O$ is at rest, while the one on the right assumes that $O$ has some velocity to the right

independent of the motion of the source or observer of light. To determine the distance to an object, an inertial observer $O$ can bounce a pulse of light off the object, noting the time of emission $t_1$ and time of reception $t_2$ of the pulse. The distance to the object and the time of reception of the pulse by the object (as measured by the observer) are then

$$x = \frac{1}{2}c(t_2 - t_1), \qquad t = \frac{1}{2}(t_1 + t_2). \tag{11.7}$$

The observer, who feels at rest, will draw the spacetime diagram shown on the left of Fig. 11.3 to describe this measurement, and it is obvious why $O$ would choose to define the time of the reception of the pulse as $(t_1 + t_2)/2$. However, if $O$ is moving with respect to an inertial observer and we want to describe $O$'s act of measurement with respect to that observer, then we would use the spacetime diagram on the right to describe that measurement.

### 11.2.1 The k Factor

If there are two inertial observers, they can communicate with each other and compare their descriptions of the spacetime around them. The basic requirement of relativity is that the descriptions of these two inertial observers agree under the relativity principles that we have described above. Imagine that observer $O$ communicates with observer $O'$ by sending two light pulses separated by a time interval $T$. Observer $O'$ will receive those two light pulses, but since the observers can be moving relative

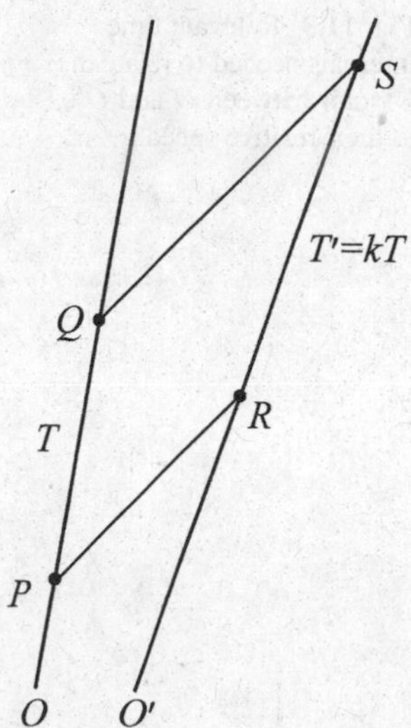


**Fig. 11.4** Definition of the $k$ factor relating two inertial observers, $O$ and $O'$, who are communicating by light pulses. Events $P$ and $R$, and $Q$ and $S$ are connected by pulses of light

to one another, the time separation of the received pulses can differ from $T$; let's call it $T'$, as shown in Fig. 11.4. The eponymous $k$ **factor** of the $k$ calculus is defined so that it relates $T$ and $T'$ through

$$T' = kT\,. \tag{11.8}$$

Note that if the inertial observers are at rest relative to one another (although they may be spatially separated), then $k = 1$. Since the only thing different between $O$ and $O'$ is their relative velocity $u$, it follows that $k$ is a function of $u$ (or more precisely $u/c$) alone. We will derive this relationship in Example 11.1. Based on this relationship, it follows that if $k$ is the factor relating time intervals for two inertial observers *moving away from another* with relative speed $u$, then $k^{-1}$ is the factor relating time intervals for two observers *approaching one another* with the same relative speed.

***Example 11.1*** Show that if inertial observer $O'$ is moving away from inertial observer $O$ with relative speed $u$, then

$$k = \sqrt{\frac{1+u/c}{1-u/c}}\,. \tag{11.9}$$

Consider the spacetime diagram shown in Fig. 11.5. By definition of the $k$ factor, $T' = kT$ and $t_2 = kT' = k^2T$. Thus, in the time interval between 0 and

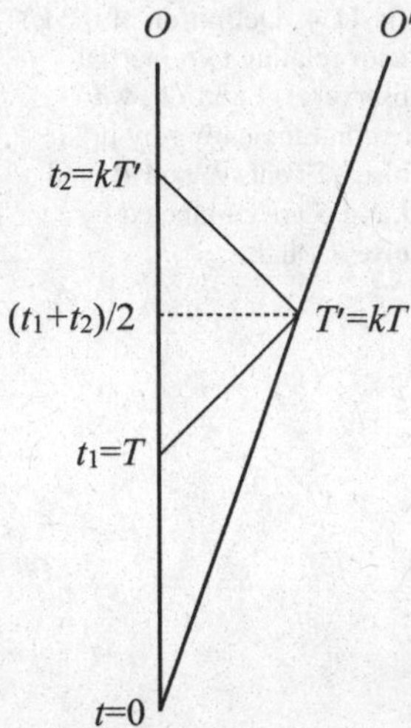


**Fig. 11.5** Relevant time intervals needed to relate the $k$ factor between $O$ and $O'$ to their relative speed $u$

$$t \equiv \frac{1}{2}(t_1 + t_2) = \frac{1}{2}T(1 + k^2)\,, \tag{11.10}$$

observer $O$ measures that observer $O'$ has moved a distance

$$x = ut = u\frac{1}{2}T(1 + k^2)\,. \tag{11.11}$$

But according to (11.7), we also have

$$x = \frac{1}{2}c(t_2 - t_1) = \frac{1}{2}cT(k^2 - 1)\,. \tag{11.12}$$

So equating these last two expressions for $x$, we can conclude that

$$\frac{u}{c} = \frac{k^2 - 1}{k^2 + 1}\,. \tag{11.13}$$

Then by simply cross-multiplying and rearranging terms in this last expression, we can solve for $k$ in terms of $u$, which is just (11.9). □

***Example 11.2*** Now, suppose there are three inertial observers $O$, $O'$, and $O''$. Observer $O$ sees $O'$ moving away in the positive $x$-direction at relative speed $u$, while observer $O'$ sees $O''$ moving away in the positive $x$-direction at relative speed $u'$. We will use the $k$ calculus to determine the relative speed $w$ at which observer $O$ sees $O''$ moving.

So let $k_u$ and $k_{u'}$ denote the $k$-factors relating $O$ and $O'$, and $O'$ and $O''$, respectively. Then the $k$-factor relating $O$ and $O''$ is simply the product

$$k = k_u k_{u'} \,. \tag{11.14}$$

From (11.13), we know that

$$\begin{aligned} \frac{w}{c} &= \frac{k^2-1}{k^2+1} = \frac{k_u^2 k_{u'}^2 - 1}{k_u^2 k_{u'}^2 + 1} = \frac{\left(\frac{1+u/c}{1-u/c}\right)\left(\frac{1+u'/c}{1-u'/c}\right) - 1}{\left(\frac{1+u/c}{1-u/c}\right)\left(\frac{1+u'/c}{1-u'/c}\right) + 1} \\ &= \frac{(1+u/c)(1+u'/c) - (1-u/c)(1-u'/c)}{(1+u/c)(1+u'/c) + (1-u/c)(1-u'/c)} = \frac{(u+u')/c}{1+uu'/c^2}\,, \end{aligned} \tag{11.15}$$

and thus we recover (11.5)—the relativistic velocity addition formula,

$$w = \frac{u+u'}{1+uu'/c^2}\,, \tag{11.16}$$

which we derived earlier (Exercise 11.1) using the Lorentz transformation equations. □

**Exercise 11.2** Show that

$$k + \frac{1}{k} = 2\gamma \quad \text{and} \quad k - \frac{1}{k} = 2\gamma\, u/c\,, \tag{11.17}$$

where $\gamma \equiv 1/\sqrt{1-u^2/c^2}$. Note that $\gamma \geq 1$ for all $u$.

**Exercise 11.3** Derive the Lorentz transformation equations relating $(ct, x)$ and $(ct', x')$ using the $k$-calculus formalism.
*Hint*: Consider an arbitrary point $P$ in spacetime with coordinates $(ct, x)$ and $(ct', x')$ with respect to inertial observers $O$ and $O'$, respectively. Let $t = t' = 0$ correspond to the intersection of the two world lines. As shown in Fig. 11.6, a pulse of light emitted from the spatial origin of $O$ at time $t_1 = t - x/c$ will be reflected by event $P$ and return to the origin of $O$ at time $t_2 = t + x/c$. The same pulse will be emitted and received by the spatial origin of $O'$ at $t_1' = t' - x'/c$ and $t_2' = t' + x'/c$, respectively. Relate the times $t - x/c$, $t' - x'/c$ and $t + x/c$, $t' + x'/c$ with the appropriate $k$-factors; then solve these equations for $(ct', x')$ in terms of $(ct, x)$, eliminating $k$ in favor of the relative speed $u$.

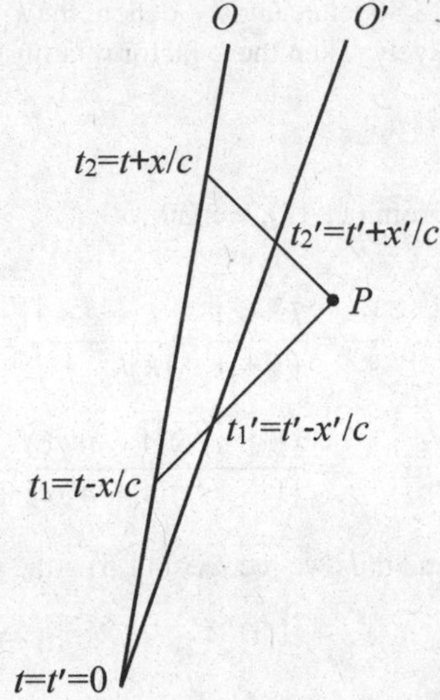


**Fig. 11.6** Spacetime diagram relating time and space measurements of an event $P$ as seen by two different inertial observers $O$ and $O'$

## 11.3 Some Consequences of Special Relativity

Because the Lorentz transformations involve both the spatial and temporal coordinates in spacetime, adopting the principles of relativity and the constancy of the speed of light introduces some consequences that differ from our more common-sense Galilean view of the universe. Here, we will highlight some of these consequences, which follow from simple application of either the Lorentz transformation equations (11.3) or the $k$-calculus formalism.

### *11.3.1 Lack of Absolute Simultaneity*

We can show qualitatively in Fig. 11.7 that different inertial observers $O$ and $O'$ will disagree on whether or not two events are *simultaneous*. We illustrate this in 2-dimensional Minkowski spacetime, showing the construction of the $t = \text{const}$ and $t' = \text{const}$ lines using the $k$-calculus radar method. Clearly, the dotted line showing events that all occur at $t = \text{const}$ for $O$ does not match the $t' = \text{const}$ line. For example, Fig. 11.8 shows three events that are simultaneous with respect to $O$ (since $t_P = t_Q = t_R$), but are not simultaneous with respect to $O'$ (since $t'_P > t'_Q > t'_R$). Thus, different observers can disagree about the temporal ordering of different events in spacetime.

Note that the constant time and constant space lines are *symmetric* with respect to the 45° lines on which light travels; this is true for all inertial observers. Note also that in the full 4-dimensional spacetime, the $t = \text{const}$ and $t' = \text{const}$ lines become

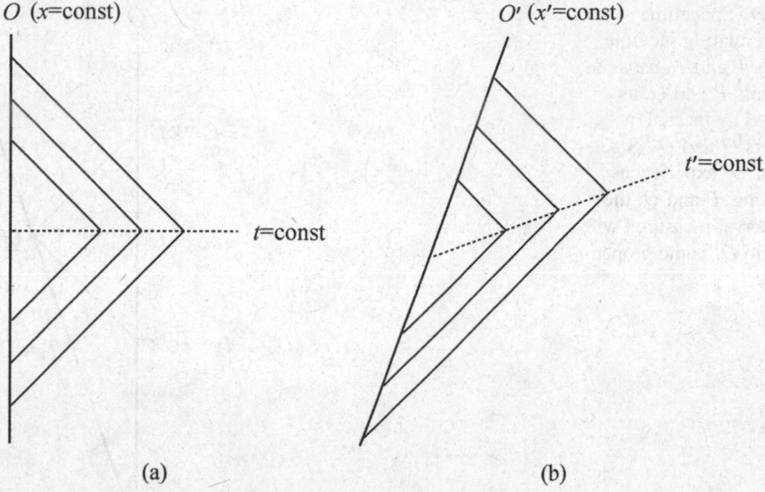


**Fig. 11.7** Lack of absolute simultaneity for two different inertial observers $O$ and $O'$ in 2-d Minkowski spacetime. The $t$ = const and $t'$ = const lines are different for the two inertial observers

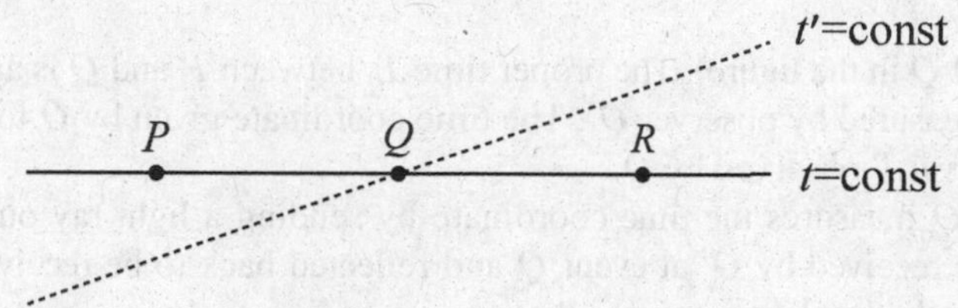


**Fig. 11.8** Three events, $P$, $Q$, $R$, which are simultaneous with respect to inertial observer $O$, but which are not simultaneous with respect to inertial observer $O'$

3-dimensional *surfaces*, which again do not agree with one another. We will return in Sect. 11.5.1.1 to the problem of simultaneity in relativity after we look in detail at how distances and times are described in different reference frames.

## 11.3.2 Time Dilation

To discuss time dilation, we first need to introduce a clock into the discussion. We can think of a clock as a stand-alone device that measures the rate at which time passes for an individual observer at rest with respect to the clock. This time is called the **proper time**. Now, let's consider a clock at rest with respect to an inertial observer $O'$, who is moving with speed $u$ relative to another inertial observer $O$. We want to determine the rate of time on the moving clock as measured by $O$. We construct the spacetime diagram shown in Fig. 11.9 to describe a time interval between two

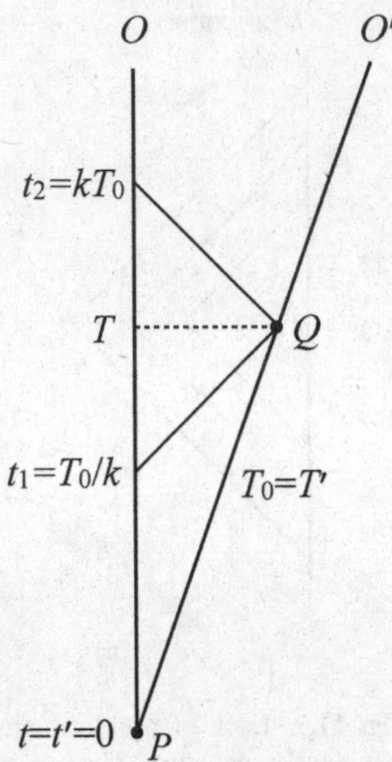


**Fig. 11.9** Spacetime diagram relating the time intervals $T$ and $T_0$ between two events $P$ and $Q$, as measured by inertial observers $O$ and $O'$. Since observer $O'$ experiences both events $P$ and $Q$, the time interval measured with respect to $O'$ is the proper time $T_0$

events ($P$ and $Q$ in the figure). The proper time $T_0$ between $P$ and $Q$ is also the time interval $T'$ measured by observer $O'$. The time coordinate given by $O$ to event $Q$ is the time interval $T$ observed by $O$.

Observer $O$ measures the time coordinate by sending a light ray out at time $t_1$ such that it is received by $O'$ at event $Q$ and reflected back to be received by $O$ at time $t_2$. Using the $k$ calculus, we see that

$$T_0 = T' = kt_1 \quad \Rightarrow \quad t_1 = T_0/k\,, \tag{11.18}$$

and

$$t_2 = kT_0\,. \tag{11.19}$$

Thus,

$$T = \frac{1}{2}(t_1 + t_2) = \frac{T_0}{2}\left(\frac{1}{k} + k\right)\,. \tag{11.20}$$

Using the result of Exercise 11.2, we find

$$T = \gamma T_0 > T_0\,, \tag{11.21}$$

where $\gamma \equiv \sqrt{1 - u^2/c^2}$. Since $\gamma \geq 1$, it follows that $T \geq T_0$. We interpret this result as showing that a clock moving with respect to an observer runs *slower* than a clock at rest with respect to that observer.

**Exercise 11.4** Rederive (11.21) for time dilation, but this time using the Lorentz transformation equations (11.3).

**Exercise 11.5** A clock is moving at speed 0.7 $c$ with respect to an inertial observer. After one year has passed on the observer's clock, how much time has passed on the moving clock?

### *11.3.3 Length Contraction*

Once we introduce an object into our spacetime, we also introduce special reference frames, called **proper frames** in which the object is at rest. If the object is not subject to any forces, then its proper frame is also an inertial frame. Here we are interested in determining how the measured length of an object depends on the choice of reference frame (or relative velocity between the object and an observer). So let's consider a rod oriented in the $x$-direction at rest with respect to inertial observer $O'$, who is moving with speed $u$ relative to inertial observer $O$. The **proper length** $L_0 \equiv L'$ is the length of the rod as measured by $O'$. The length $L$ of the rod as measured by $O$ is defined as the distance between the two ends of the rod measured at the *same time* by $O$.

We can determine the relationship between $L$ and $L_0$ through the $k$ calculus. We begin by drawing the relevant spacetime diagram, shown in Fig. 11.10. The rod, which is at rest with respect to $O'$, is represented as the thick black line along a line of constant $t' = 0$. One end of the rod remains at the spatial origin of $O'$ while the other end traces a parallel line through the spacetime. The length of the rod as measured by $O'$ is found from (11.7) to be

$$L_0 \equiv L' = \frac{1}{2}c(t_2' - t_1') . \tag{11.22}$$

The length of the rod as measured by $O$ is the distance between the two endpoints of the rod, measured at the *same* time in $O$'s reference frame, i.e., at

$$\frac{1}{2}(t_2 + t_3) = \frac{1}{2}(t_1 + t_4) . \tag{11.23}$$

From Fig. 11.10, we see that this is

$$L \equiv \frac{1}{2}c(t_4 - t_1) - \frac{1}{2}c(t_3 - t_2) . \tag{11.24}$$

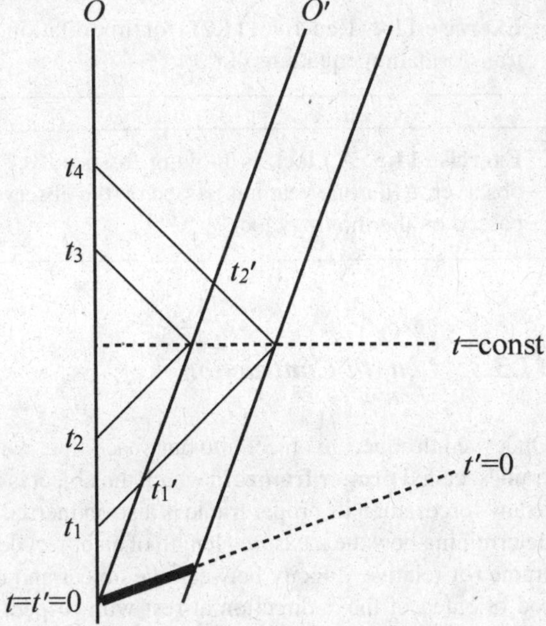


**Fig. 11.10** Spacetime diagram showing a rod (thick black line) at rest with respect to inertial observer $O'$, moving with speed $u$ relative to inertial observer $O$. $L_0$ is the length of the rod as measured with respect to $O'$

Now, we can use the following $k$-factor relations between the different time intervals:

$$t_1' = kt_1\,, \qquad t_4 = kt_2'\,, \qquad t_3 = k^2 t_2\,, \tag{11.25}$$

to obtain

$$t_2 = \left(\frac{1}{1+k^2}\right)(t_1 + t_4) \quad \Rightarrow \quad t_3 - t_2 = \left(\frac{k^2-1}{1+k^2}\right)(t_1 + t_4)\,. \tag{11.26}$$

Thus,

$$\begin{aligned}
L &= \frac{1}{2}c\left[(t_4 - t_1) - \left(\frac{k^2-1}{1+k^2}\right)(t_1 + t_4)\right] \\
&= \frac{1}{2}c\left(\frac{1}{1+k^2}\right)\left[(1+k^2)(t_4 - t_1) - (k^2-1)(t_1 + t_4)\right] \\
&= \left(\frac{1}{1+k^2}\right)c\left[t_4 - k^2 t_1\right] = \left(\frac{1}{1+k^2}\right)c\left[kt_2' - k^2 t_1'/k\right] \\
&= \left(\frac{2k}{1+k^2}\right)\frac{1}{2}c\,(t_2' - t_1') = \gamma^{-1} L_0\,,
\end{aligned} \tag{11.27}$$

where we used the result of Exercise 11.2 to get the last equality. So, we arrive at

$$L = \gamma^{-1} L_0 < L_0 \,. \tag{11.28}$$

Thus, *moving rods are contracted*—they are physically shorter when observed from a moving reference frame. (Distances perpendicular to the direction of the relative motion are not changed, however.)

**Exercise 11.6** Rederive (11.28) for length contraction, but this time using the Lorentz transformation equations (11.3).

### *11.3.4 Relativistic Doppler Effect*

We are familiar with the Doppler effect for sound, but when we include the relativistic effects on time, we find that the Doppler effect for electromagnetic waves (which travel at the speed of light) must be modified. To determine this modification, we consider a source of monochromatic electromagnetic radiation at rest with respect to an inertial observer $O$. If $f$ is the frequency of the radiation and $T \equiv 1/f$ is its period, as measured by $O$ in the proper frame of the source, then the frequency of the radiation as measured by an inertial observer $O'$, who is moving away from $O$ with relative speed $u$, is given by

$$f' \equiv \frac{1}{T'} = \frac{1}{kT} = f\sqrt{\frac{1-u/c}{1+u/c}}\,, \tag{11.29}$$

where we used (11.8) to relate the time intervals measured in the two inertial reference frames and (11.9) to express $k$ in terms of $u$. Thus, radiation is *red-shifted* (i.e., shifted to a *lower* frequency) when observed from an inertial reference frame moving away from the source with relative speed $u$. The relation

$$f' = f\sqrt{\frac{1-u/c}{1+u/c}} \tag{11.30}$$

is called the **relativistic Doppler effect**.

The relativistic Doppler effect also includes a **transverse relativistic Doppler effect**, which is attributable to time dilation. For motion *perpendicular* to the propagation of the electromagnetic wave, think of the source as a "clock" that ticks with period $T$, as measured by $O$. Then the period of the wave as measured by the moving observer $O'$ is *time dilated* relative to the source according to (11.21), leading to

$$f' = \frac{1}{T'} = \frac{1}{\gamma T} = \gamma^{-1} f = f\sqrt{1 - u^2/c^2}\,. \tag{11.31}$$

We will return to the relativistic Doppler effect at the end of Sect. 11.6 after defining the 4-momentum vector for a photon.

## 11.4 Lorentz Transformations and the Poincaré Group

The Lorentz transformation equations given in (11.3) correspond to a **boost** of an inertial reference frame in the $x$-direction. More generally, a Lorentz transformation can be either a boost in an *arbitrary* direction, see e.g., (11.47) below, or a spatial *rotation* around some axis $\hat{\mathbf{n}}$ in space. The set of such transformations form a 6-parameter group, called the **Lorentz group**, with three parameters needed to specify a boost (e.g., the components of the boost velocity $\mathbf{u}$) and three parameters needed to specify a rotation (e.g., the Euler angles $(\phi, \theta, \psi)$, or an axis $\hat{\mathbf{n}}$ and rotation angle $\Psi$). Note that we need to include both boosts and spatial rotations as elements of the Lorentz group, since the composition of two boosts in arbitrary directions is not another boost, but rather a boost followed by a spatial rotation (see Sect. 11.4.1 and Problem 11.7 for more details).

In addition to being invariant under boosts, the laws of physics are also invariant under spatial rotations and spacetime *translations*, with the translations simply shifting the spatial origin and/or the zero of the time coordinate:

$$\begin{aligned} t' &= t - t_0\,, \\ x' &= x - x_0\,, \\ y' &= y - y_0\,, \\ z' &= z - z_0\,. \end{aligned} \tag{11.32}$$

The set of spacetime translations form a 4-parameter group. Together, the set of Lorentz transformations and spacetime translations form a 10-parameter group called the **Poincaré group** or **inhomogeneous Lorentz group**.

***Example 11.3*** Here we show that Lorentz boosts in spacetime are analogous to rotations in ordinary Euclidean space.

Recall from (6.12) that (passive) rotations through an angle $\theta$ in the $xy$-plane can be written as

$$\begin{aligned} x' &= x \cos\theta + y \sin\theta, \\ y' &= -x \sin\theta + y \cos\theta. \end{aligned} \tag{11.33}$$

But instead of using the angle $\theta$ to describe the rotation, we can also introduce the **slope parameter**

$$m \equiv \tan\theta. \tag{11.34}$$

Thus,

$$\cos\theta = \frac{1}{\sqrt{1+m^2}}, \qquad \sin\theta = \frac{m}{\sqrt{1+m^2}}, \tag{11.35}$$

and so the rotation can be written in terms of $m$ as

$$\begin{aligned} x' &= \frac{1}{\sqrt{1+m^2}}(x+my), \\ y' &= \frac{1}{\sqrt{1+m^2}}(-mx+y). \end{aligned} \tag{11.36}$$

Now compare these expressions to the transformation for a Lorentz boost in the $x$-direction

$$\begin{aligned} x' &= \gamma(x-ut), \\ t' &= \gamma(-ux/c^2+t), \end{aligned} \tag{11.37}$$

where

$$\gamma \equiv \frac{1}{\sqrt{1-u^2/c^2}}. \tag{11.38}$$

If we use $ct$ and $ct'$ for the time coordinate in the two-dimensional $x$-$ct$ plane in spacetime, and introduce the dimensionless parameter

$$\beta \equiv \frac{u}{c}, \tag{11.39}$$

which gives the ratio of the boost velocity $u$ to the speed of light $c$, then the Lorentz boost can be written as

$$\begin{aligned} x' &= \frac{1}{\sqrt{1-\beta^2}}(x-\beta ct), \\ ct' &= \frac{1}{\sqrt{1-\beta^2}}(-\beta x+ct). \end{aligned} \tag{11.40}$$

Thus, $\beta$ is analogous to the slope parameter for rotations, with (11.36) and (11.40) having the same form (except for a few minus signs).

There is a more direct analogy between spatial rotations and boosts if we use hyperbolic trigonometry in spacetime and define the **hyperbolic angle** $\chi$ (also known as the **rapidity**) by

$$\beta \equiv \tanh \chi \,. \tag{11.41}$$

Then

$$\cosh \chi = \frac{1}{\sqrt{1-\beta^2}} = \gamma \,, \qquad \sinh \chi = \frac{\beta}{\sqrt{1-\beta^2}} = \beta\gamma \,, \tag{11.42}$$

so

$$\begin{aligned} x' &= x \cosh \chi - ct \sinh \chi \,, \\ ct' &= -x \sinh \chi + ct \cosh \chi \,. \end{aligned} \tag{11.43}$$

This is similar to the transformation equations (11.33) for a rotation through an angle $\theta$. The only difference is the use of hyperbolic trig functions in place of circular trig functions, and a hyperbolic angle $\chi$ in place of the circular angle $\theta$.[2]

The analogy can be carried further if we look at the slope factor for the combination of two rotations through an angle $\theta_1$ followed by an angle $\theta_2$. If $m_1 = \tan\theta_1$ and $m_2 = \tan\theta_2$, then

$$m \equiv \tan(\theta_1 + \theta_2) = \frac{\tan\theta_1 + \tan\theta_2}{1 - \tan\theta_1\, \tan\theta_2} = \frac{m_1 + m_2}{1 - m_1 m_2} \,. \tag{11.44}$$

The velocity addition formula for boosts is

$$u = \frac{u_1 + u_2}{1 + u_1 u_2/c^2} \quad \Rightarrow \quad \beta = \frac{\beta_1 + \beta_2}{1 + \beta_1\beta_2} \,, \tag{11.45}$$

which is equivalent to

$$\frac{\tanh\chi_1 + \tanh\chi_2}{1 + \tanh\chi_1\, \tanh\chi_2} = \tanh\left(\chi_1 + \chi_2\right) \,. \tag{11.46}$$

Thus, the composition of two boosts in the same direction is equivalent to two hyperbolic rotations through successive hyperbolic angles $\chi_1$ and $\chi_2$. □

[2] There is also an additional minus sign that comes from the fact that $\cosh^2\chi - \sinh^2\chi = 1$, while $\cos^2\theta + \sin^2\theta = 1$.

### 11.4.1 Boosts in an Arbitrary Direction

The Lorentz boost transformations produce length contraction in the direction of motion as well as time dilation for all moving inertial frames, but they leave lengths unchanged that are perpendicular to the direction of motion. Thus, for a boost in an arbitrary direction, we need to distinguish between components of a vector that are *parallel to* or *perpendicular to* the boost velocity $\mathbf{u}$. We define these as $\mathbf{r}_\parallel$ and $\mathbf{r}_\perp$, respectively. We can then write the transformation equations for a boost of velocity $\mathbf{u}$ in an arbitrary direction as

$$\begin{aligned} ct' &= \gamma(ct - \mathbf{u}\cdot\mathbf{r}/c)\,, \\ \mathbf{r}' &= \gamma(\mathbf{r}_\parallel - \mathbf{u}t) + \mathbf{r}_\perp\,, \end{aligned} \tag{11.47}$$

where

$$\mathbf{r}_\parallel \equiv \hat{\mathbf{u}}(\hat{\mathbf{u}}\cdot\mathbf{r})\,, \qquad \mathbf{r}_\perp \equiv \mathbf{r} - \mathbf{r}_\parallel\,, \qquad \hat{\mathbf{u}} \equiv \mathbf{u}/u\,, \tag{11.48}$$

and $\mathbf{r}$, $\mathbf{r}'$ are the position vectors of a point $P$ with cordinates $(x, y, z)$, $(x', y', z')$ with respect to the two inertial frames. The above transformation generalizes (11.3) for a boost by $u$ in the $x$-direction. Using the above definitions of $\mathbf{r}_\parallel$ and $\mathbf{r}_\perp$, the Lorentz transformation equations can also be written as

$$\begin{aligned} ct' &= \gamma(ct - \mathbf{u}\cdot\mathbf{r}/c)\,, \\ \mathbf{r}' &= \mathbf{r} + (\gamma - 1)\frac{(\mathbf{u}\cdot\mathbf{r})\mathbf{u}}{u^2} - \gamma\mathbf{u}t\,. \end{aligned} \tag{11.49}$$

We can write the Lorentz transformation equations in matrix form

$$\mathsf{X}' = \Lambda\mathsf{X} \quad \Leftrightarrow \quad x^{\alpha'} = \sum_\alpha \Lambda^{\alpha'}{}_\alpha x^\alpha\,, \tag{11.50}$$

where $x^\alpha \equiv (x^0, x^1, x^2, x^3) \equiv (ct, x, y, z)$ and similarly for $x^{\alpha'}$. Expanding the components of $\Lambda^{\alpha'}{}_\alpha$, we have

$$\begin{bmatrix} ct' \\ x' \\ y' \\ z' \end{bmatrix} = \begin{bmatrix} \gamma & -\gamma u_x/c & -\gamma u_y/c & -\gamma u_z/c \\ -\gamma u_x/c & 1+(\gamma-1)\frac{u_x^2}{u^2} & (\gamma-1)\frac{u_x u_y}{u^2} & (\gamma-1)\frac{u_x u_z}{u^2} \\ -\gamma u_y/c & (\gamma-1)\frac{u_x u_y}{u^2} & 1+(\gamma-1)\frac{u_y^2}{u^2} & (\gamma-1)\frac{u_y u_z}{u^2} \\ -\gamma u_z/c & (\gamma-1)\frac{u_x u_z}{u^2} & (\gamma-1)\frac{u_y u_z}{u^2} & 1+(\gamma-1)\frac{u_z^2}{u^2} \end{bmatrix} \begin{bmatrix} ct \\ x \\ y \\ z \end{bmatrix}, \tag{11.51}$$

for a Lorentz boost in an arbitrary direction.

As we showed earlier, the composition of two boosts by velocities $\mathbf{u}$ and $\mathbf{u}'$ in the *same direction* (i.e., $\mathbf{u} \parallel \mathbf{u}'$) is another Lorentz boost by velocity

$$\mathbf{w} = \frac{\mathbf{u} + \mathbf{u}'}{1 + \mathbf{u} \cdot \mathbf{u}'/c^2} . \tag{11.52}$$

If the velocities $\mathbf{u}$ and $\mathbf{u}'$ are not parallel, then the composition of two boosts in these directions is not another boost, but rather a Lorentz tranformation consisting of a boost *followed* by a rotation about an axis perpendicular to the plane spanned by $\mathbf{u}$ and $\mathbf{u}'$. This rotation is called a **Wigner rotation** or **Thomas rotation**, and it is investigated in Problem 11.7.

**Exercise 11.7** Show that (11.47) reduces to (11.3) for $\mathbf{u} = u\hat{\mathbf{x}}$.

## 11.4.2 Transformation of the Velocity Vector

With the full set of coordinate transformations in hand, we can now look at how two observers can compare the velocities that they measure of moving objects. Let's consider a boost in the $x$-direction with speed $u$, given by (11.3). The velocity is defined by each observer as usual, so $\mathbf{v} \equiv \mathrm{d}\mathbf{r}/\mathrm{d}t$ and $\mathbf{v}' \equiv \mathrm{d}\mathbf{r}'/\mathrm{d}t'$. In our simple case,

$$\begin{aligned}
\mathrm{d}t' &= \frac{\partial t'}{\partial x}\mathrm{d}x + \frac{\partial t'}{\partial t}\mathrm{d}t , \\
\mathrm{d}x' &= \frac{\partial x'}{\partial x}\mathrm{d}x + \frac{\partial x'}{\partial t}\mathrm{d}t , \\
\mathrm{d}y' &= \mathrm{d}y , \\
\mathrm{d}z' &= \mathrm{d}z .
\end{aligned} \tag{11.53}$$

Thus, the transformation equations for the components of the ordinary 3-velocity vector $\mathbf{v} \equiv \mathrm{d}\mathbf{r}/\mathrm{d}t$ are given by

$$\begin{aligned}
v^{x'} &= \frac{\mathrm{d}x'}{\mathrm{d}t'} = \frac{v^x - u}{1 - uv^x/c^2} , \\
v^{y'} &= \frac{\mathrm{d}y'}{\mathrm{d}t'} = \frac{v^y}{\gamma(1 - uv^x/c^2)} , \\
v^{z'} &= \frac{\mathrm{d}z'}{\mathrm{d}t'} = \frac{v^z}{\gamma(1 - uv^x/c^2)} .
\end{aligned} \tag{11.54}$$

Note that the tranformation equation for $v^{x'}$ is equivalent that derived earlier in the context of the composition of two boosts in the same direction.

**Exercise 11.8** Verify the transformation equations (11.54) for the velocity vector $\mathbf{v} = \mathrm{d}\mathbf{r}/\mathrm{d}t$.

## 11.5 Spacetime Line Element and 4-Vectors

In ordinary 3-dimensional Euclidean space, the line element $\mathrm{d}s^2$ is the infinitesimal squared displacement between two nearby points. In Cartesian coordinates $x^i \equiv (x, y, z)$, it has the form

$$\mathrm{d}s^2 = \mathrm{d}x^2 + \mathrm{d}y^2 + \mathrm{d}z^2 . \tag{11.55}$$

This expression is invariant under rotations, since rotations preserve distances in Euclidean space. For the 4-dimensional spacetime of special relativity, the **spacetime line element** (also denoted $\mathrm{d}s^2$) specifies proper distances and proper times (called **spacetime intervals**) between two nearby events. These spacetime intervals should be preserved under Lorentz transformations, just as spatial distances are preserved under ordinary rotations. In Cartesian coordinates $x^\alpha \equiv (ct, x, y, z)$, the spacetime line element that satifies this condition has the form

$$\mathrm{d}s^2 = -c^2\mathrm{d}t^2 + \mathrm{d}x^2 + \mathrm{d}y^2 + \mathrm{d}z^2 = \sum_{\alpha,\beta} \eta_{\alpha\beta}\mathrm{d}x^\alpha \mathrm{d}x^\beta , \tag{11.56}$$

where

$$\eta_{\alpha\beta} \quad \Leftrightarrow \quad \eta = \mathrm{diag}(-1, 1, 1, 1) = \begin{bmatrix} -1 & 0 & 0 & 0 \\ 0 & 1 & 0 & 0 \\ 0 & 0 & 1 & 0 \\ 0 & 0 & 0 & 1 \end{bmatrix} . \tag{11.57}$$

The minus sign in front of $c^2\mathrm{d}t^2$ means that the geometry is **non-Euclidean**. Despite the notation, the spacetime line element $\mathrm{d}s^2$ can be either positive, negative, or zero. Spacetime with this line element is called **Minkowski spacetime**, and the geometry is said to be **Lorentzian**. The matrix elements $\eta_{\alpha\beta}$ are the components of the **Minkowski metric** in Cartesian coordinates. The Minkowski metric $\eta_{\alpha\beta} = \mathrm{diag}(-1, 1, 1, 1)$ is the spacetime analog of the 3-dimensional Kronecker delta $\delta_{ij} = \mathrm{diag}(1, 1, 1)$.

As anticipated, the spacetime line element is invariant under Lorentz transformations and spacetime translations—i.e., if $(ct, x, y, z)$ and $(ct', x', y', z')$ are Cartesian coordinates associated with two different inertial reference frames, then

$$\mathrm{d}s^2 = -c^2\mathrm{d}t^2 + \mathrm{d}x^2 + \mathrm{d}y^2 + \mathrm{d}z^2 = -c^2\mathrm{d}t'^2 + \mathrm{d}x'^2 + \mathrm{d}y'^2 + \mathrm{d}z'^2 . \tag{11.58}$$

This is easy to show for spacetime translations (11.32), and we already know that spatial rotations (which have $t' = t$) preserve the spatial part of the line element. For a Lorentz boost, the invariance of the spacetime line element is equivalent to

$$\sum_{\alpha',\beta'} \eta_{\alpha'\beta'} \Lambda^{\alpha'}{}_{\alpha} \Lambda^{\beta'}{}_{\beta} = \eta_{\alpha\beta} \tag{11.59}$$

where the matrix elements $\Lambda^{\alpha'}{}_{\alpha}$ are given by (11.51).

**Exercise 11.9** Verify (11.59).

**Exercise 11.10** In 2-dimensional Euclidean space, the set of points that are equidistant from the origin form a series of concentric circles $x^2 + y^2 = \text{const}$. Show that the set of points equidistant (with respect to the spacetime line element) from the origin of 2-d Minkowski spacetime are hyperbolae, e.g., $-c^2t^2 + x^2 = \text{const}$.

### *11.5.1 Causal Structure*

The spacetime line element (11.56) allows us to define the **causal structure** of Minkowski spacetime. It specifies, for example, whether one event can physically influence another. There are three possibilities:

(i) Two nearby events in spacetime are said to be **null related** if

$$\mathrm{d}s^2 = 0 . \tag{11.60}$$

Thus, $0 = -c^2\mathrm{d}t^2 + |\mathrm{d}\mathbf{r}|^2$, or, equivalently, $|\mathrm{d}\mathbf{r}/\mathrm{d}t| = c$, which means that an object moving along the infinitesimal world line connecting the two nearby events must travel at the speed of light. Since null-related events are connected by a light ray, we can also say that such events are **lightlike related**.

(ii) Two nearby events in spacetime are said to be **spacelike related** if

$$\mathrm{d}s^2 > 0 . \tag{11.61}$$

For this case, an object traveling along the infinitesimal world line connecting the two nearby events must travel *faster* than the speed of light. Such an object is called a **tachyon**. Since no such object has ever been observed, spacelike-related events *cannot* physically influence one another.

(iii) Two nearby events in spacetime are said to be **timelike related** if

$$ds^2 < 0\,. \tag{11.62}$$

For this case, an object moving along the infinitesimal world line connecting the two nearby events would move at *less* than the speed of light. A world line for which $ds^2 < 0$ everywhere along it is said to be a **timelike world line**. (Note that this world line need not be straight.) Actual physical particles trace out timelike world lines in spacetime.

#### 11.5.1.1 Light Cones

At each event $P$ in spacetime, one can draw the future and past **light cones** of $P$. The sides of a light cone are generated by 45° lines along which $ds^2 = 0$. For 4-dimensional spacetime, the $t =$ const cross-sections of a light cone are 2-spheres. But since at least one spatial dimension is supressed when drawing a spacetime diagram, the cross-sections of a light cone are typically represented by circles (e.g., 1-spheres) as shown in Fig. 11.11.

Since every inertial observer agrees that all events in the future light cone of $P$ occur after $P$, the future light cone of $P$ is called the **absolute future** of $P$. Similarly, every inertial observer agrees that all events in the past light cone of $P$ occur before $P$, so the past light cone of $P$ is called the **absolute past** of $P$. Events that lie outside

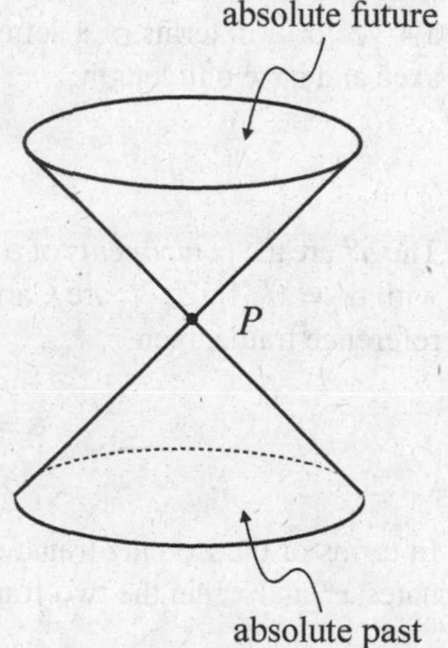


**Fig. 11.11** Spacetime diagram for the light cone for event $P$. Time increases vertically upward; space extends horizontally. Since one spatial dimension is suppressed, $t =$ const cross-sections of the light cone are represented here as cirles; in 4-dimensional spacetime, these circles are actually 2-spheres

the future and past light cones of $P$ are said to lie in the **relative past** or **relative future** of $P$, since different inertial observers will disagree on whether such events occurred before, after, or simultaneous with $P$ (See Sect. 11.3.1).

As information cannot travel faster than the speed of light (in vacuum), event $P$ can only influence events inside or on its future light cone. For the same reason, event $P$ can only have been influenced by events inside or on its past light cone. Events outside the future or past light cone of $P$ cannot influence $P$, nor can $P$ influence them. For this reason, the region outside of the future and past light cones of $P$ is also called the **absolute elsewhere** for event $P$. The set of future and past light cones at *all* points in spacetime fully specify the causal structure of spacetime.

### *11.5.2 4-Vectors and Inner Product*

Much like a 3-dimensional displacement vector is a directed line segment joining two points in Euclidean space, a **4-vector** in Minkowski spacetime is simply a directed line segment connecting two spacetime events $P$ and $Q$. We will use a san-serif symbol, like $\mathsf{a}$, or index notation, like $a^\alpha$, to denote 4-vectors. We will continue to denote ordinary 3-dimensional vectors by a boldface symbol $\mathbf{a}$, or by its components $a^i$. We can add two 4-vectors, multiply a 4-vector by a scalar, subtract two 4-vectors, ..., just like we do with ordinary 3-dimensional vectors in Euclidean space using the parallelogram rule, etc. (See Appendix D for information about vectors in arbitrary $n$-dimensional vector spaces.) The only point of concern is that we should distinguish between 4-vectors with superscripts $a^\alpha$ and 4-vectors with subscripts $a_\alpha$, since the Minkowski metric has as $-1$ in its time-time component with respect to an inertial frame, $\eta_{\alpha\beta} = \mathrm{diag}(-1, 1, 1, 1)$. See Appendix A.2 for more details about the difference between such vectors.

If $x^\alpha \equiv (ct, x, y, z)$, with $\alpha = 0, 1, 2, 3$, denote Cartesian coordinates associated with an inertial reference frame in Minkowski spacetime, then we can decompose a 4-vector $\mathsf{a}$ in terms of a set of basis vectors $\mathsf{e}_\alpha$, which point along the coordinate axes and have unit length:

$$\mathsf{a} = \sum_\alpha a^\alpha \mathsf{e}_\alpha \,. \tag{11.63}$$

The $a^\alpha$ are the *components* of $\mathsf{a}$ with respect to the basis $\{\mathsf{e}_\alpha\}$. If $x^{\alpha'} \equiv (ct', x', y', z')$, with $\alpha' = 0', 1', 2', 3'$, are Cartesian coordinates associated with a different inertial reference frame, then

$$\mathsf{a} = \sum_{\alpha'} a^{\alpha'} \mathsf{e}_{\alpha'} = \sum_\alpha a^\alpha \mathsf{e}_\alpha \,. \tag{11.64}$$

In terms of the Lorentz transformation matrices $\Lambda^{\alpha'}{}_\alpha$ and $\Lambda^\alpha{}_{\alpha'}$ relating the coordinates $x^\alpha$ and $x^{\alpha'}$ in the two frames, we have

$$a^{\alpha'} = \sum_{\alpha} \Lambda^{\alpha'}{}_{\alpha} a^{\alpha} , \qquad \mathsf{e}_{\alpha'} = \sum_{\alpha} \mathsf{e}_{\alpha} \Lambda^{\alpha}{}_{\alpha'} . \tag{11.65}$$

This follows as a simple consequence of $\Lambda^{\alpha'}{}_{\alpha}$ and $\Lambda^{\alpha}{}_{\alpha'}$ being inverses of one another:

$$\delta^{\alpha}{}_{\beta} = \sum_{\alpha'} \Lambda^{\alpha}{}_{\alpha'} \Lambda^{\alpha'}{}_{\beta} , \qquad \delta^{\alpha'}{}_{\beta'} = \sum_{\alpha} \Lambda^{\alpha'}{}_{\alpha} \Lambda^{\alpha}{}_{\beta'} . \tag{11.66}$$

Note that the first equation in (11.65) relates the components of the spacetime vector $\mathsf{a}$ with respect to two different bases, while the second equation in (11.65) relates the two different sets of basis vectors themselves.

**Exercise 11.11** Show that for a Lorentz boost with speed $u$ in the $x$-direction, the vector components $a^{\alpha'}$ and $a^{\alpha}$ are related by

$$\begin{aligned} a^{0'} &= \gamma(a^0 - ua^1/c) , \\ a^{1'} &= \gamma(a^1 - ua^0/c) , \\ a^{2'} &= a^2 , \\ a^{3'} &= a^3 , \end{aligned} \tag{11.67}$$

where $\gamma \equiv 1/\sqrt{1-u^2/c^2}$. Note that these are the same as the transformation equations (11.3) relating the coordinates $x^{\alpha}$ to $x^{\alpha'}$.

#### 11.5.2.1 Inner Product

As discussed in Appendix D.3, an **inner product** is a mapping that takes two vectors $\mathsf{a}$ and $\mathsf{b}$ and returns a scalar $\mathsf{a} \cdot \mathsf{b}$ subject to the conditions of commutativity, positivity, and linearity, (D.24a), (D.24b), and (D.24c). For Minkowski spacetime, we require that the inner product of the infinitesimal displacement vector $\mathsf{dx}$ with itself equal the spacetime interval, i.e.,

$$\mathsf{dx} \cdot \mathsf{dx} = \mathrm{d}s^2 = -c^2 \mathrm{d}t^2 + \mathrm{d}x^2 + \mathrm{d}y^2 + \mathrm{d}z^2 . \tag{11.68}$$

Using linearity of the inner product, the above requirement is equivalent to

$$\mathsf{e}_{\alpha} \cdot \mathsf{e}_{\beta} = \mathrm{diag}(-1, 1, 1, 1) = \eta_{\alpha\beta} , \tag{11.69}$$

which equates the inner product of the Cartesian basis vectors with the Minkowski metric components $\eta_{\alpha\beta}$. To prove this result, write $\mathsf{dx} = \sum_\alpha \mathrm{d}x^\alpha\, \mathsf{e}_\alpha$. Then linearity of the inner product gives

$$\mathsf{dx}\cdot\mathsf{dx} = \left(\sum_\alpha \mathrm{d}x^\alpha\, \mathsf{e}_\alpha\right)\cdot\left(\sum_\beta \mathrm{d}x^\beta\, \mathsf{e}_\beta\right) = \sum_{\alpha,\beta} \mathrm{d}x^\alpha \mathrm{d}x^\beta \left(\mathsf{e}_\alpha\cdot\mathsf{e}_\beta\right). \qquad (11.70)$$

But according to (11.56),

$$\mathrm{d}s^2 = -c^2\mathrm{d}t^2 + \mathrm{d}x^2 + \mathrm{d}y^2 + \mathrm{d}z^2 = \sum_{\alpha,\beta} \eta_{\alpha\beta}\mathrm{d}x^\alpha \mathrm{d}x^\beta . \qquad (11.71)$$

So $\mathsf{dx}\cdot\mathsf{dx} = \mathrm{d}s^2$ if and only if $\mathsf{e}_\alpha\cdot\mathsf{e}_\beta = \eta_{\alpha\beta}$ as claimed.

Proceeding in a similar fashion, it is also easy to show that

$$\mathsf{a}\cdot\mathsf{b} = \sum_{\alpha,\beta} \eta_{\alpha\beta} a^\alpha b^\beta = -a^0b^0 + a^1b^1 + a^2b^2 + a^3b^3 . \qquad (11.72)$$

for any two spacetime 4-vectors $\mathsf{a}$ and $\mathsf{b}$. As usual, we will say that two 4-vectors $\mathsf{a}$ and $\mathsf{b}$ are **orthogonal** if $\mathsf{a}\cdot\mathsf{b} = 0$. The "squared" **norm** of a 4-vector $\mathsf{a}$ is defined to be $\mathsf{a}\cdot\mathsf{a}$. Since this quantity may be positive, negative, or zero, the "squared" terminology should not be taken in its usual sense for real numbers. Along these same lines, we will say that a 4-vector $\mathsf{a}$ is *normalized* if $\mathsf{a}\cdot\mathsf{a} = \pm 1$.

Just like the line element, the inner product of 4-vectors is invariant under a Lorentz transformation, so

$$\mathsf{a}\cdot\mathsf{b} = -a^0b^0 + a^1b^1 + a^2b^2 + a^3b^3 = -a^{0'}b^{0'} + a^{1'}b^{1'} + a^{2'}b^{2'} + a^{3'}b^{3'} , \qquad (11.73)$$

or, equivalently,

$$\mathsf{e}_\alpha\cdot\mathsf{e}_\beta = \mathrm{diag}(-1,1,1,1) = \mathsf{e}_{\alpha'}\cdot\mathsf{e}_{\beta'} , \qquad (11.74)$$

for any two sets of basis vectors $\{\mathsf{e}_\alpha\}$ and $\{\mathsf{e}_{\alpha'}\}$ corresponding to different inertial reference frames.

**Exercise 11.12** Prove (11.73) using (11.59) and (11.65).

### *11.5.3 Proper Time*

We have already defined proper time as the time measured by a clock at rest with respect to an inertial observer. We can extend this to non-inertial observers by defining the **proper time** $\mathrm{d}\tau$ between two infinitesimally nearby *timelike-related* events to be

$$c\,\mathrm{d}\tau \equiv \sqrt{-\mathrm{d}s^2}\,. \tag{11.75}$$

Since the events are timelike related, the value of line element $\mathrm{d}s^2 < 0$, so taking the above square-root is not a problem. Expanding $\mathrm{d}s^2$ and factoring out $c\,\mathrm{d}t$, we have

$$c\,\mathrm{d}\tau = \sqrt{c^2\mathrm{d}t^2 - \mathrm{d}x^2 - \mathrm{d}y^2 - \mathrm{d}z^2} = c\,\mathrm{d}t\,\sqrt{1-(v_x^2+v_y^2+v_z^2)/c^2} = \gamma^{-1}\,c\,\mathrm{d}t\,, \tag{11.76}$$

where $v^i \equiv \mathrm{d}x^i/\mathrm{d}t$ are the components of an ordinary 3-velocity vector $\mathbf{v}$ and $\gamma \equiv 1/\sqrt{1-v^2/c^2}$. Thus,

$$\mathrm{d}t = \gamma\,\mathrm{d}\tau\,, \tag{11.77}$$

which is just the infinitesimal version of the time dilation formula (11.21). This result implies that the proper time interval $\mathrm{d}\tau$ between two nearby timelike-related events is always less than (or equal to) the coordinate time interval $\mathrm{d}t$ between them. As the proper time is constructed from the invariant spacetime interval $\mathrm{d}s^2$, *all* observers agree on its value—i.e., it is a spacetime invariant.

The proper time along a timelike world line $x^\alpha(\sigma)$, which is parameterized by $\sigma$ and connects spacetime events $P \equiv x^\alpha(\sigma_P)$ and $Q \equiv x^\alpha(\sigma_Q)$, is defined to be

$$\tau_{PQ}[x^\alpha] \equiv \int_P^Q \mathrm{d}\tau = c^{-1}\int_{\sigma_P}^{\sigma_Q} \mathrm{d}\sigma\,\sqrt{-\sum_{\alpha,\beta}\eta_{\alpha\beta}\frac{\mathrm{d}x^\alpha}{\mathrm{d}\sigma}\frac{\mathrm{d}x^\beta}{\mathrm{d}\sigma}}\,. \tag{11.78}$$

This is just the sum of infinitesimal proper time intervals $\mathrm{d}\tau$ along the world line from $P$ to $Q$ as shown in Fig. 11.12. Physically, proper time is the time interval that a clock would read if carried along the world line from $P$ to $Q$.

**Exercise 11.13** Show that the proper time between two events $P$ and $Q$ can be made arbitrarily close to zero by traveling along timelike world lines that are arbitrarily close to being lightlike.

**Exercise 11.14** Show that the proper time between two events $P$ and $Q$ is *greatest* if measured by an inertial observer who passes through both events—i.e., for a straight world line connecting the events. (*Hint*: Use calculus of variations (Appendix C) to find the stationary points of the proper time functional (11.78). Then use the result of Exercise 11.13 to conclude that this stationary point corresponds to a local maximum.) Thus, straight world lines in spacetime have the *largest* proper time! Contrast this to straight lines in ordinary 3-dimensional Euclidean space, which have the shortest distance between them.

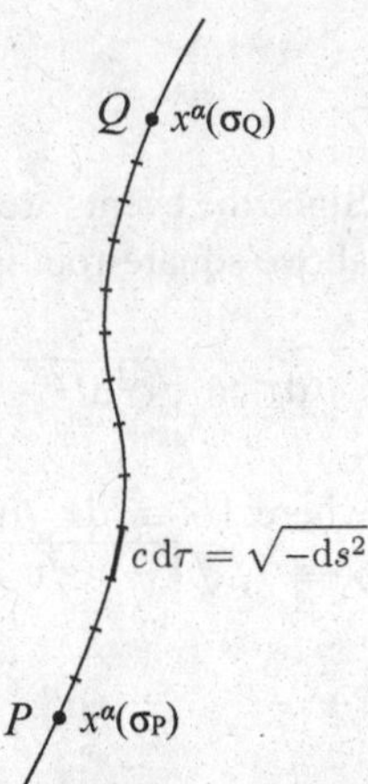


**Fig. 11.12** Timelike world line connecting two spacetime events $P$ and $Q$. The proper time along the world line $x^\alpha(\sigma)$ from $P$ to $Q$ is just the sum of infinitesimal proper time intervals $d\tau$ along the curve

## 11.6 Relativistic Kinematics

In Newtonian mechanics, the velocity, acceleration, and momentum of a particle are key quantities for describing its motion. This is also the case in relativistic mechanics. But since time and space are coupled in special relativity, velocity, acceleration, and momentum become 4-vectors, which are needed to describe world lines in spacetime. Since coordinate time is no longer a universal constant for all observers, we use the proper time of the particle for measuring rates of change.

### *11.6.1 4-Velocity*

The **4-velocity** $\mathsf{u}$ of a particle is defined to be the tangent vector to the particle's (timelike) world line parametrized by proper time $\tau$. With respect to an inertial frame having Cartesian coordinates $x^\alpha \equiv (ct, x, y, z)$, the components of the 4-velocity vector $\mathsf{u}$ are

$$u^\alpha \equiv \frac{dx^\alpha(\tau)}{d\tau}, \qquad \alpha = 0, 1, 2, 3. \tag{11.79}$$

From its definition, it immediately follows that

$$\mathsf{u} \cdot \mathsf{u} = -c^2, \tag{11.80}$$

which is somewhat surprising as it implies that all objects move through spacetime with a *constant* timelike 4-velocity. In addition,

$$u^\alpha = \gamma(c, v^i) \quad \Leftrightarrow \quad \mathsf{u} = [\gamma c, \gamma \mathbf{v}]^T, \tag{11.81}$$

where $\mathbf{v} \equiv \mathrm{d}\mathbf{x}/\mathrm{d}t$ is the ordinary 3-dimensional velocity vector and $\gamma \equiv 1/\sqrt{1 - v^2/c^2}$. Note that the 3-dimensional velocities that show up in $\gamma$ are the velocities of the particle rather than the relative velocities between inertial frames, so these velocities can be variable.

**Exercise 11.15** Verify (11.80) and (11.81). (*Hint*: You will need to use $c\,\mathrm{d}\tau = \sqrt{-\mathrm{d}s^2}$ and $\gamma = \mathrm{d}t/\mathrm{d}\tau$ to obtain these two results.)

### *11.6.2 4-Acceleration*

The **4-acceleration** $\mathsf{a}$ of a particle is defined to be the (proper) time derivative of the 4-velocity of the particle:

$$\mathsf{a} \equiv \frac{\mathrm{d}\mathsf{u}}{\mathrm{d}\tau}. \tag{11.82}$$

Since $\mathsf{u}$ is normalized to $-c$, it follows that

$$\mathsf{a} \cdot \mathsf{u} = \frac{\mathrm{d}\mathsf{u}}{\mathrm{d}\tau} \cdot \mathsf{u} = \frac{1}{2}\frac{\mathrm{d}}{\mathrm{d}\tau}(\mathsf{u} \cdot \mathsf{u}) = \frac{1}{2}\frac{\mathrm{d}}{\mathrm{d}\tau}\left(-c^2\right) = 0. \tag{11.83}$$

Since the 4-velocity $\mathsf{u}$ is a timelike vector, this means that the 4-acceleration $\mathsf{a}$ is a spacelike vector. In terms of components with respect to an inertial frame with Cartesian coordinates $x^\alpha \equiv (ct, x, y, z)$, one can show that

$$a^\alpha = \gamma^2 \left(\gamma^2 \mathbf{v} \cdot \mathbf{a}/c,\ a^i + \gamma^2 (\mathbf{v} \cdot \mathbf{a}) v^i/c^2\right), \tag{11.84}$$

where $\mathbf{v}$ and $\mathbf{a}$ are the ordinary 3-dimensional velocity and acceleration vectors, and $\gamma$ is as before. In the (non-inertial) co-moving reference frame of the accelerating particle, the 4-acceleration always has (instantaneously) just spatial components with no time component.

**Exercise 11.16** Verify (11.84) using (11.81).

**Exercise 11.17** Calculate the components $u^\alpha$ and $a^\alpha$ of the 4-velocity and 4-acceleration for a particle moving along the timelike world line

$$\begin{aligned} ct &= a^{-1}c^2 \sinh(a\tau/c)\,, \\ x &= a^{-1}c^2 (\cosh(a\tau/c) - 1)\,, \\ y &= \text{const}\,, \\ z &= \text{const}\,. \end{aligned} \tag{11.85}$$

What is the magnitude of the 4-acceleration vector?

## *11.6.3 4-Momentum*

The **4-momentum** $\mathsf{p}$ of a particle of mass $m$ is defined to be

$$\mathsf{p} \equiv m\mathsf{u}\,, \tag{11.86}$$

where $\mathsf{u}$ is the 4-velocity of the particle. Using (11.81) it immediately follows that in an inertial frame having Cartesian coordinates $x^\alpha \equiv (ct, x, y, z)$, the components of $\mathsf{p}$ are given by

$$p^\alpha = \left(E/c,\, p^i\right)\,, \tag{11.87}$$

where

$$E \equiv \gamma mc^2\,, \qquad \mathbf{p} \equiv \gamma m\mathbf{v}\,, \tag{11.88}$$

are the **relativistic energy** and **relativistic 3-momentum** of the particle, as measured in this frame. Given the form of this decomposition, the 4-momentum vector is also called the **energy-momentum 4-vector**. Note that the relativistic 3-momentum differs from the 3-momentum $m\mathbf{v}$ in Newtonian mechanics by a factor of $\gamma \equiv 1/\sqrt{1 - v^2/c^2}$.

The above expression for $E$ is the *total* relativistic energy of the particle. Note that the particle has energy even when it is at rest. This is called the **rest energy** of the particle and is just $mc^2$. We can define the **relativistic kinetic energy** as the difference between the total energy and the rest energy:

$$T \equiv E - mc^2 = mc^2(\gamma - 1)\,. \tag{11.89}$$

Since

$$\gamma \approx 1 + \frac{1}{2}\frac{v^2}{c^2} \qquad (\text{for } v \ll c)\,, \tag{11.90}$$

it follows that

$$T \approx \frac{1}{2}mv^2 \qquad (\text{for } v \ll c)\,. \tag{11.91}$$

Thus, the relativistic kinetic energy (11.89) reduces to the familiar Newtonian expression for the kinetic energy in the limit of particle velocities that are small compared to the speed of light.

> **Exercise 11.18** Relating relativistic energy, momentum, and rest mass:
>
> (a) Show that
>
> $$\mathsf{p}\cdot\mathsf{p} = -m^2c^2\,. \tag{11.92}$$
>
> (b) Using the result from part (a), show that with respect to an inertial frame
>
> $$E^2 = m^2c^4 + |\mathbf{p}|^2c^2\,. \tag{11.93}$$

#### 11.6.3.1 Photons (or Zero-rest-mass Particles)

Because of the $\gamma$-factor, both the relativistic energy and momentum approach infinity as the speed of the particle approaches the speed of light. Thus, we will need to develop a different approach for dealing with photons. The definitions of the 4-velocity, 4-acceleration, and 4-momentum given above have assumed that the particle moves along a *timelike* world line, for which $\mathrm{d}s^2 < 0$ for all nearby points on the world line. Since light (i.e., photons) moves along a null world line (for which $\mathrm{d}\tau = 0$), we cannot define a 4-velocity vector for light using $u^\alpha \equiv \mathrm{d}x^\alpha/\mathrm{d}\tau$. But we can still define a tangent vector $\mathrm{d}x^\alpha/\mathrm{d}\lambda$ for the photon's path, where $\lambda$ is some parameter along the null world line. This tangent vector is not unique, since we cannot normalize it to have a fixed (non-zero) length. But we can restrict the choice of $\lambda$ to be such that

$$\frac{\mathrm{d}^2x^\alpha(\lambda)}{\mathrm{d}\lambda^2} = 0\,. \tag{11.94}$$

Such a $\lambda$ is said to be an **affine parameter** along the photon's world line. It is easy to see that if $\lambda$ is an affine parameter, then so is $\lambda' \equiv a\lambda + b$, where $a$ and $b$ are constants. Hence, affine parameters form a restricted class of parameters, but they are not unique.

Nonetheless, the 4-momentum p of a photon can be *uniquely* defined by taking advantage of the known expressions for the energy and momentum of a photon from quantum mechanics. With respect to an inertial reference frame having Cartesian coordinates $x^\alpha \equiv (ct, x, y, z)$, we define

$$p^\alpha \equiv (\hbar\omega/c,\ \hbar k^i)\,, \tag{11.95}$$

where $\mathbf{k}$ is the 3-dimensional **wave vector**—i.e., $\mathbf{k}$ points in the direction of propagation of the photon and has magnitude $|\mathbf{k}| = \omega/c$, where $\omega$ is the angular frequency of the light as measured by the observer.[3] Since $|\mathbf{k}| = \omega/c$, it follows that $\mathsf{p}\cdot\mathsf{p} = 0$. Hence, with respect to (11.92), we see that a photon is a **zero-rest-mass particle**.

**Exercise 11.19** Show that a photon propagating in the $xy$-plane making an angle $\theta$ with the $x$-direction has 4-momentum

$$p^\alpha = \frac{\hbar\omega}{c}(1, \cos\theta, \sin\theta, 0)\,, \tag{11.96}$$

where $\omega$ is the angular frequency of the light as measured in this frame.

***Example 11.4*** Our definition (11.95) of the 4-momentum for a photon depends on the frequency of the photon. But we have seen in Sect. 11.3.4 that the relativistic Doppler effect can actually change the observed frequency. Here we consider the consequences of the Doppler effect in terms of the 4-momentum vector of the photon.

We will consider, as before, two inertial observers $O$ and $O'$, with $O'$ boosted in the $x$-direction with speed $u$. Let's assume that we have a source of light, at rest in $O$, which emits photons with angular frequency $\omega$ in the $xy$-plane making an angle $\theta$ with respect to the $x$-axis. Then (11.96) gives the components of the 4-momentum of the photon as measured with respect to $O$. With respect to $O'$, the components will have a similar form

$$p^{\alpha'} = \frac{\hbar\omega'}{c}(1, \cos\theta', \sin\theta', 0)\,, \tag{11.97}$$

where $\omega'$ and $\theta'$ are the angular frequency and direction of photon propagation as measured with respect to $O'$. To relate the primed quantities $\omega'$ and $\theta'$ to the unprimed quantities $\omega$ and $\theta$, we use the transformation equations (11.67) for the components of a 4-vector under a Lorentz boost in the $x$-direction:

[3]With respect to any other inertial reference frame $x^{\alpha'} \equiv (ct', x', y', z')$, we use the transformation law for vector components to get $p^{\alpha'}$.

$$
\begin{aligned}
p^{0'} &= \gamma(p^0 - \beta p^1)\,, \\
p^{1'} &= \gamma(p^1 - \beta p^0)\,, \\
p^{2'} &= p^2\,, \\
p^{3'} &= p^3\,,
\end{aligned}
\tag{11.98}
$$

where $\beta \equiv u/c$ and $\gamma \equiv 1/\sqrt{1-\beta^2}$. Together with (11.96) and (11.97), these equations imply

$$
\begin{aligned}
\omega' &= \gamma\omega\,(1 - \beta\cos\theta)\,, \\
\omega'\cos\theta' &= \gamma\omega(\cos\theta - \beta)\,, \\
\omega'\sin\theta' &= \omega\sin\theta\,.
\end{aligned}
\tag{11.99}
$$

Solving these equations for $\cos\theta'$, we find:

$$
\cos\theta' = \frac{\cos\theta - \beta}{1 - \beta\cos\theta}\,. \tag{11.100}
$$

Note that the first equation in (11.99) already gives $\omega'$ in terms of $\omega$ and $\cos\theta$; in terms of the $\cos\theta'$, we have

$$
\omega' = \omega\frac{\sqrt{1-\beta^2}}{1+\beta\cos\theta'}\,. \tag{11.101}
$$

Specializing to the case $\theta' = 0 = \theta$ (so that the direction of photon propagation is in the same direction as the boost), we have

$$
\omega' = \omega\sqrt{\frac{1-\beta}{1+\beta}}\,, \tag{11.102}
$$

which is consistent with (11.30). Specializing to the case $\theta' = \pi/2$ (so that the direction of photon propagation is perpendicular to the direction of the boost as measured by $O'$), we have

$$
\omega' = \omega\sqrt{1-\beta^2}\,, \tag{11.103}
$$

which is consistent with (11.31) for the transverse relativistic Doppler effect. □

**Exercise 11.20** Verify (11.101), and then show that (11.100) and (11.101) imply $\omega'\sin\theta' = \omega\sin\theta$, which is the third equation in (11.99).

## 11.7 Relativistic Dynamics

Just as Newton's second law $\mathbf{F} = m\mathbf{a}$ cannot be derived, its generalization to special relativity must also be postulated. We must choose some form for the force equation, whose ultimate justification comes from agreement with experiment. The generalized equation must also be invariant under Lorentz transformations (so that the laws of mechanics take the same form in all inertial reference frames), and reduce to the non-relativistic (Newtonian) equation when velocities are small.

We can satisfy Lorentz invariance by postulating a 4-vector equation of the form

$$\mathsf{f} = m\mathsf{a}, \tag{11.104}$$

where $\mathsf{a}$ is the 4-acceleration of (11.82), and where $\mathsf{f}$ is the **4-force**, whose relation to the 3-dimensional force $\mathbf{F}$ we must determine. Note that when $\mathsf{f} = 0$, we recover the equation of motion for a *free* particle, as we should:

$$\mathsf{a} \equiv \frac{\mathrm{d}\mathsf{u}}{\mathrm{d}\tau} = 0 \quad \Leftrightarrow \quad \frac{\mathrm{d}^2 x^\alpha}{\mathrm{d}\tau^2} = 0\,. \tag{11.105}$$

If $x^\alpha \equiv (ct, x, y, z)$ are Cartesian coordinates associated with some inertial frame, then the right-hand side of (11.104) has components

$$ma^\alpha = \frac{\mathrm{d}p^\alpha}{\mathrm{d}\tau} = \left(\frac{1}{c}\frac{\mathrm{d}E}{\mathrm{d}\tau}, \frac{\mathrm{d}p^i}{\mathrm{d}\tau}\right) = \gamma\left(\frac{1}{c}\frac{\mathrm{d}E}{\mathrm{d}t}, \frac{\mathrm{d}p^i}{\mathrm{d}t}\right), \tag{11.106}$$

where $E \equiv \gamma mc^2$ and $\mathbf{p} \equiv \gamma m\mathbf{v}$ are the relativistic energy and relativistic 3-momentum of the particle. (As usual, $\gamma \equiv 1/\sqrt{1 - v^2/c^2}$.) The spatial part of the above equation suggests generalizing Newton's 2nd law as

$$\mathbf{F} = \frac{\mathrm{d}\mathbf{p}}{\mathrm{d}t}, \quad \text{where} \quad \mathbf{p} \equiv \gamma m\mathbf{v}\,. \tag{11.107}$$

This equation reduces to the standard form of Newton's 2nd law $\mathbf{F} = m\mathbf{a}$ in the limit when $v \ll c$. The equation also has the nice property that

$$\frac{1}{c}\frac{\mathrm{d}E}{\mathrm{d}t} = \mathbf{F}\cdot\mathbf{v}/c\,, \tag{11.108}$$

which recovers the work energy theorem in Newtonian mechanics with $E$ replaced by the kinetic energy $\frac{1}{2}mv^2$ and $\mathbf{F}$ replaced by $m\mathbf{a}$. Thus, $f^\alpha = ma^\alpha$ for

$$f^\alpha = \gamma(\mathbf{F}\cdot\mathbf{v}/c, F^i)\,. \tag{11.109}$$

This relates the components of the 4-force f with the forces **F** that we measure with respect to an inertial frame.

**Exercise 11.21** Verify (11.108) for $E \equiv \gamma mc^2$ and $\mathbf{F} = \mathrm{d}\mathbf{p}/\mathrm{d}t$ with $\mathbf{p} \equiv \gamma m\mathbf{v}$.

**Exercise 11.22** Show that (11.107) becomes

$$\mathbf{F} = \gamma m \left[\mathbf{a} + \gamma^2(\mathbf{v}\cdot\mathbf{a})\mathbf{v}/c^2\right], \qquad (11.110)$$

when written in terms of the ordinary 3-dimensional acceleration **a**.

## 11.8 Relativistic Lagrangian Formalism

As we saw in Chaps. 2 and 3, Newtonian mechanics can be cast in terms of a variational principle. This is also the case for the relativistic form of mechanics that we have developed in the previous sections.

### *11.8.1 Free Particle*

As you showed in Exercise 11.14, the relativistic equation of motion for a *free* particle ($\mathrm{d}\mathsf{u}/\mathrm{d}\tau = 0$) can be found by maximizing the proper time:

A free particle moves between two timelike-separated events in such a way that its world line is a stationary point of the proper-time functional between the two events.

The action is then just the proper-time integral

$$\tau_{PQ}[x^\alpha] = \int_P^Q \mathrm{d}\tau = c^{-1}\int_{\sigma_P}^{\sigma_Q} \mathrm{d}\sigma \sqrt{-\sum_{\alpha,\beta}\eta_{\alpha\beta}\frac{\mathrm{d}x^\alpha}{\mathrm{d}\sigma}\frac{\mathrm{d}x^\beta}{\mathrm{d}\sigma}}, \qquad (11.111)$$

where $\sigma$ is any parameter along the particle's world line, $x^\alpha \equiv x^\alpha(\sigma)$. The corresponding Lagrangian is the integrand

$$\bar{L}\left(x^\alpha, \mathrm{d}x^\alpha/\mathrm{d}\sigma, \sigma\right) \equiv \frac{\mathrm{d}\tau}{\mathrm{d}\sigma} = c^{-1}\sqrt{-\sum_{\alpha,\beta}\eta_{\alpha\beta}\frac{\mathrm{d}x^\alpha}{\mathrm{d}\sigma}\frac{\mathrm{d}x^\beta}{\mathrm{d}\sigma}}. \qquad (11.112)$$

Note that the coordinate time $x^0 = ct$ is treated on the same footing as the spatial coordinates $x^i$ in this form of the Lagrangian.

**Exercise 11.23** Prove that the Euler-Lagrange equations

$$\frac{\mathrm{d}}{\mathrm{d}\sigma}\left(\frac{\partial \bar{L}}{\partial(\mathrm{d}x^\alpha/\mathrm{d}\sigma)}\right) - \frac{\partial \bar{L}}{\partial x^\alpha} = 0\,, \tag{11.113}$$

for the Lagrangian (11.112) yield the equations of motion $\mathrm{d}^2x^\alpha/\mathrm{d}\tau^2 = 0$ for a free particle.

If one takes $\sigma = t$ and multiplies the Lagrangian by $-mc^2$ (which doesn't change the equations of motion), we obtain a new Lagrangian

$$L(x^i, \dot{x}^i, t) \equiv -mc^2\frac{d\tau}{dt} = -mc^2\gamma^{-1} = -mc^2\sqrt{1 - v^2/c^2}\,, \tag{11.114}$$

where $t$ is now the independent variable. Note that in the limit $v \ll c$:

$$L \approx -mc^2 + \frac{1}{2}mv^2 = T + \text{const}\,, \tag{11.115}$$

which is the Lagrangian for a free particle in Newtonian mechanics. The momentum $\pi_i$ conjugate to $x^i$ is

$$\pi_i \equiv \frac{\partial L}{\partial \dot{x}^i} = \gamma m v_i\,, \tag{11.116}$$

which is just the expression for the relativistic momentum $\mathbf{p} \equiv \gamma m \mathbf{v}$. Thus, the free particle equations of motion for this Lagrangian are

$$\frac{\mathrm{d}\mathbf{p}}{\mathrm{d}t} = \mathbf{0}\,, \tag{11.117}$$

as to be expected from (11.107) for $\mathbf{F} = \mathbf{0}$.

### 11.8.2 Forces Derivable from a Potential

To include forces $\mathbf{F}$ which are derivable from a potential $U$, we can simply subtract the potential from the free-particle Lagrangian, redefining $L$ to be

$$L(x^i, \dot{x}^i, t) = -mc^2\sqrt{1 - v^2/c^2} - U(x^i, \dot{x}^i, t)\,. \tag{11.118}$$

If the potential is independent of velocities, then $\pi_i = p_i$, and the Euler-Lagrange equations become

$$\frac{\mathrm{d}p_i}{\mathrm{d}t} = -\frac{\partial U}{\partial x^i} = F_i . \tag{11.119}$$

Again, this is to be expected from (11.107) for $\mathbf{F} = -\nabla U$.

If the potential $U$ depends on the velocities, then the force can be described as being mediated by a field that can carry energy and momentum. Consequently, the canonical momentum $\pi_i$ conjugate to $x^i$ will also include terms corresponding to the momentum in this field. For example, a charged particle $q$ in an electromagnetic field has

$$U(\mathbf{r}, \dot{\mathbf{r}}, t) = q\,[\Phi(\mathbf{r}, t) - \mathbf{A}(\mathbf{r}, t) \cdot \dot{\mathbf{r}}] , \tag{11.120}$$

for which

$$\pi_i \equiv \frac{\partial L}{\partial \dot{x}^i} = \gamma m v_i + q A_i , \tag{11.121}$$

and

$$\frac{\mathrm{d}\pi_i}{\mathrm{d}t} = -\frac{\partial U}{\partial x^i} . \tag{11.122}$$

Following the same procedure used in Exercise 2.14, these equations can then be written in the standard form:

$$\frac{\mathrm{d}\mathbf{p}}{\mathrm{d}t} = q(\mathbf{E} + \mathbf{v} \times \mathbf{B}) , \tag{11.123}$$

where $\mathbf{p} \equiv \gamma m \mathbf{v}$ and

$$\mathbf{E} \equiv -\nabla\Phi - \frac{\partial \mathbf{A}}{\partial t} , \qquad \mathbf{B} \equiv \nabla \times \mathbf{A} . \tag{11.124}$$

This is just the Lorentz force law, with $\mathbf{F} = \mathrm{d}\mathbf{p}/\mathrm{d}t$, for a point charge $q$ moving in an electromagnetic field.

### 11.8.3 Relativistic Field Theory

The Lagrangian and Hamiltonian formulations of continuous systems and fields (discussed in Chap. 10) can also be applied to relativistic *fields*, e.g., the electromagnetic field or the gravitational field in Einstein's theory of general relativity. But since relativistic field theory is worthy of its own (semester-long or longer) course, we will stop our presentation here and refer the interested reader to Landau and Lifshitz (1975), which is a standard text on the subject. We've included a few problems on relativistic fields in the "Additional Problems" section of this chapter, for those students who have also read Chap. 10. But these problems just scratch the surface of what you would be doing in a proper course on the subject.

## Suggested References

*Full references are given in the bibliography at the end of the book.*

Bondi (1962): A popular book about special relativity; extremely well-written. It has a nice discussion of the $k$ calculus, appropriate for the layperson.

Einstein (1905): Albert Einstein's original 1905 paper on special relativity. An English translation of the paper is available at http://einsteinpapers.press.princeton.edu/vol2-trans/154.

Hartle (2003): An excellent textbook on general relativity, appropriate for advanced undergraduates or beginning graduate students. It takes a "physics first" approach to the subject, emphasizing physical concepts before the mathematics. Chapters 4 and 5 discuss special relativity.

Landau and Lifshitz (1975): Standard graduate-level text on the classical theory of fields, with special attention paid to electromagnetic and gravitational fields. Natural extension of the material discussed in this and the previous chapter.

Mermin (2005): A popular book about special relativity from one of the best popularizers of special relativity and quantum mechanics.

Schutz (2009): A solid introduction to the general theory of relativity, appropriate for advanced undegraduates or beginning graduate students. Chapters 1-4 discuss special relativity.

Taylor and Wheeler (1992): A very thorough introduction to special relativity, appropriate for undergraduates. Emphasizes geometrical understanding.

## Additional Problems

**Problem 11.1** The **twin paradox** in special relativity is the following: Alice and Bob are twins. Alice stays home, at rest on Earth, while Bob travels at constant speed to a nearby star and then immediately turns around and returns to Earth. According to Alice, Bob's biological clock ticks more slowly as he travels to the star and then returns, in accord with our finding that moving clocks run slowly. But according to Bob, *he* is at rest and Alice is moving relative to him, so *her* biological clock must be ticking more slowly than his. But how can they both be aging more slowly than the other? Who, if any one of them, is older when Bob returns? (*Hint*: Use the spacetime diagram shown in Fig. 11.13 to help answer this question. Note that Bob is not in a single inertial frame during his roundtrip to the star and back.)

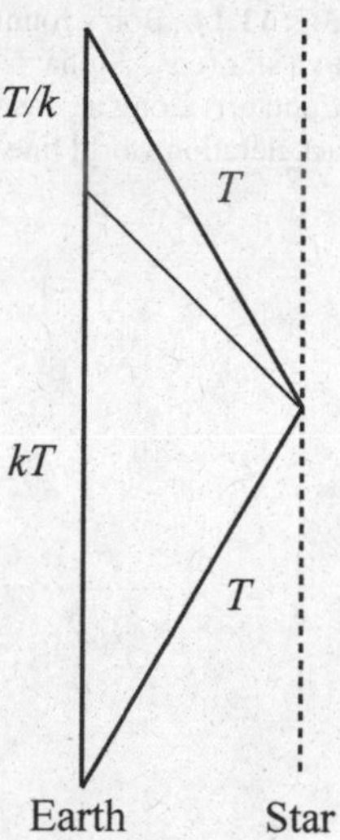


**Fig. 11.13** Spacetime diagram for the "twin paradox" problem, Problem 11.1. Alice stays on Earth, while Bob travels at constant speed to a nearby star, and then immediately turns around and returns to Earth

**Problem 11.2** In Problem 11.1, we assumed that Bob traveled to a nearby star at constant speed, and then *immediately* turned around and returned to Earth at the same constant speed as before. Obviously, this simplifying assumption is not physically realistic, as Bob would experience *infinite* acceleration when turning around. To remedy this situation, assume instead that Bob travels to the star and back, first along a world line that has constant acceleration $a = g = 9.8\ \mathrm{m/s^2}$ (out to $D/2$, half the distance to the star), then along a world line with $a = -g$ (which gets him out to $D$ and back to $D/2$), and then finally along a world line with $a = g$, to bring him back to Earth, as shown in Fig. 11.14. For concreteness, let's assume that the star is Alpha Centauri, which is at a distance $D = 4.4$ lyr from Earth.

(a) Calculate the total elapsed time for Bob's trip to Alpha Centauri as measured by both Alice and Bob. (*Hint*: Use the results of Exercise 11.17.) You should find that 7.2 yr has elapsed for Bob and 12.1 yr for Alice.
(b) Repeat the above calculation, but this time for Bob to make a roundtrip to the center of the Milky Way galaxy, a distance $D \sim 10$ kpc from Earth. (Recall that 1 kpc $= 3.26$ lyr.) You should find that Bob ages by 40 yr, while Alice ages by 65,200 yr!

**Problem 11.3** Consider a pole vaulter, carrying a 20-foot pole, running so fast in the direction of its length that the pole appears to be only 10 feet long as measured by you, at rest with respect to the ground. Suppose that the pole vaulter approaches a shed that is also 10 feet long, with its front door and back door both open, so that he can easily run right through. According to you, there is a time when the pole is completely within the shed. But according the pole vaulter, the shed is only 5 feet long, while his pole is 20 feet long; hence it is impossible for the pole to ever be

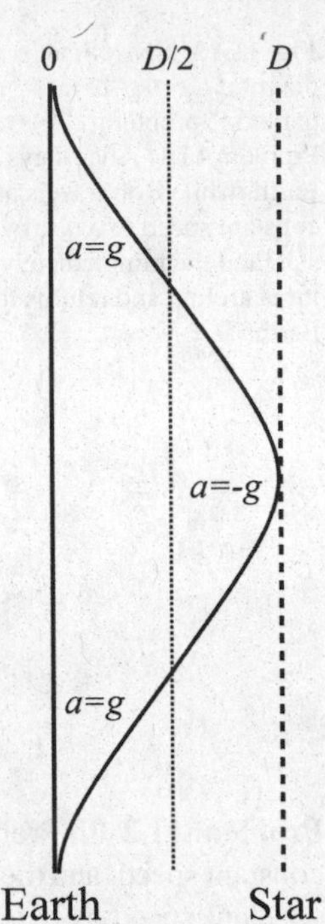


**Fig. 11.14** Bob's roundtrip to a star (e.g., Alpha Centauri) along a constant acceleration world line

completely within the shed. Who is right? Draw a spacetime diagram to resolve the apparent paradox.

**Problem 11.4** A particle is moving along the $x$-axis with velocity

$$\frac{\mathrm{d}x}{\mathrm{d}t} = \frac{gt}{\sqrt{1 + g^2t^2/c^2}}\,, \tag{11.125}$$

where $g$ is a constant with units of acceleration.

(a) Calculate the $\gamma$-factor for the particle's motion.
(b) Calculate the components of the particle's 4-velocity $u^\alpha = (c\mathrm{d}t/\mathrm{d}\tau, \mathrm{d}x/\mathrm{d}\tau)$.
(c) Find an expression for the proper time $\tau$ in terms of the coordinate time $t$ by integrating $\gamma = \mathrm{d}t/\mathrm{d}\tau$.
(d) Show that in terms of the proper time $\tau$:

$$t(\tau) = \frac{c}{g}\,\sinh(g\tau/c)\,, \qquad x(\tau) = \frac{c^2}{g}\,\cosh(g\tau/c)\,. \tag{11.126}$$

(e) Using the above expressions for $x$ and $t$ as functions of $\tau$, calculate the components of the 4-force $f^\alpha = ma^\alpha = m\mathrm{d}u^\alpha/\mathrm{d}\tau$.
(f) Find an expression for the 3-force $F$ defined via $f^\alpha = \gamma(Fv/c, F)$. (*Hint*: If you do the calculation correctly, the expression for $F$ should look *very* familiar.)

**Problem 11.5** **(Stellar aberration)** Suppose you want to observe a star that is located directly overhead at midnight in a direction that lies in the plane of the Earth's orbit around the Sun. (For simplicity, assume that the Earth's orbit is circular, and that the Earth's rotational axis is perpendicular to the orbital plane.) At what angle from the vertical must you point a telescope in order to see the star? (*Hint*: The problem is analogous to running in the rain with a cylindrical tube, tilted in such a way that the rain drops don't hit the inner wall of the tube.)

**Problem 11.6** **(Relativistic beaming)** Consider a monochromatic source of electromagnetic radiation that emits isotropically (i.e., equally well in all directions) in its rest frame $O$. Denote the number flux of photons associated with the radiation by $\mathscr{F}$, with units [photons/(m$^2$·s)]. (This flux is independent of direction in the source frame.) Suppose that the source moves with speed $v$ in the $x'$-direction of an observer's reference frame $O'$.

(a) Consider a photon with frequency $\omega$ propagating in direction $\theta$ with respect the $x'$-axis as measured in source frame $O$. Show that the corresponding frequency and direction as measured in the observer's frame $O'$ are given by

$$\omega' = \gamma\omega(1 + \beta\cos\theta)\,, \qquad \cos\theta' = \frac{\cos\theta + \beta}{1 + \beta\cos\theta}\,, \tag{11.127}$$

where $\beta \equiv v/c$ and $\gamma \equiv 1/\sqrt{1-\beta^2}$. The inverse relations are obtained by interchanging primed and unprimed quantities and replacing $\beta$ with $-\beta$. (*Hint*: Recall Example 11.4.)

(b) Now consider two photons that are emitted in direction $\theta'$, a time interval $\Delta t'_{\rm e}$ apart. They are received by a distant observer in direction $\theta'$ with a time interval $\Delta t'_{\rm o}$. (Both time intervals are measured with respect to the observer's frame $O'$.) Show that

$$\Delta t'_{\rm o} = (1 - \beta\cos\theta')\Delta t'_{\rm e}\,. \tag{11.128}$$

In the source frame, $\Delta t_{\rm o} = \Delta t_{\rm e}$.

(c) Show that $\Delta t'_{\rm e}$ and $\Delta t_{\rm e}$ are related by $\Delta t'_{\rm e}/\Delta t_{\rm e} = \gamma$.

(d) Show that the number flux of photons $\mathscr{F}'(\theta')$ as seen in the observer's frame is given by

$$\mathscr{F}'(\theta') = \mathscr{F}\left[\gamma(1 - \beta\cos\theta')\right]^{-3}\,. \tag{11.129}$$

(*Hint*: The number of photons passing through a given solid angle in a given interval of time should be the same in both frames.)

(e) Finally, calculate the *beaming factor* for radiation propagating in the forward direction, $\mathscr{F}'(0)/\mathscr{F}$. You should find

$$\frac{\mathscr{F}'(0)}{\mathscr{F}} = \gamma^3(1+\beta)^3\,, \tag{11.130}$$

which grows very rapidly as $\beta \to 1$. Thus, *radiation from a relativistic particle is highly beamed in the direction of its motion.*

**Problem 11.7** Determine the rotation (called a **Wigner rotation** or **Thomas rotation**) associated with the composition of two Lorentz boosts, the first in the $x$-direction by $u$, and the second in the transformed $y'$-direction by $u'$.

**Problem 11.8** (**Terrell rotation**) In Sect. 11.3.3, we showed that a moving rod of proper length $L_0$ will appear to be *contracted* in the direction of its motion, as measured by a stationary observer. The contracted length is $L_0\sqrt{1-\beta^2}$, where $v \equiv \beta c$ is the speed of the moving rod with respect to the observer. Since lengths perpendicular to the direction of motion are not affected, a moving square with proper side length $L_0$ would appear to a stationary observer as a *rectangle* with height $L_0$ and width $L_0\sqrt{1-\beta^2}$. But now suppose that we have a *cube* of proper edge length $L_0$, moving with speed $v$ relative to a stationary observer. Assume that the front and back faces of the cube are parallel to the velocity of the cube as shown in Fig. 11.15. Show that if the observer *photographs* the cube as it moves by, it will appear to be *rotated* by an angle $\sin^{-1}\beta$, and not length contracted as one might expect from the above discussion. This unexpected outcome was first reported in Terrell (1959), more than 50 years after the formulation of special relativity. (*Hint*: Light rays entering the camera from the back edge $D'$ of the moving cube had to leave $L_0/c$ earlier than those from the front edges $A'$ and $B'$ because they had farther to travel the camera. Thus, the photograph records back edge $D'$ at distance $vL_0/c = \beta L_0$ behind front edge $A'$.)

**Problem 11.9** For collisions (or disintegrations) of particles in special relativity, the total relativistic momentum and total relativistic energy are conserved:

$$\sum_I \gamma_I m_I \mathbf{v}_I = \text{const}\,, \qquad \sum_I \gamma_I m_I c^2 = \text{const}\,. \tag{11.131}$$

If the collision is *elastic*, one additionally has conservation of total relativistic kinetic energy

$$\sum_I (\gamma_I - 1) m_I c^2 = \text{const} \qquad \text{(elastic collision)}\,. \tag{11.132}$$

Note that, using (11.131), conservation of total relativistic kinetic energy is equivalent to conservation of total mass, $\sum_I m_I = \text{const}$, so the total mass of a system

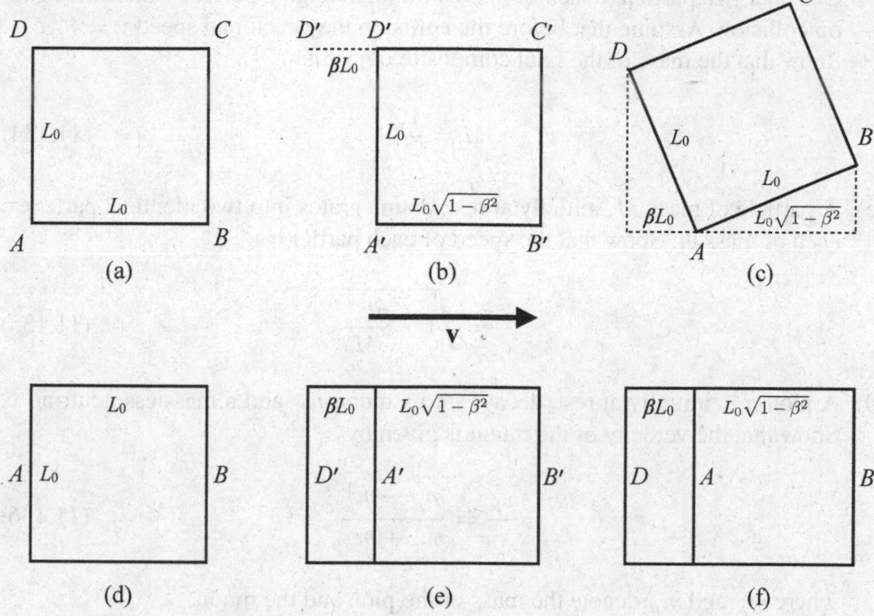


**Fig. 11.15** Photographing a moving cube. The top three panels show the top view of the cube. The bottom three panels show photographs of the front view of the cube. Panels (a) and (d) correspond to the cube in its rest frame. Panels (b), (c), (e), (f) correspond to the moving cube as seen by the stationary observer. The front edges of the cube are $A$, $B$ or $A'$, $B'$; back edges are $C$, $D$ or $C'$, $D'$. Panel (b) shows that light entering the camera from back edge $D'$ left the cube at an earlier time, when the cube was located a distance $\beta L_0$ to the left. Panel (c) shows an interpretation of the photograph in panel (e) as a rotation of an uncontracted cube. (*Note* $\beta = 5/13$ was used for these figures.)

is conserved in special relativity only for elastic collisions. This is in contrast to Newtonian mechanics, where the total mass of a system is *always* conserved. In Newtonian mechanics, if the kinetic energy of a system increases (decreases), there is a corresponding decrease (increase) in the internal energy of the system.

(a) Show that (11.131) can be written more compactly as

$$\sum_I p_I^\alpha = \text{const}\,, \tag{11.133}$$

where $p_I^\alpha$ are the components of the energy-momentum 4-vector with respect to an inertial coordinate system $x^\alpha = (ct, x, y, z)$.

(b) Consider two particles, each of mass $m$, that undergo a perfectly inelastic head-on collision. Assume that before the collision they each had speed $v = (4/5)c$. Show that the mass of the final composite object is

$$M = \frac{10}{3} m\,. \tag{11.134}$$

(c) A particle of mass $M$, initially at rest, disintegrates into two identical particles, each of mass $m$. Show that the speed of each particle is

$$\frac{v}{c} = \sqrt{1 - \frac{4m^2}{M}}\,. \tag{11.135}$$

(d) A pion $\pi^-$, initially at rest, decays into a muon $\mu^-$ and a massless neutrino $\nu$. Show that the velocity of the muon is given by

$$\frac{v_\mu}{c} = \frac{m_\pi^2 - m_\mu^2}{m_\pi^2 + m_\mu^2}\,, \tag{11.136}$$

where $m_\pi$ and $m_\mu$ denote the mass of the pion and the muon.

**NOTE**: *The following problems are related to the electromagnetic field, which is the classic example of a relativistic field theory. To solve these problems requires that you have read the material from Chaps.* 9 and 10 (*the wave equation, and the Lagrangian and Hamiltonian formulations of field theory*). *If you have skipped these chapters, you should probably skip these problems.*

**Problem 11.10** Maxwell's equations for the electric and magnetic fields **E** and **B** in the presence of sources $\rho$ and **J** (the electric charge density and current density) are given in (11.6).

(a) Show that if we write **E** and **B** in terms of potentials $\Phi$ and **A** via

$$\mathbf{E} = -\nabla\Phi - \frac{\partial \mathbf{A}}{\partial t}\,, \qquad \mathbf{B} = \nabla \times \mathbf{A}\,, \tag{11.137}$$

then the two Maxwell equations with no source terms are automatically satisfied.

(b) Show that the above equations relating **E**, **B** to $\Phi$, **A** do not uniquely determine $\Phi$ and **A**. Do this by verifying that the transformations

$$\mathbf{A} \rightarrow \mathbf{A}' \equiv \mathbf{A} + \nabla\Lambda\,, \qquad \Phi \rightarrow \Phi' \equiv \Phi - \frac{\partial \Lambda}{\partial t}\,, \tag{11.138}$$

leave **E** and **B** unchanged. (Here $\Lambda = \Lambda(\mathbf{r}, t)$ is an arbitrary function.) The above transformation is called a **gauge transformation**.

(c) Show that Maxwell's equations with sources can be written in terms of the potentials $\Phi$ and $\mathbf{A}$ as

$$\Box^2\Phi + \frac{\partial}{\partial t}\left(\nabla \cdot \mathbf{A} + \frac{1}{c^2}\frac{\partial \Phi}{\partial t}\right) = -\frac{\rho}{\varepsilon_0}\,,$$
$$\Box^2\mathbf{A} - \nabla\left(\nabla \cdot \mathbf{A} + \frac{1}{c^2}\frac{\partial \Phi}{\partial t}\right) = -\mu_0\mathbf{J}\,, \tag{11.139}$$

where

$$\Box^2 \equiv \nabla^2 - \frac{1}{c^2}\frac{\partial^2}{\partial t^2} \tag{11.140}$$

is the *D'Alembertian* (or wave operator).

(d) Show that it is always possible to make a gauge transformation for which the term in parentheses vanishes—i.e.,

$$\nabla \cdot \mathbf{A} + \frac{1}{c^2}\frac{\partial \Phi}{\partial t} = 0\,. \tag{11.141}$$

This is called the **Lorentz gauge**. To do this, show that if

$$\nabla \cdot \mathbf{A} + \frac{1}{c^2}\frac{\partial \Phi}{\partial t} \equiv \alpha \neq 0\,, \tag{11.142}$$

then a gauge transformation to a new set of scalar and vector potentials $\Phi'$ and $\mathbf{A}'$ will satisfy

$$\nabla \cdot \mathbf{A}' + \frac{1}{c^2}\frac{\partial \Phi'}{\partial t} = 0 \quad \Leftrightarrow \quad \Box^2\Lambda = -\alpha\,. \tag{11.143}$$

Since this last equation always admits a solution (it is just an ordinary wave equation with source term), then it is always possible to satisfy the gauge condition.

Hence, in the Lorentz gauge, Maxwell's equations for the scalar and vector potential $\Phi$ and $\mathbf{A}$ are simply

$$\Box^2\Phi = -\frac{\rho}{\varepsilon_0}\,, \qquad \Box^2\mathbf{A} = -\mu_0\mathbf{J}\,. \tag{11.144}$$

**Problem 11.11** In this problem, we continue working with Maxwell's equations in terms of the scalar and vector potentials $\Phi$ and $\mathbf{A}$ (See Problem 11.10). We start by constructing spacetime 4-vectors from the scalar and vector potentials, and from the electric charge and current densities:

$$A^\alpha \equiv (\Phi/c, \mathbf{A})\,, \qquad J^\alpha \equiv (\rho c, \mathbf{J})\,, \tag{11.145}$$

with respect to inertial coordinates $x^\alpha \equiv (ct, x, y, z)$. We also define the **electromagnetic field tensor**

$$F_{\alpha\beta} \equiv \partial_\alpha A_\beta - \partial_\beta A_\alpha \,, \tag{11.146}$$

where[4]

$$A_\alpha \equiv \sum_\beta \eta_{\alpha\beta} A^\beta = (-\Phi/c, \mathbf{A}) \,. \tag{11.147}$$

Note that the electromagnetic field tensor $F_{\alpha\beta}$ is anti-symmetric under interchange of $\alpha$ and $\beta$.

(a) Show that **E** and **B** are related to $F_{\alpha\beta}$ via

$$E_i = -cF_{0i} \,, \qquad B_i = \frac{1}{2}\sum_{j,k} \varepsilon_{ijk} F_{jk} \,. \tag{11.148}$$

(b) Show that the gauge transformation (11.138) can be written in terms of the spacetime 4-vector $A_\alpha$ as

$$A_\alpha \to A'_\alpha = A_\alpha + \partial_\alpha \Lambda \,, \tag{11.149}$$

and that the field tensor is invariant under such a transformation—i.e., $F'_{\alpha\beta} = F_{\alpha\beta}$.

(c) Recall from Problem 11.10 that the two Maxwell equations without source terms are automatically satisfied by working with the potentials $\Phi$ and **A**. Show that in terms of $F_{\alpha\beta}$ the source-free equations are contained in the identity

$$\partial_\alpha F_{\beta\gamma} + \partial_\beta F_{\gamma\alpha} + \partial_\gamma F_{\alpha\beta} = 0 \,, \tag{11.150}$$

which is automatically satisfied by the definition (11.146).

(d) Show that the two Maxwell's equations with source terms can be written as

$$\sum_\alpha \partial_\alpha F^{\alpha\beta} = -\mu_0 J^\beta \,, \tag{11.151}$$

where $F^{\alpha\beta} \equiv \sum_{\gamma,\delta} \eta^{\alpha\gamma}\eta^{\beta\delta} F_{\gamma\delta}$ (i.e., $F^{00} = F_{00}$, $F^{0i} = -F_{0i}$, $F^{ij} = F_{ij}$).

[4] Recall that spacetime vectors with up indices $A^\alpha$ are not identical to spacetime vectors with down indices $A_\alpha$; they are related by application of $\eta_{\alpha\beta}$ or $\eta^{\alpha\beta}$. See Appendix A.2 for more details.

(e) Show that the continuity equation (conservation of electric charge)

$$\frac{\partial \rho}{\partial t} + \nabla \cdot \mathbf{J} = 0 \quad \Leftrightarrow \quad \sum_\alpha \partial_\alpha J^\alpha = 0 \tag{11.152}$$

follows from (11.151) using the anti-symmetry properties of $F_{\alpha\beta}$.

Hence, in terms of the electromagnetic field tensor, Maxwell's equations are simply

$$\partial_\alpha F_{\beta\gamma} + \partial_\beta F_{\gamma\alpha} + \partial_\gamma F_{\alpha\beta} = 0\,, \qquad \sum_\alpha \partial_\alpha F^{\alpha\beta} = -\mu_0 J^\beta\,. \tag{11.153}$$

**Problem 11.12** Show that Maxwell's equations with sources (11.151) are derivable from the Lagrangian density

$$\mathscr{L} = -\frac{1}{4}\sum_{\alpha,\beta} F^{\alpha\beta} F_{\alpha\beta} + \mu_0 \sum_\alpha J^\alpha A_\alpha \tag{11.154}$$

with respect to variations in the 4-vector potential $A_\alpha$.

# Appendix A
# Vector Calculus

Vector calculus is an indispensable mathematical tool for classical mechanics. It provides a geometric (i.e., coordinate-independent) framework for formulating the laws of mechanics, and for solving for the motion of a particle, or a system of particles, subject to external forces. In this appendix, we summarize several key results of differential and integral vector calculus, which are used repeatedly throughout the text. For a more detailed introduction to these topics, including proofs, we recommend, e.g., Schey (1996), Boas (2006), and Griffiths (1999), on which we've based our discussion. Since the majority of the calculations in classical mechanics involve working with ordinary three-dimensional spatial vectors, e.g., position, velocity, acceleration, etc., we focus attention on such vectors in this appendix. Extensions to four-dimensional vectors, which arise in the context of relativistic mechanics, are discussed in Chap. 11, and the calculus of *differential forms* is discussed in Appendix B.

## A.1 Vector Algebra

In addition to adding two vectors, $\mathbf{A}+\mathbf{B}$, and multiplying a vector by a scalar, $a\mathbf{A}$, we can form various products of vectors: (i) The **dot product** (also called the **scalar product** or **inner product**) of two vectors $\mathbf{A}$ and $\mathbf{B}$ is defined by

$$\mathbf{A}\cdot\mathbf{B} \equiv AB\cos\theta\,, \qquad (A.1)$$

where $A \equiv |\mathbf{A}|$, $B \equiv |\mathbf{B}|$ are the magnitudes (or norms) of $\mathbf{A}$, $\mathbf{B}$, and $\theta$ is the angle between the two vectors. (ii) The **cross product** (also called the **vector product** or **exterior product**) of $\mathbf{A}$ and $\mathbf{B}$ is defined by

$$\mathbf{A}\times\mathbf{B} \equiv AB\sin\theta\,\hat{\mathbf{n}}\,, \qquad (A.2)$$



M.J. Benacquista and J.D. Romano, *Classical Mechanics*, Undergraduate Lecture Notes in Physics, https://doi.org/10.1007/978-3-319-68780-3

where $\theta$ is as before (assumed to be between 0° and 180°), and $\hat{\mathbf{n}}$ is a unit vector perpendicular to the plane spanned by $\mathbf{A}$ and $\mathbf{B}$, whose direction is given by the right-hand rule.[1] Note that if $\mathbf{A}$ and $\mathbf{B}$ are parallel, then $\mathbf{A} \times \mathbf{B} = \mathbf{0}$, while if $\mathbf{A}$ and $\mathbf{B}$ are perpendicular, $\mathbf{A} \cdot \mathbf{B} = 0$.

The dot product and cross product of $\mathbf{A}$ and $\mathbf{B}$ can also be written rather simply in terms of the components $A_i$, $B_i$ ($i = 1, 2, 3$) of $\mathbf{A}$ and $\mathbf{B}$ with respect to an orthonormal basis $\{\hat{\mathbf{e}}_1, \hat{\mathbf{e}}_2, \hat{\mathbf{e}}_3\}$. (See Appendix D.2.1 for a general review about decomposing a vector into its components with respect to a basis.) Expanding $\mathbf{A}$ and $\mathbf{B}$ as

$$\mathbf{A} = \sum_i A_i \, \hat{\mathbf{e}}_i \,, \qquad \mathbf{B} = \sum_i B_i \, \hat{\mathbf{e}}_i \,, \tag{A.3}$$

it follows that

$$\mathbf{A} \cdot \mathbf{B} = \sum_{i,j} \delta_{ij} A_i B_j = \sum_i A_i B_i \tag{A.4}$$

and

$$(\mathbf{A} \times \mathbf{B})_i = \sum_{j,k} \varepsilon_{ijk} A_j B_k \,, \tag{A.5}$$

where

$$\delta_{ij} \equiv \begin{cases} 1 & i = j \\ 0 & i \neq j \end{cases} \tag{A.6}$$

is the **Kronecker delta**, and[2]

$$\varepsilon_{ijk} \equiv \begin{cases} 1 & \text{if } ijk \text{ is an even permutation of 123} \\ -1 & \text{if } ijk \text{ is an odd permutation of 123} \\ 0 & \text{otherwise} \end{cases} \tag{A.7}$$

is the **Levi-Civita symbol**. We note that the above component expressions for dot product and cross product are valid with respect to *any* orthonormal basis, and not just for Cartesian coordinates.

Geometrically, the dot product of two vectors is the projection of one vector onto the direction of the other vector, times the magnitude of the other vector. Thus, by

[1] Point the fingers of your right hand in the direction of $\mathbf{A}$, and then curl them toward your palm in the direction of $\mathbf{B}$. Your thumb then points in the direction of $\hat{\mathbf{n}}$.

[2] An odd (even) permutation of 123 corresponds to an odd (even) number of interchanges of two of the numbers. For example, 213 is an odd permutation of 123, while 231 is an even permutation.

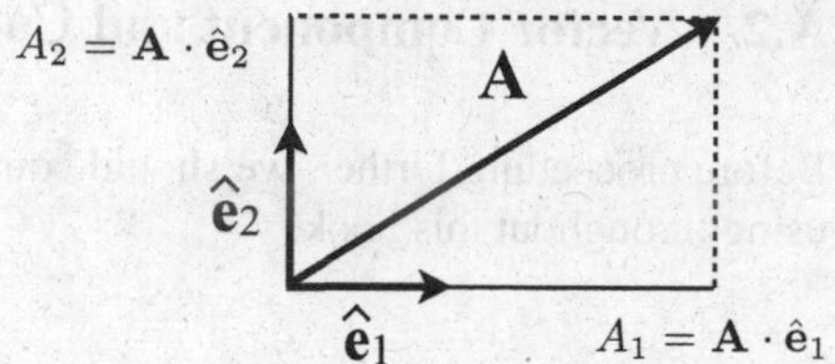


**Fig. A.1** Components of a 2-dimensional vector **A** in terms of dot products with the orthonormal basis vectors $\hat{\mathbf{e}}_1$, $\hat{\mathbf{e}}_2$

taking the dot product of **A** with the orthonormal basis vectors $\hat{\mathbf{e}}_i$, we obtain the components of **A** with respect to this basis, i.e., $A_i = \mathbf{A} \cdot \hat{\mathbf{e}}_i$. This is shown in Fig. A.1.

> **Exercise A.1** Prove that the geometric and component expressions for both the dot product, (A.1) and (A.4), and cross product, (A.2) and (A.5), are equivalent to one another, choosing a convenient coordinate system to do the calculation.

A key identity relating the Kronecker delta and Levi-Civita symbol is

$$\sum_i \varepsilon_{ijk}\varepsilon_{ilm} = \delta_{jl}\delta_{km} - \delta_{jm}\delta_{kl} \,. \tag{A.8}$$

Using this identity and the component forms of the dot product and cross product, one can prove the following three results:
Scalar triple product:

$$\mathbf{A} \cdot (\mathbf{B} \times \mathbf{C}) = \mathbf{B} \cdot (\mathbf{C} \times \mathbf{A}) = \mathbf{C} \cdot (\mathbf{A} \times \mathbf{B}) \tag{A.9}$$

Vector triple product:

$$\mathbf{A} \times (\mathbf{B} \times \mathbf{C}) = \mathbf{B}(\mathbf{A} \cdot \mathbf{C}) - \mathbf{C}(\mathbf{A} \cdot \mathbf{B}) \tag{A.10}$$

Jacobi identity:

$$\mathbf{A} \times (\mathbf{B} \times \mathbf{C}) + \mathbf{B} \times (\mathbf{C} \times \mathbf{A}) + \mathbf{C} \times (\mathbf{A} \times \mathbf{B}) = 0 \tag{A.11}$$

> **Exercise A.2** Prove the above three identities.

## A.2 Vector Component and Coordinate Notation

Before proceeding further, we should comment on the index notation that we'll be using throughout this book.

### *A.2.1 Contravariant and Covariant Vectors*

In general, one should distinguish between vectors with components $A^i$ (so-called **contravariant vectors**) and vectors with components $A_i$ (so-called **covariant vectors** or **dual vectors**). By definition these components transform *inversely* to one another under a change of basis or coordinate system. For example, if $A^i$ and $A_i$ denote the components of a contravariant and covariant vector with respect to a coordinate basis (See Appendix A.4.1), then under a coordinate transformation $x^i \to x^{i'} = x^{i'}(x^i)$:

$$A^{i'} = \sum_i \frac{\partial x^{i'}}{\partial x^i} A^i \,, \qquad A_{i'} = \sum_i \frac{\partial x^i}{\partial x^{i'}} A_i \,. \tag{A.12}$$

But since most of the calculations that we will perform involve quantities in ordinary 3-dimensional Euclidean space with components defined with respect to an *orthonormal* basis, then

$$\hat{\mathbf{e}}_i \cdot \hat{\mathbf{e}}_j = \delta_{ij} = \text{diag}(1, 1, 1)\,, \tag{A.13}$$

and the two sets of components $A^i$ and $A_i$ can be mapped to one another using the Kronecker delta:

$$A_i = \sum_j \delta_{ij} A^j \quad \Leftrightarrow \quad A_1 = A^1\,, \quad A_2 = A^2\,, \quad A_3 = A^3\,. \tag{A.14}$$

Hence, $A_i = A^i$, so it doesn't matter where we place the index. For simplicity of notation, we will typically use the subscript notation, which is the standard notation in the classical mechanics literature.

This equality between covariant and contravariant components will not hold, however, when we discuss spacetime 4-vectors in the context of special relativity (Chap. 11). This is because a set of orthonormal spacetime basis vectors satisfies (See Sect. 11.5.2.1)

$$\mathbf{e}_\alpha \cdot \mathbf{e}_\beta = \eta_{\alpha\beta} = \mathrm{diag}(-1, 1, 1, 1)\,, \tag{A.15}$$

where $\alpha = 0, 1, 2, 3$ labels the spacetime coordinates $x^\alpha \equiv (ct, x, y, z)$ of an inertial reference frame. Thus, the two sets of components $A^\alpha$ and $A_\alpha$ are related by

$$A_\alpha = \sum_\beta \eta_{\alpha\beta} A^\beta \quad \Leftrightarrow \quad A_0 = -A^0\,, \quad A_1 = A^1\,, \quad A_2 = A^2\,, \quad A_3 = A^3\,. \tag{A.16}$$

So for relativistic mechanics, $A^0$ and $A_0$ differ by a minus sign, and hence it will be important to distinguish between the components of contravariant and covariant vectors. Whether an index is a superscript or a subscript *does* make a difference in special relativity.

> **Exercise A.3** Show that under a coordinate transformation $x^i \to x^{i'}(x^i)$, the components $A^i$ of a contravariant vector transform like the coordinate differentials $\mathrm{d}x^i$, while the components $A_i$ of a covariant vector transform like the partial derivative operators $\partial/\partial x^i$.

### A.2.2 Coordinate Notation

Regarding coordinates, we will generally use superscripts (as we have above) to denote the collection of coordinates as a whole, e.g.,

(i) $x^i \equiv (x^1, x^2, x^3) = (x, y, z)$, $(r, \theta, \phi)$, or $(\rho, \phi, z)$ for ordinary 3-dimensional Euclidean space,
(ii) $x^\alpha \equiv (ct, x, y, z)$ for Minkowski spacetime,
(iii) $q^a \equiv (q^1, q^2, \ldots, q^n)$ for generalized coordinates defining the configuration of a system of particles having $n$ degrees of freedom.

Note that the superscripts just label the different coordinates, e.g., $x^2$ and $x^3$, and do *not* correspond to the square or cube of a single coordinate $x$. We choose this notation because tangents to curves in these spaces naturally define contravariant vectors, e.g.,

$$v^i \equiv \frac{\mathrm{d}x^i(\lambda)}{\mathrm{d}\lambda}\,, \tag{A.17}$$

and partial derivatives of scalars with respect to the coordinates naturally define covariant vectors, e.g.,

$$\omega_i \equiv \frac{\partial \varphi}{\partial x^i}\,, \tag{A.18}$$

with the placement of the superscript or subscript indices matching on both sides of the equation.[3] This is valid even for spaces that are not Euclidean and for coordinates that are not Cartesian, e.g., the angular coordinates describing the configuration of a planar double pendulum (See e.g., Problem 1.4).

### A.2.3 *Other Indices*

All other types of indices that we might need to use, e.g., to label different functions, basis vectors, or particles in a system, etc., will be placed as either superscripts or subscripts in whichever way is most notationally convenient for the discussion at hand. There is no "transformation law" associated with changes in these types of indices, so there is no standard convention for their placement.

## A.3 Differential Vector Calculus

To do calculus with vectors, we need *fields*—both **scalar fields**, which assign a real number to each position in space, and **vector fields**, which assign a three-dimensional vector to each position. An example of a scalar field is the gravitational potential $\Phi(\mathbf{r})$ for a stationary mass distribution, written as a function of the spatial location $\mathbf{r}$. An example of a vector field is the velocity $\mathbf{v}(\mathbf{r}, t)$ of a fluid at a fixed time $t$, which is a function of position $\mathbf{r}$ within the fluid.

Given a scalar field $U(\mathbf{r})$ and vector field $\mathbf{A}(\mathbf{r})$, we can define the following derivatives:

**(i) Gradient**:

$$(\nabla U) \cdot \hat{\mathbf{t}} \equiv \lim_{\Delta s \to 0} \left[ \frac{U(\mathbf{r}_2) - U(\mathbf{r}_1)}{\Delta s} \right], \tag{A.19}$$

where $\mathbf{r}_1$ and $\mathbf{r}_2$ are the endpoints (i.e., the 'boundary') of the vector displacement $\Delta\mathbf{s} \equiv \Delta s\, \hat{\mathbf{t}}$. Thus, $(\nabla U) \cdot \hat{\mathbf{t}}$ measures the change in $U$ in the direction of $\hat{\mathbf{t}}$. This is the **directional derivative** of the scalar field $U$. The direction of $\nabla U$ is *perpendicular* to the contour lines $U(\mathbf{r}) = $ const, since the right-hand side is zero for points that

[3] If we had swapped the notation and denoted the components of contravariant vectors with subscripts and the components of covariant vectors with superscripts, then to match indices would require denoting the collection of coordinates with subscripts, like $x_i$. In retrospect, this might have been a less confusing notation for coordinates (e.g., no chance of confusing the second coordinate $x_2$ with $x$-squared, etc.). But we will stick with the coordinate index notation that we have adopted above since it is the standard notation in the literature.

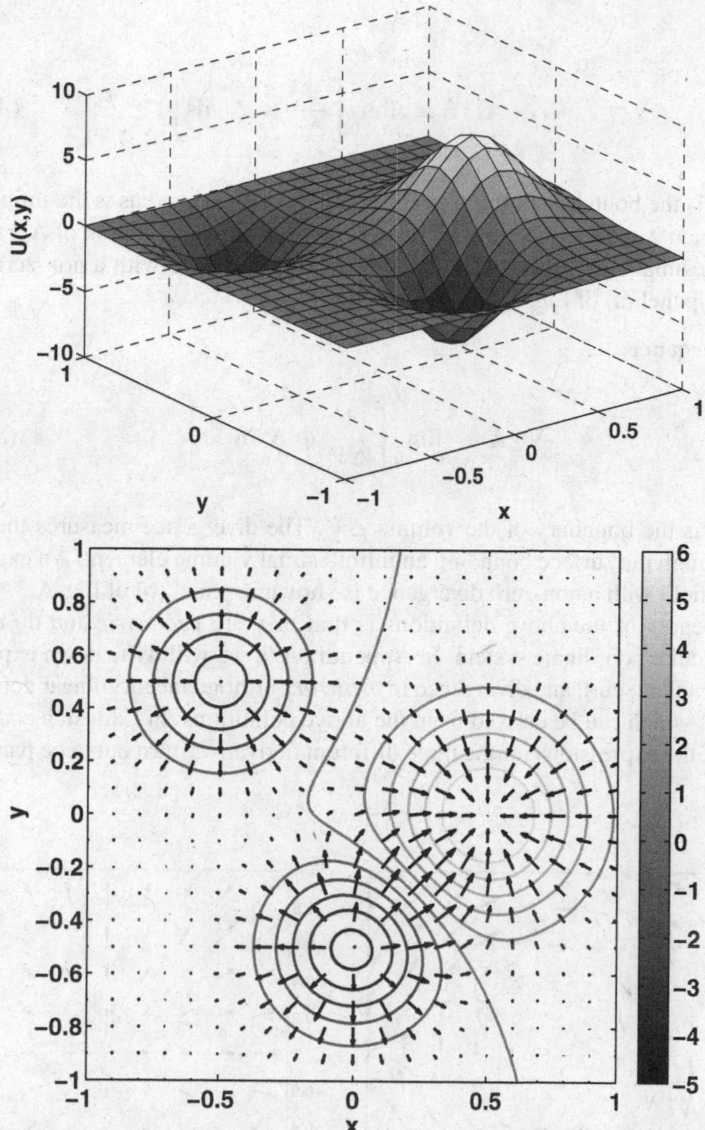


**Fig. A.2** Top panel: Function $U(x, y)$ displayed as a 2-dimensional surface. Bottom panel: Contour plot (lines of constant $U$, lighter lines corresponding to larger values) with gradient vector field $\nabla U$ superimposed. Note that the direction of $\nabla U$ is perpendicular to the $U(x, y) = \text{const}$ lines and is largest in magnitude where the change in $U$ is greatest

lie along a contour. Hence the gradient $\nabla U$ points in the direction of *steepest ascent* of the function $U$. This is illustrated graphically in Fig. A.2 for a function of two variables $U(x, y)$.

**(ii) Curl:**

$$(\nabla \times \mathbf{A}) \cdot \hat{\mathbf{n}} \equiv \lim_{\Delta a \to 0} \left[ \frac{1}{\Delta a} \oint_C \mathbf{A} \cdot d\mathbf{s} \right], \qquad (A.20)$$

where $C$ is the boundary of the area element $\Delta\mathbf{a} \equiv \hat{\mathbf{n}}\, \Delta a$, and $d\mathbf{s}$ is the infinitesimal displacement vector tangent to $C$. The curl measures the circulation of $\mathbf{A}(\mathbf{r})$ around an infinitesimal closed curve. An example of a vector field with a non-zero curl is shown in panel (a) of Fig. A.3.

**(iii) Divergence:**

$$\nabla \cdot \mathbf{A} \equiv \lim_{\Delta V \to 0} \left[ \frac{1}{\Delta V} \oint_S \mathbf{A} \cdot \hat{\mathbf{n}}\, da \right], \qquad (A.21)$$

where $S$ is the boundary of the volume $\Delta V$. The divergence measures the flux of $\mathbf{A}(\mathbf{r})$ through the surface bounding an infinitesimal volume element. An example of a vector field with a non-zero divergence is shown in panel (b) of Fig. A.3.

The beauty of the above definitions is that they are *geometric* and do not refer to a particular coordinate system. In Appendix A.5, we will write down expressions for the gradient, curl, and divergence in *arbitrary* orthogonal curvilinear coordinates $(u, v, w)$, which can be derived from the above definitions. In Cartesian coordinates $(x, y, z)$, the expressions for the three different derivatives turn out to be particularly simple:

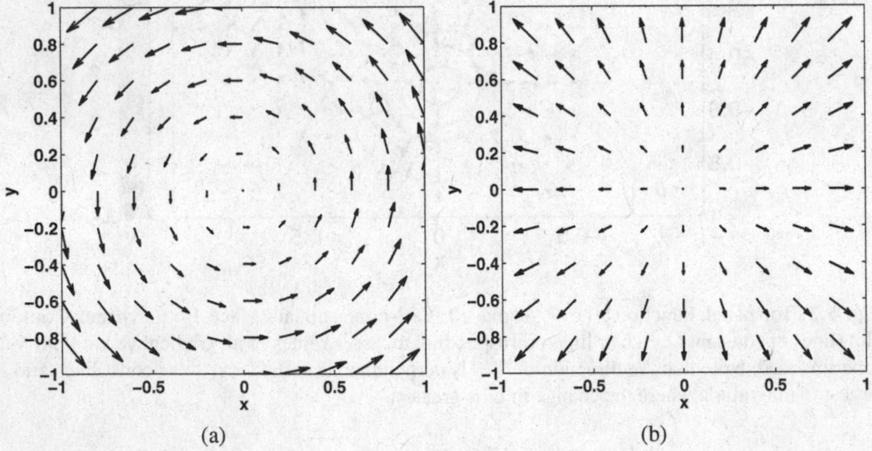


**Fig. A.3** Panel (a) Example of a vector field, $\mathbf{A}(\mathbf{r}) = -y\,\hat{\mathbf{x}} + x\,\hat{\mathbf{y}}$, with a non-zero curl, $\nabla \times \mathbf{A} = 2\hat{\mathbf{z}}$. Panel (b) Example of a vector field $\mathbf{A}(\mathbf{r}) = x\,\hat{\mathbf{x}} + y\,\hat{\mathbf{y}} + z\,\hat{\mathbf{z}}$, with a non-zero divergence, $\nabla \cdot \mathbf{A} = 3$. In both cases, just the $z = 0$ values of the vector fields are shown in these figures

$$\begin{aligned}
\nabla U &= \frac{\partial U}{\partial x}\,\hat{\mathbf{x}} + \frac{\partial U}{\partial y}\,\hat{\mathbf{y}} + \frac{\partial U}{\partial z}\,\hat{\mathbf{z}}\,, \\
\nabla \times \mathbf{A} &= \left(\frac{\partial A_z}{\partial y} - \frac{\partial A_y}{\partial z}\right)\hat{\mathbf{x}} + \left(\frac{\partial A_x}{\partial z} - \frac{\partial A_z}{\partial x}\right)\hat{\mathbf{y}} + \left(\frac{\partial A_y}{\partial x} - \frac{\partial A_x}{\partial y}\right)\hat{\mathbf{z}}\,, \\
\nabla \cdot \mathbf{A} &= \frac{\partial A_x}{\partial x} + \frac{\partial A_y}{\partial y} + \frac{\partial A_z}{\partial z}\,.
\end{aligned} \tag{A.22}$$

In more compact form,

$$(\nabla U)_i = \partial_i U\,, \quad (\nabla \times \mathbf{A})_i = \sum_{j,k} \varepsilon_{ijk}\partial_j A_k\,, \quad \nabla \cdot \mathbf{A} = \sum_i \partial_i A_i\,, \tag{A.23}$$

where $\partial_i$ is shorthand for the partial derivative $\partial/\partial x^i$, where $x^i \equiv (x, y, z)$.

We conclude this subsection by noting that the curl and divergence of a vector field $\mathbf{A}(\mathbf{r})$, although important derivative operations, *do not completely capture how a vector field changes as you move from point to point*. A simple counting argument shows that, in three-dimensions, we need $3 \times 3 = 9$ components to completely specify how a vector field changes from point to point (three components of $\mathbf{A}$ times the three directions in which to take the derivative). The curl and divergence supply $3 + 1 = 4$ of those components. So we are missing 5 components, which turn out to have the geometrical interpretation of **shear** (See, e.g., Romano and Price 2012). The shear can be calculated in terms of the **directional derivative of a vector field**, which we shall discuss in Appendix A.4. Figure A.4 shows an example of a vector field that has zero curl and zero divergence, but is clearly not a constant. This is an example of a *pure-shear* field (See Exercise A.11).

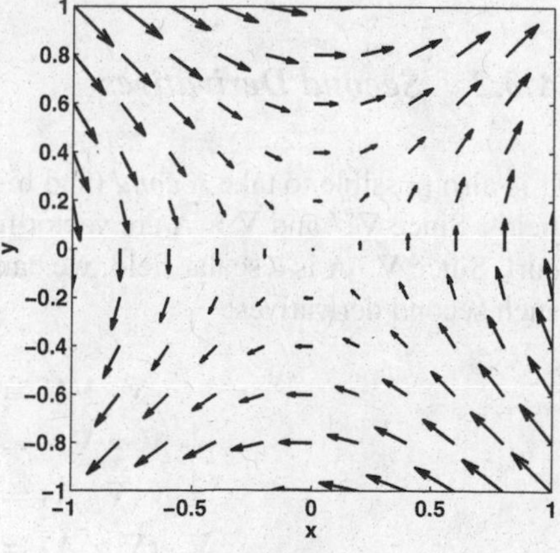


**Fig. A.4** Example of a vector field $\mathbf{A}(\mathbf{r}) = y\,\hat{\mathbf{x}} + x\,\hat{\mathbf{y}}$ that has both $\nabla \times \mathbf{A} = 0$ and $\nabla \cdot \mathbf{A} = 0$, but non-zero shear

### A.3.1 Product Rules

It turns out that the product rule

$$\frac{\mathrm{d}}{\mathrm{d}x}(fg) = \frac{\mathrm{d}f}{\mathrm{d}x}g + f\frac{\mathrm{d}g}{\mathrm{d}x} \tag{A.24}$$

for ordinary functions of one variable, $f(x)$ and $g(x)$, extends to the gradient, curl, and divergence operations, although the resulting expressions are more complicated. Since there are four different ways of combining a pair of scalar and/or vector fields (i.e., $fg$, $\mathbf{A}\cdot\mathbf{B}$, $f\mathbf{A}$, $\mathbf{A}\times\mathbf{B}$) and two different ways of taking derivatives of vector fields (either curl or divergence), there are six different product rules: here are six different product

$$\begin{aligned}
\nabla(fg) &= (\nabla f)g + f(\nabla g)\,, && \text{(A.25a)}\\
\nabla(\mathbf{A}\cdot\mathbf{B}) &= \mathbf{A}\times(\nabla\times\mathbf{B}) + \mathbf{B}\times(\nabla\times\mathbf{A}) + (\mathbf{A}\cdot\nabla)\mathbf{B} + (\mathbf{B}\cdot\nabla)\mathbf{A}\,, && \text{(A.25b)}\\
\nabla\times(f\mathbf{A}) &= (\nabla f)\times\mathbf{A} + f\nabla\times\mathbf{A}\,, && \text{(A.25c)}\\
\nabla\times(\mathbf{A}\times\mathbf{B}) &= (\mathbf{B}\cdot\nabla)\mathbf{A} - (\mathbf{A}\cdot\nabla)\mathbf{B} + \mathbf{A}(\nabla\cdot\mathbf{B}) - \mathbf{B}(\nabla\cdot\mathbf{A})\,, && \text{(A.25d)}\\
\nabla\cdot(f\mathbf{A}) &= (\nabla f)\cdot\mathbf{A} + f\nabla\cdot\mathbf{A}\,, && \text{(A.25e)}\\
\nabla\cdot(\mathbf{A}\times\mathbf{B}) &= (\nabla\times\mathbf{A})\cdot\mathbf{B} - \mathbf{A}\cdot(\nabla\times\mathbf{B})\,. && \text{(A.25f)}
\end{aligned}$$

We will discuss some of these product rules in more detail in Appendix A.4.

**Exercise A.4** Prove the above product rules. (*Hint*: Do the calculations in Cartesian coordinates where the expressions for gradient, curl, and divergence are the simplest.)

### A.3.2 Second Derivatives

It is also possible to take *second* (and higher-order) derivatives of scalar and vector fields. Since $\nabla U$ and $\nabla\times\mathbf{A}$ are vector fields, we can take either their divergence or curl. Since $\nabla\cdot\mathbf{A}$ is a scalar field, we can take only its gradient. Thus, there are five such second derivatives:

$$\begin{aligned}
\nabla\cdot\nabla U &\equiv \nabla^2 U\,, && \text{(A.26a)}\\
\nabla\times\nabla U &= 0\,, && \text{(A.26b)}\\
\nabla(\nabla\cdot\mathbf{A}) &= \text{a vector field}\,, && \text{(A.26c)}\\
\nabla\cdot(\nabla\times\mathbf{A}) &= 0\,, && \text{(A.26d)}\\
\nabla\times(\nabla\times\mathbf{A}) &\equiv \nabla(\nabla\cdot\mathbf{A}) - \nabla^2\mathbf{A}\,. && \text{(A.26e)}
\end{aligned}$$

Note that the curl of a gradient, $\nabla \times \nabla U$, and the divergence of a curl, $\nabla \cdot (\nabla \times \mathbf{A})$, are both identically zero. The divergence of a gradient defines the **Laplacian** of a scalar field, $\nabla^2 U$, and the curl of a curl defines the Laplacian of a vector field, $\nabla^2 \mathbf{A}$ (second term on the right-hand side of (A.26e)). In Cartesian coordinates $x^i \equiv (x, y, z)$, the scalar and vector Laplacians are given by

$$\begin{aligned} \nabla^2 U &= \frac{\partial^2 U}{\partial x^2} + \frac{\partial^2 U}{\partial y^2} + \frac{\partial^2 U}{\partial z^2}, \\ \left(\nabla^2 \mathbf{A}\right)_i &= \frac{\partial^2 A_i}{\partial x^2} + \frac{\partial^2 A_i}{\partial y^2} + \frac{\partial^2 A_i}{\partial z^2}, \qquad i = 1, 2, 3. \end{aligned} \tag{A.27}$$

The gradient of a divergence is a non-zero vector field in general, but it has no special name, as it does not appear as frequently as the Laplacian operator.

***Example A.1*** Prove that $\nabla \times \nabla U = \mathbf{0}$ and $\nabla \cdot (\nabla \times \mathbf{A}) = 0$.
*Solution*: Since these are vector equations, we can do the proof in any coordinate system. For simplicity, we will use Cartesian coordinates $x^i \equiv (x, y, z)$ where the gradient, curl, and divergence are given by (A.23). Then

$$[\nabla \times \nabla U]_i = \sum_{j,k} \varepsilon_{ijk} \partial_j \partial_k U = 0, \tag{A.28}$$

since partial derivatives commute and $\varepsilon_{ijk}$ is totally anti-symmetric. Similarly,

$$\nabla \cdot (\nabla \times \mathbf{A}) = \sum_i \partial_i \left( \sum_{j,k} \varepsilon_{ijk} \partial_j A_k \right) = 0, \tag{A.29}$$

again since partial derivatives commute and $\varepsilon_{ijk}$ is totally anti-symmetric. □

**Exercise A.5** Verify $\nabla \times (\nabla \times \mathbf{A}) = \nabla(\nabla \cdot \mathbf{A}) - \nabla^2 \mathbf{A}$ in Cartesian coordinates.

## A.4 Directional Derivatives

You might have noticed that the right-hand sides of (A.25b) and (A.25d) for $\nabla(\mathbf{A} \cdot \mathbf{B})$ and $\nabla \times (\mathbf{A} \times \mathbf{B})$ involve quantities of the form $(\mathbf{B} \cdot \nabla)\mathbf{A}$, which are *not* gradients, curls, or divergences of a vector or scalar field. Geometrically, $(\mathbf{B} \cdot \nabla)\mathbf{A}$ represents the directional derivative of the vector field **A** in the direction of **B**, which generalizes the definition of the directional derivative of a scalar field. To calculate $(\mathbf{B}\cdot\nabla)\mathbf{A}$, we need to evaluate the directional derivatives of the components $A_i$ with respect to a basis $\hat{\mathbf{e}}_i$, as well as the directional derivatives of the *basis vectors* themselves. But before doing that calculation, it is worthwhile to remind ourselves about directional derivatives of scalar fields, and also how to calculate **coordinate basis vectors** in arbitrary curvilinear coordinates $(u, v, w)$.

### A.4.1 Directional Derivative of a Function; Coordinate Basis Vectors

Suppose we are given a curve $x^i = x^i(\lambda)$ parametrized by $\lambda$, with tangent vector $v^i \equiv \mathrm{d}x^i/\mathrm{d}\lambda$ defined along the curve. Then the directional derivative of a (scalar) function $f$ evaluated at any point along the curve is given by

$$\frac{\mathrm{d}f}{\mathrm{d}\lambda} \equiv \sum_i \frac{\mathrm{d}x^i}{\mathrm{d}\lambda}\frac{\partial f}{\partial x^i} = \sum_i v^i \frac{\partial f}{\partial x^i} \equiv \mathbf{v}(f)\,. \tag{A.30}$$

The notation $\mathbf{v}(f)$ should be thought of as $\mathbf{v}$ "acting on" $f$. Note that if we abstract away the function $f$, we have $\mathbf{v} = \sum_i v^i\partial/\partial x^i$, with the partial derivative operators $\partial/\partial x^i$ playing the role of **coordinate basis vectors**. Denoting these basis vectors by the boldface symbol $\boldsymbol{\partial}_i$, we have[4]

$$\mathbf{v} = \sum_i v^i \boldsymbol{\partial}_i\,. \tag{A.31}$$

Thus, the directional derviative of a scalar field sets up a one-to-one correspondence between vectors $\mathbf{v}$ and directional derivative operators $\sum_i v^i\partial/\partial x^i$.

One nice feature about this correspondence between vectors and directional derivative operators is that it suggests how to calculate the coordinate basis vectors for arbitrary curvilinear coordinates $(u, v, w)$ in terms of the Cartesian basis vectors $\hat{\mathbf{x}}, \hat{\mathbf{y}}, \hat{\mathbf{z}}$. One simply takes the chain rule

$$\frac{\partial}{\partial u} = \frac{\partial x}{\partial u}\frac{\partial}{\partial x} + \frac{\partial y}{\partial u}\frac{\partial}{\partial y} + \frac{\partial z}{\partial u}\frac{\partial}{\partial z}\,, \qquad \text{etc.}\,, \tag{A.32}$$

---

[4] A particular coordinate basis vector $\boldsymbol{\partial}_i$ points along the $x^i$ coordinate line, with all other coordinates (i.e., $x^j$ with $j \neq i$) constant.

and formally converts it to a vector equation

$$\boldsymbol{\partial}_u = \frac{\partial x}{\partial u}\boldsymbol{\partial}_x + \frac{\partial y}{\partial u}\boldsymbol{\partial}_y + \frac{\partial z}{\partial u}\boldsymbol{\partial}_z\,, \qquad \text{etc.}\,, \tag{A.33}$$

with partial derivative operators replaced everywhere by coordinate basis vectors. But since the coordinate basis vectors in Cartesian coordinates are orthogonal and have unit norm, with $\boldsymbol{\partial}_x = \hat{\mathbf{x}}$, etc., it follows that

$$\begin{aligned}
\boldsymbol{\partial}_u &= \frac{\partial x}{\partial u}\,\hat{\mathbf{x}} + \frac{\partial y}{\partial u}\,\hat{\mathbf{y}} + \frac{\partial z}{\partial u}\,\hat{\mathbf{z}}\,,\\
\boldsymbol{\partial}_v &= \frac{\partial x}{\partial v}\,\hat{\mathbf{x}} + \frac{\partial y}{\partial v}\,\hat{\mathbf{y}} + \frac{\partial z}{\partial v}\,\hat{\mathbf{z}}\,,\\
\boldsymbol{\partial}_w &= \frac{\partial x}{\partial w}\,\hat{\mathbf{x}} + \frac{\partial y}{\partial w}\,\hat{\mathbf{y}} + \frac{\partial z}{\partial w}\,\hat{\mathbf{z}}\,.
\end{aligned} \tag{A.34}$$

The norms of these coordinate basis vectors are then given by

$$\begin{aligned}
N_u &\equiv |\boldsymbol{\partial}_u| = \sqrt{\left(\frac{\partial x}{\partial u}\right)^2 + \left(\frac{\partial y}{\partial u}\right)^2 + \left(\frac{\partial z}{\partial u}\right)^2}\,,\\
N_v &\equiv |\boldsymbol{\partial}_v| = \sqrt{\left(\frac{\partial x}{\partial v}\right)^2 + \left(\frac{\partial y}{\partial v}\right)^2 + \left(\frac{\partial z}{\partial v}\right)^2}\,,\\
N_w &\equiv |\boldsymbol{\partial}_w| = \sqrt{\left(\frac{\partial x}{\partial w}\right)^2 + \left(\frac{\partial y}{\partial w}\right)^2 + \left(\frac{\partial z}{\partial w}\right)^2}\,,
\end{aligned} \tag{A.35}$$

which we can then use to calculate *unit* vectors

$$\hat{\mathbf{u}} = N_u^{-1}\boldsymbol{\partial}_u\,, \qquad \hat{\mathbf{v}} = N_v^{-1}\boldsymbol{\partial}_v\,, \qquad \hat{\mathbf{w}} = N_w^{-1}\boldsymbol{\partial}_w\,. \tag{A.36}$$

In general, these unit vectors will *not* be orthogonal, although they will be for several common coordinate systems, including spherical coordinates $(r, \theta, \phi)$ and cylindrical coordinates $(\rho, \phi, z)$ (See Appendix A.5 for details). We will use the above results in the next section when calculating the directional derivative of a vector field in non-Cartesian coordinates.

### A.4.2 Directional Derivative of a Vector Field

Let's return now to the problem of calculating $(\mathbf{B}\cdot\nabla)\mathbf{A}$, which started this discussion of directional derivatives. In Cartesian coordinates, it is natural to define $(\mathbf{B}\cdot\nabla)\mathbf{A}$ in terms of its components via

$$(\mathbf{B}\cdot\nabla)\mathbf{A} \equiv \sum_{i,j}\left[(B_i\partial_i)A_j\right]\hat{\mathbf{e}}_j \tag{A.37}$$

since the orthonormal basis vectors $\hat{\mathbf{e}}_i = \{\hat{\mathbf{x}}, \hat{\mathbf{y}}, \hat{\mathbf{z}}\}$ are *constant* vector fields. In non-Cartesian coordinates, where the coordinate basis vectors change from point to point, we would need to make the appropriate coordinate transformations for both the vector components and the partial derivative operators. Although straightforward, this is usually a rather long and tedious process.

A simpler method for calculating $(\mathbf{B}\cdot\nabla)\mathbf{A}$ in non-Cartesian coordinates $x^i$ is to expand both $\mathbf{A}$ and $\mathbf{B}$ in terms of the orthonormal basis vectors $\hat{\mathbf{e}}_i$,

$$(\mathbf{B}\cdot\nabla)\mathbf{A} = \sum_{i,j}(B_i\hat{\mathbf{e}}_i\cdot\nabla)(A_j\hat{\mathbf{e}}_j) = \sum_{i,j}B_i(\nabla_{\hat{\mathbf{e}}_i}A_j)\hat{\mathbf{e}}_j + \sum_{i,j}B_iA_j(\nabla_{\hat{\mathbf{e}}_i}\hat{\mathbf{e}}_j)\,, \tag{A.38}$$

and then evaluate $\nabla_{\hat{\mathbf{e}}_i}\hat{\mathbf{e}}_j$ by further expanding $\hat{\mathbf{e}}_j$ as a linear combination of the Cartesian basis vectors $\hat{\mathbf{e}}_{i'} = \{\hat{\mathbf{x}}, \hat{\mathbf{y}}, \hat{\mathbf{z}}\}$. (Here we are using the notation $\nabla_{\hat{\mathbf{e}}_i} \equiv \hat{\mathbf{e}}_i\cdot\nabla$, and we are using a prime to distinguish the Cartesian basis vectors $\hat{\mathbf{e}}_{i'}$ from the non-Cartesian basis vectors $\hat{\mathbf{e}}_i$.) This leads to

$$\nabla_{\hat{\mathbf{e}}_i}\hat{\mathbf{e}}_j = \nabla_{\hat{\mathbf{e}}_i}\left(\sum_{k'}\Lambda_{jk'}\hat{\mathbf{e}}_{k'}\right) \equiv \sum_{k'}(\nabla_{\hat{\mathbf{e}}_i}\Lambda_{jk'})\hat{\mathbf{e}}_{k'}\,, \tag{A.39}$$

where we have applied the derivatives only to the expansion coefficients $\Lambda_{jk'}$, since the Cartesian basis vectors $\hat{\mathbf{e}}_{k'}$ are constants. For example, for the spherical coordinate basis vectors $\hat{\mathbf{e}}_i = \{\hat{\mathbf{r}}, \hat{\boldsymbol{\theta}}, \hat{\boldsymbol{\phi}}\}$, we have

$$\begin{aligned}
\Lambda &= \begin{bmatrix} \sin\theta\cos\phi & \sin\theta\sin\phi & \cos\theta \\ \cos\theta\cos\phi & \cos\theta\sin\phi & -\sin\theta \\ -\sin\phi & \cos\phi & 0 \end{bmatrix}, \\
\Lambda^{-1} &= \begin{bmatrix} \cos\phi\sin\theta & \cos\phi\cos\theta & -\sin\phi \\ \sin\phi\sin\theta & \sin\phi\cos\theta & \cos\phi \\ \cos\theta & -\sin\theta & 0 \end{bmatrix},
\end{aligned} \tag{A.40}$$

for the matrix of expansion coefficients $\Lambda_{jk'}$ and its inverse $(\Lambda^{-1})_{k'l}$ (See Example A.2 for details). If we then re-express the Cartesian basis vectors in terms of the original non-Cartesian basis vectors using the inverse transformation matrix $(\Lambda^{-1})_{k'l}$, we obtain

$$\nabla_{\hat{\mathbf{e}}_i}\hat{\mathbf{e}}_j = \sum_{k'}(\nabla_{\hat{\mathbf{e}}_i}\Lambda_{jk'})\sum_l(\Lambda^{-1})_{k'l}\hat{\mathbf{e}}_l = \sum_l C_{ijl}\hat{\mathbf{e}}_l\,, \tag{A.41}$$

where

$$C_{ijl} \equiv \sum_{k'}(\nabla_{\hat{\mathbf{e}}_i}\Lambda_{jk'})(\Lambda^{-1})_{k'l}\,. \tag{A.42}$$

The $C_{ijl}$ are often called **connection coefficients**. Thus,

$$(\mathbf{B} \cdot \nabla)\mathbf{A} = \sum_{i,j} B_i (\nabla_{\hat{\mathbf{e}}_i} A_j)\hat{\mathbf{e}}_j + \sum_{i,j,l} B_i A_j C_{ijl} \hat{\mathbf{e}}_l \,. \tag{A.43}$$

Finally, if the non-Cartesian coordinates $x^i$ are *orthogonal*, as is the case for spherical coordinates $(r, \theta, \phi)$ and cylindrical coordinates $(\rho, \phi, z)$, then $\nabla_{\hat{\mathbf{e}}_i} = N_i^{-1}\partial/\partial x^i$, where $N_i$ is a normalization factor relating the (in general, unnormalized) coordinate basis vectors $\boldsymbol{\partial}_i$ to the orthonormal basis vectors $\hat{\mathbf{e}}_i$. For example, in spherical coordinates $\hat{\mathbf{r}} = \boldsymbol{\partial}_r$, $\hat{\boldsymbol{\theta}} = r^{-1}\boldsymbol{\partial}_\theta$, and $\hat{\boldsymbol{\phi}} = (r \sin\theta)^{-1}\boldsymbol{\partial}_\phi$.

Although this might seem like a complicated procedure when discussed abstractly, in practice it is relatively easy to carry out, as the following example shows.

***Example A.2*** Calculate $\nabla_{\hat{\mathbf{e}}_i} \hat{\mathbf{e}}_j$ in spherical coordinates $(r, \theta, \phi)$.
*Solution*: Recall that spherical coordinates $(r, \theta, \phi)$ are related to Cartesian coordinates $(x, y, z)$ via

$$x = r \sin\theta \cos\phi \,, \qquad y = r \sin\theta \sin\phi \,, \qquad z = r \cos\theta \,. \tag{A.44}$$

Using the chain rule to relate partial derivatives, e.g.,

$$\frac{\partial}{\partial r} = \frac{\partial x}{\partial r}\frac{\partial}{\partial x} + \frac{\partial y}{\partial r}\frac{\partial}{\partial y} + \frac{\partial z}{\partial r}\frac{\partial}{\partial z} \,, \qquad \text{etc.,} \tag{A.45}$$

it follows that

$$\begin{aligned} \hat{\mathbf{r}} &= \boldsymbol{\partial}_r = \sin\theta\cos\phi\,\hat{\mathbf{x}} + \sin\theta\sin\phi\,\hat{\mathbf{y}} + \cos\theta\,\hat{\mathbf{z}} \,, \\ \hat{\boldsymbol{\theta}} &= r^{-1}\boldsymbol{\partial}_\theta = \cos\theta\cos\phi\,\hat{\mathbf{x}} + \cos\theta\sin\phi\,\hat{\mathbf{y}} - \sin\theta\,\hat{\mathbf{z}} \,, \\ \hat{\boldsymbol{\phi}} &= (r\sin\theta)^{-1}\boldsymbol{\partial}_\phi = -\sin\phi\,\hat{\mathbf{x}} + \cos\phi\,\hat{\mathbf{y}} \,, \end{aligned} \tag{A.46}$$

using the one-to-one correspondence between vectors and directional derivative operators discussed in Appendix A.4.1. The inverse transformation is given by

$$\begin{aligned} \hat{\mathbf{x}} &= \boldsymbol{\partial}_x = \cos\phi\sin\theta\,\hat{\mathbf{r}} + \cos\phi\cos\theta\,\hat{\boldsymbol{\theta}} - \sin\phi\,\hat{\boldsymbol{\phi}} \,, \\ \hat{\mathbf{y}} &= \boldsymbol{\partial}_y = \sin\phi\sin\theta\,\hat{\mathbf{r}} + \sin\phi\cos\theta\,\hat{\boldsymbol{\theta}} + \cos\phi\,\hat{\boldsymbol{\phi}} \,, \\ \hat{\mathbf{z}} &= \boldsymbol{\partial}_z = \cos\theta\,\hat{\mathbf{r}} - \sin\theta\,\hat{\boldsymbol{\theta}} \,. \end{aligned} \tag{A.47}$$

Performing the derivatives as described above, we find

$$\begin{aligned}
\nabla_{\hat{\mathbf{r}}}\hat{\mathbf{r}} &= \partial_r\left(\sin\theta\cos\phi\,\hat{\mathbf{x}}+\sin\theta\sin\phi\,\hat{\mathbf{y}}+\cos\theta\,\hat{\mathbf{z}}\right)=0\,,\\
\nabla_{\hat{\boldsymbol{\theta}}}\hat{\mathbf{r}} &= r^{-1}\partial_\theta\left(\sin\theta\cos\phi\,\hat{\mathbf{x}}+\sin\theta\sin\phi\,\hat{\mathbf{y}}+\cos\theta\,\hat{\mathbf{z}}\right)\\
&= r^{-1}\left(\cos\theta\cos\phi\,\hat{\mathbf{x}}+\cos\theta\sin\phi\,\hat{\mathbf{y}}-\sin\theta\,\hat{\mathbf{z}}\right)=r^{-1}\,\hat{\boldsymbol{\theta}}\,,\\
\nabla_{\hat{\boldsymbol{\phi}}}\hat{\mathbf{r}} &= (r\sin\theta)^{-1}\partial_\phi\left(\sin\theta\cos\phi\,\hat{\mathbf{x}}+\sin\theta\sin\phi\,\hat{\mathbf{y}}+\cos\theta\,\hat{\mathbf{z}}\right)\\
&= (r\sin\theta)^{-1}\left(-\sin\theta\sin\phi\,\hat{\mathbf{x}}+\sin\theta\cos\phi\,\hat{\mathbf{y}}\right)=r^{-1}\,\hat{\boldsymbol{\phi}}\,.
\end{aligned} \tag{A.48}$$

Continuing in this fashion:

$$\begin{aligned}
&\nabla_{\hat{\mathbf{r}}}\hat{\mathbf{r}}=0\,,\quad \nabla_{\hat{\boldsymbol{\theta}}}\hat{\mathbf{r}}=r^{-1}\,\hat{\boldsymbol{\theta}}\,,\quad \nabla_{\hat{\boldsymbol{\phi}}}\hat{\mathbf{r}}=r^{-1}\,\hat{\boldsymbol{\phi}}\,,\\
&\nabla_{\hat{\mathbf{r}}}\hat{\boldsymbol{\theta}}=0\,,\quad \nabla_{\hat{\boldsymbol{\theta}}}\hat{\boldsymbol{\theta}}=-r^{-1}\,\hat{\mathbf{r}}\,,\quad \nabla_{\hat{\boldsymbol{\phi}}}\hat{\boldsymbol{\theta}}=r^{-1}\cot\theta\,\hat{\boldsymbol{\phi}}\,,\\
&\nabla_{\hat{\mathbf{r}}}\hat{\boldsymbol{\phi}}=0\,,\quad \nabla_{\hat{\boldsymbol{\theta}}}\hat{\boldsymbol{\phi}}=0\,,\quad \nabla_{\hat{\boldsymbol{\phi}}}\hat{\boldsymbol{\phi}}=-r^{-1}\,\hat{\mathbf{r}}-r^{-1}\cot\theta\,\hat{\boldsymbol{\theta}}\,.
\end{aligned} \tag{A.49}$$

□

**Exercise A.6** Calculate $\nabla_{\hat{\mathbf{e}}_i}\hat{\mathbf{e}}_j$ in cylindrical coordinates $(\rho,\phi,z)$. You should find

$$x=\rho\cos\phi\,,\qquad y=\rho\sin\phi\,,\qquad z=z\,. \tag{A.50}$$

Relation between basis vectors:

$$\begin{aligned}
\hat{\boldsymbol{\rho}}&=\boldsymbol{\partial}_\rho=\cos\phi\,\hat{\mathbf{x}}+\sin\phi\,\hat{\mathbf{y}}\,,\\
\hat{\boldsymbol{\phi}}&=\rho^{-1}\boldsymbol{\partial}_\phi=-\sin\phi\,\hat{\mathbf{x}}+\cos\phi\,\hat{\mathbf{y}}\,,\\
\hat{\mathbf{z}}&=\boldsymbol{\partial}_z=\hat{\mathbf{z}}\,,
\end{aligned} \tag{A.51}$$

with inverse relations:

$$\begin{aligned}
\hat{\mathbf{x}}&=\cos\phi\,\hat{\boldsymbol{\rho}}-\sin\phi\,\hat{\boldsymbol{\phi}}\,,\\
\hat{\mathbf{y}}&=\sin\phi\,\hat{\boldsymbol{\rho}}+\cos\phi\,\hat{\boldsymbol{\phi}}\,,\\
\hat{\mathbf{z}}&=\hat{\mathbf{z}}\,,
\end{aligned} \tag{A.52}$$

and directional derivatives:

$$\begin{aligned}
&\nabla_{\hat{\boldsymbol{\rho}}}\hat{\boldsymbol{\rho}}=0\,,\quad \nabla_{\hat{\boldsymbol{\phi}}}\hat{\boldsymbol{\rho}}=\rho^{-1}\,\hat{\boldsymbol{\phi}}\,,\quad \nabla_{\hat{\mathbf{z}}}\hat{\boldsymbol{\rho}}=0\,,\\
&\nabla_{\hat{\boldsymbol{\rho}}}\hat{\boldsymbol{\phi}}=0\,,\quad \nabla_{\hat{\boldsymbol{\phi}}}\hat{\boldsymbol{\phi}}=-\rho^{-1}\,\hat{\boldsymbol{\rho}}\,,\quad \nabla_{\hat{\mathbf{z}}}\hat{\boldsymbol{\phi}}=0\,,\\
&\nabla_{\hat{\boldsymbol{\rho}}}\hat{\mathbf{z}}=0\,,\quad \nabla_{\hat{\boldsymbol{\phi}}}\hat{\mathbf{z}}=0\,,\quad \nabla_{\hat{\mathbf{z}}}\hat{\mathbf{z}}=0\,.
\end{aligned} \tag{A.53}$$

## A.5 Orthogonal Curvilinear Coordinates

In this section we derive expressions for the gradient, curl, and divergence in general orthogonal curvilinear coordinates $(u, v, w)$. Our starting point will be the definitions of gradient, curl, and divergence given in (A.19), (A.20), and (A.21). Examples of orthogonal curvilinear coordinates include Cartesian coordinates $(x, y, z)$, spherical coordinates $(r, \theta, \phi)$, and cylindrical coordinates $(\rho, \phi, z)$. These are the three main coordinate systems that we will be using most in this text.

Recall that in Cartesian coordinates $(x, y, z)$, the **line element** or infinitesimal squared distance between two nearby points is given by

$$ds^2 = dx^2 + dy^2 + dz^2 \qquad \text{(Cartesian)}. \tag{A.54}$$

Using the transformation equations (A.44) and (A.50), it is fairly easy to show that in spherical coordinates $(r, \theta, \phi)$ and in cylindrical coordinates $(\rho, \phi, z)$:

$$\begin{aligned} ds^2 &= dr^2 + r^2\, d\theta^2 + r^2 \sin^2\theta\, d\phi^2 \qquad &\text{(spherical)},\\ ds^2 &= d\rho^2 + \rho^2\, d\phi^2 + dz^2 \qquad &\text{(cylindrical)}. \end{aligned} \tag{A.55}$$

More generally, in orthogonal curvilinear coordinates $(u, v, w)$:

$$ds^2 = f^2\, du^2 + g^2\, dv^2 + h^2\, dw^2\,, \tag{A.56}$$

where $f$, $g$, and $h$ are functions of $(u, v, w)$ in general. The fact that there are no cross terms, like $du\, dv$, is a consequence of the coordinates being *orthogonal*. Note that $f = 1$, $g = r$, and $h = r \sin\theta$ for spherical coordinates and $f = 1$, $g = \rho$, and $h = 1$ for cylindrical coordinates. These results are summarized in Table A.1.

For completely arbitrary curvilinear coordinates $x^i \equiv (x^1, x^2, x^3)$, the line element, (A.56), has the more general form

$$ds^2 = \sum_{i,j} g_{ij}\, dx^i dx^j\,, \tag{A.57}$$

**Table A.1** Coordinates $(u, v, w)$ and functions $f, g, h$ for different orthogonal curvilinear coordinate systems

| Coordinates | $u$ | $v$ | $w$ | $f$ | $g$ | $h$ |
|---|---|---|---|---|---|---|
| Cartesian | $x$ | $y$ | $z$ | 1 | 1 | 1 |
| Spherical | $r$ | $\theta$ | $\phi$ | 1 | $r$ | $r \sin\theta$ |
| Cylindrical | $\rho$ | $\phi$ | $z$ | 1 | $\rho$ | 1 |

where $g_{ij} \equiv g_{ij}(x^1, x^2, x^3)$. The quantities $g_{ij}$ are called the components of the **metric**, and they can be represented in this case by a $3\times3$ matrix (See Appendix D.4.3 for a general discussion of matrix calculations). The metric components arise, for example, when finding the *geodesic* curves (shortest distance paths) between two points in a general curved space (See e.g., (C.14) and Exercise (C.7) in the context of the calculus of variations, Appendix C).

To make connection with the definitions of gradient, curl, and divergence given in (A.19), (A.20), and (A.21), we need expressions for the infinitesimal displacement vector, volume element, and area elements in general orthogonal curvilinear coordinates $(u, v, w)$. From the line element (A.56), we can conclude that the infinitesimal displacement vector $\mathbf{ds}$ connecting two nearby points is

$$\mathbf{ds} = f\,\mathrm{d}u\,\hat{\mathbf{u}} + g\,\mathrm{d}v\,\hat{\mathbf{v}} + h\,\mathrm{d}w\,\hat{\mathbf{w}}\,. \tag{A.58}$$

This is illustrated graphically in panel (a) of Fig. A.5. It is also easy to see from this figure that the infinitesimal volume element $\mathrm{d}V$ is given by

$$\mathrm{d}V = fgh\,\mathrm{d}u\,\mathrm{d}v\,\mathrm{d}w\,. \tag{A.59}$$

The infinitesimal area elements are

$$\hat{\mathbf{n}}\,\mathrm{d}a = \begin{cases} \pm\hat{\mathbf{u}}\,gh\,\mathrm{d}v\,\mathrm{d}w \\ \pm\hat{\mathbf{v}}\,hf\,\mathrm{d}w\,\mathrm{d}u \\ \pm\hat{\mathbf{w}}\,fg\,\mathrm{d}u\,\mathrm{d}v \end{cases} \tag{A.60}$$

with the $\pm$ sign depending on whether the unit normals to the area elements point in the direction of increasing (or decreasing) coordinate value. One such area element is illustrated graphically in panel (b) of Fig. A.5.

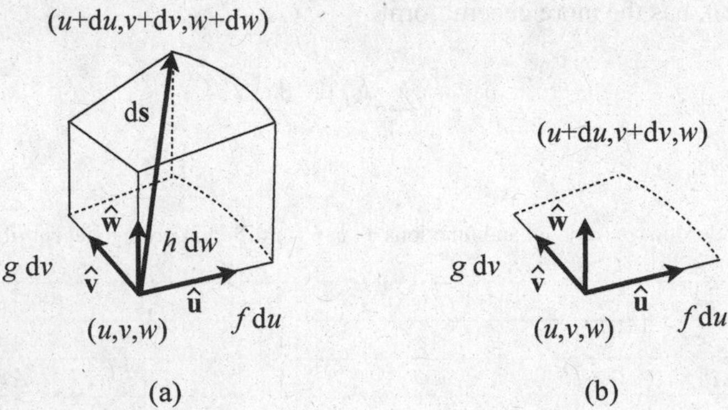


**Fig. A.5** Panel (a) Infinitesimal displacement vector $\mathbf{ds}$ and volume element $\mathrm{d}V = fgh\,\mathrm{d}u\,\mathrm{d}v\,\mathrm{d}w$ in general orthogonal curvilinear coordinates. Panel (b) The infinitesimal area element corresponding to the bottom ($w = \text{const}$) surface of the volume element shown in panel (a)

For arbitrary curvilinear coordinates $x^i \equiv (x^1, x^2, x^3)$ with non-zero off-diagonal terms, we note that the above expressions generalize to

$$dV = \sqrt{\det \mathsf{g}}\, dx^1\, dx^2\, dx^3 \,, \tag{A.61}$$

where $\det \mathsf{g}$ is the determinant of the matrix $\mathsf{g}$ of metric components $g_{ij}$ (See Appendix D.4.3.2), and

$$\hat{\mathbf{n}}\, da = \begin{cases} \pm\hat{\mathbf{n}}_1 \sqrt{g_{22}g_{33} - (g_{23})^2}\, dx^2\, dx^3 \\ \pm\hat{\mathbf{n}}_2 \sqrt{g_{33}g_{11} - (g_{31})^2}\, dx^3\, dx^1 \\ \pm\hat{\mathbf{n}}_3 \sqrt{g_{11}g_{22} - (g_{12})^2}\, dx^1\, dx^2 \end{cases} \tag{A.62}$$

where

$$\hat{\mathbf{n}}_1 = \frac{\boldsymbol{\partial}_2 \times \boldsymbol{\partial}_3}{|\boldsymbol{\partial}_2 \times \boldsymbol{\partial}_3|} \,, \quad \text{etc.} \tag{A.63}$$

Of particular relevance for both arbitrary curvilinear coordinates $(x^1, x^2, x^3)$ and orthogonal curvilinear coordinates $(u, v, w)$ is the distinction between the 3-dimensional and 2-dimensional *coordinate* volume and area elements, e.g., $d^3x \equiv dx^1\, dx^2\, dx^3$ and $d^2x \equiv dx^2\, dx^3$, etc. (which are just products of coordinate differentials), and the *invariant* volume and area elements, $dV$ and $\hat{\mathbf{n}}\, da$, which include the appropriate factors of the metric components $g_{ij}$.

**Exercise A.7** Verify (A.61) and (A.62).

## A.5.1 Gradient

Using the definition (A.19), it follows that

$$(\nabla U) \cdot d\mathbf{s} = dU \,. \tag{A.64}$$

From (A.58), the left-hand side of the above equation can be written as

$$(\nabla U) \cdot d\mathbf{s} = (\nabla U)_u\, f\, du + (\nabla U)_v\, g\, dv + (\nabla U)_w\, h\, dw \,, \tag{A.65}$$

while the right-hand side can be written as

$$dU = \frac{\partial U}{\partial u}\, du + \frac{\partial U}{\partial v}\, dv + \frac{\partial U}{\partial w}\, dw \,. \tag{A.66}$$

By equating these last two equations, we can read off the components of the gradient, from which we obtain

$$\nabla U = \frac{1}{f}\frac{\partial U}{\partial u}\,\hat{\mathbf{u}} + \frac{1}{g}\frac{\partial U}{\partial v}\,\hat{\mathbf{v}} + \frac{1}{h}\frac{\partial U}{\partial w}\,\hat{\mathbf{w}}\,. \tag{A.67}$$

**Exercise A.8** Consider a particle of mass $m$ moving in the potential

$$U(x, y, z) = \frac{1}{2}k(x^2 + y^2) + mgz\,. \tag{A.68}$$

Calculate the force $\mathbf{F} = -\nabla U$ in (a) spherical coordinates and (b) cylindrical coordinates. You should find:

$$\begin{aligned}
&\text{(a)}\quad \mathbf{F} = (-kr\sin^2\theta - mg\cos\theta)\,\hat{\mathbf{r}} + (-kr\sin\theta\cos\theta + mg\sin\theta)\,\hat{\boldsymbol{\theta}}\,,\\
&\text{(b)}\quad \mathbf{F} = -k\rho\,\hat{\boldsymbol{\rho}} - mg\,\hat{\mathbf{z}}\,.
\end{aligned} \tag{A.69}$$

### A.5.2 Curl

Using the definition (A.20), it follows that

$$(\nabla\times\mathbf{A})\cdot\hat{\mathbf{n}}\,\mathrm{d}a = \oint_C \mathbf{A}\cdot\mathrm{d}\mathbf{s}\,, \tag{A.70}$$

where $C$ is the *infinitesimal* closed curve bounding the area element $\hat{\mathbf{n}}\,\mathrm{d}a$, with orientation given by the right-hand rule relative to $\hat{\mathbf{n}}$. To calculate the components of $\nabla\times\mathbf{A}$, we take (in turn) the three different infinitesimal area elements given in (A.60). Starting with $\hat{\mathbf{n}}\,\mathrm{d}a = \hat{\mathbf{u}}\,gh\,\mathrm{d}v\,\mathrm{d}w$, the left-hand side of (A.70) becomes

$$(\nabla\times\mathbf{A})\cdot\hat{\mathbf{n}}\,\mathrm{d}a = (\nabla\times\mathbf{A})_u\,gh\,\mathrm{d}v\,\mathrm{d}w\,. \tag{A.71}$$

Since this area element lies in a $u =$ const surface, $\mathrm{d}u = 0$, for which

$$\mathbf{A}\cdot\mathrm{d}\mathbf{s} = A_v\,g\,\mathrm{d}v + A_w\,h\,\mathrm{d}w\,. \tag{A.72}$$

Integrating this around the corresponding boundary curve $C$ shown in Fig. A.6, we find that the right-hand side of (A.70) becomes

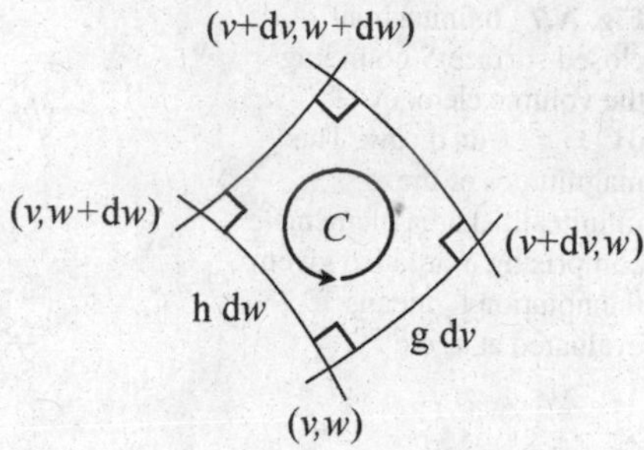


**Fig. A.6** Infinitesimal closed curve $C$ in the $u = \text{const}$ surface bounding the area element $\hat{\mathbf{n}}\,\mathrm{d}a = \hat{\mathbf{u}}\,gh\,\mathrm{d}v\,\mathrm{d}w$

$$\oint_C \mathbf{A}\cdot\mathrm{d}\mathbf{s} = (A_v g)|_w\,\mathrm{d}v + (A_w h)|_{v+dv}\,\mathrm{d}w - (A_v g)|_{w+dw}\,\mathrm{d}v - (A_w h)|_v\,\mathrm{d}w$$
$$= \frac{\partial}{\partial v}(A_w h)\,\mathrm{d}v\,\mathrm{d}w - \frac{\partial}{\partial w}(A_v g)\,\mathrm{d}w\,\mathrm{d}v\,. \tag{A.73}$$

Thus,

$$(\nabla\times\mathbf{A})_u = \frac{1}{gh}\left[\frac{\partial}{\partial v}(A_w\,h) - \frac{\partial}{\partial w}(A_v\,g)\right]. \tag{A.74}$$

Repeating the above calculation for the other two components yields

$$(\nabla\times\mathbf{A})_v = \frac{1}{hf}\left[\frac{\partial}{\partial w}(A_u\,f) - \frac{\partial}{\partial u}(A_w\,h)\right], \tag{A.75}$$

and

$$(\nabla\times\mathbf{A})_w = \frac{1}{fg}\left[\frac{\partial}{\partial u}(A_v\,g) - \frac{\partial}{\partial v}(A_u\,f)\right]. \tag{A.76}$$

### A.5.3 Divergence

Using the definition (A.21), it follows that

$$(\nabla\cdot\mathbf{A})\,\mathrm{d}V = \oint_S \mathbf{A}\cdot\hat{\mathbf{n}}\,\mathrm{d}a\,, \tag{A.77}$$

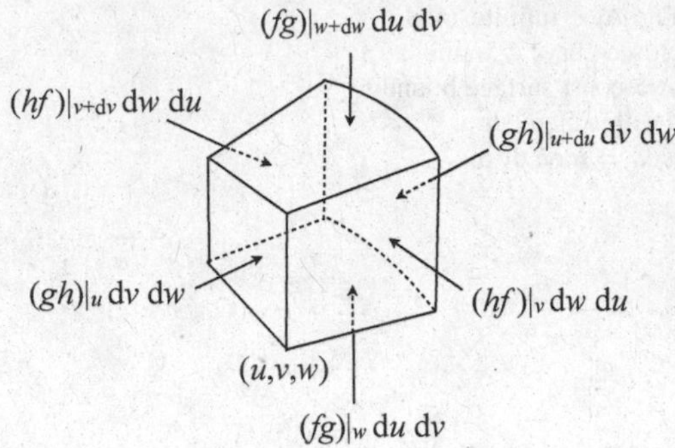


**Fig. A.7** Infinitesimal closed surface $S$ bounding the volume element $\mathrm{d}V = fgh\,\mathrm{d}u\,\mathrm{d}v\,\mathrm{d}w$. The magnitudes of the infinitesimal area elements comprising $S$ are also given; the notation $\big|_u$ means evaluated at $u$, etc

where $S$ is the *infinitesimal* closed surface bounding the volume element $\mathrm{d}V$, with outward pointing normal $\hat{\mathbf{n}}$. From (A.59), the left-hand side of the above equation can be written as

$$(\nabla \cdot \mathbf{A})\,\mathrm{d}V = (\nabla \cdot \mathbf{A})\, fgh\,\mathrm{d}u\,\mathrm{d}v\,\mathrm{d}w\,. \tag{A.78}$$

From (A.60), the integrand of the right-hand side of (A.77) contains the terms

$$\mathbf{A} \cdot \hat{\mathbf{n}}\,\mathrm{d}a = \begin{cases} \pm A_u\, gh\,\mathrm{d}v\,\mathrm{d}w \\ \pm A_v\, hf\,\mathrm{d}w\,\mathrm{d}u\,. \\ \pm A_w\, fg\,\mathrm{d}u\,\mathrm{d}v \end{cases} \tag{A.79}$$

Integrating over the boundary surface $S$ shown in Fig. A.7, we obtain

$$\begin{aligned} \oint_S \mathbf{A} \cdot \hat{\mathbf{n}}\,\mathrm{d}a = \; & (A_u\, gh)|_{u+du}\; \mathrm{d}v\,\mathrm{d}w - (A_u\, gh)|_u\; \mathrm{d}v\,\mathrm{d}w \\ & +(A_v\, hf)|_{v+dv}\; \mathrm{d}w\,\mathrm{d}u - (A_v\, hf)|_v\; \mathrm{d}w\,\mathrm{d}u \\ & + (A_w\, fg)|_{w+dw}\; \mathrm{d}u\,\mathrm{d}v - (A_w\, fg)|_w\; \mathrm{d}u\,\mathrm{d}v \\ = & \left[\frac{\partial}{\partial u}(A_u\, gh) + \frac{\partial}{\partial v}(A_v\, hf) + \frac{\partial}{\partial w}(A_w\, fg)\right] \mathrm{d}u\,\mathrm{d}v\,\mathrm{d}w\,. \end{aligned} \tag{A.80}$$

Thus,

$$\nabla \cdot \mathbf{A} = \frac{1}{fgh}\left[\frac{\partial}{\partial u}(A_u\, gh) + \frac{\partial}{\partial v}(A_v\, hf) + \frac{\partial}{\partial w}(A_w\, fg)\right]. \tag{A.81}$$

### A.5.4 Laplacian

Since the Laplacian of a scalar field is defined as the divergence of the gradient, it immediately follows from (A.67) and (A.81) that

$$\nabla^2 U = \frac{1}{fgh}\left[\frac{\partial}{\partial u}\left(\frac{gh}{f}\frac{\partial U}{\partial u}\right) + \frac{\partial}{\partial v}\left(\frac{hf}{g}\frac{\partial U}{\partial v}\right) + \frac{\partial}{\partial w}\left(\frac{fg}{h}\frac{\partial U}{\partial w}\right)\right]. \tag{A.82}$$

**Exercise A.9** Show that in spherical coordinates $(r, \theta, \phi)$:

$$\nabla U = \frac{\partial U}{\partial r}\hat{\mathbf{r}} + \frac{1}{r}\frac{\partial U}{\partial \theta}\hat{\boldsymbol{\theta}} + \frac{1}{r\sin\theta}\frac{\partial U}{\partial \phi}\hat{\boldsymbol{\phi}},$$

$$\nabla \times \mathbf{A} = \frac{1}{r\sin\theta}\left[\frac{\partial}{\partial \theta}(A_\phi \sin\theta) - \frac{\partial A_\theta}{\partial \phi}\right]\hat{\mathbf{r}} + \frac{1}{r}\left[\frac{1}{\sin\theta}\frac{\partial A_r}{\partial \phi} - \frac{\partial}{\partial r}(A_\phi r)\right]\hat{\boldsymbol{\theta}} + \frac{1}{r}\left[\frac{\partial}{\partial r}(A_\theta r) - \frac{\partial A_r}{\partial \theta}\right]\hat{\boldsymbol{\phi}},$$

$$\nabla \cdot \mathbf{A} = \frac{1}{r^2}\frac{\partial}{\partial r}\left(r^2 A_r\right) + \frac{1}{r\sin\theta}\frac{\partial}{\partial \theta}\left(\sin\theta\, A_\theta\right) + \frac{1}{r\sin\theta}\frac{\partial A_\phi}{\partial \phi},$$

$$\nabla^2 U = \frac{1}{r^2}\frac{\partial}{\partial r}\left(r^2\frac{\partial U}{\partial r}\right) + \frac{1}{r^2\sin\theta}\frac{\partial}{\partial \theta}\left(\sin\theta\frac{\partial U}{\partial \theta}\right) + \frac{1}{r^2\sin^2\theta}\frac{\partial^2 U}{\partial \phi^2}. \tag{A.83}$$

**Exercise A.10** Show that in cylindrical coordinates $(\rho, \phi, z)$:

$$\nabla U = \frac{\partial U}{\partial \rho}\hat{\boldsymbol{\rho}} + \frac{1}{\rho}\frac{\partial U}{\partial \phi}\hat{\boldsymbol{\phi}} + \frac{\partial U}{\partial z}\hat{\mathbf{z}},$$

$$\nabla \times \mathbf{A} = \left[\frac{1}{\rho}\frac{\partial A_z}{\partial \phi} - \frac{\partial A_\phi}{\partial z}\right]\hat{\boldsymbol{\rho}} + \left[\frac{\partial A_\rho}{\partial z} - \frac{\partial A_z}{\partial \rho}\right]\hat{\boldsymbol{\theta}} + \frac{1}{\rho}\left[\frac{\partial}{\partial \rho}(A_\phi \rho) - \frac{\partial A_\rho}{\partial \phi}\right]\hat{\mathbf{z}},$$

$$\nabla \cdot \mathbf{A} = \frac{1}{\rho}\frac{\partial}{\partial \rho}\left(\rho A_\rho\right) + \frac{1}{\rho}\frac{\partial A_\phi}{\partial \phi} + \frac{\partial A_z}{\partial z},$$

$$\nabla^2 U = \frac{1}{\rho}\frac{\partial}{\partial \rho}\left(\rho\frac{\partial U}{\partial \rho}\right) + \frac{1}{\rho^2}\frac{\partial^2 U}{\partial \phi^2} + \frac{\partial^2 U}{\partial z^2}. \tag{A.84}$$

## A.6 Integral Theorems of Vector Calculus

Using the definitions of gradient, curl, and divergence given above, one can prove the following fundamental theorems of integral vector calculus:

**Theorem A.1 Fundamental theorem for gradients.**

$$\int_C (\nabla U) \cdot d\mathbf{s} = U(\mathbf{r}_b) - U(\mathbf{r}_a)\,, \tag{A.85}$$

*where* $\mathbf{r}_a$, $\mathbf{r}_b$ *are the endpoints of* $C$.

**Theorem A.2 Stokes' theorem.**

$$\int_S (\nabla \times \mathbf{A}) \cdot \hat{\mathbf{n}}\, da = \oint_C \mathbf{A} \cdot d\mathbf{s}\,, \tag{A.86}$$

*where* $C$ *is the closed curved bounding the surface* $S$, *with orientation given by the right-hand rule relative to* $\hat{\mathbf{n}}$.

**Theorem A.3 Divergence theorem.**

$$\int_V (\nabla \cdot \mathbf{A})\, dV = \oint_S \mathbf{A} \cdot \hat{\mathbf{n}}\, da\,, \tag{A.87}$$

*where* $S$ *is the closed surface bounding the volume* $V$, *with outward pointing normal* $\hat{\mathbf{n}}$.

For infinitesimal volume elements, area elements, and path lengths, the proofs of these theorems follow trivially from the definitions given in (A.19), (A.20), and (A.21). For *finite* size volumes, areas, and path lengths, one simply adds together the contribution from infinitesimal elements. The neighboring surfaces, edges, and endpoints of these infinitesimal elements have *oppositely-directed* normals, tangent vectors, etc., and hence yield terms that cancel out when forming the sum. For detailed proofs, we recommend Schey (1996), Boas (2006), or Griffiths (1999).

## A.7 Some Additional Theorems for Vector Fields

Here we state (without proof) some additional theorems for vector fields. These make use of the identities $\nabla \times \nabla U = 0$ and $\nabla \cdot (\nabla \times \mathbf{A}) = 0$, which we derived earlier (See Example A.1), and the integral theorems of vector calculus from the previous subsection.

**Theorem A.4** *Any vector field* **F** *can be written in the form*

$$\mathbf{F} = -\nabla U + \nabla \times \mathbf{W}. \tag{A.88}$$

Note that this decomposition is not unique as the transformations

$$\begin{aligned} U &\to U + C\,, \quad \text{where } C = \text{const}\,, \\ \mathbf{W} &\to \mathbf{W} + \nabla \Lambda\,, \end{aligned} \tag{A.89}$$

leave **F** unchanged.

**Theorem A.5** *Curl-free vector fields:*

$$\nabla \times \mathbf{F} = 0 \quad \Leftrightarrow \quad \mathbf{F} = -\nabla U \quad \Leftrightarrow \quad \oint_C \mathbf{F} \cdot d\mathbf{s} = 0\,. \tag{A.90}$$

**Theorem A.6** *Divergence-free vector fields:*

$$\nabla \cdot \mathbf{F} = 0 \quad \Leftrightarrow \quad \mathbf{F} = \nabla \times \mathbf{W} \quad \Leftrightarrow \quad \oint_S \mathbf{F} \cdot \hat{\mathbf{n}}\, da = 0\,. \tag{A.91}$$

Both of the above theorems require that: (i) **F** be differentiable, and (ii) the region of interest be simply-connected (i.e., that there are not any holes; see the discussion in Appendix B.2). We will assume that both of these conditions are always satisfied.

Theorem A.5 is particularly relevant in the context of **conservative forces**, which we encounter often in the main text. Recall that **F** is conservative if and only if the work done by **F** in moving a particle from $\mathbf{r}_a$ to $\mathbf{r}_b$ is *independent* of the path connecting the two points. But path-independence is equivalent to the condition that $\oint_C \mathbf{F} \cdot d\mathbf{s} = 0$ for any closed curve $C$. Thus, from Theorem A.5, we can conclude that a conservative force is curl-free, i.e., $\nabla \times \mathbf{F} = 0$, and that it can always be written as the gradient of a scalar field.

**Exercise A.11** In two dimensions, consider the vector fields

$$\mathbf{A} = x\,\hat{\mathbf{x}} + y\,\hat{\mathbf{y}}\,, \qquad \mathbf{B} = -y\,\hat{\mathbf{x}} + x\,\hat{\mathbf{y}}\,, \qquad \mathbf{C} = y\,\hat{\mathbf{x}} + x\,\hat{\mathbf{y}}\,. \tag{A.92}$$

(a) Show that $\nabla \times \mathbf{A} = 0$, $\nabla \cdot \mathbf{B} = 0$, and $\nabla \times \mathbf{C} = 0$, $\nabla \cdot \mathbf{C} = 0$.
(b) Make plots of these vector fields.
(c) Show that $\mathbf{A} = \rho\,\hat{\boldsymbol{\rho}}$ and $\mathbf{B} = \rho\,\hat{\boldsymbol{\phi}}$, where $(\rho, \phi)$ are plane polar coordinates related to $(x, y)$ via $x = \rho\cos\phi$ and $y = \rho\sin\phi$.

(d) Show that $\mathbf{C} = \frac{1}{2}\nabla V$, where $(U, V)$ are orthogonal hyperbolic coordinates on the plane defined by $U \equiv x^2 - y^2$ and $V \equiv 2xy$.

A non-constant vector field like **C**, which is both curl-free and divergence-free, is said to be a **pure-shear** vector field. The "shearing pattern" of the **C** will look like that in Fig. A.4.

## A.8 Dirac Delta Function

The **Dirac delta function** $\delta(\mathbf{r} - \mathbf{r}_0)$ is a mathematical representation of a "spike"—i.e., a quantity that is zero at all points except at the spike, where it is infinite,

$$\delta(\mathbf{r} - \mathbf{r}_0) = \begin{cases} 0\,, & \text{if } \mathbf{r} \neq \mathbf{r}_0 \\ \infty\,, & \text{if } \mathbf{r} = \mathbf{r}_0 \end{cases} \tag{A.93}$$

and such that[5]

$$\int_V \mathrm{d}V\, \delta(\mathbf{r} - \mathbf{r}_0) = 1 \tag{A.94}$$

for any volume $V$ containing the spike. The Dirac delta function is not an ordinary mathematical function. It is what mathematicians call a **generalized function** or **distribution**. An example of a Dirac delta function is the mass density of an idealized point particle, $\mu(\mathbf{r}) = m\delta(\mathbf{r} - \mathbf{r}_0)$.

In one dimension, the Dirac delta function $\delta(x - x_0)$ can be represented as the limit of a sequence of functions $f_n(x)$ all of which have unit area, but which get narrower and higher as $n \to \infty$. Some simple example sequences are:

(i) A sequence of top-hat functions centered at $x_0$ with width $2/n$:

$$f_n(x) = \begin{cases} n/2 & x_0 - 1/n < x < x_0 + 1/n \\ 0 & \text{otherwise} \end{cases} \tag{A.95}$$

(ii) A sequence of Gaussian probability distributions with mean $\mu = x_0$ and standard deviation $\sigma = 1/n$:

[5]We are adopting here the standard "physicist's" definition of a 3-dimensional Dirac delta function (See e.g., Griffiths 1999), where we integrate it against the 3-dimensional volume element $\mathrm{d}V$. But note that we could also define a 3-dimensional Dirac delta function $\tilde{\delta}(\mathbf{r} - \mathbf{r}_0)$ with respect to the *coordinate* volume element $\mathrm{d}^3x$ via $\int_V \mathrm{d}^3x\, \tilde{\delta}(\mathbf{r} - \mathbf{r}_0) = 1$. (Recall that $\mathrm{d}^3x = \mathrm{d}u\,\mathrm{d}v\,\mathrm{d}w$ while $\mathrm{d}V = fgh\,\mathrm{d}u\,\mathrm{d}v\,\mathrm{d}w$ for orthogonal curvilinear coordinates $(u, v, w)$.) The difference between these two definitions of the Dirac delta function shows up in their transformation properties under a coordinate transformation, see Footnote 7.

$$f_n(x) = \frac{n}{\sqrt{2\pi}} e^{-n^2(x-x_0)^2/2} \tag{A.96}$$

(iii) A sequence of sinc functions[6] centered at $x_0$ of the form:

$$f_n(x) = \frac{n}{\pi} \operatorname{sinc}[n(x - x_0)] \tag{A.97}$$

Equivalently, the Dirac delta function can be defined in terms of its action on a set of **test functions** $f(x)$, which are infinitely differentiable and which vanish as $x \to \pm\infty$. The defining property of a 1-dimensional Dirac delta function is then

$$\int_a^b \mathrm{d}x\, f(x)\delta(x - x') = \begin{cases} f(x') & a < x' < b \\ 0 & \text{otherwise} \end{cases} \tag{A.98}$$

for any test function $f(x)$.

**Exercise A.12** Prove the following properties of the 1-dimensional Dirac delta function, which follow from the defining property (A.98):

$$\begin{aligned} \delta(x-a) &= \frac{\mathrm{d}}{\mathrm{d}x}\left(u(x-a)\right), \\ \delta'(-x) &= -\delta'(x), \\ \delta(-x) &= \delta(x), \\ \delta(ax) &= \frac{1}{|a|}\delta(x), \\ \delta[f(x)] &= \sum_i \frac{\delta(x-x_i)}{|f'(x_i)|}. \end{aligned} \tag{A.99}$$

In the above expressions, $u(x)$ is the unit step function,

$$u(x) = \begin{cases} 0, & x < 0 \\ 1, & x \geq 0 \end{cases} \tag{A.100}$$

and $f(x)$ is such that $f(x_i) = 0$ and $f'(x_i) \neq 0$. The last two properties indicate that the one-dimensional Dirac delta function $\delta(x)$ transforms like a *density* under a change of variables—i.e., $\delta(x)\,\mathrm{d}x = \delta(y)\,\mathrm{d}y$.

[6] The sinc function, $\operatorname{sinc} x$, is defined by $\operatorname{sinc} x \equiv \sin x / x$.

In three dimensions, the defining property of the Dirac delta function is

$$\int_V \mathrm{d}V\, f(\mathbf{r})\delta(\mathbf{r}-\mathbf{r}') = \begin{cases} f(\mathbf{r}') & \text{if } \mathbf{r}' \in V \\ 0 & \text{otherwise} \end{cases} \tag{A.101}$$

for any test function $f(\mathbf{r})$. Note that this definition implies

$$\begin{aligned} \delta(\mathbf{r}-\mathbf{r}') &= \delta(x-x')\delta(y-y')\delta(z-z'), \\ \delta(\mathbf{r}-\mathbf{r}') &= \frac{1}{r^2\sin\theta}\delta(r-r')\delta(\theta-\theta')\delta(\phi-\phi'), \\ \delta(\mathbf{r}-\mathbf{r}') &= \frac{1}{\rho}\delta(\rho-\rho')\delta(\phi-\phi')\delta(z-z'), \end{aligned} \tag{A.102}$$

in order that

$$\begin{aligned} \int \mathrm{d}V\, \delta(\mathbf{r}-\mathbf{r}') &= \int\int\int \mathrm{d}x\, \mathrm{d}y\, \mathrm{d}z\, \delta(x-x')\delta(y-y')\delta(z-z') \\ &= \int\int\int \mathrm{d}r\, \mathrm{d}\theta\, \mathrm{d}\phi\, \delta(r-r')\delta(\theta-\theta')\delta(\phi-\phi') \\ &= \int\int\int \mathrm{d}\rho\, \mathrm{d}\phi\, \mathrm{d}z\, \delta(\rho-\rho')\delta(\phi-\phi')\delta(z-z') \end{aligned} \tag{A.103}$$

be *independent* of the choice of coordinates. For general orthogonal curvilinear coordinates $(u, v, w)$,

$$\delta(\mathbf{r}-\mathbf{r}') = \frac{1}{fgh}\delta(u-u')\delta(v-v')\delta(w-w') \tag{A.104}$$

as a consequence of $\mathrm{d}V = fgh\, \mathrm{d}u\, \mathrm{d}v\, \mathrm{d}w$.[7]

There is also an integral representation of the 1-dimensional Dirac delta function, which can be heuristically "derived" by taking a limit of sinc functions:

$$\delta(x) = \lim_{L\to\infty} \frac{L}{\pi}\mathrm{sinc}(Lx) = \lim_{L\to\infty} \frac{1}{2\pi}\int_{-L}^{L} \mathrm{d}k\, \mathrm{e}^{ikx} = \frac{1}{2\pi}\int_{-\infty}^{\infty} \mathrm{d}k\, \mathrm{e}^{ikx}, \tag{A.105}$$

where we used (A.97). Thus,

$$\delta(x-x') = \frac{1}{2\pi}\int_{-\infty}^{\infty} \mathrm{d}k\, \mathrm{e}^{\pm ik(x-x')}. \tag{A.106}$$

[7] If we used the alternative definition of the 3-dimensional Dirac delta function $\tilde{\delta}(\mathbf{r}-\mathbf{r}_0)$ discussed in Footnote 5, then $\tilde{\delta}(\mathbf{r}-\mathbf{r}') = \delta(u-u')\delta(v-v')\delta(w-w')$, without the factor of $fgh$.

Similarly, in 3-dimensions,

$$\delta(\mathbf{r}-\mathbf{r}') = \frac{1}{(2\pi)^3}\int_{\text{all space}} \mathrm{d}V_{\mathbf{k}}\, \mathrm{e}^{\pm i\mathbf{k}\cdot(\mathbf{r}-\mathbf{r}')}\,, \tag{A.107}$$

where $\mathrm{d}V_{\mathbf{k}}$ is the 3-dimensional volume element in **k**-space.

***Example A.3*** Recall that in Newtonian gravity the gravitational potential $\Phi(\mathbf{r}, t)$ satisfies Poisson's equation

$$\nabla^2\Phi(\mathbf{r}, t) = 4\pi G\mu(\mathbf{r}, t)\,, \tag{A.108}$$

where $G$ is Newton's constant and $\mu(\mathbf{r}, t)$ is the mass density of the source distribution. Note that the left-hand side of the above equation is just the Laplacian of $\Phi$. We now show that for a stationary point source $\mu(\mathbf{r}, t) = m\delta(\mathbf{r}-\mathbf{r}_0)$, the potential is given by the well-known formula

$$\Phi(\mathbf{r}) = -\frac{Gm}{|\mathbf{r}-\mathbf{r}_0|}\,. \tag{A.109}$$

We begin by noting that

$$\boldsymbol{\nabla}\cdot\left(\frac{\hat{\mathbf{r}}}{r^2}\right) = 0\,, \qquad \text{for } r \neq 0\,, \tag{A.110}$$

which follows from the expression for the divergence in spherical coordinates (See Exercise A.9). To determine its behavior at $r = 0$, we consider the volume integral of $\boldsymbol{\nabla}\cdot(\hat{\mathbf{r}}/r^2)$ over a spherical volume of radius $R$ centered at the origin. Using the divergence theorem (A.87), we obtain

$$\int_V \boldsymbol{\nabla}\cdot\left(\frac{\hat{\mathbf{r}}}{r^2}\right)\mathrm{d}V = \oint_S \frac{\hat{\mathbf{r}}}{r^2}\cdot\hat{\mathbf{n}}\,\mathrm{d}a = \int_{\phi=0}^{2\pi}\int_{\theta=0}^{\pi}\frac{1}{R^2}R^2\sin\theta\,\mathrm{d}\theta\,\mathrm{d}\phi = 4\pi\,, \tag{A.111}$$

*independent* of the radius $R$. Thus, by comparison with the definition of the Dirac delta function, (A.94), we can conclude that

$$\boldsymbol{\nabla}\cdot\left(\frac{\hat{\mathbf{r}}}{r^2}\right) = 4\pi\delta(\mathbf{r})\,. \tag{A.112}$$

But since

$$\boldsymbol{\nabla}\left(\frac{1}{r}\right) = -\frac{\hat{\mathbf{r}}}{r^2}\,, \tag{A.113}$$

we can also write

$$\nabla^2\left(\frac{1}{r}\right) = -4\pi\delta(\mathbf{r})\,. \tag{A.114}$$

Finally, by simpling shifting the origin, we have

$$\nabla\cdot\left(\frac{\mathbf{r}-\mathbf{r}'}{|\mathbf{r}-\mathbf{r}'|^3}\right) = 4\pi\delta(\mathbf{r}-\mathbf{r}')\,,\quad \nabla^2\left(\frac{1}{|\mathbf{r}-\mathbf{r}'|}\right) = -4\pi\delta(\mathbf{r}-\mathbf{r}')\,. \tag{A.115}$$

Thus, $\Phi(\mathbf{r}) = -Gm/|\mathbf{r}-\mathbf{r}_0|$ as claimed. □

**Exercise A.13** Show that

$$\begin{aligned} \nabla\times(r^n\hat{\mathbf{r}}) &= 0 \quad \text{for all } n\,, \\ \nabla\cdot(r^n\hat{\mathbf{r}}) &= (n+2)r^{n-1} \quad \text{for } n\neq -2\,. \end{aligned} \tag{A.116}$$

Thus, it is only for $n=-2$ that we get a Dirac delta function.

## Suggested References

*Full references are given in the bibliography at the end of the book.*

Boas (2006): Chapter 6 is devoted to vector algebra and vector calculus, especially suited for undergraduates.

Griffiths (1999): Chapter 1 provides an excellent review of vector algebra and vector calculus, at the same level as this appendix. Our discussion of orthogonal curvilinear coordinates in Appendix A.5 is a summary of Appendix A in Griffiths, which has more detailed derivations and discussion.

Schey (1996): An excellent introduction to vector calculus emphasizing the geometric nature of the divergence, gradient, and curl operations.

# Appendix B
# Differential Forms

Although we will not need to develop the full machinery of tensor calculus for the applications to classical mechanics covered in this book, the concept of a **differential form** and the associated operations of **exterior derivative** and **wedge product** will come in handy from time to time. For example, they are particulary useful for determining whether certain differential equations or constraints on a mechanical system (See e.g., Sect. 2.2.3) are *integrable* or not. They are also helpful in understanding the geometric structure underlying Poisson brackets (Sect. 3.5). More generally, differential forms are actually the quantities that you integrate on a manifold, with the integral theorems of vector calculus (Appendix A.6) being special cases of a more general (differential-form version) of Stokes' theorem.

In broad terms, the exterior derivative is a generalization of the total derivative (or gradient) of a function, and the curl of a vector field in three dimensions. The wedge product is a generalization of the cross-product of two vectors. And differential forms are quantities constructed from a sum of wedge products of coordinate differentials $\mathrm{d}x^i$. Readers interested to learn more about differential forms and related topics should see e.g., Flanders (1963) and Schutz (1980).

## B.1 Definitions

Since we have not developed a general framework for working with tensors, our presentation of differential forms will be somewhat heuristic, starting with familar examples for 0-forms and 1-forms, and then adding mathematical operations as needed (e.g., wedge product and exterior derivative) to construct higher-order differential forms. To keep things sufficiently general, we will consider an $n$-dimensional manifold $M$ with coordinates $x^i \equiv (x^1, x^2, \ldots, x^n)$. From time to time we will consider ordinary 3-dimensional space to make connection with more familiar mathematical objects and operations.

 
M.J. Benacquista and J.D. Romano, *Classical Mechanics*, Undergraduate Lecture Notes in Physics, https://doi.org/10.1007/978-3-319-68780-3

### B.1.1 0-Forms, 1-Forms, and Exterior Derivative

To begin, a **0-form** is just a function

$$\alpha \equiv \alpha(x^1, x^2, \ldots, x^n)\,, \tag{B.1}$$

while a **1-form** is a linear combination of the coordinate differentials,

$$\beta \equiv \sum_i \beta_i \, \mathrm{d}x^i\,, \tag{B.2}$$

for which the components $\beta_i \equiv \beta_i(x^1, x^2, \ldots, x^n)$ transform according to

$$\beta_{i'} = \sum_i \frac{\partial x^i}{\partial x^{i'}} \beta_i\,, \qquad i' = 1', 2', \ldots, n'\,, \tag{B.3}$$

under a coordinate transformation $x^i \to x^{i'}(x^i)$. We impose this requirement on the components in order that

$$\beta \equiv \sum_i \beta_i \, \mathrm{d}x^i = \sum_{i'} \beta_{i'} \, \mathrm{d}x^{i'} \tag{B.4}$$

be invariant under a coordinate transformation. The set $\{\mathrm{d}x^1, \mathrm{d}x^2, \ldots, \mathrm{d}x^n\}$ is a coordinate *basis* for the $n$-dimensional space of 1-forms on $M$. A simple example of a 1-form is the **exterior derivative** of a 0-form $\alpha$,

$$\mathrm{d}\alpha \equiv \sum_i (\partial_i \alpha) \, \mathrm{d}x^i\,, \tag{B.5}$$

where $\partial_i \alpha \equiv \partial \alpha / \partial x^i$. Note that the exterior derivative of a 0-form is just the usual total differential (or gradient) of a function.

> **Exercise B.1** Verify (B.4) using (B.3) and the transformation property of the coordinate differentials $\mathrm{d}x^i$.

### B.1.2 2-Forms and Wedge Product

To construct a **2-form** from two 1-forms, we introduce the **wedge product** of two forms. We require this product to be *anti-symmetric*,

$$\mathrm{d}x^i \wedge \mathrm{d}x^j = -\mathrm{d}x^j \wedge \mathrm{d}x^i\,, \tag{B.6}$$

and *linear* with respect to its arguments,

$$\alpha \wedge (f\beta + g\gamma) = f(\alpha \wedge \beta) + g(\alpha \wedge \gamma)\,, \tag{B.7}$$

where $f$ and $g$ are any two functions. Given this definition, it immediately follows that the wedge product of two 1-forms $\alpha$ and $\beta$ can be written as

$$\alpha \wedge \beta = \sum_{i,j} \alpha_i \beta_j \,\mathrm{d}x^i \wedge \mathrm{d}x^j = \sum_{i<j} (\alpha_i \beta_j - \alpha_j \beta_i)\,\mathrm{d}x^i \wedge \mathrm{d}x^j\,, \tag{B.8}$$

where we used the anti-symmetry of $\mathrm{d}x^i \wedge \mathrm{d}x^j$ to get the last equality. Note that in three dimensions

$$\alpha_i \beta_j - \alpha_j \beta_i = \sum_k \varepsilon_{ijk} (\boldsymbol{\alpha} \times \boldsymbol{\beta})_k\,, \tag{B.9}$$

where on the right-hand side we are treating $\alpha_i$ and $\beta_i$ as the components of two vectors $\boldsymbol{\alpha}$ and $\boldsymbol{\beta}$. Thus, the wedge product of two 1-forms generalizes the *cross product* of two vectors in three dimensions. The most general 2-form on $M$ will have the form

$$\gamma \equiv \sum_{i<j} \gamma_{ij}\,\mathrm{d}x^i \wedge \mathrm{d}x^j\,, \tag{B.10}$$

where the components $\gamma_{ij} \equiv \gamma_{ij}(x^1, x^2, \ldots, x^n)$ are totally anti-symmetric under interchange of $i$ and $j$ (i.e., $\gamma_{ij} = -\gamma_{ji}$ for all $i$ and $j$).

The exterior derivative can also be extended to an arbitrary 1-form $\alpha$. We simply take the exterior derivative of the components $\alpha_j$, for $j = 1, 2, \ldots, n$, and then wedge those 1-forms $\mathrm{d}\alpha_j = \sum_i (\partial_i \alpha_j)\,\mathrm{d}x^i$ with the coordinate differentials $\mathrm{d}x^j$. This leads to the 2-form

$$\mathrm{d}\alpha \equiv \sum_{i,j} (\partial_i \alpha_j)\,\mathrm{d}x^i \wedge \mathrm{d}x^j = \sum_{i<j} (\partial_i \alpha_j - \partial_j \alpha_i)\,\mathrm{d}x^i \wedge \mathrm{d}x^j\,. \tag{B.11}$$

Note that in three dimensions

$$\partial_i \alpha_j - \partial_j \alpha_i = \sum_k \varepsilon_{ijk} (\nabla \times \boldsymbol{\alpha})_k\,, \tag{B.12}$$

where on the right-hand side we are treating $\alpha_i$ as the components of a vector field $\boldsymbol{\alpha}$. So the exterior derivative of a 1-form generalizes the *curl* of a vector field in three dimensions.

### B.1.3 3-Forms and Higher-Order Forms

We can continue in this fashion to construct 3-forms, 4-forms, etc., by requiring that the wedge product be *associative*,

$$\alpha \wedge (\beta \wedge \gamma) = (\alpha \wedge \beta) \wedge \gamma = \alpha \wedge \beta \wedge \gamma \,. \tag{B.13}$$

Thus, a general $p$-form $\alpha$ can be written as

$$\alpha = \sum_{i_1 < i_2 < \cdots < i_p} \alpha_{i_1 i_2 \cdots i_p}\, \mathrm{d}x^{i_1} \wedge \mathrm{d}x^{i_2} \wedge \cdots \wedge \mathrm{d}x^{i_p} \,, \tag{B.14}$$

where the components $\alpha_{i_1 i_2 \cdots i_p} \equiv \alpha_{i_1 i_2 \cdots i_p}(x^1, x^2, \ldots, x^n)$ are totally anti-symmetric under interchange of the indices $i_1, i_2, \ldots, i_p$. Similarly, the exterior derivative of a $p$-form $\alpha$ is the $(p+1)$ form

$$\mathrm{d}\alpha = \sum_{i_1 < i_2 < \cdots < i_{p+1}} \left(\partial_{i_1} \alpha_{i_2 i_3 \cdots i_{p+1}} - \partial_{i_2} \alpha_{i_1 i_3 \cdots i_{p+1}} \cdots - \partial_{i_{p+1}} \alpha_{i_2 i_3 \cdots i_p i_1}\right) \mathrm{d}x^{i_1} \wedge \mathrm{d}x^{i_2} \wedge \cdots \wedge \mathrm{d}x^{i_p+1} \,. \tag{B.15}$$

Note that $(n+1)$ and higher-rank forms in an $n$-dimensional space are identically zero due to the anti-symmetry of the wedge product, i.e., $\mathrm{d}x^i \wedge \mathrm{d}x^j = 0$ for $i = j$.

### B.1.4 Total Anti-Symmetrization

By introducing a notation for totally anti-symmetrizing a set of indices, e.g.,

$$\begin{aligned} [ij] &\equiv \frac{1}{2!}(ij - ji)\,, \\ [ijk] &\equiv \frac{1}{3!}(ijk - ikj + jki - jik + kij - kji)\,, \\ &\text{etc.}\,, \end{aligned} \tag{B.16}$$

we can write down the general expressions for the components of the wedge product and exterior derivative in compact form:

$$(\alpha \wedge \beta)_{i_1 \cdots i_p j_1 \cdots j_q} = \frac{(p+q)!}{p!q!}\, \alpha_{[i_1 \cdots i_p} \beta_{j_1 \cdots j_q]} \tag{B.17}$$

and

$$(d\alpha)_{i_1 i_2 \cdots i_{p+1}} = (p+1)\partial_{[i_1}\alpha_{i_2 \cdots i_{p+1}]}\,, \tag{B.18}$$

where $\alpha$ and $\beta$ denote a $p$-form and $q$-form, respectively.

**Exercise B.2** Let $\alpha$ be a $p$-form and $\beta$ be a $q$-form. Show that

$$\alpha \wedge \beta = (-1)^{pq}\beta \wedge \alpha\,. \tag{B.19}$$

**Exercise B.3** Let $\alpha$ be a $p$-form and $\beta$ be a $q$-form. Show that the exterior derivative of the wedge product $\alpha \wedge \beta$ satisfies

$$d(\alpha \wedge \beta) = (d\alpha) \wedge \beta + (-1)^p \alpha \wedge (d\beta)\,, \tag{B.20}$$

as a consequence of the ordinary product rule for partial derivatives and the anti-symmetry of the differential forms.

## B.2 Closed and Exact Forms

Given the above definitions, we can introduce some additional terminology:

- A $p$-form $\alpha$ is said to be **exact** if there exists a $(p-1)$-form $\beta$ for which $\alpha = d\beta$.
- A $p$-form $\alpha$ is said to be **closed** if $d\alpha = 0$.

It is easy to show that all exact forms are closed—i.e.,

$$d(d\beta) = 0\,, \tag{B.21}$$

as a consequence of the commutativity of partial derivatives, $\partial_i\partial_j = \partial_j\partial_i$. This result is called the **Poincaré lemma**. But what about the converse? Are all closed forms also exact?

The answer is that all closed forms are *locally* exact, but *globally* this need not be true (See e.g., Schutz 1980 for a proof). Global exactness of a closed form requires that the space be *topologically trivial* (i.e., simply-connected), in the sense that the space shouldn't contain any "holes". More precisely, this means that any closed loop in the space should be (smoothly) contractible to a point. Ordinary 3-dimensional space with no points removed or the surface of a 2-sphere are examples of simply-connected spaces. The *punctured plane* ($\mathbb{R}^2$ with the origin removed) or the surface of a torus are examples of spaces that are not simply-connected. Any closed curve

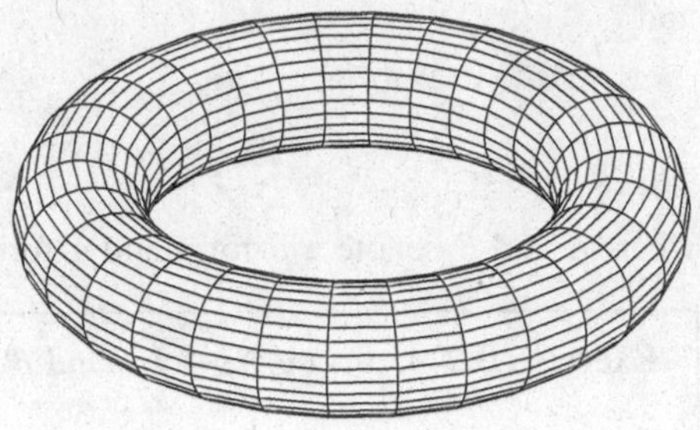

**Fig. B.1** The coordinate lines on a torus are examples of closed curves that are not contractible to a point

encircling the origin of the punctured plane, and any of the coordinates lines on the torus shown in Fig. B.1 are not contractible to a point. (See Schutz 1980 or Flanders 1963 for more details.)

**Exercise B.4** In three dimensions, show that $\mathrm{d}(\mathrm{d}\alpha) = 0$ corresponds to

(a) $\nabla \times \nabla\alpha = \mathbf{0}$ if $\alpha$ is a 0-form;
(b) $\nabla \cdot (\nabla \times \mathbf{A}) = 0$ if $\alpha$ is a 1-form $\alpha = \sum_i \alpha_i \, \mathrm{d}x^i$ with $A_i \equiv \alpha_i$.

**Exercise B.5** Consider the 1-form

$$\alpha \equiv \frac{1}{x^2 + y^2} (-y \, \mathrm{d}x + x \, \mathrm{d}y) \tag{B.22}$$

defined on the punctured plane. Show that $\alpha$ is closed, but globally is not exact. Find a function $f(x, y)$ for which $\alpha = \mathrm{d}f$ locally. (*Hint*: Plane polar coordinates $(r, \phi)$ might be useful for this.)

## B.3 Frobenius' Theorem

You may recall from a math methods class trying to determine if a 1st-order differential equation of the form

$$A(x, y) \, \mathrm{d}x + B(x, y) \, \mathrm{d}y = 0 \tag{B.23}$$

is integrable or not. You may also remember that if

$$\partial_y A = \partial_x B \,, \tag{B.24}$$

then (at least locally) there exists a function $\varphi \equiv \varphi(x, y)$ for which

$$d\varphi = A(x, y)\, dx + B(x, y)\, dy\,, \quad \text{with} \quad A = \partial_x \varphi\,, \quad B = \partial_y \varphi\,. \tag{B.25}$$

But requiring that the differential equation be exact is actually *too strong* a requirement for integrability. More generally, (B.23) is integrable if and only if there exists a function $\mu \equiv \mu(x, y)$, called an **integrating factor**, for which

$$\mu(x, y)\,[A(x, y)\, dx + B(x, y)\, dy] \tag{B.26}$$

is exact, so that[1]

$$\partial_y(\mu A) = \partial_x(\mu B)\,. \tag{B.27}$$

It turns out that in two dimensions one can *always* find such an integrating factor. Thus, *all* 1st-order differential equations of the form given in (B.23) are integrable (Exercise B.6). But explicitly finding an integrating factor in practice is not an easy task in general.

Now in three and higher dimensions not all 1st-order differential equations are integrable, so testing for integrability is a necessary and important task. Writing the differential equation in $n$ dimensions as

$$\alpha \equiv \sum_i \alpha_i\, dx^i = 0\,, \tag{B.28}$$

where $\alpha_i \equiv \alpha_i(x^1, x^2, \ldots, x^n)$, the question of integrability again becomes does there exist an integrating factor $\mu \equiv \mu(x^1, x^2, \ldots, x^n)$ for which

$$d\varphi = \mu\alpha \tag{B.29}$$

for some $\varphi$. This would imply

$$\partial_i(\mu\alpha_j) = \partial_j(\mu\alpha_i) \quad \text{for all} \quad i, j = 1, 2, \ldots, n\,, \tag{B.30}$$

which in the language of differential forms becomes

$$0 = d\mu \wedge \alpha + \mu\, d\alpha \quad \Leftrightarrow \quad d\alpha = -\mu^{-1} d\mu \wedge \alpha\,. \tag{B.31}$$

[1] Recall from thermodynamics that heat flow is described by an *inexact* differential $đQ$ (notationally, the bar on the 'd' is to indicate that it is not the total differential of a function $Q$). But $đQ$ becomes exact when multiplied by an integrating factor, i.e., $dS = đQ/T$, where $T$ is the temperature and $S$ is the entropy.

But since $\alpha \wedge \alpha = 0$, it follows that

$$\mathrm{d}\alpha \wedge \alpha = 0\,. \tag{B.32}$$

This is a *necessary* condition for (B.28) to be integrable. That it is also a *sufficient* condition was proven by Frobenius in 1877. Thus, **Frobenius' theorem** tells us that (B.32) is necessary and sufficient for the integrability of the 1st-order differential equation (B.28).

Frobenius's theorem can also be extended to the case of a *system* of 1st-order differential equations:

$$\alpha^A \equiv \sum_i \alpha_i^A \, \mathrm{d}x^i = 0\,, \qquad A = 1, 2, \ldots, M\,, \tag{B.33}$$

where $M < n$ and $\alpha_i^A \equiv \alpha_i^A(x^1, x^2, \ldots, x^n)$. We would like to know if this system is integrable in the sense of defining an $(n - M)$-dimensional hypersurface in the original $n$-dimensional space of coordinates. The necessary and sufficient condition for this to be true is the existence of an invertible transformation from the $\alpha^A$ to a set of exact 1-forms (i.e., total differentials):

$$\mathrm{d}\varphi^A = \sum_B \mu_{AB}\alpha^B \quad \Leftrightarrow \quad \alpha^A = \sum_B \left(\mu^{-1}\right)_{AB} \mathrm{d}\varphi^B\,, \tag{B.34}$$

where $\varphi^A \equiv \varphi^A(x^1, x^2, \ldots, x^n)$ is a set of functions, and $\mu_{AB} \equiv \mu_{AB}(x^1, x^2, \ldots, x^n)$ are the components of an invertible matrix (which is a generalization of the integrating factor $\mu$ for a single equation). In this context, Frobenius' theorem states that the set of constraints given by (B.33) is integrable if and only if

$$\mathrm{d}\alpha^A \wedge \alpha^1 \wedge \alpha^2 \wedge \cdots \wedge \alpha^M = 0\,, \qquad A = 1, 2, \ldots, M\,. \tag{B.35}$$

For a single differential equation, we recover (B.32).

**Exercise B.6** Using Frobenius' theorem, prove that *any* 1st-order differential equation in two dimensions,

$$\alpha \equiv A(x, y)\, \mathrm{d}x + B(x, y)\, \mathrm{d}y = 0\,, \tag{B.36}$$

is integrable.

**Exercise B.7** (*Adapted from Flanders* 1963.) Consider the 1st-order differential equation

$$\alpha \equiv yz\,\mathrm{d}x + xz\,\mathrm{d}y + \mathrm{d}z = 0\,, \tag{B.37}$$

in three dimensions.

(a) Use Frobenius' theorem to show that this equation is integrable.
(b) Verify that $\mu = \mathrm{e}^{xy}$ is an integrating factor for $\alpha$ with $\varphi = z\mathrm{e}^{xy}$.

## B.4 Integration of Differential Forms

Although you may not have thought about it this way, the things that you integrate on a manifold are really just differential forms. Indeed, the integrand $f(x)\,\mathrm{d}x$ of the familiar integral

$$\int_{x_1}^{x_2} f(x)\,\mathrm{d}x \tag{B.38}$$

from calculus is, in the language of this appendix, a 1-form. And the transformation property of $f(x)$ under a change of variables $x \to y(x)$,

$$f(x) \to f(y) \equiv \left.\frac{f(x)}{\mathrm{d}y/\mathrm{d}x}\right|_{x=x(y)}, \tag{B.39}$$

is just what you need in order for

$$f(x)\,\mathrm{d}x = f(y)\,\mathrm{d}y \tag{B.40}$$

to be *independent* of the choice of coordinates (compare these last two equations with (B.3) and (B.4)).

More generally, a 1-form field $\alpha$ on an $n$-dimensional manifold $M$ can be thought of as mapping from a 1-dimensional curve $C$ (with parameter $\lambda$ and tangent vector $\mathrm{d}x^i/\mathrm{d}\lambda$) to the value of the line integral

$$\int_C \alpha = \int_C \sum_i \alpha_i\,\mathrm{d}x^i \equiv \int_{\lambda_1}^{\lambda_2} \sum_i \alpha_i \frac{\mathrm{d}x^i}{\mathrm{d}\lambda}\,\mathrm{d}\lambda\,. \tag{B.41}$$

Similarly, a 2-form field $\beta$ can be thought of as a mapping from a 2-dimensional surface $S$ (with coordinates $(u, v)$ and tangent vectors $\partial x^i/\partial u$, $\partial x^i/\partial v$) to the value of the surface integral

$$\begin{aligned}\int_S \beta &= \int_S \sum_{i<j} \beta_{ij}\, \mathrm{d}x^i \wedge \mathrm{d}x^j \\ &\equiv \int_{u_1}^{u_2}\int_{v_1}^{v_2} \sum_{i<j} \beta_{ij} \left(\frac{\partial x^i}{\partial u}\frac{\partial x^j}{\partial v} - \frac{\partial x^j}{\partial u}\frac{\partial x^i}{\partial v}\right) \mathrm{d}u\, \mathrm{d}v \\ &= \int_{u_1}^{u_2}\int_{v_1}^{v_2} \sum_{i<j} \beta_{ij} \frac{\partial(x^i, x^j)}{\partial(u,v)}\, \mathrm{d}u\, \mathrm{d}v\,,\end{aligned} \tag{B.42}$$

where the **Jacobian** of the transformation from $(x^i, x^j)$ to $(u, v)$ is[2]

$$\frac{\partial(x^i, x^j)}{\partial(u,v)} \equiv \begin{vmatrix} \frac{\partial x^i}{\partial u} & \frac{\partial x^i}{\partial v} \\ \frac{\partial x^j}{\partial u} & \frac{\partial x^j}{\partial v} \end{vmatrix} = \frac{\partial x^i}{\partial u}\frac{\partial x^j}{\partial v} - \frac{\partial x^j}{\partial u}\frac{\partial x^i}{\partial v}\,. \tag{B.43}$$

The extension to 3-form, 4-form, $\cdots$, $n$-form fields follows by noting that the above integrands are special cases of the general result that a $p$-form $\gamma$ maps a set of $p$ vectors with components $\{A^i, B^j, \ldots, C^k\}$ to the real number

$$\sum_{i<j<\cdots<k} \gamma_{ij\cdots k} \left(A^i B^j \cdots C^k - A^j B^i \cdots C^k - \cdots - A^k B^j \cdots C^i\right). \tag{B.44}$$

By taking

$$A^i = \mathrm{d}u\, \frac{\partial x^i}{\partial u}\,, \qquad B^j = \mathrm{d}v\, \frac{\partial x^j}{\partial v}\,, \qquad \cdots, \qquad C^k = \mathrm{d}w\, \frac{\partial x^k}{\partial w}\,, \tag{B.45}$$

where $(u, v, \ldots, w)$ are the coordinates for a $p$-dimensional hypersurface, we are able to generalize (B.41) and (B.42) to arbitrary $p$-forms, with the appropriate Jacobians entering these expressions. Figure B.2 shows the tangent vectors and infinitesimal coordinate area element for a 2-dimensional surface spanned by the coordinates $(u, v)$.

Note that all of these integrals are *oriented* in the sense that swapping the order of the coordinates, e.g., $(u, v) \to (v, u)$, in the parametrization of the 2-dimensional surface $S$, changes the sign of the Jacobian and hence the sign of the integral. In addition, if one decides to change coordinates to do a particular integral, the Jacobian of the transformation enters automatically via the wedge product of the coordinate differentials. For example, in two dimensions, if one transforms from $(x, y)$ to $(u, v)$ it follows that

[2]The vertical lines in (B.43) mean you should take the *determinant* of the $2 \times 2$ matrix of partial derivatives. See Appendix D.4.3.2 for more details, if needed.

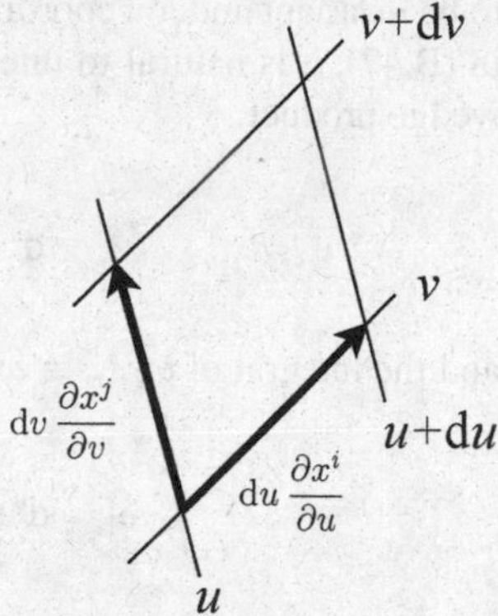


**Fig. B.2** Infinitesimal coordinate area element for a 2-dimensional surface spanned by the coordinates $(u, v)$

$$\begin{aligned}
\mathrm{d}x \wedge \mathrm{d}y &= \left(\frac{\partial x}{\partial u}\,\mathrm{d}u + \frac{\partial x}{\partial v}\,\mathrm{d}v\right) \wedge \left(\frac{\partial y}{\partial u}\,\mathrm{d}u + \frac{\partial y}{\partial v}\,\mathrm{d}v\right) \\
&= \left(\frac{\partial x}{\partial u}\frac{\partial y}{\partial v} - \frac{\partial x}{\partial v}\frac{\partial y}{\partial u}\right) \mathrm{d}u \wedge \mathrm{d}v = \frac{\partial(x, y)}{\partial(u, v)}\,\mathrm{d}u \wedge \mathrm{d}v\,,
\end{aligned} \tag{B.46}$$

where we used the chain rule and the anti-symmetry of the wedge product to get the first and second equalities above. In $n$ dimensions, for a coordinate transformation from $x^i \equiv (x^1, x^2, \ldots, x^n)$ to $x^{i'} = (x^{1'}, x^{2'}, \ldots, x^{n'})$, we have

$$\begin{aligned}
\mathrm{d}x^1 \wedge \cdots \wedge \mathrm{d}x^n &= \sum_{i'_1} \cdots \sum_{i'_n} \frac{\partial x^1}{\partial x^{i'_1}} \cdots \frac{\partial x^n}{\partial x^{i'_n}} \mathrm{d}x^{i'_1} \wedge \cdots \wedge \mathrm{d}x^{i'_n} \\
&= \sum_{i'_1} \cdots \sum_{i'_n} \frac{\partial x^1}{\partial x^{i'_1}} \cdots \frac{\partial x^n}{\partial x^{i'_n}} \varepsilon^{i'_1 \cdots i'_n}\,\mathrm{d}x^{1'} \wedge \cdots \wedge \mathrm{d}x^{n'} \\
&= \frac{\partial(x^1, \ldots, x^n)}{\partial(x^{1'}, \ldots, x^{n'})} \mathrm{d}x^{1'} \wedge \cdots \wedge \mathrm{d}x^{n'}\,,
\end{aligned} \tag{B.47}$$

where we made use of the $n$-dimensional Levi-Civita symbol and used (D.81) to get the last equality. Note that this is precisely the *inverse* transformation of the components of an $n$-form

$$\omega_{1\cdots n} = \frac{\partial(x^{1'}, \ldots, x^{n'})}{\partial(x^1, \ldots, x^n)}\,\omega_{1'\cdots n'}\,, \tag{B.48}$$

which is needed for

$$\omega = \omega_{1\cdots n}\,\mathrm{d}x^1 \wedge \cdots \wedge \mathrm{d}x^n = \omega_{1'\cdots n'}\,\mathrm{d}x^{1'} \wedge \cdots \wedge \mathrm{d}x^{n'} \tag{B.49}$$

to be invariant under a coordinate transformation. Given the presence of the Jacobian in (B.47), it is natural to interpret the $n$-dimensional coordinate element $\mathrm{d}^n x$ as the wedge product

$$\mathrm{d}^n x \equiv \mathrm{d}x^1 \wedge \mathrm{d}x^2 \wedge \cdots \wedge \mathrm{d}x^n \,, \tag{B.50}$$

and the integral of $\omega_{1\cdots n} \equiv \omega_{1\cdots n}(x^1, x^2, \ldots, x^n)$ as the integral of the $n$-form $\omega$:

$$\int \omega_{1\cdots n}\, \mathrm{d}^n x \equiv \int \omega_{1\cdots n}\, \mathrm{d}x^1 \wedge \cdots \wedge \mathrm{d}x^n = \int \omega \,. \tag{B.51}$$

### B.4.1 Stokes' Theorem for Differential Forms

Finally, to end this section, we note that the integral theorems of vector calculus (Appendix A.6) are actually special cases of an all-inclusive **Stokes' theorem**, written in terms of differential forms,

$$\int_U \mathrm{d}\alpha = \oint_{\partial U} \alpha \,, \tag{B.52}$$

where $\alpha$ is a $(p-1)$ form and $U$ is $p$-dimensional region in $M$ with boundary $\partial U$. We will not prove (B.52) here (See e.g., Flanders 1963). Rather we leave it as an exercise (Exercise B.9) to show that in three dimensions (B.52) reduces to the fundamental theorem for gradients (A.85), Stokes' theorem (A.86), and the divergence theorem (A.87), if one makes the appropriate identification of 1-forms and 2-forms with vector fields, and exterior derivative with either the gradient, curl, or divergence. Thus, (B.52) unifies the integral theorems of vector calculus.

**Exercise B.8** (a) Show explicitly by taking partial derivatives that a coordinate transformation from Cartesian coordinates $(x, y)$ to plane polar coordinates $(r, \phi)$ leads to

$$\mathrm{d}x \wedge \mathrm{d}y = r\, \mathrm{d}r \wedge \mathrm{d}\phi \,. \tag{B.53}$$

(b) Similarly, show that a coordinate transformation from Cartesian coordinates $(x, y, z)$ to spherical coordinates $(r, \theta, \phi)$ leads to

$$\mathrm{d}x \wedge \mathrm{d}y \wedge \mathrm{d}z = r^2 \sin\theta\, \mathrm{d}r \wedge \mathrm{d}\theta \wedge \mathrm{d}\phi \,. \tag{B.54}$$

**Exercise B.9** Show that in three dimensions (B.52) reduces to the fundamental theorem for gradients (A.85), Stokes' theorem (A.86), and the divergence theorem (A.87), by making the following identifications:

(a) For $p = 1$, identify the 0-form $\alpha$ with the function $U$, and the exterior derivative $\mathrm{d}\alpha$ with the gradient $\nabla U$. Also, identify

$$\frac{\mathrm{d}x^i}{\mathrm{d}s}\,\mathrm{d}s \tag{B.55}$$

with the line element $\mathrm{d}\mathbf{s}$, where $\mathrm{d}x^i/\mathrm{d}s$ is the tangent vector to the curve $C$ parameterized by the arc length $s$.

(b) For $p = 2$, identify the 1-form $\alpha$ with the vector field $A_i \equiv \alpha_i$, and use (B.12) to identify $\mathrm{d}\alpha$ with $\nabla \times \mathbf{A}$. Also, identify

$$\sum_{j<k} \varepsilon_{ijk} \frac{\partial(x^j, x^k)}{\partial(u, v)}\,\mathrm{d}u\,\mathrm{d}v \tag{B.56}$$

with the area element $\hat{\mathbf{n}}\,\mathrm{d}a$, where $(u, v)$ are coordinates on $S$.

(c) For $p = 3$, identifty the 2-form $\alpha$ with the vector field $A_i \equiv \sum_{j<k} \varepsilon_{ijk}\alpha_{jk}$, and the exterior derivative $\mathrm{d}\alpha$ with $\varepsilon_{ijk}\nabla \cdot \mathbf{A}$. Also, identify

$$\sum_{i<j<k} \varepsilon_{ijk} \frac{\partial(x^i, x^j, x^k)}{\partial(u, v, w)}\,\mathrm{d}u\,\mathrm{d}v\,\mathrm{d}w \tag{B.57}$$

with the volume element $\mathrm{d}V$, where $(u, v, w)$ are coordinates in $V$.

## Suggested References

*Full references are given in the bibliography at the end of the book.*

Flanders (1963): A classic text about differential forms, appropriate for graduate students or advanced undergraduates comfortable with abstract mathematics.

Schutz (1980): A introduction to differential geometry, including tensor calculus and differential forms, with an emphasis on geometrical methods. Appropriate for graduate students or advanced undergraduates comfortable with abstract mathematics.

# Appendix C
# Calculus of Variations

The calculus of variations is an extension of the standard procedure for finding the **extrema** (i.e., maxima and minima) of a function $f(x)$ of a single real variable $x$. But instead of extremizing a function $f(x)$, we extremize a **functional** $I[y]$, which is a "function of a function" $y = f(x)$. Classic problems that can be solved using the calculus of variations are: (i) finding the curve connecting two points in the plane that has the shortest distance (a *geodesic* problem), (ii) finding the shape of a closed curve of fixed length that encloses the maximum area (an *isoperimetric* problem), (iii) finding the shape of a wire joining two points such that a bead will slide along the wire under the influence of gravity in the shortest amount of time (the famous *brachistochrone* problem of Johann Bernoulli). The calculus of variations also provides an alternative way of obtaining the equations of motion for a particle, or a system of particles, in classical mechanics. In this appendix, we derive the Euler equations, discuss ways of solving these equations in certain simplified scenarios, and extend the formalism to deal with integral constraints. For a more thorough introduction to the calculus of variations, see, e.g., Boas (2006), Gelfand and Fomin (1963), and Lanczos (1949). Specific applications to classical mechanics will be given in Chap. 3.

## C.1 Functionals

In its simplest form, a **functional** $I = I[y]$ is a mapping from some specified set of functions $\{y = f(x)\}$ to the set of real numbers $\mathbb{R}$. For the types of problems that we will be most interested in, the functions $y = f(x)$ are defined on some finite interval $x \in [x_1, x_2]$; they are single-valued and have continuous first derivatives; and they have *fixed endpoints* $\wp_1 \equiv (x_1, y_1)$, $\wp_2 \equiv (x_2, y_2)$. (Curves that cannot be described by a single-valued function can be put in parametric form, $x = x(t)$, $y = y(t)$, which we will discuss in detail in Appendix C.6.) A simple concrete example of a functional is the arc length of the curve traced out by a function $y = f(x)$ that connects $\wp_1$ and $\wp_2$:





M.J. Benacquista and J.D. Romano, *Classical Mechanics*, Undergraduate Lecture Notes in Physics, https://doi.org/10.1007/978-3-319-68780-3

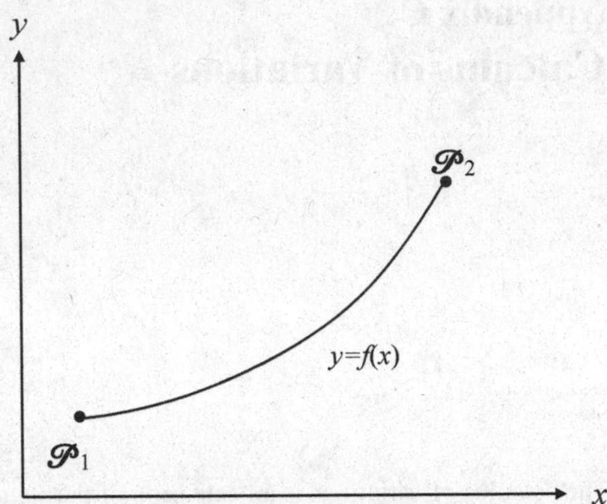


**Fig. C.1** The arc length of the curve traced out by the function $y = f(x)$ between $\wp_1$ and $\wp_2$ can be thought of as the value of a functional $I[y]$ evaluated for this particular function $y = f(x)$

$$I[y] \equiv \int_{\wp_1}^{\wp_2} \mathrm{d}s = \int_{\wp_1}^{\wp_2} \sqrt{\mathrm{d}x^2 + \mathrm{d}y^2} = \int_{x_1}^{x_2} \sqrt{1 + y'^2}\, \mathrm{d}x\,, \tag{C.1}$$

as shown in Fig. C.1. The corresponding calculus of variations problem is then to find the function $y = f(x)$ that minimizes the arc length between the two endpoints. We know the answer to this problem is a straight line, but to actually *prove* it requires some work. We will do this explicitly using the formalism of the calculus of variations in Example C.1 below.

More generally, we will consider functionals of the form

$$I[y] \equiv \int_{x_1}^{x_2} F(y, y', x)\, \mathrm{d}x\,, \tag{C.2}$$

where $x$ is the independent variable and the set of functions $\{y = f(x)\}$ is as before, but the integrand $F$ is now an *arbitrary* function of the three variables $(y, y', x)$. (For the arc-length functional defined previously, $F(y, y', x) = \sqrt{1 + y'^2}$, which is independent of $x$ and $y$, but that does not have to be the case in general.) Note that to do the integral over $x$, we need to express both $y$ and $y'$ in terms of $f(x)$, but for the variational calculations that follow, we simply treat $F$ as an ordinary function of three *independent* variables. The fact that $y$ and $y'$ are related to one another only shows up later on, when we need to relate the variation $\delta y'$ to $\delta y$, cf. (C.6).

Toward the end of this appendix, in Appendix C.7, we will extend our definition of a functional to $n$-degrees of freedom:

$$I[y_1, y_2, \ldots, y_n] \equiv \int_{x_1}^{x_2} F(y_1, y_2, \ldots, y_n; y_1', y_2', \ldots, y_n'; x)\, \mathrm{d}x\,, \tag{C.3}$$

where $y_i \equiv f_i(x)$, $i = 1, 2, \dots, n$, are $n$ functions of the independent variable $x \in [x_1, x_2]$, which is a form more appropriate for classical mechanics problems with $x$ replaced by the time $t$. But for most of this appendix, we will work with the simpler functional given by (C.2).

## C.2 Deriving the Euler Equation

Given $I[y]$, we now want to find its extrema—i.e., those functions $y = f(x)$ for which $I[y]$ has a local maximum or minimum. Similar to ordinary calculus, a *necessary* (but not sufficient) condition for $y$ to be an extremum is that the *1st-order* change in $I[y]$ vanish for arbitrary variations to $y = f(x)$ that preserve the boundary conditions. We define such a variation to $y = f(x)$ by

$$\delta y \equiv \bar{f}(x) - f(x)\,, \tag{C.4}$$

where $\bar{f}(x)$ is a function that differs infinitesimally from $f(x)$ at each value of $x$ in the domain $[x_1, x_2]$ (See Fig. C.2). In terms of $\delta y$, the variation of the functional is then given by $\delta I[y] \equiv I[y+\delta y] - I[y]$, where we ignore all terms that are 2nd-order or higher in $\delta y$, $\delta y'$. The condition $\delta I[y] = 0$ determines the **stationary values** of the functional. These include maxima and minima, but also points of inflection or *saddle points*. To check if a stationary value is an extremum, we need to calculate the change in $I[y]$ to *2nd-order* in $\delta y$. If the 2nd-order contribution $\delta^2 I[y]$ is positive, then we have a minimum; if it is negative, a maximum; and if it is zero, a saddle point. However, in the calculations that follow, we will stop at 1st order, as it will usually be obvious from the context of the problem whether our stationary solution is a maximum or minimum, without having to explicitly carry out the 2nd-order variation.

Given definition (C.2) of the functional $I[y]$, it follows that

$$\begin{aligned}\delta I[y] &\equiv I[y+\delta y] - I[y] \\ &= \int_{x_1}^{x_2} F(y+\delta y, y'+\delta y', x)\,\mathrm{d}x - \int_{x_1}^{x_2} F(y, y', x)\,\mathrm{d}x \\ &= \int_{x_1}^{x_2} \left\{ \left(\frac{\partial F}{\partial y}\right)\delta y + \left(\frac{\partial F}{\partial y'}\right)\delta y' \right\} \mathrm{d}x\,,\end{aligned} \tag{C.5}$$

where we ignored all 2nd-order terms to the get the last line. As mentioned earlier, the variations $\delta y$ and $\delta y'$ are not independent of one another, but are related by

$$\delta y' \equiv \delta\left(\frac{\mathrm{d}y}{\mathrm{d}x}\right) = \frac{\mathrm{d}}{\mathrm{d}x}\delta y\,. \tag{C.6}$$

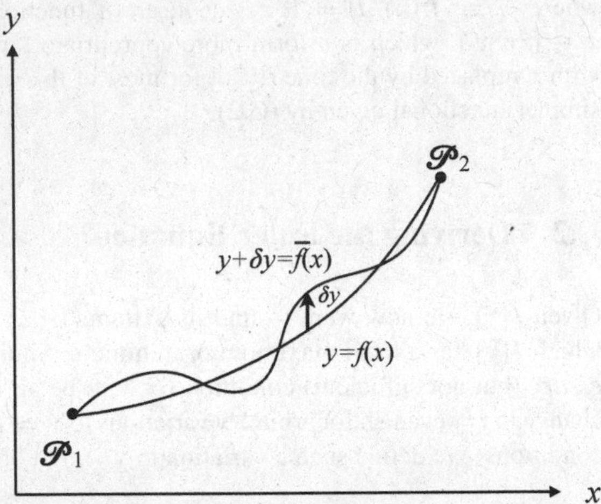


**Fig. C.2** Graphical illustration of a variation $\delta y \equiv \bar{f}(x) - f(x)$ to the function $y = f(x)$. Note that the variation must vanish at the end points in order to preserve the boundary conditions

Making this substitution and then integrating the term involving $\delta y'$ by parts, we find

$$\delta I[y] = \left(\frac{\partial F}{\partial y'}\right)\delta y\bigg|_{x_1}^{x_2} + \int_{x_1}^{x_2}\left\{\left(\frac{\partial F}{\partial y}\right) - \frac{\mathrm{d}}{\mathrm{d}x}\left(\frac{\partial F}{\partial y'}\right)\right\}\delta y\,\mathrm{d}x\,. \tag{C.7}$$

But since the variation $\delta y$ must vanish at the endpoints, i.e.,

$$\delta y|_{x_1} = 0\,, \quad \delta y|_{x_2} = 0\,, \tag{C.8}$$

the first term on the right-hand side of (C.7) is zero. Then, since the variation $\delta y$ is otherwise arbitrary, it follows that

$$\delta I[y] = 0 \quad \Leftrightarrow \quad \frac{\partial F}{\partial y} - \frac{\mathrm{d}}{\mathrm{d}x}\left(\frac{\partial F}{\partial y'}\right) = 0\,. \tag{C.9}$$

The equation on the right-hand side is called the **Euler equation**. (In the context of classical mechanics, where $F$ is the Lagrangian of the system, the above equation is called the **Euler-Lagrange equation**. See Chap. 3 for details.)

***Example C.1*** Using the Euler equation, show that the curve that minimizes the distance between two fixed points in a plane is a straight line.

*Proof* From (C.1) we have $F(y, y', x) = \sqrt{1 + y'^2}$, which is independent of both $x$ and $y$. Thus, the Euler equation (C.9) simplifies to

$$\frac{\mathrm{d}}{\mathrm{d}x}\left(\frac{\partial F}{\partial y'}\right) = 0 \quad \Leftrightarrow \quad \frac{\partial F}{\partial y'} = \text{const}\,. \tag{C.10}$$

Performing the derivative for our particular $F$, we get

$$\frac{\partial F}{\partial y'} = \frac{y'}{\sqrt{1 + y'^2}} = \text{const}\,, \tag{C.11}$$

which, after rearranging, gives $y' = A$ (another constant). So $y = Ax + B$, which is the equation of a straight line. The integration constants $A$ and $B$ are determined by the fixed endpoint conditions $y_1 = f(x_1)$ and $y_2 = f(x_2)$. □

A straight line is an example of a **geodesic**—i.e., the shortest distance path between two points. In the following two exercises, you are asked to determine the geodesics on the surface of the cylinder and the surface of a sphere. Recall that the line element on the surface of a cylinder of radius $R$ is

$$\mathrm{d}s^2 = R^2\,\mathrm{d}\phi^2 + \mathrm{d}z^2\,, \tag{C.12}$$

and the line element on the surface of a sphere of radius $R$ is

$$\mathrm{d}s^2 = R^2\,(\mathrm{d}\theta^2 + \sin^2\theta\,\mathrm{d}\phi^2)\,. \tag{C.13}$$

In general, for an $n$-dimensional space with coordinates $x^i \equiv (x^1, x^2, \ldots, x^n)$, the line element can be written as

$$\mathrm{d}s^2 = \sum_{i,j=1}^{n} g_{ij}\,\mathrm{d}x^i\,\mathrm{d}x^j\,, \tag{C.14}$$

where $g_{ij} \equiv g_{ij}(x^1, x^2, \ldots, x^n)$.

**Exercise C.1** Show that a geodesic on the surface of a cylinder of radius $\rho = R$ is a *helix*—i.e., $z = A\phi + B$, where $A$, $B$ are constants, determined by the location of the endpoints $\wp_1 = (\phi_1, z_1)$ and $\wp_2 = (\phi_2, z_2)$. This result should not be surprising given that the surface of a cylinder is *intrinsically flat*, just like a plane in two dimensions.

**Exercise C.2** Show that a geodesic on the surface of a sphere is an arc of a *great circle*—i.e., the intersection of the surface of the sphere with a plane passing through the center of the sphere, $A\cos\phi + B\sin\phi + \cot\theta = 0$, where $A$ and $B$ are constants, determined by the end points of the curve. (*Hint*: Take $\theta$ as the independent variable for this calculation.)

**Exercise C.3** Consider the functionals

$$I_1[y] \equiv \int_{x_1}^{x_2} F(y, y', x)\,\mathrm{d}x\,, \qquad I_2[y] \equiv \int_{x_1}^{x_2} F^2(y, y', x)\,\mathrm{d}x\,, \tag{C.15}$$

where $F(y, y', x)$ is everywhere positive. As usual, assume that the functions $y = f(x)$ are fixed at the end points $x_1$ and $x_2$. Show that the Euler equations for $I_1[y]$ and $I_2[y]$ agree if and only if

$$\frac{\mathrm{d}F}{\mathrm{d}x} = 0 \quad \text{or} \quad \frac{\partial F}{\partial y'} = 0\,. \tag{C.16}$$

Thus, unlike varying an ordinary function $f(x) > 0$ for which the stationary values of $f(x)$ and $g(x) \equiv f^2(x)$ are identical, the stationary values of the functionals defined by $F(y, y', x)$ and $F^2(y, y', x)$ differ in general.

## C.3 A More Formal Discussion of the Variational Process

We can give a more formal derivation of Euler's equation and the associated variational process by writing the variation $\delta y$ of the function $y = f(x)$ as

$$\delta y(x) \equiv \bar{f}(x) - f(x) = \varepsilon\eta(x)\,, \tag{C.17}$$

where $\eta(x)$ is a function satisfying the appropriate boundary conditions (e.g., it vanishes at the endpoints), and $\varepsilon$ is a real variable, which we take to be infinitesimal for $\delta y$ to represent an infinitesimal variation of $y$. The variation of a functional $I[y]$ resulting from the above variation of $y$ is then

$$\delta I[y] \equiv I[y + \varepsilon\eta] - I[y]\,. \tag{C.18}$$

Note that since $I[y+\varepsilon\eta]$ is an *ordinary* function of the real variable $\varepsilon$, we can Taylor expand $I[y + \varepsilon\eta]$ or take its derivatives with respect to $\varepsilon$ in the usual way. For our applications, we will be particularly interested in the first derivative of $I[y + \varepsilon\eta]$ with respect to $\varepsilon$ evaluated at $\varepsilon = 0$:

$$\left. \frac{dI[y+\varepsilon\eta]}{d\varepsilon} \right|_{\varepsilon=0} \equiv \lim_{\varepsilon\to 0} \frac{I[y+\varepsilon\eta]-I[y]}{\varepsilon} . \tag{C.19}$$

In terms of this derivative, we can define the **functional derivative** of $I$, denoted $\delta I[y]/\delta y(x)$ or more simply $\delta I[y]/\delta y$, as[1]

$$\left. \frac{dI[y+\varepsilon\eta]}{d\varepsilon} \right|_{\varepsilon=0} = \int_{x_1}^{x_2} dx \, \frac{\delta I[y]}{\delta y(x)} \eta(x) , \tag{C.20}$$

where the integration is over the domain $x \in [x_1, x_2]$ of the functions $y = f(x)$ on which the functional $I[y]$ is defined. Note that the above definition is the *functional analogue* of the definition of the *directional derivative* of a function $\varphi(x^1, x^2, \ldots, x^n)$ in the direction of $\boldsymbol{\eta}$:

$$\left. \frac{d\varphi(\mathbf{x}+\varepsilon\boldsymbol{\eta})}{d\varepsilon} \right|_{\varepsilon=0} \equiv \lim_{\varepsilon\to 0} \frac{\varphi(\mathbf{x}+\varepsilon\boldsymbol{\eta})-\varphi(\mathbf{x})}{\varepsilon} = \sum_{i=1}^{n} \frac{\partial\varphi}{\partial x^i}\eta^i , \tag{C.21}$$

where integration over the continuous variable $x$ in (C.20) replaces the summation over the discrete index $i$ in (C.21); see also Appendix A.4.1 and (A.30).

If the functional $I[y]$ has the form given in (C.2), i.e.,

$$I[y] \equiv \int_{x_1}^{x_2} dx \, F(y, y', x) , \tag{C.22}$$

where the functions $y = f(x)$ are fixed at $x_1$ and $x_2$, then the variational procedure described above leads to the same results that we found in the previous section, namely

$$\left. \frac{dI[y+\varepsilon\eta]}{d\varepsilon} \right|_{\varepsilon=0} = \left( \frac{\partial F}{\partial y'} \right) \eta \Big|_{x_1}^{x_2} + \int_{x_1}^{x_2} \left\{ \left( \frac{\partial F}{\partial y} \right) - \frac{d}{dx} \left( \frac{\partial F}{\partial y'} \right) \right\} \eta \, dx . \tag{C.23}$$

But since the function $\eta(x)$ vanishes at the endpoints, the above expression simplifies to

$$\left. \frac{dI[y+\varepsilon\eta]}{d\varepsilon} \right|_{\varepsilon=0} = \int_{x_1}^{x_2} \left\{ \frac{\partial F}{\partial y} - \frac{d}{dx} \left( \frac{\partial F}{\partial y'} \right) \right\} \eta \, dx , \tag{C.24}$$

for which

[1] *A word of caution.* The functional derivative $\delta I[y]/\delta y(x)$ is a *density* in $x$, being defined *inside* an integral, (C.20). As such, the dimensions of $dx \, \delta I[y]/\delta y(x)$ are the same as the dimensions of $I$ divided by the dimensions of $y$. For example, if $I[y]$ is the arc length functional, then $\delta I[y]/\delta y(x)$ has dimension of 1/length.

$$\frac{\delta I[y]}{\delta y} = \frac{\partial F}{\partial y} - \frac{\mathrm{d}}{\mathrm{d}x}\left(\frac{\partial F}{\partial y'}\right). \tag{C.25}$$

Thus, in terms of a functional derivative, the Euler equation (C.9) can be written as $\delta I[y]/\delta y(x) = 0$.

**Exercise C.4** Let $I[y]$ be a functional that depends only on the value of $y$ at a particular value of $x$, e.g.,

$$I[y] \equiv y(x_0). \tag{C.26}$$

Show that for this case the functional derivative is the Dirac delta function:

$$\frac{\delta I[y]}{\delta y(x)} = \delta(x - x_0). \tag{C.27}$$

## C.4 Alternate Form of the Euler Equation

When $F$ does not explicitly depend upon $x$, it is convenient to work with the Euler equation in an alternative form. If we simply take the total derivative of $F$ with respect to $x$ we have

$$\frac{\mathrm{d}F}{\mathrm{d}x} = \frac{\partial F}{\partial y}y' + \frac{\partial F}{\partial y'}y'' + \frac{\partial F}{\partial x}. \tag{C.28}$$

But since

$$\frac{\mathrm{d}}{\mathrm{d}x}\left(y'\frac{\partial F}{\partial y'}\right) = y''\frac{\partial F}{\partial y'} + y'\frac{\mathrm{d}}{\mathrm{d}x}\left(\frac{\partial F}{\partial y'}\right), \tag{C.29}$$

we can rewrite (C.28) as

$$\frac{\mathrm{d}F}{\mathrm{d}x} = \frac{\partial F}{\partial x} + \frac{\mathrm{d}}{\mathrm{d}x}\left(y'\frac{\partial F}{\partial y'}\right) + y'\left[\frac{\partial F}{\partial y} - \frac{\mathrm{d}}{\mathrm{d}x}\left(\frac{\partial F}{\partial y'}\right)\right]. \tag{C.30}$$

Thus, using the Euler equation, (C.9), we have

$$\delta I[y] = 0 \quad \Leftrightarrow \quad 0 = \frac{\partial F}{\partial x} + \frac{\mathrm{d}}{\mathrm{d}x}\left(y'\frac{\partial F}{\partial y'} - F\right). \tag{C.31}$$

## C.5 Possible Simplifications

The Euler equation (C.9) or its alternate form (C.31) is a 2nd-order ordinary differential equation with respect to the independent variable $x$. This equation may simplify depending on the form of $F(y, y', x)$:

1. If $F$ is independent of $y$, then $\partial F/\partial y = 0$ and the Euler equation (C.9) can be integrated to yield
$$\frac{\partial F}{\partial y'} = \text{const}\,. \tag{C.32}$$

2. If $F$ does not depend explicitly on $x$, then $\partial F/\partial x = 0$ and the alternate form of the Euler equation (C.31) can be integrated to yield
$$y'\frac{\partial F}{\partial y'} - F = \text{const}\,. \tag{C.33}$$

It turns out that simplification (2) is equivalent to making a change of the independent variable in the integrand of the functional from $x$ to $y$ using

$$\mathrm{d}x = x'\,\mathrm{d}y \quad \Leftrightarrow \quad y' = 1/x'\,, \tag{C.34}$$

so that

$$I[y] = \int_{x_1}^{x_2} F(y, y')\,\mathrm{d}x = \int_{y_1}^{y_2} F(y, 1/x')\,x'\,\mathrm{d}y \equiv \int_{y_1}^{y_2} \tilde{F}(x, x', y)\,\mathrm{d}y \equiv \tilde{I}[x]\,. \tag{C.35}$$

But since

$$\tilde{F}(x, x', y) \equiv x'\,F(y, 1/x') \tag{C.36}$$

is independent of $x$, the Euler equation for $\tilde{I}[x]$ simplifies to $\partial\tilde{F}/\partial x' = \text{const}$. But note that

$$\frac{\partial \tilde{F}}{\partial x'} = \frac{\partial}{\partial x'}\left[x'\,F(y, 1/x')\right] = F + x'\frac{\partial F}{\partial y'}\frac{\partial(1/x')}{\partial x'} = F - \frac{1}{x'}\frac{\partial F}{\partial y'} = F - y'\frac{\partial F}{\partial y'}\,, \tag{C.37}$$

which means that

$$\frac{\partial \tilde{F}}{\partial x'} = \text{const} \quad \Leftrightarrow \quad y'\frac{\partial F}{\partial y'} - F = \text{const} \tag{C.38}$$

as claimed. (Note: The change of independent variables from $x$ to $y$ assumes that the function $y = f(x)$ is invertible, so that $x = f^{-1}(y)$ and $\tilde{F}(x, x', y)$ are well-defined.)

***Example C.2*** A soap film is suspended between two circular loops of wire, as shown in Fig. C.3. Ignoring the effects of gravity, the soap film takes the shape of a surface of revolution, which has *minimimum* surface area. Thus, in terms of the function $y = f(x)$, the functional that we need to minimize is the surface area of revolution

$$I[y] = \int_{\wp_1}^{\wp_2} 2\pi y \, ds = 2\pi \int_{x_1}^{x_2} y\sqrt{1 + y'^2}\, dx \,, \tag{C.39}$$

which has

$$F(y, y', x) = 2\pi y\sqrt{1 + y'^2} \,. \tag{C.40}$$

But since $F$ does not depend explicitly on $x$, we can use simplification (2) to write

$$y'\frac{\partial F}{\partial y'} - F = y'\frac{2\pi y y'}{\sqrt{1 + y'^2}} - 2\pi y\sqrt{1 + y'^2} = -\frac{2\pi y}{\sqrt{1 + y'^2}} = \text{const} \,. \tag{C.41}$$

Rewriting this constant as $-2\pi A$ and solving for $y'$ yields

$$y' \equiv \frac{dy}{dx} = \frac{1}{A}\sqrt{y^2 - A^2} \,. \tag{C.42}$$

This is a separable equation, which can be integrated using the hyperbolic trig substitutions $y = A\cosh u$, recalling that $\cosh^2 u - \sinh^2 u = 1$, and $d\cosh u = \sinh u \, du$. Thus,

$$x = A \int \frac{dy}{\sqrt{y^2 - A^2}} + B = A \int \frac{A \sinh u \, du}{A \sinh u} + B = Au + B = A\cosh^{-1}(y/A) + B \,, \tag{C.43}$$

or, equivalently,

$$y = A\cosh\left(\frac{x - B}{A}\right) \,. \tag{C.44}$$

Such a curve is called a **catenary**. As usual, the integration constants $A$ and $B$ can be determined by the boundary conditions for $y = f(x)$, which are related to the radii of the two circular loops of wire. Unfortunately, solving for $A$ and $B$ involves solving a transcendental equation. See Chap. 17 of Arfken (1970) for a discussion of special cases of this problem. □

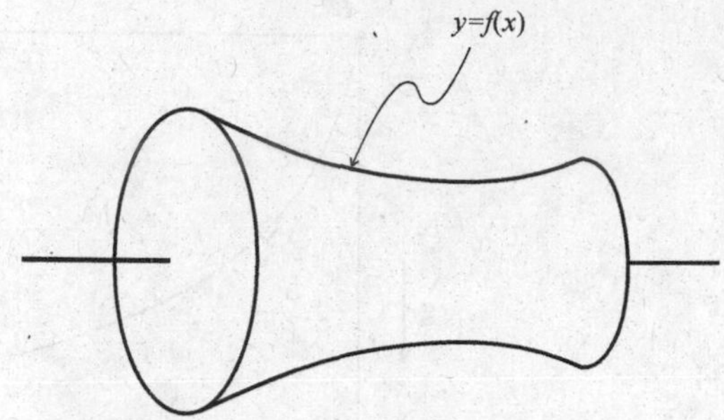


**Fig. C.3** Soap film suspended between two circular loops of wire. If we ignore the effects of gravity, the soap film takes the shape of a surface of revolution about the horizontal axis, which has minimum surface area

**Exercise C.5** Find the shape of a wire joining two points such that a bead will slide along the wire under the influence of gravity (without friction) in the shortest amount of time. (See Fig. C.4.) Assume that the bead is released from rest at $y = 0$. Such a curve is called a **brachistochrone**, which in Greek means "shortest time."

*Hint*: You should extremize the functional

$$I[y] = \int_{\wp_1}^{\wp_2} \frac{\mathrm{d}s}{v} = \int_{x_1}^{x_2} \mathrm{d}x \, \frac{\sqrt{1+y'^2}}{\sqrt{2gy}}\,, \tag{C.45}$$

where conservation of energy

$$\frac{1}{2}mv^2 - mgy = 0 \quad \Rightarrow \quad v = \sqrt{2gy} \tag{C.46}$$

was used to yield an expression for the speed $v$ in terms of $y$. By using simplification (2) or changing the independent variable of the functional from $x$ to $y$, you should find

$$y' \equiv \frac{\mathrm{d}y}{\mathrm{d}x} = \sqrt{\frac{1-Ay}{Ay}}\,. \tag{C.47}$$

which has solution

$$x = \frac{1}{2A}(\theta - \sin\theta)\,, \qquad y = \frac{1}{2A}(1 - \cos\theta)\,. \tag{C.48}$$

This is the parametric representation of a **cycloid** (i.e., the path traced out by a point on the rim of a wheel as it rolls without slipping across a horizontal surface). See Fig. C.5.

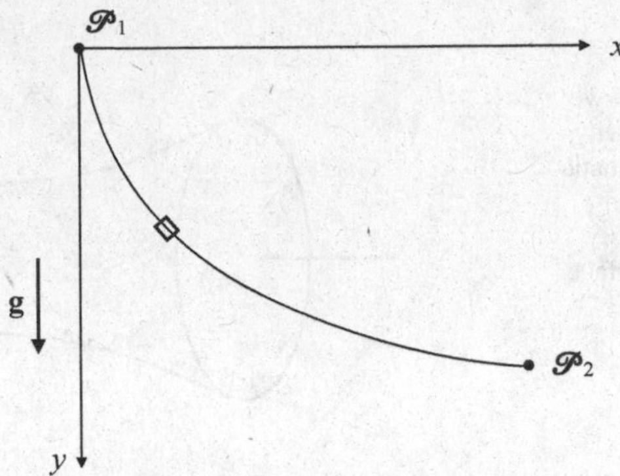


**Fig. C.4** Geometrical set-up for the brachistochrone problem, Exercise C.5. The goal is to find the shape of the wire connecting points $\wp_1$ and $\wp_2$ such that a bead slides along the wire under the influence of gravity (and in the absence of friction) in the shortest amount of time. Note that we have chosen the $y$-axis to increase in the downward direction

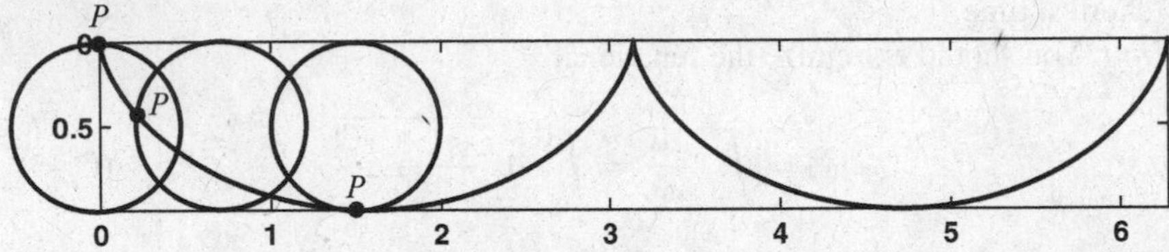


**Fig. C.5** A cycloid is the path traced out by a point on the rim of a wheel as it rolls without slipping across a flat surface. It is also the shape of the wire that solves the brachistochrone problem, Exercise C.5. To be consistent with the geometry of Exercise C.5 shown in Fig. C.4, we are considering the wheel as rolling to the right in contact with the *top* horizontal surface

**Exercise C.6** (*Adapted from Kuchăr* 1995.) Consider a two-dimensional surface of revolution obtained by rotating the curve $z = f(\rho)$ around the $z$-axis, where $\rho \equiv \sqrt{x^2 + y^2}$. An example of such a surface is shown in panel (a) of Fig. C.6, which is a *paraboloid*, defined by $f(\rho) = \rho^2$. Surfaces of revolution are most conveniently described by embedding equations

$$x = \rho \cos\phi \,, \qquad y = \rho \sin\phi \,, \qquad z = f(\rho) \,, \tag{C.49}$$

where $\phi$ is the standard azimuthal angle in the $xy$-plane.

(a) Write down the line element $\mathrm{d}s^2$ on the surface of revolution in terms of the coordinates $(\rho, \phi)$ by simply substituting the embedding equations into the 3-dimensional line element $\mathrm{d}x^2 + \mathrm{d}y^2 + \mathrm{d}z^2$.

(b) By varying the arc length functional, obtain the geodesic equation for a curve $\rho = \rho(\phi)$ on the surface of revolution, and show that it can be solved via quadratures,

$$\phi - \phi_0 = c_1 \int_{\rho_0}^{\rho} \frac{d\rho}{\rho} \sqrt{\frac{1 + [f'(\rho)]^2}{\rho^2 - c_1^2}}, \tag{C.50}$$

where $c_1$ is a constant.

(c) Evaluate the above integral for the case of a surface of a cone with half-angle $\alpha$, which is defined by $f(\rho) = \rho \cot \alpha$. (See panel (b) of Fig. C.6.) You should find

$$\rho = \frac{c_1}{\cos(\phi \sin \alpha + c_2)}, \tag{C.51}$$

where $c_1$, $c_2$ are constants determined by the boundary conditions.

(d) Show that the above solution is equivalent to a straight line in a flat 2-dimensional space with Cartesian coordinates

$$\bar{x} \equiv \rho \cos(\phi \sin \alpha), \qquad \bar{y} \equiv \rho \sin(\phi \sin \alpha). \tag{C.52}$$

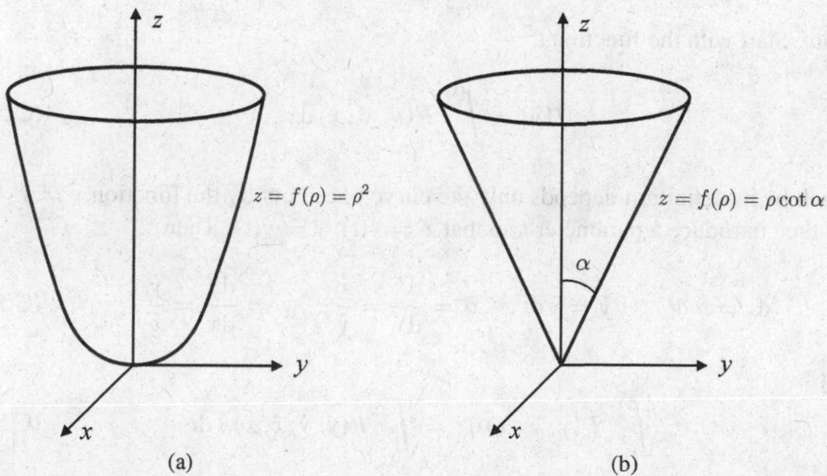


**Fig. C.6** Examples of surfaces of revolution, obtained by rotating a curve $z = f(\rho)$ around the $z$-axis. Panel (a) A *paraboloid* defined by $f(\rho) \equiv \rho^2$. Panel (b) A *cone* with half-angle $\alpha$, defined by $f(\rho) = \rho \cot \alpha$

## C.6 Variational Problem in Parametric Form

Although we have been considering functionals of the form given by (C.2), where the curve is explicitly described by the function $y = f(x)$, there may be cases where it is more convenient (or even necessary) to describe the curve in parametric form, e.g., $x = x(t)$, $y = y(t)$. Such an example is the variational problem to find the shape of a *closed* curve of fixed length that encloses the greatest area. (We will revisit this in more detail in Appendix C.8.) Here we derive the necessary and sufficient conditions for a functional to depend only on the curve in the $xy$-plane and not on the choice of parametric representation of the curve. The relevant theorem is:

**Theorem C.1** *The necessary and sufficient conditions for the functional*

$$I[x, y] = \int_{t_1}^{t_2} G(x, y, \dot{x}, \dot{y}, t)\, dt \tag{C.53}$$

*to depend only on the curve in the $xy$-plane and not on the choice of parametric representation of the curve is that $G$ not depend explicitly on $t$ and be a positive-homogeneous function of degree one in $\dot{x}$ and $\dot{y}$—i.e.,*

$$G(x, y, \lambda\dot{x}, \lambda\dot{y}) = \lambda G(x, y, \dot{x}, \dot{y}) \tag{C.54}$$

*for all $\lambda > 0$.*

*Proof* Start with the functional[2]

$$I[y] = \int_{x_1}^{x_2} F(y, y', x)\, \mathrm{d}x\,, \tag{C.55}$$

which by its definition depends only the curve traced out by the function $y = f(x)$. We then introduce a parameter $t$ so that $x = x(t)$, $y = y(t)$. Then

$$\mathrm{d}x = \dot{x}\,\mathrm{d}t\,, \quad \mathrm{d}y = \dot{y}\,\mathrm{d}t\,, \quad x' = \frac{\mathrm{d}x}{\mathrm{d}y} = \frac{\dot{x}}{\dot{y}}\,, \quad y' = \frac{\mathrm{d}y}{\mathrm{d}x} = \frac{\dot{y}}{\dot{x}}\,, \tag{C.56}$$

and

$$\int_{x_1}^{x_2} F(y, y', x)\, \mathrm{d}x = \int_{t_1}^{t_2} F(y, \dot{y}/\dot{x}, x)\dot{x}\, \mathrm{d}t\,. \tag{C.57}$$

Thus,

[2]This derivation closely follows that given in Sect. 10 of Gelfand and Fomin (1963).

$$I[y] = \int_{x_1}^{x_2} F(y, y', x)\,\mathrm{d}x = \int_{t_1}^{t_2} G(x, y, \dot{x}, \dot{y})\,\mathrm{d}t \equiv I[x, y]\,, \tag{C.58}$$

where

$$G(x, y, \dot{x}, \dot{y}) = F(y, \dot{y}/\dot{x}, x)\dot{x}\,. \tag{C.59}$$

Note that $G$ has the properties that it does not depend explicitly on $t$, and that

$$G(x, y, \lambda\dot{x}, \lambda\dot{y}) = \lambda G(x, y, \dot{x}, \dot{y}) \tag{C.60}$$

for all $\lambda > 0$. Thus, $G$ is a *positive-homogeneous function of degree one* in $\dot{x}$ and $\dot{y}$.

Conversely, suppose that we have a functional of the form

$$I[x, y] = \int_{t_1}^{t_2} G(x, y, \dot{x}, \dot{y})\,\mathrm{d}t\,, \tag{C.61}$$

where $G$ does not explicitly depend on $t$ and is a positive-homogeneous function of degree one in $\dot{x}$ and $\dot{y}$. Then if we change the parametrization from $t$ to a new parameter $\tau$, we have

$$\mathrm{d}t = \frac{\mathrm{d}t}{\mathrm{d}\tau}\,\mathrm{d}\tau\,, \quad \dot{x} = \frac{\mathrm{d}x}{\mathrm{d}\tau}\frac{\mathrm{d}\tau}{\mathrm{d}t}\,, \quad \dot{y} = \frac{\mathrm{d}y}{\mathrm{d}\tau}\frac{\mathrm{d}\tau}{\mathrm{d}t}\,, \tag{C.62}$$

and

$$G(x, y, \dot{x}, \dot{y}) = G\left(x, y, \frac{\mathrm{d}x}{\mathrm{d}\tau}\frac{\mathrm{d}\tau}{\mathrm{d}t}, \frac{\mathrm{d}y}{\mathrm{d}\tau}\frac{\mathrm{d}\tau}{\mathrm{d}t}\right) = G\left(x, y, \frac{\mathrm{d}x}{\mathrm{d}\tau}, \frac{\mathrm{d}y}{\mathrm{d}\tau}\right)\frac{\mathrm{d}\tau}{\mathrm{d}t}\,, \tag{C.63}$$

where the last equality used the positive-homogeneous-of-degree-one property of $G$. Thus,

$$I[x, y] = \int_{t_1}^{t_2} G(x, y, \dot{x}, \dot{y})\,\mathrm{d}t = \int_{\tau_1}^{\tau_2} G\left(x, y, \frac{\mathrm{d}x}{\mathrm{d}\tau}, \frac{\mathrm{d}y}{\mathrm{d}\tau}\right)\mathrm{d}\tau\,, \tag{C.64}$$

which shows that the functional is independent of the parametric representation of the curve. □

***Example C.3*** Here we show explicitly that the Euler equations obtained from the parametrized functional

$$I[x,y] \equiv \int_{t_1}^{t_2} G(x,y,\dot{x},\dot{y})\,\mathrm{d}t = \int_{t_1}^{t_2} F(y,\dot{y}/\dot{x},x)\dot{x}\,\mathrm{d}t \tag{C.65}$$

reduce to the standard Euler equation (C.9) obtained from

$$I[y] = \int_{x_1}^{x_2} F(y,y',x)\,\mathrm{d}x\,. \tag{C.66}$$

*Proof* The Euler equations obtained from (C.65) by varying both $x$ and $y$ are

$$\frac{\partial G}{\partial x} - \frac{\mathrm{d}}{\mathrm{d}t}\left(\frac{\partial G}{\partial \dot{x}}\right) = 0\,, \tag{C.67a}$$

$$\frac{\partial G}{\partial y} - \frac{\mathrm{d}}{\mathrm{d}t}\left(\frac{\partial G}{\partial \dot{y}}\right) = 0\,. \tag{C.67b}$$

Since $G$ does not depend explicitly on $t$, then by an extension of (C.33) to two variables (See also Appendix C.7.3), we also have

$$\dot{x}\frac{\partial G}{\partial \dot{x}} + \dot{y}\frac{\partial G}{\partial \dot{y}} - G = \text{const}\,. \tag{C.68}$$

Differentiating this last equation with respect to $t$ yields

$$\dot{x}\left[\frac{\mathrm{d}}{\mathrm{d}t}\left(\frac{\partial G}{\partial \dot{x}}\right) - \frac{\partial G}{\partial x}\right] + \dot{y}\left[\frac{\mathrm{d}}{\mathrm{d}t}\left(\frac{\partial G}{\partial \dot{y}}\right) - \frac{\partial G}{\partial y}\right] = 0\,, \tag{C.69}$$

where we have cancelled out the terms involving $\ddot{x}$ and $\ddot{y}$. This last equation shows that the two equations (C.67a) and (C.67b) are *not* independent, but follow one from the other. So, without loss of generality, let's consider (C.67b). Then by writing $G$ in terms of $F$ and performing the derivatives, we find

$$\begin{aligned}
0 &= \frac{\partial G}{\partial y} - \frac{\mathrm{d}}{\mathrm{d}t}\left(\frac{\partial G}{\partial \dot{y}}\right)\\
&= \frac{\partial}{\partial y}[F(y,\dot{y}/\dot{x},x)\dot{x}] - \frac{\mathrm{d}x}{\mathrm{d}t}\frac{\mathrm{d}}{\mathrm{d}x}\left(\frac{\partial}{\partial \dot{y}}[F(y,\dot{y}/\dot{x},x)\dot{x}]\right)\\
&= \frac{\partial F}{\partial y}\dot{x} - \dot{x}\frac{\mathrm{d}}{\mathrm{d}x}\left(\frac{\partial F}{\partial y'}\frac{1}{\dot{x}}\dot{x}\right)\\
&= \dot{x}\left[\frac{\partial F}{\partial y} - \frac{\mathrm{d}}{\mathrm{d}x}\left(\frac{\partial F}{\partial y'}\right)\right],
\end{aligned} \tag{C.70}$$

which is proportional to the standard Euler equation (C.9). □

**Exercise C.7** Consider the parametrized form of the arc length functional in two dimensions,

$$I[x_1, x_2] \equiv \int_{\wp_1}^{\wp_2} \mathrm{d}s = \int_{t_1}^{t_2} G(x_1, x_2, \dot{x}_1, \dot{x}_2)\,\mathrm{d}t\,, \tag{C.71}$$

where

$$G(x_1, x_2, \dot{x}_1, \dot{x}_2) = \frac{\mathrm{d}s}{\mathrm{d}t} = \sqrt{\sum_{i,j} g_{ij}\,\dot{x}_i\,\dot{x}_j}\,. \tag{C.72}$$

Note that by writing the arc length in this form (See (C.14)), we are allowing for the possibility that the 2-dimensional space be curved (e.g., the surface of a sphere) and that the coordinates need not be Cartesian, so $g_{ij} \equiv g_{ij}(x_1, x_2)$ in general.

(a) Show that the Euler equations for this functional are

$$\frac{\mathrm{d}}{\mathrm{d}t}\left(\sum_j g_{ij}\dot{x}_j\right) - \frac{1}{2}\sum_{j,k}\frac{\partial g_{jk}}{\partial x_i}\dot{x}_j\dot{x}_k = \frac{1}{G}\left(\frac{\mathrm{d}G}{\mathrm{d}t}\right)\sum_j g_{ij}\dot{x}_j\,. \tag{C.73}$$

These are the geodesic equations in an *arbitrary* parametrization.

(b) Show that if we choose the parameter $t$ to be linearly related to the arc length $s$ along the curve,

$$t = as + b\,, \qquad a, b = \text{const}\,, \tag{C.74}$$

then $G = \text{const}$, and the geodesic equation simplifies to

$$\frac{\mathrm{d}}{\mathrm{d}t}\left(\sum_j g_{ij}\dot{x}_j\right) - \frac{1}{2}\sum_{j,k}\frac{\partial g_{jk}}{\partial x_i}\dot{x}_j\dot{x}_k = 0\,. \tag{C.75}$$

Such a parametrization of the curve is called an **affine parametrization.**

(c) Show that one obtains the same simplified form of the geodesic equation by varying instead the *kinetic energy functional,*

$$J[x_1, x_2] \equiv \int_{t_1}^{t_2} K(x_1, x_2, \dot{x}_1, \dot{x}_2)\,\mathrm{d}t\,, \tag{C.76}$$

where

$$K(x_1, x_2, \dot{x}_1, \dot{x}_2) \equiv \frac{1}{2}\sum_{i,j} g_{ij}\,\dot{x}_i\,\dot{x}_j = \frac{1}{2}G^2(x_1, x_2, \dot{x}_1, \dot{x}_2)\,. \tag{C.77}$$

The equivalence of these two approaches follows from the fact that $G$ and $K$ differ by an overall multiplicative constant when $t$ is an affine parameter. If $t$ is *not* an affine parameter, then the two functionals $I$ and $J$ lead to the different equations of motion, consistent with the results of Exercise C.3. (Note that these results hold, in general, in $n$ dimensions.)

## C.7 Generalizations

The standard calculus of variations problem (C.2) discussed in the preceding sections can be extended in several ways. Here we describe three such extensions.

### C.7.1 Functionals that Depend on Higher-Order Derivatives

Consider a functional of the form

$$I[y] = \int_{x_1}^{x_2} F(y, y', y'', x)\,\mathrm{d}x\,, \tag{C.78}$$

where the set of functions $\{y = f(x)\}$ is now restricted so that *both* $y$ and $y'$ are fixed at the endpoints $\wp_1$ and $\wp_2$. Then proceeding in a manner similar to that in Appendix C.2, we find

$$\delta I[y] = 0 \quad \Leftrightarrow \quad \frac{\partial F}{\partial y} - \frac{\mathrm{d}}{\mathrm{d}x}\left(\frac{\partial F}{\partial y'}\right) + \frac{\mathrm{d}^2}{\mathrm{d}x^2}\left(\frac{\partial F}{\partial y''}\right) = 0\,. \tag{C.79}$$

Note that the Euler equation for this case may contain *3rd* or even *4th-order* derivatives of $y = f(x)$. Although most problems in classical mechanics involve *2nd-order* differential equations, 3rd or higher-order differential equations have applications in certain areas of chaos theory (See, e.g., Goldstein et al. 2002).

### C.7.2 Allowing Variations with Free Endpoints

The standard variational problem (C.2) with fixed endpoints can be generalized by allowing the variations $\delta y$ to be *non-zero* at either one or both endpoints $\wp_1$, $\wp_2$. The derivation given in Appendix C.2 then leads to

$$\delta I[y] = 0 \quad \Leftrightarrow \quad \frac{\partial F}{\partial y} - \frac{\mathrm{d}}{\mathrm{d}x}\left(\frac{\partial F}{\partial y'}\right) = 0\,, \quad \left.\frac{\partial F}{\partial y'}\right|_{x_1} = 0\,, \quad \left.\frac{\partial F}{\partial y'}\right|_{x_2} = 0\,. \tag{C.80}$$

The conditions

$$\left.\frac{\partial F}{\partial y'}\right|_{x_1} = 0\,, \qquad \left.\frac{\partial F}{\partial y'}\right|_{x_2} = 0\,, \tag{C.81}$$

are sometimes called **natural boundary conditions** for the curve. In the context of classical mechanics, the natural boundary conditions for a particle moving in response to a velocity-independent conservative force correspond to *zero velocity* at the endpoints. The only solution to the equations of motion that satisfies these boundary conditions at *both* endpoints is the trivial solution, where the particle just sits at one location forever. Imposing the natural boundary condition at just *one* endpoint and fixed boundary conditions at the other allows for non-trivial solutions, in general.

**Exercise C.8** Redo the brachistochrone problem (Exercise C.5), but this time allowing the second endpoint at $x_2$ to be free—i.e., $\delta y|_{x_2} \neq 0$. You should find that the solution is again a cycloid, but which intersects the line $x = x_2$ at a *right angle*.

### C.7.3 Generalization to Several Dependent Variables

The derivation of the Euler equation given in Appendix C.2 can be easily be extended to functionals of the form

$$I[y_1, \ldots, y_n] \equiv \int_{x_1}^{x_2} F(y_1, \ldots, y_n; y_1', \ldots, y_n'; x)\,\mathrm{d}x\,, \tag{C.82}$$

where $y_i \equiv f_i(x)$, $i = 1, 2, \ldots, n$, are $n$ functions of the independent variable $x \in [x_1, x_2]$, which we require to be fixed at the endpoints $\wp_1$ and $\wp_2$. The corresponding Euler equations obtained from the variational principle $\delta I[y_1, \ldots, y_n] = 0$ with respect to variations $\delta y_i$ that vanish at the endpoints are

$$\frac{\partial F}{\partial y_i} - \frac{\mathrm{d}}{\mathrm{d}x}\left(\frac{\partial F}{\partial y_i'}\right) = 0\,, \qquad i = 1, 2, \ldots, n\,. \tag{C.83}$$

These equations form a system of $n$ 2nd-order ordinary differential equations for the functions $y_i = f_i(x)$ with respect to the independent variable $x$.

The simplifications discussed in Appendix C.5 carry over to the general case of $n$ degrees of freedom:

1. If $F$ is independent of a particular $y_{\underline{i}}$, then $\partial F/\partial y_{\underline{i}} = 0$ and the corresponding Euler equation can be integrated to yield

$$\frac{\partial F}{\partial y'_{\underline{i}}} = \text{const}\,. \tag{C.84}$$

2. If $F$ does not depend explicitly on $x$, then

$$h \equiv \sum_{i=1}^{n} y'_i \frac{\partial F}{\partial y'_i} - F = \text{const}\,. \tag{C.85}$$

To prove the second result above:

$$\frac{\mathrm{d}h}{\mathrm{d}x} = \sum_{i=1}^{n} y''_i \frac{\partial F}{\partial y'_i} + \sum_{i=1}^{n} y'_i \frac{\mathrm{d}}{\mathrm{d}x}\left(\frac{\partial F}{\partial y'_i}\right) - \sum_{i=1}^{n}\left[\frac{\partial F}{\partial y_i} y'_i + \frac{\partial F}{\partial y'_i} y''_i\right] = 0\,, \tag{C.86}$$

where we assumed that $F$ does not depend explicitly on $x$ (i.e., $\partial F/\partial x = 0$) and used the Euler equations (C.83) to get the last equality.

There is also an additional simplification if $F$ does not depend on the derivative of one of the variables:

3. If $F$ is independent of a particular derivative, which we will take (without loss of generality) to be $y'_n$, then the Euler equation for $y_n$ becomes $\partial F/\partial y_n = 0$, which can be solved algebraically for $y_n$ in terms of all of the other variables and their derivatives (assuming $\partial^2 F/\partial y_n^2 \neq 0$). The Euler equations for all the other variables can then be obtained from the *reduced* functional

$$\underline{I}[y_1, \ldots, y_{n-1}] \equiv \int_{x_1}^{x_2} \underline{F}(y_1, \ldots, y_{n-1}; y'_1, \ldots, y'_{n-1}; x)\,\mathrm{d}x\,, \tag{C.87}$$

where

$$\begin{aligned}&\underline{F}(y_1, \ldots, y_{n-1}; y'_1, \ldots, y'_{n-1}; x)\\ &\quad\equiv F(y_1, \ldots, y_n; y'_1, \ldots, y'_{n-1}; x)\big|_{y_n = y_n(y_1, \ldots, y_{n-1}; y'_1, \ldots, y'_{n-1}; x)}\end{aligned} \tag{C.88}$$

**Exercise C.9** Verify simplification (3) above by showing that

$$\frac{\partial \underline{F}}{\partial y_i} - \frac{\mathrm{d}}{\mathrm{d}x}\left(\frac{\partial \underline{F}}{\partial y_i'}\right) = 0 \quad \Rightarrow \quad \frac{\partial F}{\partial y_i} - \frac{\mathrm{d}}{\mathrm{d}x}\left(\frac{\partial F}{\partial y_i'}\right) = 0\,, \tag{C.89}$$

for $i = 1, 2, \ldots, n-1$ as a consequence of $\partial F/\partial y_n = 0$.

## C.8 Isoperimetric Problems

So far, we've been considering variational problems of the form $\delta I[y] = 0$, where the functions $y = f(x)$ have been subject only to *boundary conditions*, e.g., fixed at the endpoints $x = x_1$ and $x_2$. But there might also exist situations where the functions $y = f(x)$ are subject to an *integral constraint*

$$J[y] \equiv \int_{x_1}^{x_2} G(y, y', x)\,\mathrm{d}x = J_0\,, \tag{C.90}$$

where $J_0$ is a constant. Such problems are called **isoperimetric problems**, since the classic example of such a problem is to find the shape of a closed curve of *fixed* length (perimeter) $\ell$ that encloses the maximum area. Due to the constraint, the variations of $y$ in $\delta I = 0$ are not free, but are subject to the condition that $\delta J[y] = 0$. We can incorporate this condition into the variational problem by using the method of *Lagrange multipliers* (See Sect. 2.4). This amounts to adding to $\delta I = 0$ a multiple of $\delta J = 0$,

$$\delta I[y] + \lambda \delta J[y] = 0\,, \tag{C.91}$$

where $\lambda$ is an undetermined constant (the Lagrange multiplier for this problem). Note that (C.91) can be recast as finding the stationary values of the functional

$$\bar{I}[y, \lambda] \equiv I[y] + \lambda\,(J[y] - J_0) = \int_{x_1}^{x_2} (F + \lambda G)\,\mathrm{d}x - \lambda\,J_0 \tag{C.92}$$

with respect to *unconstrained* variations of both $y$ and $\lambda$. (The variation with respect to $\lambda$ recovers the integral constraint $J[y] - J_0 = 0$.) Performing the variations give rise to two equations, which can be solved for the two unknowns $y = f(x)$ and $\lambda$ (if desired).

***Example C.4*** Here we will find the shape of a closed curve of fixed length $\ell$ that encloses the largest area. Since the curve is closed, we will not be able to represent it as a single-valued function $y = f(x)$ or $x = g(y)$. Instead, we have to represent it *parametrically*—i.e., by $x = x(t)$, $y = y(t)$, where $t \in [t_1, t_2]$ is a parameter along the curve. Without loss of generality, we can orient the curve in the $xy$-plane so that

$$x|_{t_1,t_2} = 0\,, \quad y|_{t_1,t_2} = 0\,, \quad \dot{x}|_{t_1,t_2} = -v\,, \quad \dot{y}|_{t_1,t_2} = 0\,, \tag{C.93}$$

with the derivatives so chosen as to avoid a kink at the origin. (See Fig. C.7.)

The functional that we want to extremize is the area under the curve

$$I[x, y] = \oint y\,\mathrm{d}x = \int_{t_1}^{t_2} y\dot{x}\,\mathrm{d}t\,, \tag{C.94}$$

subject to the constraint

$$J[x, y] = \oint \mathrm{d}s = \int_{t_1}^{t_2} \sqrt{\dot{x}^2 + \dot{y}^2}\,\mathrm{d}t = \ell\,. \tag{C.95}$$

The curve is traversed clockwise so that the area obtained is enclosed by the curve.

Using the method of Lagrange multipliers discussed above, we extremize the combined functional

$$\bar{I}[x, y, \lambda] = \int_{t_1}^{t_2} (F + \lambda G)\,\mathrm{d}t - \lambda\ell\,, \tag{C.96}$$

where

$$F + \lambda G = y\dot{x} + \lambda\sqrt{\dot{x}^2 + \dot{y}^2}\,. \tag{C.97}$$

Note that $F + \lambda G$ does not explicitly depend on $t$ and is a positive-homogeneous function of degree one in $\dot{x}$ and $\dot{y}$. Thus, from the discussion of Appendix C.6, the solution to the variational problem will depend only on the curve in the $xy$-plane and not on a particular parametric representation of the curve.

The Euler equations obtained from $F + \lambda G$ by varying $x$ and $y$ are

$$\frac{\mathrm{d}}{\mathrm{d}t}\left(y + \frac{\lambda\dot{x}}{\sqrt{\dot{x}^2 + \dot{y}^2}}\right) = 0 \quad \Rightarrow \quad y + \frac{\lambda\dot{x}}{\sqrt{\dot{x}^2 + \dot{y}^2}} = A\,, \tag{C.98}$$

$$\frac{\mathrm{d}}{\mathrm{d}t}\left(\frac{\lambda\dot{y}}{\sqrt{\dot{x}^2 + \dot{y}^2}}\right) - \dot{x} = 0 \quad \Rightarrow \quad \frac{\lambda\dot{y}}{\sqrt{\dot{x}^2 + \dot{y}^2}} - x = B\,, \tag{C.99}$$

where $A$ and $B$ are constants. These equations can be simplified if we switch the parametric representation from $t$ to arc length $s$, noting that

$$\frac{\mathrm{d}t}{\mathrm{d}s} = \frac{1}{\sqrt{\dot{x}^2 + \dot{y}^2}}\,. \tag{C.100}$$

The right-hand sides of (C.98) and (C.99) then become

$$y + \lambda \frac{\mathrm{d}x}{\mathrm{d}s} = A\,, \qquad \lambda \frac{\mathrm{d}y}{\mathrm{d}s} - x = B\,. \tag{C.101}$$

We now apply the boundary conditions of (C.93) but in terms of arc length $s$,

$$x|_{s=0,\ell} = 0\,, \quad y|_{s=0,\ell} = 0\,, \quad \left.\frac{\mathrm{d}x}{\mathrm{d}s}\right|_{s=0,\ell} = -1\,, \quad \left.\frac{\mathrm{d}y}{\mathrm{d}s}\right|_{s=0,\ell} = 0\,, \tag{C.102}$$

which lead to $B = 0$ and $A = -\lambda$. The first equation in (C.101) can then be solved for $y$ in terms of $\mathrm{d}x/\mathrm{d}s$,

$$y = -\lambda \left(1 + \frac{\mathrm{d}x}{\mathrm{d}s}\right), \tag{C.103}$$

and substituted back into the second equation. This leads to

$$\frac{\mathrm{d}^2 x}{\mathrm{d}s^2} = -\lambda^{-2} x\,, \tag{C.104}$$

with solution

$$x(s) = -|\lambda| \sin(s/|\lambda|)\,, \tag{C.105}$$

which satisfies the boundary conditions above. Using (C.103) we have

$$y(s) = -\lambda\,(1 - \cos(s/|\lambda|))\,. \tag{C.106}$$

Note that these last two equations are the parametric representation of a circle

$$x^2 + (y + \lambda)^2 = \lambda^2\,, \tag{C.107}$$

with radius $R = |\lambda|$ and center $(0, -\lambda)$, In order that the circle lie above the $x$-axis, as suggested by Fig. C.7, we need $\lambda = -R$. Finally, since $2\pi R = \ell$ is the length of the curve, the Lagrange multiplier $\lambda = -\ell/2\pi$. Thus,

$$x = -\frac{\ell}{2\pi} \sin(2\pi s/l)\,, \qquad y = \frac{\ell}{2\pi}(1 - \cos(2\pi s/l))\,. \tag{C.108}$$

The area enclosed by the circle is $A = \pi R^2 = \ell^2/4\pi$, which is the largest area enclosed by a closed curve of fixed length $\ell$. □

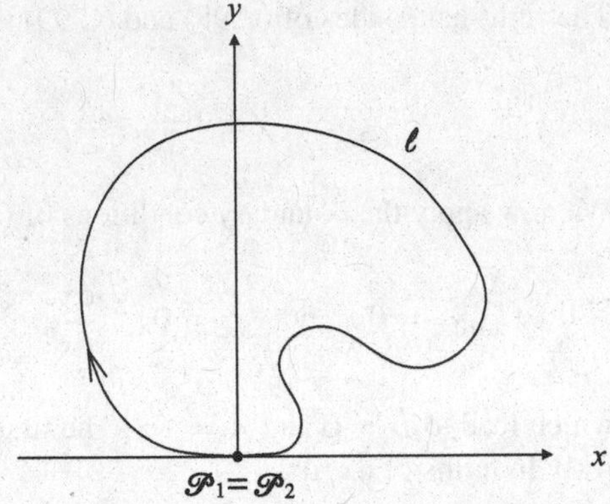


**Fig. C.7** Closed curve of fixed length $\ell$ in the $xy$-plane. The coordinates and parametrization are chosen so that the curve starts and stops at the origin and is tangent to the $x$-axis at the origin. The curve is traversed in the clockwise direction so that the area calculated in (C.94) is that *enclosed* by the curve

**Exercise C.10** Find the shape of the curve of fixed length $\ell$ and *free* endpoint $\wp_2$ on the $x$-axis that encloses the largest area between it and the $x$-axis. (See Fig. C.8.) You should find:

$$x = \frac{\ell}{\pi}\left(1 - \cos(\pi s/l)\right), \qquad y = -\frac{\ell}{\pi}\sin(\pi s/l), \tag{C.109}$$

which is the parametric representation of a *semi-circle* of radius $R = \ell/\pi$, length $\ell$, and center $(\ell/\pi, 0)$. Note that the area enclosed by this curve and the $x$-axis is $\ell^2/2\pi$, which is *twice* as large as that calculated in Example C.4 for the closed-curve variational problem.

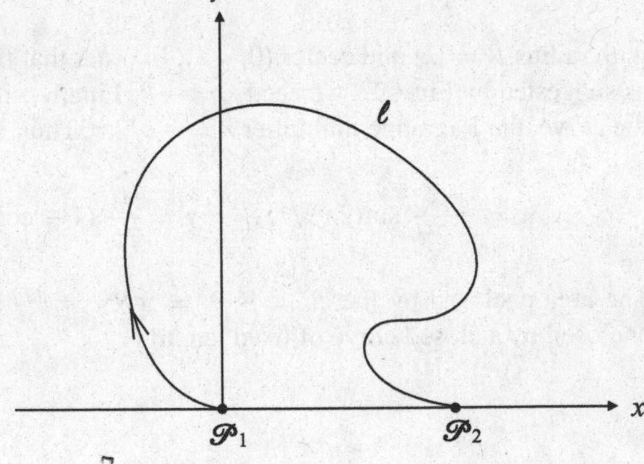


**Fig. C.8** Curve of fixed length $\ell$, having one endpoint $\wp_1$ fixed at the origin and the other endpoint $\wp_2$ *freely-variable* in the $x$-direction

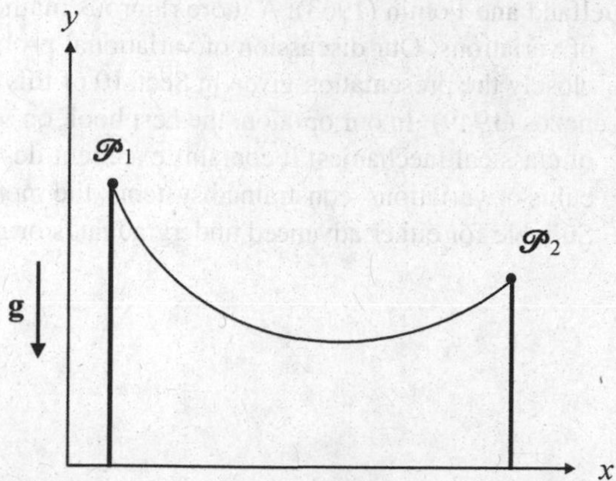


**Fig. C.9** Flexible hanging cable of fixed length $\ell$ supported at endpoints $\wp_1$ and $\wp_2$. A uniform gravitational field **g** points downward

**Exercise C.11** Find the shape of a flexible hanging cable of fixed length $\ell$, which is supported at endpoints $\wp_1$ and $\wp_2$ in a uniform gravitational field **g** pointing downward. (See Fig. C.9.)
*Hint*: The shape minimizes the gravitational potential energy of the cable

$$I[x] = \int_{\wp_1}^{\wp_2} \mathrm{d}m\, gy = \int_{\wp_1}^{\wp_2} \mu \mathrm{d}s\, gy = \mu g \int_{x_1}^{x_2} y\sqrt{1+y'^2}\,\mathrm{d}x\,, \qquad \text{(C.110)}$$

where $\mu$ is the mass-per-unit-length of the cable (assumed constant), subject to the constraint that the cable has fixed length $\ell$, i.e.,

$$J[y] = \int_{\wp_1}^{\wp_2} \mathrm{d}s = \int_{x_1}^{x_2} \sqrt{1+y'^2}\,\mathrm{d}x = \ell\,. \qquad \text{(C.111)}$$

You should find that the solution has the form of a catenary, $y \sim \cosh x$.

## Suggested References

*Full references are given in the bibliography at the end of the book.*


Boas (2006): Chapter 9 is devoted solely to the calculus of variations; an excellent introduction to the topic well-suited for undergraduates with many examples and problems. Solutions to several of the problems presented in this appendix can be found in Boas (2006).

Gelfand and Fomin (1963): A more rigorous mathematical treament of the calculus of variations. Our discussion of variational problems in parametric form follows closely the presentation given in Sect. 10 of this book.

Lanczos (1949): In our opinion, the best book on variational methods in the context of classical mechanics. It contains excellent descriptions/explanations of the calculus of variations, constrained systems, the method of Lagrange multipliers, etc. Suitable for either advanced undergraduates or graduate students.

# Appendix D
# Linear Algebra

Linear algebra can be thought of as an extension of the mathematical structure of ordinary (3-dimensional) vectors and matrices to an *arbitrary* number of dimensions. The general mathematical framework of linear algebra is particularly relevant for the matrix calculations required for describing rigid-body motion (Chaps. 6 and 7), and for calculating the normal modes associated with small oscillations (Chap. 8). Although not strictly necessary for classical mechanics, we will allow our vectors and matrices to be *complex-valued*, since this generalization requires limited additional work, and it turns out to be extremely useful for quantum mechanics, where complex numbers are the rule, not the exception. We will, however, restrict ourselves to a *finite* number of dimensions $n$, although it is also possible to have infinite-dimensional vector spaces (e.g., function spaces).

Since we will only be summarizing key results here and not giving detailed proofs, we encourage readers to refer to other texts, e.g., Boas (2006); Dennery and Kryzwicki (1967); Griffiths (2005); Halmos (1958) to fill in the missing details. Our approach in this appendix is similar to that of Appendix A in (Griffiths, 2005).

## D.1 Vector Space

An abstract vector space consists of two types of objects (**vectors** and **scalars**) and two types of operations (**vector addition** and **scalar multiplication**), which interact with one another and are subject to certain properties (enumerated below). We will denote vectors by boldface symbols, $\mathbf{A}$, $\mathbf{B}$, $\mathbf{C}$, $\cdots$, and scalars (which we will take to be complex numbers) by italicized symbols, $a, b, c, \cdots$. Vector addition will be denoted by a $+$ sign between two vectors, e.g., $\mathbf{A} + \mathbf{B}$, and scalar multiplication by juxtaposition of a scalar and a vector, e.g., $a\mathbf{A}$. The properties obeyed by these operations are as follows.



M.J. Benacquista and J.D. Romano, *Classical Mechanics*, Undergraduate Lecture Notes in Physics, https://doi.org/10.1007/978-3-319-68780-3

### D.1.1 Vector Addition

1. Closure: The addition of two vectors is also a vector:

$$\mathbf{A} + \mathbf{B} = \mathbf{C} \tag{D.1}$$

2. Commutativity:

$$\mathbf{A} + \mathbf{B} = \mathbf{B} + \mathbf{A} \tag{D.2}$$

3. Associativity:

$$(\mathbf{A} + \mathbf{B}) + \mathbf{C} = \mathbf{A} + (\mathbf{B} + \mathbf{C}) \tag{D.3}$$

4. Zero vector:

$$\exists \mathbf{0} \quad \text{such that} \quad \mathbf{A} + \mathbf{0} = \mathbf{A} \quad \forall \mathbf{A} \tag{D.4}$$

5. Inverse vector:

$$\forall \mathbf{A} \quad \exists - \mathbf{A} \quad \text{such that} \quad \mathbf{A} + (-\mathbf{A}) = \mathbf{0} \tag{D.5}$$

### D.1.2 Scalar Multiplication

1. Closure: The multiplication of a scalar and a vector is also a vector:

$$a\mathbf{A} = \mathbf{B} \tag{D.6}$$

2. Identity: The scalar 1 is the identity operator on vectors:

$$1\mathbf{A} = \mathbf{A} \tag{D.7}$$

3. Scalar multiplication is distributive with respect to scalar addition:

$$(a + b)\mathbf{A} = a\mathbf{A} + b\mathbf{A} \tag{D.8}$$

4. Scalar multiplication is distributive with respect to vector addition:

$$a(\mathbf{A} + \mathbf{B}) = a\mathbf{A} + a\mathbf{B} \tag{D.9}$$

5. Scalar multiplication is associative with respect to scalar multiplication:

$$a(b\mathbf{A}) = (ab)\mathbf{A} \tag{D.10}$$

Using the various properties given above, it is easy to see that $0\mathbf{A} = \mathbf{0}$ and $(-1)\mathbf{A} = -\mathbf{A}$, since

$$(\mathbf{A} + 0\mathbf{A}) = (1\mathbf{A} + 0\mathbf{A}) = (1+0)\mathbf{A} = 1\mathbf{A} = \mathbf{A}\,, \tag{D.11}$$

and

$$(\mathbf{A} + (-1)\mathbf{A}) = (1\mathbf{A} + (-1)\mathbf{A}) = (1+(-1))\mathbf{A} = 0\mathbf{A} = \mathbf{0}\,. \tag{D.12}$$

The above definitions and properties allow us to extend the familiar properties of ordinary 3-dimensional vectors and real numbers to other sets of objects. Although *most* properties of 3-dimensional vectors carry over to these higher-dimensional abstract vector spaces, some do not, such as the cross (vector) product of two vectors, e.g., $\mathbf{A} \times \mathbf{B}$, which is defined in 3-dimensions by (A.2). (But see the *wedge product* of differential forms described in Appendix B.)

## D.2 Basis Vectors

A key concept when working with vectors is that of a *basis*. But in order to define what we mean by a basis, we must first introduce some terminology:

- A **linear combination** of vectors $\mathbf{A}, \mathbf{B}, \cdots$ is any vector of the form

$$a\mathbf{A} + b\mathbf{B} + \cdots \tag{D.13}$$

  where $a, b, \cdots$ are scalars.
- A vector $\mathbf{C}$ is **linearly independent** of the set of vectors $\{\mathbf{A}, \mathbf{B}, \cdots\}$ if and only if $\mathbf{C}$ cannot be written as a linear combination of the vectors in the set—i.e.,

$$\neg\exists a, b, \cdots \quad \text{such that} \quad \mathbf{C} = a\mathbf{A} + b\mathbf{B} + \cdots \tag{D.14}$$

- A *set* of vectors $\{\mathbf{A}, \mathbf{B}, \cdots\}$ is a **linearly independent set** if and only if each vector in the set is linearly independent of all the other vectors in the set.
- A set of vectors $\{\mathbf{A}, \mathbf{B}, \cdots\}$ **spans** the vector space if and only if any vector $\mathbf{C}$ in the vector space can be written as a linear combinaton of the vectors in the set—i.e.,

$$\exists a, b, \cdots \quad \text{such that} \quad \mathbf{C} = a\mathbf{A} + b\mathbf{B} + \cdots \tag{D.15}$$

In terms of the above definitions, a **basis** for a vector space is defined to be any set of vectors which is (i) linearly independent and (ii) spans the vector space. The number of basis vectors is defined as the **dimension** of the vector space. Thus, an $n$-dimensional vector space has a basis consisting of $n$ vectors,

$$\{\mathbf{e}_1, \mathbf{e}_2, \ldots, \mathbf{e}_n\}\,. \tag{D.16}$$

This is, of course, consistent with what we know about the space of ordinary 3-dimensional vectors. The set of unit vectors $\{\hat{\mathbf{x}}, \hat{\mathbf{y}}, \hat{\mathbf{z}}\}$ is a basis for that space. (More about what *unit* means for a more general vector space in just a bit.)

### *D.2.1 Components of a Vector*

Given a set of basis vectors (D.16), we can write

$$\mathbf{A} = A_1\mathbf{e}_1 + A_2\mathbf{e}_2 + \cdots + A_n\mathbf{e}_n \equiv \sum_i A_i\mathbf{e}_i \,. \tag{D.17}$$

The scalars $A_1, A_2, \ldots, A_n$ are called the **components** of $\mathbf{A}$ with respect to the basis $\{\mathbf{e}_1, \mathbf{e}_2, \ldots, \mathbf{e}_n\}$. The *decomposition* of $\mathbf{A}$ into its components is *unique* for a *given* basis as shown in Exercise D.1 below. But for a *different* set of basis vectors, e.g., $\{\mathbf{e}'_1, \mathbf{e}'_2, \ldots, \mathbf{e}'_n\}$, the components of $\mathbf{A}$ will be different.

**Exercise D.1** Prove that the decomposition (D.17) of $\mathbf{A}$ into its components $A_1, A_2, \ldots, A_n$ is unique. *Hint*: Use proof by contradiction—i.e., assume that there exist *other* components $A'_1, A'_2, \ldots, A'_n$, for which

$$\mathbf{A} = \sum_i A'_i\mathbf{e}_i \,. \tag{D.18}$$

Then show that this leads to a contradiction regarding the linear independence of the basis vectors unless $A'_i = A_i$ for all $i$.

Vector addition and scalar multiplication are what you might expect in terms of the components of the vectors. That is, if we denote the correspondence between vectors and components by[1]

$$\mathbf{A} \quad \leftrightarrow \quad \mathsf{A} \equiv [A_1, A_2, \ldots, A_n]^T \equiv \begin{bmatrix} A_1 \\ A_2 \\ \vdots \\ A_n \end{bmatrix}, \tag{D.19}$$

then it is easy to show that

[1]Our notation is such that $\mathbf{A}$ denotes the abstract vector, $A_i$ its $i$th component with respect to a basis, and $\mathsf{A}$ the collection of components $A_1, A_2, \ldots, A_n$ represented as an $n \times 1$ column matrix. The superscript $T$ denotes transpose, which converts a row matrix into a colum matrix, and vice versa.

$$\mathbf{A} + \mathbf{B} \quad \leftrightarrow \quad \mathsf{A} + \mathsf{B} = [A_1 + B_1, A_2 + B_2, \ldots, A_n + B_n]^T \,, \tag{D.20}$$

and

$$a\mathbf{A} \quad \leftrightarrow \quad a\mathsf{A} = [aA_1, aA_2, \ldots, aA_n]^T \,. \tag{D.21}$$

Thus, vector addition corresponds to ordinary addition of the (scalar) components of the vectors, and scalar multiplication of a vector corresponds to multiplying each of the components of the vector by that scalar.

**Exercise D.2** Show that the zero vector **0** and inverse vector $-\mathbf{A}$ can be written in terms of components as

$$\mathbf{0} \quad \leftrightarrow \quad \mathsf{0} = [0, 0, \ldots, 0]^T \,, \tag{D.22}$$

and

$$-\mathbf{A} \quad \leftrightarrow \quad -\mathsf{A} = [-A_1, -A_2, \ldots, -A_n]^T . \tag{D.23}$$

## D.3 Inner Product

An abstract $n$-dimensional vector space as defined above generalizes several key properties of vectors in ordinary 3-dimensional space. But by itself, the definition of a vector space does not specify how to calculate the *length* of a vector, or the *angle* that one vector makes with another vector.[2] In order to extend these concepts to higher-dimensional vector spaces, we need to introduce an additional mathematical structure, called an **inner product** on the space of vectors. As we shall see below, this inner product generalizes the notion of the ordinary *dot product* (or scalar product) $\mathbf{A} \cdot \mathbf{B}$ of 3-dimensional vectors, (A.1), and hence will allow us to talk about *unit* vectors and *orthogonality* of two vectors.

**Definition**: Given two vectors **A** and **B**, the **inner product** of **A** and **B** (denoted $\mathbf{A} \cdot \mathbf{B}$) is a scalar (i.e., a complex number) that satisfies the following three properties:

$$\mathbf{A} \cdot \mathbf{B} = (\mathbf{B} \cdot \mathbf{A})^* \tag{D.24a}$$

$$\mathbf{A} \cdot \mathbf{A} \geq 0 \,, \quad \mathbf{A} \cdot \mathbf{A} = 0 \text{ if and only if } \mathbf{A} = \mathbf{0} \tag{D.24b}$$

$$\mathbf{A} \cdot (b\mathbf{B} + c\mathbf{C}) = b(\mathbf{A} \cdot \mathbf{B}) + c(\mathbf{A} \cdot \mathbf{C}) \tag{D.24c}$$

[2] As anybody who has taken freshman physics knows, length and angle are *key* concepts for ordinary (3-dimensional) vectors, which are sometimes *defined* as "arrows" having magnitude and direction!

Mathematicians call a vector space with the additional structure of an inner product an **inner product space**.

> **Exercise D.3** Show that if $\mathbf{C} = b\mathbf{B}$ then $\mathbf{C} \cdot \mathbf{A} = b^* (\mathbf{B} \cdot \mathbf{A})$.

The fact that $\mathbf{A} \cdot \mathbf{A} \geq 0$ allows us to interpret $\mathbf{A} \cdot \mathbf{A}$ as the (squared) length or *norm* (or magnitude) of the vector $\mathbf{A}$. We will denote the norm of $\mathbf{A}$ as either $|\mathbf{A}|$ or $A$, so that

$$|\mathbf{A}| \equiv A \equiv \sqrt{\mathbf{A} \cdot \mathbf{A}}\,. \tag{D.25}$$

This leads to following definitions:

- A vector $\mathbf{A}$ is said to have *unit norm* (or to be **normalized**) if and only if $|\mathbf{A}| = 1$.
- Two vectors $\mathbf{A}$ and $\mathbf{B}$ are said to be **orthogonal** if and only if the inner product of $\mathbf{A}$ and $\mathbf{B}$ vanishes—i.e., $\mathbf{A} \cdot \mathbf{B} = 0$.
- A set of vectors $\{\mathbf{A}_1, \mathbf{A}_2, \cdots\}$ is said to be **orthonormal** if and only if $\mathbf{A}_i \cdot \mathbf{A}_j = \delta_{ij}$ for all $i, j = 1, 2, \ldots, n$, where $\delta_{ij}$ is the Kronecker delta symbol (which equals one if $i = j$, and equals zero otherwise).
- An **orthonormal basis** is an orthonormal set of basis vectors, which we will typically denote with hats, $\hat{\mathbf{e}}_1, \hat{\mathbf{e}}_2, \ldots, \hat{\mathbf{e}}_n$. These vectors are thus linearly independent, span the vector space, and satisfy

$$\hat{\mathbf{e}}_i \cdot \hat{\mathbf{e}}_j = \delta_{ij}\,. \tag{D.26}$$

### D.3.1 Gram-Schmidt Orthonormalization Procedure

As we already know from doing calculations with ordinary vectors in 3-dimensions, it is often convenient to work with a set of orthonormal basis vectors, e.g., $\{\hat{\mathbf{x}}, \hat{\mathbf{y}}, \hat{\mathbf{z}}\}$. Hence, it is good to know that there exists a general procedure, called the **Gram-Schmidt orthonormalization procedure**, for taking an arbitrary set of basis vectors in $n$ dimensions and converting it into an orthonormal set of basis vectors. As must be the case, this new set of basis vectors is formed by taking appropriate linear combinations of the original (non-orthonormal) basis vectors.

We start with a set of basis vectors

$$\{\mathbf{e}_1, \mathbf{e}_2, \ldots, \mathbf{e}_n\}\,, \tag{D.27}$$

which we will assume is not orthonormal. (If the basis vectors were already orthonormal, then there would be nothing that you need to do!) Take $\mathbf{e}_1$ and simply divide by its norm. The result is a unit vector that points in the same direction as $\mathbf{e}_1$,

$$\hat{\mathbf{f}}_1 \equiv \frac{\mathbf{e}_1}{|\mathbf{e}_1|}. \tag{D.28}$$

Now take $\mathbf{e}_2$, and subtract off its component in the direction of $\hat{\mathbf{f}}_1$:

$$\mathbf{f}_2 \equiv \mathbf{e}_2 - (\hat{\mathbf{f}}_1 \cdot \mathbf{e}_2)\,\hat{\mathbf{f}}_1\,. \tag{D.29}$$

This makes $\mathbf{f}_2$ orthogonal to $\hat{\mathbf{f}}_1$ as one can easily check,

$$\hat{\mathbf{f}}_1 \cdot \mathbf{f}_2 = \hat{\mathbf{f}}_1 \cdot \mathbf{e}_2 - (\hat{\mathbf{f}}_1 \cdot \mathbf{e}_2)(\hat{\mathbf{f}}_1 \cdot \hat{\mathbf{f}}_1) = 0\,. \tag{D.30}$$

Then normalize $\mathbf{f}_2$,

$$\hat{\mathbf{f}}_2 \equiv \frac{\mathbf{f}_2}{|\mathbf{f}_2|}. \tag{D.31}$$

Thus, both $\hat{\mathbf{f}}_1$ and $\hat{\mathbf{f}}_2$ have unit norm and they are orthogonal to one another. For $\hat{\mathbf{f}}_3$ we proceed in a similar fashion:

$$\mathbf{f}_3 \equiv \mathbf{e}_3 - (\hat{\mathbf{f}}_1 \cdot \mathbf{e}_3)\,\hat{\mathbf{f}}_1 - (\hat{\mathbf{f}}_2 \cdot \mathbf{e}_3)\,\hat{\mathbf{f}}_2\,, \tag{D.32}$$

and

$$\hat{\mathbf{f}}_3 \equiv \frac{\mathbf{f}_3}{|\mathbf{f}_3|}. \tag{D.33}$$

Continue as above for $\hat{\mathbf{f}}_4, \hat{\mathbf{f}}_5, \ldots, \hat{\mathbf{f}}_n$.

Note that this procedure does not produce a *unique* orthonormal basis. The resulting set of orthonormal basis vectors depends on the ordering of the basis vectors as illustrated in the following exercise.

**Exercise D.4** (a) Use the Gram-Schmidt orthonormalization procedure to construct an orthonormal basis starting from

$$\mathbf{e}_1 \equiv \hat{\mathbf{x}}\,, \quad \mathbf{e}_2 \equiv \hat{\mathbf{x}} + \hat{\mathbf{y}}\,, \quad \mathbf{e}_3 \equiv \hat{\mathbf{x}} + \hat{\mathbf{y}} + \hat{\mathbf{z}}\,, \tag{D.34}$$

where $\hat{\mathbf{x}}, \hat{\mathbf{y}}, \hat{\mathbf{z}}$ are the standard orthonormal basis vectors in ordinary 3-dimensional space. (b) Repeat the procedure, but this time with the basis vectors enumerated in the reverse order,

$$\mathbf{e}_1 \equiv \hat{\mathbf{x}} + \hat{\mathbf{y}} + \hat{\mathbf{z}}\,, \qquad \mathbf{e}_2 \equiv \hat{\mathbf{x}} + \hat{\mathbf{y}}\,, \qquad \mathbf{e}_3 \equiv \hat{\mathbf{x}}\,. \tag{D.35}$$

Do you get the same result as in part (a)?

### D.3.2 Component Form of the Inner Product

To illustrate that the inner product defined above generalizes the *dot* product of ordinary 3-dimensional vectors, it is simplest to show that we can recover the form of the dot product given in (A.4), which we rewrite here as

$$\mathbf{A} \cdot \mathbf{B} = A_1 B_1 + A_2 B_2 + A_3 B_3 = \sum_i A_i B_i \,, \tag{D.36}$$

where $A_i$, $B_i$ with $i = 1, 2, 3$ are the components of the ordinary 3-dimensional vectors $\mathbf{A}$, $\mathbf{B}$ with respect to some orthonormal basis $\{\hat{\mathbf{e}}_1, \hat{\mathbf{e}}_2, \hat{\mathbf{e}}_3\}$.

So let's work now in an arbitrary $n$-dimensional vector space equipped with an inner product, and let $\{\hat{\mathbf{e}}_1, \hat{\mathbf{e}}_2, \ldots, \hat{\mathbf{e}}_n\}$ denote an orthonormal basis for this space. Then according to (D.17), we can write

$$\mathbf{A} = \sum_i A_i \hat{\mathbf{e}}_i \,, \qquad \mathbf{B} = \sum_i B_i \hat{\mathbf{e}}_i \,, \tag{D.37}$$

for any two vectors $\mathbf{A}$, $\mathbf{B}$, where $A_i$, $B_i$ with $i = 1, 2, \ldots, n$ are the components of these vectors with respect to the given orthonormal basis. Since the basis vectors are orthonormal, it is easy to show that

$$A_i = \hat{\mathbf{e}}_i \cdot \mathbf{A} \,. \tag{D.38}$$

The proof is simply

$$\hat{\mathbf{e}}_i \cdot \mathbf{A} = \hat{\mathbf{e}}_i \cdot \left( \sum_j A_j \hat{\mathbf{e}}_j \right) = \sum_j A_j (\hat{\mathbf{e}}_i \cdot \hat{\mathbf{e}}_j) = \sum_j A_j \delta_{ij} = A_i \,, \tag{D.39}$$

where we used the linearity property (D.24c) of the inner product and the orthonormality (D.26) of the basis vectors to obtain the second and third equalities.[3] Using these results, it is then fairly straightforward to show that

$$\mathbf{A} \cdot \mathbf{B} = A_1^* B_1 + A_2^* B_2 + \cdots + A_n^* B_n = \sum_i A_i^* B_i \,, \tag{D.40}$$

and, as a consequence,

[3]Note that $A_i \neq \mathbf{A} \cdot \hat{\mathbf{e}}_i$ in general, since the components of a vector can be complex. Using Exercise (D.3), it follows that $\mathbf{A} \cdot \hat{\mathbf{e}}_i = A_i^*$.

$$|\mathbf{A}|^2 = \sum_i |A_i|^2 \,. \tag{D.41}$$

Note that (D.40) does indeed generalize the dot product (D.36) to arbitrary dimensions. Setting $n = 3$ and taking our vector components to be real-valued, we see that (D.40) reduces to (D.36).

**Exercise D.5** Prove the component form (D.40) of the inner product.

### D.3.3 *Schwarz Inequality*

Recall that for ordinary vectors in 3-dimensions, the dot product $\mathbf{A} \cdot \mathbf{B}$ can also be written as

$$\mathbf{A} \cdot \mathbf{B} = AB \cos\theta \,, \tag{D.42}$$

where $A \equiv |\mathbf{A}| \equiv \sqrt{\mathbf{A} \cdot \mathbf{A}}$ and $B \equiv |\mathbf{B}| \equiv \sqrt{\mathbf{B} \cdot \mathbf{B}}$ are the magnitudes of the two vectors, and $\theta$ is the angle between them; see (A.1). If we rewrite the above equation as

$$\cos\theta = \frac{\mathbf{A} \cdot \mathbf{B}}{|\mathbf{A}||\mathbf{B}|} \,, \tag{D.43}$$

then it can be thought of as the *definition* of the angle between the two vectors in terms of their dot products and their magnitudes. This suggests a way of generalizing the concept of "angle between two vectors" to an arbitrary $n$-dimensional vector space. Namely, simply interpret the expressions on the right-hand side of (D.43) in terms of the inner product defined by (D.24a), (D.24b), (D.24c). Unfortunately this won't work since $\mathbf{A} \cdot \mathbf{B}$ is a *complex number* in general, so the angle $\theta$ would not be real. But it turns out that there is a simple solution, which amounts to taking the *absolute value* of the right-hand side,

$$\cos\theta = \frac{|\mathbf{A} \cdot \mathbf{B}|}{|\mathbf{A}||\mathbf{B}|} = \sqrt{\frac{|\mathbf{A} \cdot \mathbf{B}|^2}{(\mathbf{A} \cdot \mathbf{A})(\mathbf{B} \cdot \mathbf{B})}} \,. \tag{D.44}$$

So this is now a real quantity, but the fact that this equation actually gives us something that we can interpret as an angle is thanks to the **Schwarz inequality**

$$|\mathbf{A} \cdot \mathbf{B}|^2 \leq (\mathbf{A} \cdot \mathbf{A})(\mathbf{B} \cdot \mathbf{B}) \,, \tag{D.45}$$

which guarantees that the right-hand side of (D.44) has a value $\leq 1$.

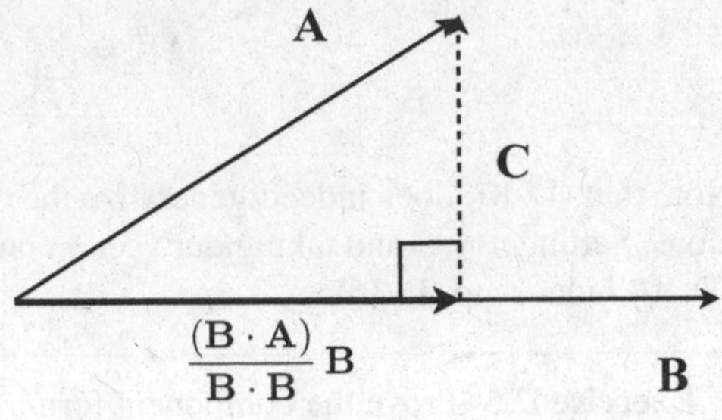


**Fig. D.1** Graphical illustration of the vector **C** used in the proof of the Schwarz inequality. Note that $((\mathbf{B}\cdot\mathbf{A})/\mathbf{B}\cdot\mathbf{B})\mathbf{B}$ is the projection of **A** onto **B**, so (by construction) **C** is orthogonal to **B**. Note that $\mathbf{C} = \mathbf{0}$ if and only if **A** and **B** are proportional to one another, which corresponds to the equal sign in the Schwarz inequality

**Exercise D.6** Prove the Schwarz inequality. (*Hint*: Consider the vector

$$\mathbf{C} \equiv \mathbf{A} - \frac{(\mathbf{B}\cdot\mathbf{A})}{\mathbf{B}\cdot\mathbf{B}}\mathbf{B}\,, \tag{D.46}$$

and then use $|\mathbf{C}|^2 \equiv \mathbf{C}\cdot\mathbf{C} \geq 0$. (See Fig. D.1.) Note that the Schwarz inequality becomes an equality when **A** and **B** are proportional (i.e., parallel or anti-parallel) to one another.)

## D.4 Linear Transformations

Given our abstract $n$-dimensional vector space, we would now like to define a certain class of operations (called *linear transformations*), which map vectors to other vectors in such a way that they preserve the linear property of vector addition and scalar multiplication of vectors. Rotations of ordinary 3-dimensional vectors, which play an important role in *all* branches of physics, are just one example of linear transformations.

**Definition**: A **linear transformation T** is a mapping that takes a vector **A** to another vector $\mathbf{A}' \equiv \mathbf{TA}$ such that

$$\mathbf{T}(a\mathbf{A} + b\mathbf{B}) = a(\mathbf{TA}) + b(\mathbf{TB})\,. \tag{D.47}$$

This method of introducing additional structure on a space in such a way that it interacts "naturally" with other structures in the space (in this case scalar multiplication and vector addition) is common practice in mathematics.

Note that multiplying every vector in the space by the same scalar $c$ (i.e., $\mathbf{A} \mapsto c\mathbf{A}$) is an example of a linear transformation since

$$\begin{aligned} c(a\mathbf{A} + b\mathbf{B}) &= c(a\mathbf{A}) + c(b\mathbf{B}) \\ &= (ca)\mathbf{A} + (cb)\mathbf{B} \\ &= (ac)\mathbf{A} + (bc)\mathbf{B} \\ &= a(c\mathbf{A}) + b(c\mathbf{B}) \,, \end{aligned} \tag{D.48}$$

where we used the distributive property of scalar multiplication with respect vector addition (D.9); the associative property of scalar multiplication of vectors (D.10); the commutative property for multiplication of two scalars (complex numbers); and the associative property of scalar multiplication of vectors (again) to get the successive equalities above. But adding a constant vector **C** to every vector in the space is *not* an example of a linear transformation as you are asked to show in the following exercise.

**Exercise D.7** Prove that adding a constant vector **C** to every vector in the space, i.e., $\mathbf{A} \mapsto \mathbf{A} + \mathbf{C}$, is *not* an example of a linear transformation.

The set of linear transformations, by itself, has an interesting mathematical structure. You can define (for any vector **A**):

(i) addition of two linear transformations:

$$(\mathbf{S} + \mathbf{T})\mathbf{A} \equiv \mathbf{S}\mathbf{A} + \mathbf{T}\mathbf{A} \tag{D.49}$$

(ii) multiplication of a linear transformation by a scalar:

$$(a\mathbf{T})\mathbf{A} \equiv a(\mathbf{T}\mathbf{A}) \tag{D.50}$$

(iii) multiplication (or **composition**) of two linear transformations:

$$(\mathbf{S}\mathbf{T})\mathbf{A} \equiv \mathbf{S}(\mathbf{T}\mathbf{A}) \tag{D.51}$$

With the first two operations, (D.49) and (D.50), the space of linear transformations has the structure of an $n^2$-dimensional vector space over the complex numbers (Exercise D.8). Note that multiplication of linear transformations is not commutative, however, since

$$\mathbf{S}\mathbf{T} \neq \mathbf{T}\mathbf{S} \,, \tag{D.52}$$

in general. (Think of rotations in 3-dimensions; see, e.g., Fig. 6.5.) We will return to the multiplicative structure of linear transformations later in this section, after we develop the connection between linear transformations and matrices.

**Exercise D.8** Show that with the above definitions of addition of linear transformations and multiplication of a linear transformation by a scalar, the set of linear transformations has the structure of a vector space over the complex numbers. Note that you will need to verify that these operations satisfy the properties given in (D.1)–(D.5) and (D.6)–(D.10).

### *D.4.1 Component Form of a Linear Transformation*

The real beauty of the linearity property (D.47) is that once you know what a linear transformation $\mathbf{T}$ does to a set of basis vectors $\{\mathbf{e}_1, \mathbf{e}_2, \ldots, \mathbf{e}_n\}$, you can easily determine what it does to *any* vector $\mathbf{A}$. To see that this is the case, let's begin by writing

$$\mathbf{T}\mathbf{e}_1 = T_{11}\mathbf{e}_1 + T_{21}\mathbf{e}_2 + \cdots + T_{n1}\mathbf{e}_n = \sum_i T_{i1}\mathbf{e}_i \,, \tag{D.53}$$

which follows from the fact that any vector (in this case $\mathbf{T}\mathbf{e}_1$) can be written as a linear combination of the basis vectors). Similarly,

$$\begin{aligned} \mathbf{T}\mathbf{e}_2 &= T_{12}\mathbf{e}_1 + T_{22}\mathbf{e}_2 + \cdots + T_{n2}\mathbf{e}_n = \sum_i T_{i2}\mathbf{e}_i \,, \\ &\vdots \\ \mathbf{T}\mathbf{e}_n &= T_{1n}\mathbf{e}_1 + T_{2n}\mathbf{e}_2 + \cdots + T_{nn}\mathbf{e}_n = \sum_i T_{in}\mathbf{e}_i \,. \end{aligned} \tag{D.54}$$

Thus, we see that the action of $\mathbf{T}$ on the $n$ basis vectors is completely captured by the $n \times n$ numbers $T_{ij}$, where

$$\mathbf{T}\mathbf{e}_j = \sum_i T_{ij}\mathbf{e}_i \,, \qquad j = 1, 2, \ldots, n \,. \tag{D.55}$$

We will write these components as an $n \times n$ matrix:

$$\mathbf{T} \leftrightarrow \mathsf{T} = \begin{bmatrix} T_{11} & T_{12} & \cdots & T_{1n} \\ T_{21} & T_{22} & \cdots & T_{2n} \\ \vdots & \vdots & \ddots & \vdots \\ T_{n1} & T_{n2} & \cdots & T_{nn} \end{bmatrix} , \tag{D.56}$$

where the doubleheaded arrow $\leftrightarrow$ reminds us that $\mathsf{T}$ are the components of $\mathbf{T}$ with respect to a particular basis. (With respect to a different basis, the matrix components will change in a manner that we will investigate shortly.) If the basis is orthonormal, then we can write

$$T_{ij} = \mathbf{e}_i \cdot (\mathbf{T}\mathbf{e}_j) \qquad \text{(orthonomal basis)}, \tag{D.57}$$

which follows immediately from (D.55).

Returning now to our claim that the action of **T** on any vector **A** is completely determined once we know the action of **T** on the basis vectors, we can write

$$\mathbf{TA} = \mathbf{T}(\sum_j A_j \mathbf{e}_j) = \sum_j A_j(\mathbf{T}\mathbf{e}_j) = \sum_j A_j \sum_i T_{ij}\mathbf{e}_i = \sum_i \left( \sum_j T_{ij} A_j \right) \mathbf{e}_i \,, \tag{D.58}$$

where we used the linearity property (D.47) to get the second equality and the action of **T** on the basis vectors (D.55) to get the third. Thus, knowing the action of **T** on the basis vectors (i.e., the components $T_{ij}$), we can determine the action of **T** on any vector **A**. If we denote **TA** as $\mathbf{A}'$ and the components of $\mathbf{A}'$ with respect to the basis $\{\mathbf{e}_1, \mathbf{e}_2, \cdots, \mathbf{e}_n\}$ as $A'_i$, then (D.58) becomes

$$\mathbf{A}' = \mathbf{TA} \quad \Leftrightarrow \quad A'_i = \sum_j T_{ij} A_j \,, \tag{D.59}$$

which is equivalent to the matrix equation $\mathsf{A}' = \mathsf{TA}$:

$$\begin{bmatrix} A'_1 \\ A'_2 \\ \vdots \\ A'_n \end{bmatrix} = \begin{bmatrix} T_{11} & T_{12} & \cdots & T_{1n} \\ T_{21} & T_{22} & \cdots & T_{2n} \\ \vdots & \vdots & \ddots & \vdots \\ T_{n1} & T_{n2} & \cdots & T_{nn} \end{bmatrix} \begin{bmatrix} A_1 \\ A_2 \\ \vdots \\ A_n \end{bmatrix}. \tag{D.60}$$

Thus, the mathematical structure of linear transformations on an $n$-dimensional vector space is equivalent to that of $n \times n$ matrices.

**Exercise D.9** Show that addition of linear transformations, multiplication of a linear transformation by a scalar, and multiplication (composition) of two linear transformations, defined by (D.49), (D.50), and (D.51), become simply:

$$S_{ij} + T_{ij} \,, \qquad aT_{ij} \,, \qquad \sum_k S_{ik} T_{kj} \,, \tag{D.61}$$

in terms of the corresponding components. Note that the last expression is just the component form of ordinary matrix multiplication of two matrices.

### D.4.2 Change of Basis

Before discussing properties of matrices in general, let's first determine how the components of a vector **A** and the components (i.e., matrix elements) of a linear transformation **T** transform under a change of basis. The classic example of a change of basis is given by a *rotation* of the coordinate basis vectors $\hat{\mathbf{x}}$, $\hat{\mathbf{y}}$, $\hat{\mathbf{z}}$ to a new set of basis vectors $\hat{\mathbf{x}}'$, $\hat{\mathbf{y}}'$, $\hat{\mathbf{z}}'$.

So let's denote the two sets of basis vectors by

$$\{\mathbf{e}_1, \mathbf{e}_2, \ldots, \mathbf{e}_n\}, \qquad \{\mathbf{e}_{1'}, \mathbf{e}_{2'}, \ldots, \mathbf{e}_{n'}\}, \tag{D.62}$$

where we use primed indices to distinguish between the two bases. We will assume, for now, that these are arbitrary bases—i.e., we do not require that they be orthonormal. Since any vector can be expanded in terms of either set of basis vectors, we can write

$$\begin{aligned}
\mathbf{e}_1 &= S_{1'1}\mathbf{e}_{1'} + S_{2'1}\mathbf{e}_{2'} + \cdots + S_{n'1}\mathbf{e}_{n'} = \sum_{j'} S_{j'1}\mathbf{e}_{j'}\,,\\
\mathbf{e}_2 &= S_{1'2}\mathbf{e}_{1'} + S_{2'2}\mathbf{e}_{2'} + \cdots + S_{n'2}\mathbf{e}_{n'} = \sum_{j'} S_{j'2}\mathbf{e}_{j'}\,,\\
&\;\;\vdots\\
\mathbf{e}_n &= S_{1'n}\mathbf{e}_{1'} + S_{2'n}\mathbf{e}_{2'} + \cdots + S_{n'2}\mathbf{e}_{n'} = \sum_{j'} S_{j'n}\mathbf{e}_{j'}\,,
\end{aligned} \tag{D.63}$$

for some set of components $S_{j'i}$. In compact form

$$\mathbf{e}_i = \sum_{j'} S_{j'i}\mathbf{e}_{j'}\,, \qquad i = 1, 2, \ldots, n\,. \tag{D.64}$$

The components $S_{j'i}$ define an $n \times n$ matrix, which is necessarily invertible (otherwise $\{\mathbf{e}_1, \mathbf{e}_2, \cdots\}$ wouldn't form a basis).

Now consider a single vector **A**, which has components $A_i$ and $A_{i'}$, respectively, with respect to the unprimed and primed bases:

$$\mathbf{A} = \sum_i A_i\mathbf{e}_i = \sum_{i'} A_{i'}\mathbf{e}_{i'}\,. \tag{D.65}$$

Then using (D.64), it follows that

$$\sum_i A_i\mathbf{e}_i = \sum_i A_i \sum_{j'} S_{j'i}\mathbf{e}_{j'} = \sum_{j'}\left(\sum_i S_{j'i}A_i\right)\mathbf{e}_{j'}\,. \tag{D.66}$$

Comparing with (D.65) we see that

$$A_{j'} = \sum_i S_{j'i} A_i \,, \qquad j' = 1', 2', \ldots, n' \,. \tag{D.67}$$

The inverse of the above equation can be written as

$$A_i = \sum_{j'} (S^{-1})_{ij'} A_{j'} \,, \tag{D.68}$$

where $(S^{-1})_{ij'}$ are the components of the inverse matrix to $S_{j'i}$:

$$\sum_{j'} (S^{-1})_{ij'} S_{j'k} = \delta_{ik} \,, \qquad \sum_i S_{j'i} (S^{-1})_{ik'} = \delta_{j'k'} \,. \tag{D.69}$$

Now take a linear transformation **T**, which maps **A** to $\mathbf{B} \equiv \mathbf{TA}$. Using (D.67), (D.59) and (D.68), it follows that

$$\begin{aligned} B_{i'} &= \sum_k S_{i'k} B_k = \sum_k S_{i'k} \sum_l T_{kl} A_l = \sum_k S_{i'k} \sum_l T_{kl} \sum_{j'} (S^{-1})_{lj'} A_{j'} \\ &= \sum_{j'} \left( \sum_{k,l} S_{i'k} T_{kl} (S^{-1})_{lj'} \right) A_{j'} = \sum_{j'} T_{i'j'} A_{j'} \end{aligned} \tag{D.70}$$

where

$$T_{i'j'} \equiv \sum_{k,l} S_{i'k} T_{kl} (S^{-1})_{lj'} \,. \tag{D.71}$$

Noting that the products and summations on the right-hand side are exactly those for a product of matrices, we have

$$\mathsf{T}' = \mathsf{STS}^{-1} \,, \tag{D.72}$$

where $\mathsf{T}'$ is the $n \times n$ matrix of components $T_{i'j'}$. Such a transformation of matrices is called a **similarity transformation**.

### *D.4.3 Matrix Definitions and Operations*

As illustrated by the calculations in the last two subsections, matrices play a key role in linear algebra. In this section, we summarize some important definitions and operations involving matrices, which we will refer to repeatedly in the main text.

Most of the discussion will be restricted to $n \times n$ (i.e., square) matrices, although the transpose and conjugate operations (complex conjugate and Hermitian conjugate) can be defined for arbitrary $n \times m$ (i.e., rectangular) matrices.

#### D.4.3.1 Transpose and Conjugate Matrices

The **transpose**, **conjugate**, and **Hermitian conjugate** of a matrix $\mathsf{T}$ are defined by:

$$(\mathsf{T}^T)_{ij} = T_{ji}\,, \quad (\mathsf{T}^*)_{ij} = T^*_{ij}\,, \quad (\mathsf{T}^\dagger)_{ij} = T^*_{ji}\,. \tag{D.73}$$

A matrix is said to be **symmetric** if and only if

$$\mathsf{T} = \mathsf{T}^T \quad \leftrightarrow \quad T_{ij} = T_{ji}\,, \tag{D.74}$$

and **Hermitian** if and only if

$$\mathsf{T} = \mathsf{T}^\dagger \quad \leftrightarrow \quad T_{ij} = T^*_{ji}\,. \tag{D.75}$$

**Anti-symmetric** and **Anti-hermitian** matrices are defined with minus signs in the last two equations.

**Exercise D.10** Show that the transpose of a product of matrices equals the product of transposes in the *opposite* order:

$$(\mathsf{ST})^T = \mathsf{T}^T\,\mathsf{S}^T\,, \tag{D.76}$$

and similarly for the Hermitian conjugate:

$$(\mathsf{ST})^\dagger = \mathsf{T}^\dagger\,\mathsf{S}^\dagger\,. \tag{D.77}$$

Note that these relations hold in general for the product of an $m \times n$ matrix and an $n \times p$ matrix.

**Exercise D.11** Show that the inner product (D.40) of two vectors **A** and **B** can be written in terms of row and column matrices as

$$\mathbf{A}\cdot\mathbf{B} = \mathsf{A}^\dagger\mathsf{B}\,. \tag{D.78}$$

### D.4.3.2 Determinants

The **determinant** of a $2 \times 2$ matrix is defined by

$$\mathsf{T} = \begin{bmatrix} a & b \\ c & d \end{bmatrix}, \qquad \det \mathsf{T} \equiv \begin{vmatrix} a & b \\ c & d \end{vmatrix} \equiv ad - bc\,. \tag{D.79}$$

For a higher-order $n \times n$ matrix $\mathsf{T}$, we define its determinant in terms of the determinants of $(n-1) \times (n-1)$ sub-matrices of $\mathsf{T}$. (This procedure is called **Laplace development** of the determinant.) Explicitly, if we expand off of the $i$th row of $\mathsf{T}$ then

$$\det \mathsf{T} = \sum_j T_{ij}(-1)^{i+j} M_{ij} = \sum_j T_{ij} C_{ij}\,, \tag{D.80}$$

where $M_{ij}$ is the **minor** of $T_{ij}$ and $C_{ij} \equiv (-1)^{i+j} M_{ij}$ is the corresponding **cofactor**. The minor $M_{ij}$ is calculated by taking the determinant of the $(n-1) \times (n-1)$ matrix obtained from $\mathsf{T}$ by removing its $i$th row and $j$th column. Note that you get the same answer for the determinant of $\mathsf{T}$ regardless of which row you expand off of, or if you expand off of a *column* instead of a row. In addition, as you will show in part (c) of Exercise D.12 below, adding a multiple of one row (or column) of a square matrix to another row (or column) does not change the value of its determinant. Thus, a judicious choice of such elementary row (or column) operations can simplify the calculation of the determinant.

**Exercise D.12** The determinant of an $n \times n$ matrix can also be defined by

$$\det \mathsf{T} = \sum_{i_1, i_2, \cdots i_n} \varepsilon_{i_1 i_2 \cdots i_n} T_{1 i_1} T_{2 i_2} \cdots T_{n i_n}\,, \tag{D.81}$$

where $\varepsilon_{i_1 i_2 \cdots i_n}$ is the $n$-dimensional Levi-Civita symbol:

$$\varepsilon_{i_1 i_2 \cdots i_n} \equiv \begin{cases} 1 & \text{if } i_1 i_2 \cdots i_n \text{ is an even permutation of } 12 \cdots n \\ -1 & \text{if } i_1 i_2 \cdots i_n \text{ is an odd permutation of } 12 \cdots n \\ 0 & \text{otherwise} \end{cases} \tag{D.82}$$

See (A.7) for the 3-dimensional version of the Levi-Civita symbol, which enters the expression for the vector product of two 3-dimensional vectors.

(a) Work out the explicit expression for the determinant of a $3 \times 3$ matrix using the definition given in (D.81).

(b) Do the same using the earlier definition (D.80), and confirm that the two expressions you obtain agree with one another.
(c) Using the above definition (D.81), show that the determinant of an $n \times n$ matrix T is unchanged if you add a multiple of one row (or column) of T to another row (or column) before taking its determinant.

#### D.4.3.3 Unit Matrix and Inverses

The **unit** (or **identity**) **matrix** $\mathsf{1}$ has components given by the Kronecker delta $\delta_{ij}$:

$$\mathsf{1} = \begin{bmatrix} 1 & 0 & \cdots & 0 \\ 0 & 1 & \cdots & 0 \\ \vdots & \vdots & \ddots & \vdots \\ 0 & 0 & \cdots & 1 \end{bmatrix} . \tag{D.83}$$

A matrix T is said to be *invertible* if and only if there exist another matrix $\mathsf{T}^{-1}$, called the **inverse matrix** of T, such that

$$\mathsf{T}\mathsf{T}^{-1} = \mathsf{T}^{-1}\mathsf{T} = \mathsf{1} , \tag{D.84}$$

or, equivalently,

$$\sum_k T_{ik}(T^{-1})_{kj} = \sum_k (T^{-1})_{ik} T_{kj} = \delta_{ij} . \tag{D.85}$$

It turns out that a matrix is invertible if and only if its determinant is non-zero. An explicit expression for the inverse matrix is

$$\mathsf{T}^{-1} = \frac{1}{\det \mathsf{T}} \mathsf{C}^T , \tag{D.86}$$

where C is the matrix of cofactors. The inverse of a product of two invertible matrices S and T is the product of the inverse matrices in $\mathsf{S}^{-1}$ and $\mathsf{T}^{-1}$ in the opposite order,

$$(\mathsf{S}\mathsf{T})^{-1} = \mathsf{T}^{-1}\mathsf{S}^{-1} . \tag{D.87}$$

**Exercise D.13** Calculate the inverse matrices for the general $2 \times 2$ and $3 \times 3$ matrices

$$\begin{bmatrix} a & b \\ c & d \end{bmatrix}, \quad \begin{bmatrix} a & b & c \\ d & e & f \\ g & h & i \end{bmatrix}, \tag{D.88}$$

assuming that the determinants are non-zero for both.

#### D.4.3.4 Orthogonal and Unitary Matrices

A matrix is said to be **orthogonal** if and only if

$$\mathsf{T}^T = \mathsf{T}^{-1} \quad \leftrightarrow \quad \sum_k T_{ik} T_{jk} = \delta_{ij}\,, \quad \sum_k T_{ki} T_{kj} = \delta_{ij}\,. \tag{D.89}$$

A matrix is said to be **unitary** if and only if

$$\mathsf{T}^\dagger = \mathsf{T}^{-1} \quad \leftrightarrow \quad \sum_k T_{ik} T^*_{jk} = \delta_{ij}\,, \quad \sum_k T^*_{ki} T_{kj} = \delta_{ij}\,. \tag{D.90}$$

**Exercise D.14** Show that an ordinary rotation in 3-dimensions, e.g.,

$$\mathsf{R}_z(\phi) = \begin{bmatrix} \cos\phi & \sin\phi & 0 \\ -\sin\phi & \cos\phi & 0 \\ 0 & 0 & 1 \end{bmatrix}, \tag{D.91}$$

is an example of an orthogonal matrix.

#### D.4.3.5 Useful Properties of Determinants

The determinant satisfies several useful properties:

$$\det \mathsf{1} = 1\,, \quad \det(\mathsf{T}^T) = \det \mathsf{T}\,, \quad \det(\mathsf{ST}) = \det \mathsf{S}\, \det \mathsf{T}\,. \tag{D.92}$$

Applying the above results to (D.84), it follows that

$$\det(\mathsf{T}^{-1}) = \frac{1}{\det \mathsf{T}} = (\det \mathsf{T})^{-1}\,. \tag{D.93}$$

In Appendix D.4.2, we derived how the components of a linear transformation $\mathsf{T}$ change under a change of basis, (D.72). Using the above properties, it follows that

$$\det(\mathsf{STS}^{-1}) = \det \mathsf{S} \det \mathsf{T} \det(\mathsf{S}^{-1}) = \det \mathsf{S} \det \mathsf{T} (\det \mathsf{S})^{-1} = \det \mathsf{T} . \quad \text{(D.94)}$$

Thus, the determinant of a matrix is *invariant* under a similarity transformation. In other words, the value of the determinant doesn't depend on what basis we use to convert a linear transformation **T** to a matrix of components $T_{ij}$.

#### D.4.3.6 Trace

There is another operation on matrices that is invariant under a similarity transformation. It is the **trace**, which is defined as the sum of the diagonal elements of the matrix,

$$\mathrm{Tr}(\mathsf{T}) \equiv \sum_i T_{ii} . \quad \text{(D.95)}$$

Since one can show that (Exercise D.15)

$$\mathrm{Tr}(\mathsf{ST}) = \mathrm{Tr}(\mathsf{TS}) , \quad \text{(D.96)}$$

it follows trivially that

$$\mathrm{Tr}(\mathsf{STS}^{-1}) = \mathrm{Tr}(\mathsf{S}^{-1}\mathsf{ST}) = \mathrm{Tr}(\mathsf{T}) . \quad \text{(D.97)}$$

Thus, the trace of a matrix, like the determinant, is also invariant under a similarity transformation, (D.72).

**Exercise D.15** Prove property (D.96) for the trace operation.

## D.5 Eigenvectors and Eigenvalues

The last topic that we will discuss in our review of linear algebra involves **eigenvectors** and **eigenvalues** of a linear transformation **T**. Eigenvectors of **T** are special vectors, which are *effectively unchanged* by the action of **T**. By "effectively unchanged" we mean that the eigenvector need only be mapped to itself *up to an overall proportionality factor*, which is called the **eigenvalue** of the eigenvector. If we denote an eigenvector of **T** by **v** and its eigenvalue by $\lambda$, then

$$\mathbf{T}\mathbf{v} = \lambda \mathbf{v} . \tag{D.98}$$

Note that the magnitude of the eigenvector $\mathbf{v}$ is not fixed by the above equation as $\mathbf{v}' \equiv a\mathbf{v}$ is also an eigenvector of $\mathbf{T}$ with the same eigenvalue $\lambda$.

***Example D.1*** As a simple example of an eigenvector, consider the space of ordinary 3-dimensional vectors and let's take as our linear transformation a counter-clockwise rotation about some axis $\hat{\mathbf{n}}$ through the angle $\Psi$, which we will denote as $\mathbf{R}_{\hat{\mathbf{n}}}(\Psi)$. Then $\hat{\mathbf{n}}$ is trivially an eigenvector of $\mathbf{R}_{\hat{\mathbf{n}}}(\Psi)$ with eigenvalue 1, since all points on the axis of rotation are left invariant by the transformation. □

**Exercise D.16** Suppose we restrict attention to ordinary *real-valued* vectors and rotations in 2-dimensions. Do any non-zero real-valued eigenvectors exist for such transformations? If so, what rotation angles do they correspond to?

## *D.5.1 Characteristic Equation*

If we introduce a basis $\{\mathbf{e}_1, \mathbf{e}_2, \ldots, \mathbf{e}_n\}$, then we can recast (D.98) as a matrix equation

$$\mathsf{T}\mathsf{v} = \lambda \mathsf{v} , \tag{D.99}$$

where $\mathsf{v}$ and $\mathsf{T}$ are the matrix representations of $\mathbf{v}$ and $\mathbf{T}$ with respect to the basis. This last equation is equivalent to

$$(\mathsf{T} - \lambda \mathsf{1})\mathsf{v} = \mathsf{0} , \tag{D.100}$$

where the right-hand side is the zero-vector $\mathsf{0} \equiv [0, 0, \ldots, 0]^T$. Since this is a homogeneous equation, $\mathsf{v} = \mathsf{0}$ is a (trivial) solution, and it is the only solution if $(\mathsf{T} - \lambda \mathsf{1})$ is invertible. Hence, a non-zero solution to this equation requires that the matrix $(\mathsf{T} - \lambda \mathsf{1})$ *not* be invertible or, equivalently, that

$$\det(\mathsf{T} - \lambda \mathsf{1}) = 0 . \tag{D.101}$$

Expanding the determinant yields an $n$th-order polynomial equation for $\lambda$, which is called the **characteristic equation**

$$\det(\mathsf{T} - \lambda \mathsf{1}) = c_0 + c_1\lambda + c_2\lambda^2 + \cdots + c_n\lambda^n = 0\,, \tag{D.102}$$

where the coefficients $c_i$ are algebraic expressions involving the matrix elements $T_{ij}$. By the **fundamental theorem of algebra**, this equations admits $n$ *complex* roots $\lambda_i$, which might be zero or repeated multiple times,

$$(\lambda_1 - \lambda)(\lambda_2 - \lambda)\cdots(\lambda_n - \lambda) = 0\,. \tag{D.103}$$

The $n$ roots are the eigenvalues of (D.99).

Given the eigenvalues, $\lambda_1, \lambda_2, \ldots, \lambda_n$, we can now substitute them back into (D.100), one at a time, and solve for the elements of the corresponding eigenvectors $\mathsf{v}_1, \mathsf{v}_2, \ldots, \mathsf{v}_n$. Since $\det(\mathsf{T} - \lambda\mathsf{1}) = 0$, not all of the components of the individual eigenvectors $\mathsf{v}_i$ will be uniquely determined. As mentioned earlier, there is always the freedom of an overall normalization factor, which we will usually chose to make each eigenvector have unit norm.

***Example D.2*** Find the eigenvectors and eigenvalues of the matrix

$$\mathsf{T} = \begin{bmatrix} 0 & 1 \\ 1 & 0 \end{bmatrix}. \tag{D.104}$$

We start by writing down the characteristic equation

$$\det(\mathsf{T} - \lambda\mathsf{1}) = \begin{vmatrix} -\lambda & 1 \\ 1 & -\lambda \end{vmatrix} = \lambda^2 - 1 = 0\,. \tag{D.105}$$

This has two real solutions

$$\lambda_+ = 1\,, \qquad \lambda_- = -1\,. \tag{D.106}$$

Substituting the solution $\lambda_+ = 1$ back into the eigenvector-eigenvalue equation (D.100), we have

$$\begin{bmatrix} -1 & 1 \\ 1 & -1 \end{bmatrix} \begin{bmatrix} v_1 \\ v_2 \end{bmatrix} = \begin{bmatrix} 0 \\ 0 \end{bmatrix}. \tag{D.107}$$

This yields two equations

$$\begin{aligned} -v_1 + v_2 &= 0\,, \\ v_1 - v_2 &= 0\,, \end{aligned} \tag{D.108}$$

which (as expected) are linearly dependent on one another. The solution to these equations is

$$v_1 = v_2\,. \tag{D.109}$$

Using our freedom in choice of an overall multiplicative factor, we can choose $v_1 = v_2 = 1/\sqrt{2}$ for which

$$\mathsf{v}_+ = \frac{1}{\sqrt{2}} \begin{bmatrix} 1 \\ 1 \end{bmatrix}. \tag{D.110}$$

Repeating this procedure for $\lambda_- = -1$, we find

$$\mathsf{v}_- = \frac{1}{\sqrt{2}} \begin{bmatrix} 1 \\ -1 \end{bmatrix}. \tag{D.111}$$

Note that these two eigenvectors have unit norm and are orthogonal to one another,

$$\mathsf{v}_+^\dagger \mathsf{v}_- = \frac{1}{2} \begin{bmatrix} 1 & 1 \end{bmatrix} \begin{bmatrix} 1 \\ -1 \end{bmatrix} = \frac{1}{2}(1-1) = 0\,. \tag{D.112}$$

Thus, the corresponding vectors $\mathbf{v}_+$, $\mathbf{v}_-$ form an orthonormal basis for the (real-valued) 2-dimensional vector space. But as we shall explain in the next subsection, it is not always the case that the eigenvectors of an arbitrary matrix form a basis for the vector space. □

**Exercise D.17** Find the eigenvectors and eigenvalues of the 2-dimensional rotation matrix

$$\mathsf{R}(\phi) = \begin{bmatrix} \cos\phi & \sin\phi \\ -\sin\phi & \cos\phi \end{bmatrix}. \tag{D.113}$$

Note that you will need to allow complex-valued eigenvectors in general.

### D.5.2 *Diagonalizing a Matrix*

If the eigenvectors $\mathbf{v}_1, \mathbf{v}_2, \ldots, \mathbf{v}_n$ of a linear transformation $\mathbf{T}$ span the $n$-dimensional vector space, then they can be used as a *new* set of basis vectors

$$\mathbf{e}_{1'} \equiv \mathbf{v}_1\,, \qquad \mathbf{e}_{2'} \equiv \mathbf{v}_2\,, \qquad \cdots, \qquad \mathbf{e}_{n'} \equiv \mathbf{v}_n\,, \tag{D.114}$$

in place of the original basis vectors $\mathbf{e}_1, \mathbf{e}_2, \ldots, \mathbf{e}_n$. Since

$$\mathbf{T}\mathbf{e}_{i'} = \lambda_i \mathbf{e}_{i'}\,, \qquad i' = 1', 2', \ldots, n'\,, \tag{D.115}$$

it follows that the components $T_{i'j'}$ of **T** in this new basis are given by $T_{i'j'} = \lambda_i \delta_{i'j'}$, which in matrix form is

$$\mathsf{T}' = \operatorname{diag}(\lambda_1, \lambda_2, \ldots, \lambda_n) = \begin{bmatrix} \lambda_1 & 0 & \cdots & 0 \\ 0 & \lambda_2 & \cdots & 0 \\ \vdots & \vdots & \ddots & \vdots \\ 0 & 0 & \cdots & \lambda_n \end{bmatrix} . \tag{D.116}$$

Thus, the matrix $\mathsf{T}'$ is *diagonal*.

It's not too hard to show that the matrix $\mathsf{S}$, which transforms the components $T_{ij}$ to the components $T_{i'j'}$ via

$$\mathsf{T}' = \mathsf{STS}^{-1} , \tag{D.117}$$

has

$$\mathsf{S}^{-1} = \left[ \mathsf{e}_{1'}\ \mathsf{e}_{2'} \cdots \mathsf{e}_{n'} \right] = \left[ \mathsf{v}_1\ \mathsf{v}_2 \cdots \mathsf{v}_n \right] , \tag{D.118}$$

or, equivalently,

$$(S^{-1})_{ij'} = (\mathsf{e}_{j'})_i . \tag{D.119}$$

In other words, the columns of $\mathsf{S}^{-1}$ are just the eigenvectors of **T** in the original basis.

*Proof*

$$\begin{aligned}
(\mathsf{STS}^{-1})_{i'j'} &= \sum_k \sum_l S_{i'k} T_{kl} (S^{-1})_{lj'} = \sum_k \sum_l S_{i'k} T_{kl} (\mathsf{e}_{j'})_l \\
&= \sum_k S_{i'k} \lambda_j (\mathsf{e}_{j'})_k = \lambda_j \sum_k S_{i'k} (S^{-1})_{kj'} \\
&= \lambda_j \delta_{i'j'} = T_{i'j'} ,
\end{aligned} \tag{D.120}$$

where we used (D.119) twice and also $\mathsf{T}\mathsf{e}_{j'} = \lambda_j \mathsf{e}_{j'}$. □

If, in addition to spanning the vector space, the eigenvectors of **T** are *orthonormal*, then the matrix $\mathsf{S}$ also has a simple form,

$$\mathsf{S} = \begin{bmatrix} \mathsf{v}_1^\dagger \\ \mathsf{v}_2^\dagger \\ \vdots \\ \mathsf{v}_n^\dagger \end{bmatrix} = \left[ \mathsf{v}_1\ \mathsf{v}_2 \cdots \mathsf{v}_n \right]^\dagger . \tag{D.121}$$

Comparing with (D.118) we see that

$$\mathsf{S}^{\dagger} = \mathsf{S}^{-1}, \tag{D.122}$$

so the similarity matrix $\mathsf{S}$ is *unitary*, cf. (D.90).

**Exercise D.18** Using the results of Example D.2, show explicitly that

$$\mathsf{S}^{-1} = \left[\mathsf{v}_+ \ \mathsf{v}_-\right] = \frac{1}{\sqrt{2}}\begin{bmatrix} 1 & 1 \\ 1 & -1 \end{bmatrix} \tag{D.123}$$

diagonalizes

$$\mathsf{T} = \begin{bmatrix} 0 & 1 \\ 1 & 0 \end{bmatrix}. \tag{D.124}$$

In so doing, calculate $\mathsf{S}$ and show that it is unitary (actually *orthogonal* in this case, since the eigenvectors are all real-valued).

Although every linear transformation (or $n \times n$ matrix) admits (complex-valued) eigenvectors, not all $n \times n$ matrices can be diagonalized. The eigenvectors would need to span the vector space, but this is not the case in general. As a simple example, consider the $2 \times 2$ matrix

$$\mathsf{T} = \begin{bmatrix} 0 & 1 \\ 0 & 0 \end{bmatrix}. \tag{D.125}$$

Its two eigenvalues $\lambda_1, \lambda_2$ are both equal to 0, and the corresponding eigenvectors $\mathsf{v}_1$, $\mathsf{v}_2$ are both proportional to $[1, 0]^T$. Hence the eigenvectors span only a 1-dimensional subspace of the 2-dimensional vector space, and the similarity transformation $\mathsf{S}$ needed to map $\mathsf{T}$ to the diagonal matix

$$\begin{bmatrix} \lambda_1 & 0 \\ 0 & \lambda_2 \end{bmatrix} = \begin{bmatrix} 0 & 0 \\ 0 & 0 \end{bmatrix} \tag{D.126}$$

does not exist. Thus, the matrix given by (D.125) cannot be diagonalized.

Fortunately, there is a certain class of matrices that are *guaranteed* to be diagonalizable. These are *Hermitian* matrices, for which $T_{ij} = T_{ji}^*$. (For a real-valued vector space, these matrices are *symmetric*, i.e., $T_{ij} = T_{ji}$.) Not only do the eigenvectors of a Hermitian matrix span the space, but the eigenvalues are *real*, and the eigenvectors corresponding to distinct eigenvalues are *orthogonal* to one another. These results are especially relevant for quantum mechanics, where the *observables* of the theory are represented by Hermitian transformations. (For proofs of these statements regarding Hermitian matrices, and for an excellent introduction to quantum theory, see Griffiths 2005.)

**Exercise D.19** Diagonalize the Hermitian matrix

$$\mathsf{T} = \begin{bmatrix} 1 & \mathrm{i} \\ -\mathrm{i} & 1 \end{bmatrix} \tag{D.127}$$

by finding its eigenvalues and eigenvectors, etc. Verify that the similarity transformation that diagonalizes T is unitary.

### *D.5.3 Determinant and Trace in Terms of Eigenvalues*

We end this section by showing that for *any* matrix T (diagonalizable or *not*), the determinant and trace of T can be written very simply in terms of its eigenvalues:

$$\det \mathsf{T} = \prod_i \lambda_i \,, \qquad \mathrm{Tr}(\mathsf{T}) = \sum_i \lambda_i \,. \tag{D.128}$$

For a diagonalizable matrix, the above two results follow immediately from (D.116) for $\mathsf{T}'$ and the fact that the determinant and trace of a matrix are invariant under a similarity transformation, (D.94) and (D.97). For a non-diagonalizable matrix, we proceed by first equating the expansion of the characteristic equation (D.102) in terms of powers of $\lambda$ and its factorization (D.103) in terms of its eigenvalues:

$$c_0 + c_1\lambda + c_2\lambda^2 + \cdots + c_n\lambda^n = (\lambda_1 - \lambda)(\lambda_2 - \lambda)\cdots(\lambda_n - \lambda) \,. \tag{D.129}$$

From this equality we can see that the constant term $c_0$ is given by

$$c_0 = \prod_i \lambda_i \,, \tag{D.130}$$

while the factor multiplying $\lambda^{n-1}$ is given by

$$c_{n-1} = (-1)^{n-1} \sum_i \lambda_i \,. \tag{D.131}$$

Now return to (D.102),

$$\det(\mathsf{T} - \lambda \mathsf{1}) = c_0 + c_1\lambda + c_2\lambda^2 + \cdots + c_n\lambda^n \,. \tag{D.132}$$

Setting $\lambda = 0$ in this equation gives

$$c_0 = \det \mathsf{T}\,, \tag{D.133}$$

while expanding the determinant using (D.81),

$$\det(\mathsf{T} - \lambda \mathsf{1}) = \sum_{i_1, i_2, \ldots, i_n} \varepsilon_{i_1 i_2 \cdots i_n} (T_{1i_1} - \lambda\delta_{1i_1})(T_{2i_2} - \lambda\delta_{2i_2}) \cdots (T_{ni_n} - \lambda\delta_{ni_n})\,, \tag{D.134}$$

gives

$$c_n = (-1)^n\,, \qquad c_{n-1} = (-1)^{n-1}\mathrm{Tr}(\mathsf{T})\,. \tag{D.135}$$

(To see this, note that the terms proportional to $\lambda^n$ and $\lambda^{n-1}$ in (D.134) must come from the product

$$(T_{11} - \lambda)(T_{22} - \lambda) \cdots (T_{nn} - \lambda)\,, \tag{D.136}$$

which leads to (D.135).) Then by comparing (D.130) and (D.131) with (D.133) and (D.135), we get (D.128).

## Suggested References

*Full references are given in the bibliography at the end of the book.*

Boas (2006): Chapter 3 is devoted to linear algebra. The treatment is especially suited for undergraduates, with many examples and problems.

Dennery and Kryzwicki (1967): A mathematical methods book suited for advanced undergraduates and graduate students. Chapter 2 discusses finite-dimensional vector spaces; Chap. 3 extends the formalism to (infinite-dimensional) function spaces.

Griffiths (2005): Appendix A provides a review of linear algebra, especially relevant for calculations that arise in quantum mechanics. Our presentation follows that of Griffiths.

Halmos (1958): A classic text on vector spaces and linear algebra, written primarily for undergraduate students majoring in mathematics. As such, the mathematical rigor is higher than that in most mathematical methods books for scientists and engineers.

# Appendix E
# Special Functions

Special functions play an important role in the physical sciences. They often arise as power series solutions of ordinary differential equations, which in turn come from a separation-of-variables decomposition of common partial differential equations (e.g., Laplace's equation, Helmholtz's equation, the wave equation, the diffusion equation, $\cdots$). Special functions behave like vectors in an infinite-dimensional vector space, sharing many of the properties of vectors described in Appendix D. Each set of special functions is *orthogonal* with respect to an inner product defined as an appropriate integral of a product of two such functions. Special functions also form a set of *basis* functions in terms of which one can expand the general solution of the original partial differential equation.

In this appendix, we review the key properties of several special functions, with particular emphasis on those functions that appear often in classical mechanics applications. We will assume that the reader is already familiar with the general (Frobenius) method of power series solutions, which we describe very briefly in Appendix E.1. As such we will omit detailed derivations of the recursion relations for the coefficients of the various power series solutions. For those details, you should consult, e.g., Chap. 12 in Boas (2006). The definitive source for anything related to special functions is Abramowitz and Stegun (1972).

## E.1 Series Solutions of Ordinary Differential Equations

The general form of a homogeneous, linear, 2nd-order ordinary differential equation is

$$y''(x) + p(x)\,y'(x) + q(x)\,y(x) = 0\,, \tag{E.1}$$

where $p(x)$ and $q(x)$ are arbitrary functions of $x$. We are interested in **power series solutions** of the form



M.J. Benacquista and J.D. Romano, *Classical Mechanics*, Undergraduate Lecture Notes in Physics, https://doi.org/10.1007/978-3-319-68780-3

$$y(x) = \sum_{n=0}^{\infty} a_n x^n \quad \text{or} \quad y(x) = x^{\sigma} \sum_{n=0}^{\infty} a_n x^n \tag{E.2}$$

for some value of $\sigma$. We need to consider the second (more general) power series expansion (i.e., with $\sigma \neq 0$), called a **Frobenius series**, if $x = 0$ is a **regular singular point** of the differential equation—that is, if $p(x)$ or $q(x)$ is singular (i.e., infinite) at $x = 0$, but $xp(x)$ and $x^2q(x)$ are *finite* at $x = 0$. If $x = 0$ is a regular point of the differential equation, then one can simply set $\sigma = 0$ and use the first expansion.

The basic procedure for finding a power series solution to (E.1) is to differentiate the power series expansion for $y(x)$ term by term, and then substitute the expansion into the differential equation for $y(x)$. Since the resulting sum must vanish for all values of $x$, the coefficients of $x^n$ must all equal zero, leading to a **recursion relation**, which relates $a_n$ to some subset of the previous $a_r$ $(r < n)$, and a quadratic equation for $\sigma$, called the **indicial equation**. The following theorerm, called *Fuch's theorem*, tells us how to obtain the general solution of the differential equation from two Frobenius series solutions.

**Theorem E.1 Fuch's theorem**: *The general solution of the differential equation (E.1) with a regular singular point at $x = 0$ consists of of either:*

*(i) a sum of two Frobenius series $S_1(x)$ and $S_2(x)$, or*
*(ii) the sum of one Frobenius series $S_1(x)$, and a second solution of the form $S_1(x) \ln x + S_2(x)$, where $S_2(x)$ is another Frobenius series.*

*Case (ii) occurs only if the roots of the indicial equation for $\sigma$ are equal to one another or differ by an integer.*

If $x = 0$ is a regular point of the differential equation, then the general solution is simply the sum of two ordinary series solutions.

## E.2 Trigonometric and Hyperbolic Functions

Trigonometric and hyperbolic functions (e.g., $\cos\theta$, $\sin\theta$, $\cosh\chi$, $\sinh\chi$, etc.) can be defined geometrically in terms of circles and hyperbolae. For example, $\cos\theta$ is the projection onto the $x$-axis of a point $P$ on the unit circle making an angle $\theta$ with respect to the $x$-axis. Here, instead, we define these functions in terms of power series solutions to differential equations.

### *E.2.1 Trig Functions*

**Trigonometric functions** are solutions to the differential equation

$$y'' + k^2 y = 0 . \tag{E.3}$$

A power series expansion leads to the two-term recursion relation

$$a_{n+2} = \frac{-k^2}{(n+1)(n+2)} a_n . \tag{E.4}$$

Thus, there are two independent solutions, one starting with $a_0$ and the other starting with $a_1$. If $a_0 = A$ and $a_1 = B$, then the general solution to this equation is a linear superposition of sine and cosine functions,

$$y(x) = A \cos kx + B \sin kx , \tag{E.5}$$

where

$$\begin{aligned} \cos kx &\equiv 1 - \frac{(kx)^2}{2!} + \frac{(kx)^4}{4!} - \cdots , \\ \sin kx &\equiv kx - \frac{(kx)^3}{3!} + \frac{(kx)^5}{5!} - \cdots . \end{aligned} \tag{E.6}$$

These functions can be written in terms of complex exponentials using Euler's identity

$$e^{i\theta} = \cos\theta + i \sin\theta , \tag{E.7}$$

which can be inverted to yield explicit expressions for the cosine and sine functions:

$$\cos\theta = \frac{1}{2}\left(e^{i\theta} + e^{-i\theta}\right) , \qquad \sin\theta = \frac{1}{2i}\left(e^{i\theta} - e^{-i\theta}\right) . \tag{E.8}$$

The trig functions are periodic with period $2\pi$, and form an orthogonal set of functions on the interval $[-\pi, \pi]$:

$$\begin{aligned} \int_{-\pi}^{\pi} dx \, \sin(nx) \sin(mx) &= \pi \delta_{nm} , \\ \int_{-\pi}^{\pi} dx \, \cos(nx) \cos(mx) &= \pi \delta_{nm} , \\ \int_{-\pi}^{\pi} dx \, \sin(nx) \cos(mx) &= 0 . \end{aligned} \tag{E.9}$$

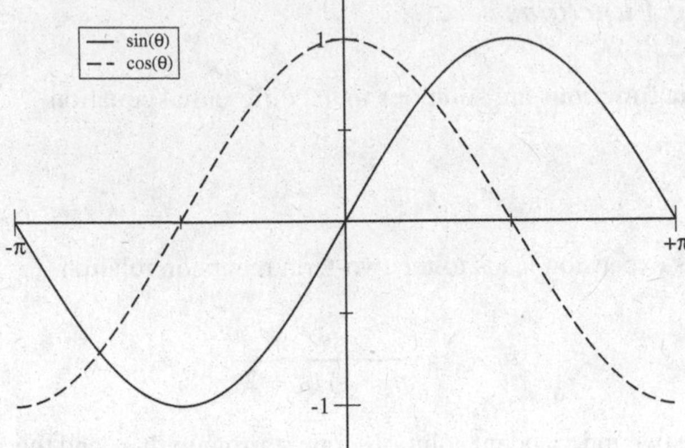


**Fig. E.1** The functions $\sin\theta$ and $\cos\theta$ plotted over the interval $-\pi$ to $\pi$

This is a key property of trig functions used in Fourier expansions of periodic functions. Plots of $\sin\theta$ and $\cos\theta$ are given in Fig. E.1. Finally, from sine and cosine we can define other trig fucntions:

$$\tan x \equiv \frac{\sin x}{\cos x} \equiv \frac{1}{\cot x}, \qquad \sec x \equiv \frac{1}{\cos x}, \qquad \csc x \equiv \frac{1}{\sin x}. \tag{E.10}$$

**Exercise E.1** Verify the recursion relation given in (E.4).

**Exercise E.2** Verify the orthogonality property of the sine and cosine functions, (E.9).

## *E.2.2 Hyperbolic Functions*

**Hyperbolic functions** are solutions to the differential equation

$$y'' - k^2 y = 0. \tag{E.11}$$

The recursion relation for this case is

$$a_{n+2} = \frac{k^2}{(n+1)(n+2)} a_n , \tag{E.12}$$

and the general solution to this equation is a linear combination of sinh and cosh functions:

$$y(x) = A\cosh kx + B\sinh kx\,, \tag{E.13}$$

where

$$\begin{aligned} \cosh kx &\equiv 1 + \frac{(kx)^2}{2!} + \frac{(kx)^4}{4!} + \cdots\,, \\ \sinh kx &\equiv kx + \frac{(kx)^3}{3!} + \frac{(kx)^5}{5!} + \cdots\,. \end{aligned} \tag{E.14}$$

Hyperbolic functions can be also be written in terms of ordinary exponentials,

$$\cosh x = \frac{1}{2}\left(\mathrm{e}^x + \mathrm{e}^{-x}\right), \qquad \sinh x = \frac{1}{2}\left(\mathrm{e}^x - \mathrm{e}^{-x}\right), \tag{E.15}$$

and trig functions,

$$\cosh x = \cos(\mathrm{i}x)\,, \qquad \sinh x = -\mathrm{i}\sin(\mathrm{i}x)\,. \tag{E.16}$$

From sinh and cosh we can define other hyperbolic functions, analogous to (E.10):

$$\tanh x \equiv \frac{\sinh x}{\cosh x} \equiv \frac{1}{\coth x}\,, \qquad \operatorname{sech} x \equiv \frac{1}{\cosh x}\,, \qquad \operatorname{csch} x \equiv \frac{1}{\sinh x}\,. \tag{E.17}$$

Plots of $\sinh x$, $\cosh x$, and $\tanh x$ are given in Fig. E.2.

**Exercise E.3** Verify (E.15) and (E.16).

## E.3 Legendre Polynomials and Associated Legendre Functions

Legendre's equation for $y(x)$ is

$$(1 - x^2)\,y'' - 2x\,y' + l(l+1)\,y = 0\,, \tag{E.18}$$

where $l$ is a constant. This ordinary differential equation arises when one uses separation of variables for Laplace's equation $\nabla^2\Phi = 0$ in spherical coordinates (here

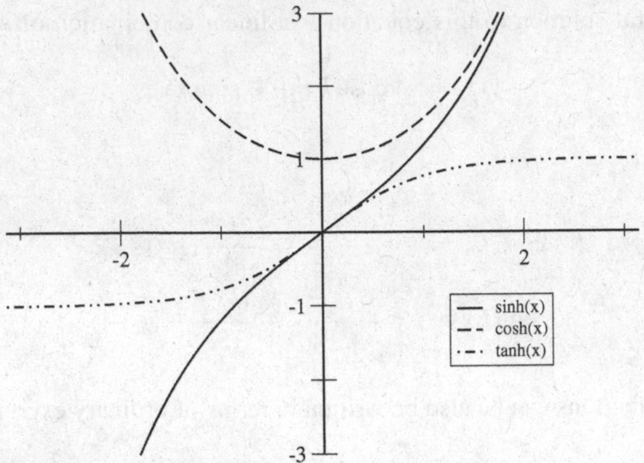


**Fig. E.2** The hyperbolic functions $\sinh x$, $\cosh x$, and $\tanh x$

$x \equiv \cos\theta$). One can show that $y(x)$ admits a regular power series solution with recursion relation

$$a_{n+2} = \frac{n(n+1) - l(l+1)}{(n+1)(n+2)} a_n , \qquad n = 0, 1, \cdots \tag{E.19}$$

Using the ratio test, it follows that the power series solution converges for $|x| < 1$. But one can also show (Exercise E.4, part (c)) that the power series solution diverges at $x = \pm 1$ (corresponding to the North and South poles of the sphere) unless the series terminates after some finite value of $n$.

**Exercise E.4** (a) Verify the recursion relation (E.19). (b) Show that for $l = 0$, the power series solution obtained by taking $a_0 = 0$ and $a_1 = 1$ is

$$y(x) = x + \frac{1}{3}x^3 + \frac{1}{5}x^5 + \cdots . \tag{E.20}$$

(c) Using the integral test, show that this solution diverges at $x = 1$ or $x = -1$.

### *E.3.1 Legendre Polynomials*

From the recursion relation (E.19), we see that if $l$ is a non-negative integer ($l = 0, 1, \cdots$), one of the power series solutions terminates (the even solution if $l$ is even, and the odd solution if $l$ is odd). The other solution can be set to zero (by hand) by choosing $a_1 = 0$ or $a_0 = 0$. The finite solutions thus obtained are *polynomials*

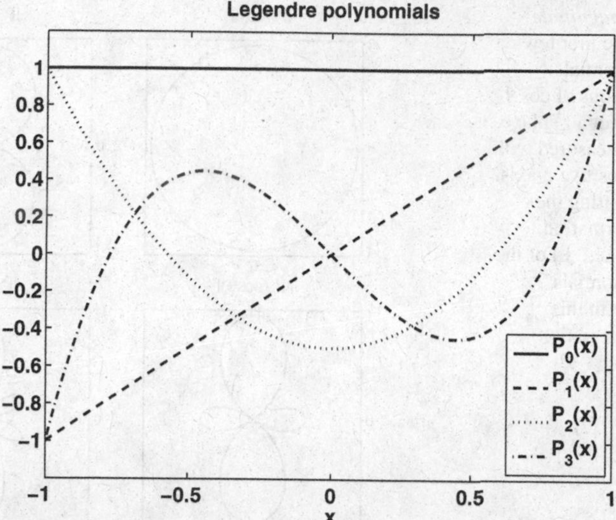


**Fig. E.3** First few Legendre polynomials $P_l(x)$ plotted as functions of $x \in [-1, 1]$

of order $l$. When appropriately normalized, they are called **Legendre polynomials**, denoted $P_l(x)$. By convention, the normalization condition is $P_l(1) = 1$. The first four Legendre polynomials are

$$\begin{aligned} P_0(x) &= 1\,, \\ P_1(x) &= x\,, \\ P_2(x) &= \frac{1}{2}(3x^2 - 1)\,, \\ P_3(x) &= \frac{1}{2}(5x^3 - 3x)\,. \end{aligned} \tag{E.21}$$

Figures E.3 and E.4 give two different graphical representations of the first few Legendre polynomials. Note that $P_l(-x) = (-1)^l P_l(x)$.

**Exercise E.5** Verify (E.21).

**Exercise E.6** (a) Show that one also obtains a polynomial solution if $l$ is a *negative integer* ($l = -1, -2, \cdots$). (b) Verify that these solutions are the *same* as those for non-negative $l$ (e.g., $l = -1$ yields the same solution as $l = 0$, and $l = -2$ yields the same solution as $l = 1$, etc.). Thus, there is no loss of generality in restricting attention to $l = 0, 1, \cdots$.

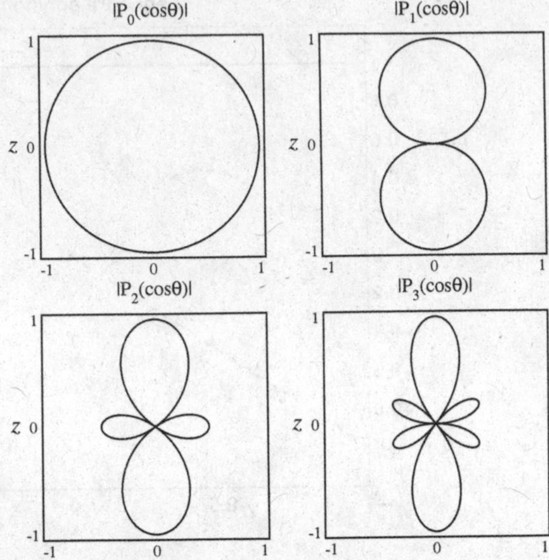


**Fig. E.4** The *magnitude* $|P_l(\cos\theta)|$ of the first few Legendre polynomials plotted as functions of $\cos\theta$ in the $xz$ plane (or $yz$) plane. The angle $\theta$ is measured with respect to the positive $z$-axis. Note that by plotting the magnitude, information about the *sign* (i.e., $\pm$) of the Legendre polynomials $P_l(\cos\theta)$ is lost in this graphical representation

## E.3.2 Some Properties of Legendre Polynomials

### E.3.2.1 Rodrigues' Formula

The Legendre polynomials can be generated using **Rodrigues' formula**:

$$P_l(x) = \frac{1}{2^l l!}\left(\frac{\mathrm{d}}{\mathrm{d}x}\right)^l (x^2-1)^l \,. \tag{E.22}$$

### E.3.2.2 Orthogonality

The Legendre polynomials for different values of $l$ are orthogonal to one another,

$$\int_{-1}^{1} \mathrm{d}x\, P_l(x)P_{l'}(x) = \frac{2}{2l+1}\delta_{ll'} \,. \tag{E.23}$$

**Exercise E.7** Prove (E.23). (*Hint*: The proof of orthogonality is simple if you write down Legendre's equation for both $P_l(x)$ and $P_{l'}(x)$; multiply these equations by $P_{l'}(x)$ and $P_l(x)$; and then subtract and integrate the result between $-1$ and 1. The derivation of the normalization constant is harder, but can be proved using mathematical induction and Rodrigues' formula for $P_l(x)$.)

#### E.3.2.3 Completeness

The Legendre polynomials are *complete* in the sense that any square-integrable function $f(x)$ defined on the interval $x \in [-1, 1]$ can be expanded in terms of Legendre polynomials:

$$f(x) = \sum_{l=0}^{\infty} A_l\, P_l(x)\,, \quad \text{where} \quad A_l = \frac{2l+1}{2}\int_{-1}^{1} \mathrm{d}x\; f(x)\, P_l(x)\,. \tag{E.24}$$

**Exercise E.8** Show that the function

$$f(x) = \begin{cases} -1\,, & -1 \le x < 0 \\ +1\,, & 0 < x \le 1 \end{cases} \tag{E.25}$$

can be expanded in terms of Legendre polynomials as

$$f(x) = \frac{3}{2}\, P_1(x) - \frac{7}{8}\, P_3(x) + \frac{11}{16}\, P_5(x) + \cdots \tag{E.26}$$

#### E.3.2.4 Generating Function

The Legendre polynomials can also be obtained as the coefficients of a power series expansion in $t$ of a so-called **generating function**

$$\frac{1}{\sqrt{1-2xt+t^2}} = \sum_{n=0}^{\infty} P_n(x)\, t^n\,. \tag{E.27}$$

With this result, one can rather easily express $1/r$ potentials using a series of Legendre polynomials

$$\frac{1}{|\mathbf{r} - \mathbf{r}'|} = \sum_{l=0}^{\infty} \frac{r_<^l}{r_>^{l+1}} P_l(\cos\gamma)\,, \tag{E.28}$$

where $r_<$ ($r_>$) is the smaller (larger) of $r$ and $r'$, and $\gamma$ is the angle between $\mathbf{r}$ and $\mathbf{r}'$,

$$\hat{\mathbf{r}} \cdot \hat{\mathbf{r}}' \equiv \cos\gamma = \cos\theta\cos\theta' + \sin\theta\sin\theta'\,\cos(\phi - \phi')\,. \tag{E.29}$$

#### E.3.2.5 Recurrence Relations

Using the generating function, one can derive the following relations, called **recurrence relations**,[1] which relate Legendre polynomials $P_n(x)$ and their derivatives $P_n'(x)$ to neighboring Legendre polynomials:

$$(n+1)\,P_{n+1} = (2n+1)x\,P_n - n\,P_{n-1}\,, \tag{E.30a}$$

$$P_n = P_{n+1}' - 2x\,P_n' + P_{n-1}'\,, \tag{E.30b}$$

$$n\,P_n = x\,P_n' - P_{n-1}'\,, \tag{E.30c}$$

$$(n+1)\,P_n = P_{n+1}' - x\,P_n'\,, \tag{E.30d}$$

$$(2n+1)\,P_n = P_{n+1}' - P_{n-1}'\,, \tag{E.30e}$$

$$(1-x^2)\,P_n' = n(P_{n-1} - x\,P_n)\,. \tag{E.30f}$$

Note that Legendre's equation

$$(1-x^2)\,P_n'' - 2x\,P_n' + n(n+1)\,P_n = 0 \tag{E.31}$$

can be obtained by differentiating (E.30f) with respect to $x$ and then using (E.30c). In addition, the normalization $P_n(1) = 1$ also follows simply from the generating function.

**Exercise E.9** Prove the above recurrence relations by differentiating the generating function with respect to $t$ and $x$ separately, and then combining the various expressions.

[1] Most authors use either "recursion relation" or "recurrence relation" exclusively, and apply it to any relation between indexed objects of different order. Here, we have decided to use "recurrence relation" when describing relationships between special functions of different order, while using "recursion relation" when describing relationships between the coefficients of the power series.

### E.3.3 *Associated Legendre Functions*

The associated Legendre equation is given by

$$(1-x^2)\,y'' - 2x\,y' + \left[l(l+1) - \frac{m^2}{(1-x^2)}\right] y = 0\,. \tag{E.32}$$

It differs from the ordinary Legendre equation, (E.18), by the extra term proportional to $m^2$. It turns out that power series solutions of this differential equation also diverge at the poles ($x = \pm 1$) unless $l = 0, 1, \cdots$ (as before) and $m = -l, -l+1, \ldots, l$. The finite solutions are called **associated Legendre functions**, $P_l^m(x)$, and are given by derivatives of the Legendre polynomials,

$$\begin{aligned} P_l^m(x) &= (-1)^m (1-x^2)^{m/2} \frac{\mathrm{d}^m}{\mathrm{d}x^m} P_l(x)\,, \quad \text{for} \quad m \geq 0\,, \\ P_l^{-m}(x) &= (-1)^m \frac{(l-m)!}{(l+m)!} P_l^m(x)\,, \quad \text{for} \quad m < 0\,. \end{aligned} \tag{E.33}$$

**Exercise E.10** Prove by direct substitution that the above expression for $P_l^m(x)$ satisfies the associated Legendre equation (E.32).

The associated Legendre functions are *not* polynomials in $x$ on account of the square root factor $(1-x^2)^{m/2}$ for odd $m$. But since we are often ultimately interested in the replacement $x \equiv \cos\theta$, these non-polynomial factors are just proportional to $\sin^m\theta$. Thus, the associated Legendre functions can be written as polynomials in $\cos\theta$ if $m$ is even, and polynomials in $\cos\theta$ multiplied by $\sin\theta$ if $m$ is odd. The first few associated Legendre functions are given by:

$\underline{l=0}$:

$$P_0^0(\cos\theta) = 1\,, \tag{E.34}$$

$\underline{l=1}$:

$$\begin{aligned} P_1^0(\cos\theta) &= \cos\theta\,, \\ P_1^1(\cos\theta) &= -\sin\theta\,, \end{aligned} \tag{E.35}$$

$\underline{l=2}$:

$$\begin{aligned} P_2^0(\cos\theta) &= \frac{1}{2}\left(3\cos^2\theta - 1\right)\,, \\ P_2^1(\cos\theta) &= -3\sin\theta\,\cos\theta\,, \\ P_2^2(\cos\theta) &= 3(1-\cos^2\theta)\,, \end{aligned} \tag{E.36}$$

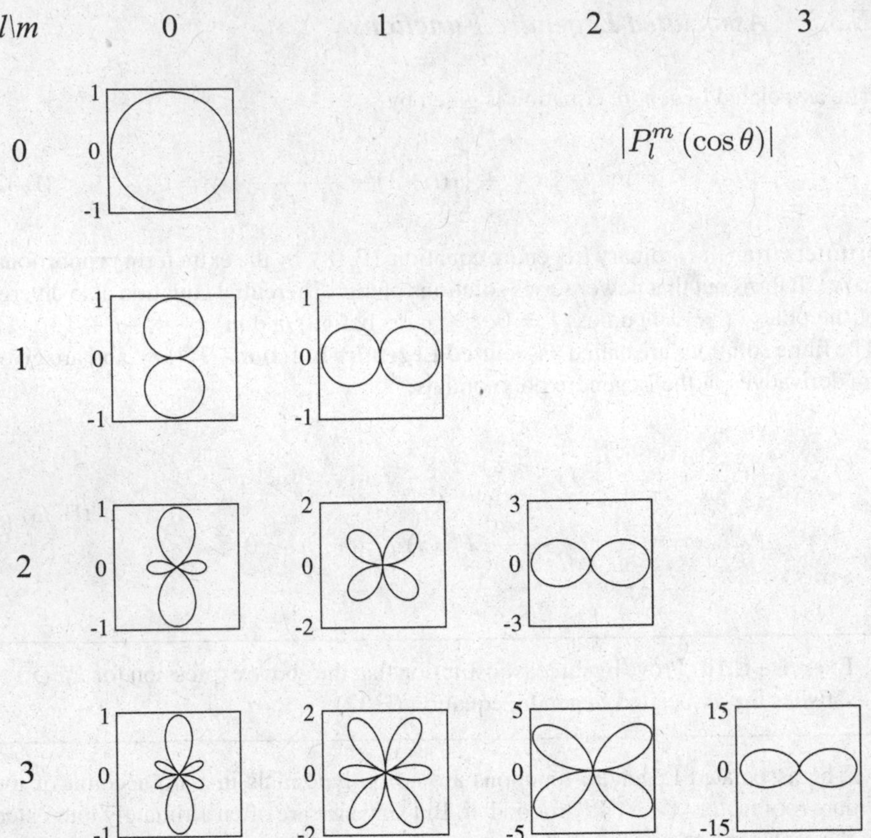


**Fig. E.5** The *magnitude* $|P_l^m(\cos\theta)|$ of the first few associated Legendre functions plotted as functions of $\cos\theta$ in the $xz$ plane (or $yz$) plane. The angle $\theta$ is measured with respect to the positive $z$-axis. Similar to the plot in Fig. E.4, the sign (i.e., $\pm$) of the associated Legendre functions $P_l^m(\cos\theta)$ is lost in this graphical representation. Note that the scale changes for larger values of $m$

$\underline{l = 3}$:

$$
\begin{aligned}
P_3^0(\cos\theta) &= \frac{1}{2}\left(5\cos^3\theta - 3\cos\theta\right) , \\
P_3^1(\cos\theta) &= -\frac{3}{2}\sin\theta\,\left(5\cos^2\theta - 1\right) , \\
P_3^2(\cos\theta) &= 15\left(\cos\theta - \cos^3\theta\right) , \\
P_3^3(\cos\theta) &= -15\sin\theta\,\left(1 - \cos^2\theta\right) .
\end{aligned}
\tag{E.37}
$$

Plots of the magnitude of the first few of these functions are given in Fig. E.5.

### *E.3.4 Some Properties of Associated Legendre Functions*

#### E.3.4.1 Rodrigues' Formula

Using Rodrigues' formula for Legendre polynomials (E.22), we can write down an analogous Rodrigues' formula for associated Legendre functions, valid for both positive and negative values of $m$:

$$P_l^m(x) = \frac{(-1)^m}{2^l l!}(1-x^2)^{m/2}\frac{\mathrm{d}^{l+m}}{\mathrm{d}x^{l+m}}(x^2-1)^l\,. \qquad \text{(E.38)}$$

#### E.3.4.2 Orthonormality

For each $m$, the associated Legendre functions are orthogonal to one another,

$$\int_{-1}^{1} \mathrm{d}x\; P_l^m(x) P_{l'}^m(x) = \frac{2}{2l+1}\frac{(l+m)!}{(l-m)!}\,\delta_{ll'}\,. \qquad \text{(E.39)}$$

#### E.3.4.3 Completeness

For each $m$, the associated Legendre functions form a complete set (in the index $l$) for square-integrable functions on $x \in [-1, 1]$:

$$f(x) = \sum_{l=0}^{\infty} A_l\, P_l^m(x)\,, \quad \text{where} \quad A_l = \frac{2l+1}{2}\frac{(l-m)!}{(l+m)!}\int_{-1}^{1} \mathrm{d}x\; f(x)\, P_l^m(x)\,. \qquad \text{(E.40)}$$

## E.4 Spherical Harmonics

**Spherical harmonics** are solutions to the $(\theta, \phi)$ part of Laplace's equation $\nabla^2\Phi = 0$ in spherical coordinates. As such they are proportional to the product of associated Legendre functions $P_l^m(\cos\theta)$ and complex exponentials $\mathrm{e}^{im\phi}$:

$$Y_{lm}(\theta,\phi) \equiv N_l^m P_l^m(\cos\theta)\mathrm{e}^{im\phi}\,, \qquad N_l^m \equiv \sqrt{\frac{2l+1}{4\pi}\frac{(l-m)!}{(l+m)!}}\,. \tag{E.41}$$

The proportionality constants have been chosen so that

$$\int_{S^2} \mathrm{d}\Omega\, Y^*_{lm}(\theta,\phi)Y_{l'm'}(\theta,\phi) = \delta_{ll'}\delta_{mm'}\,, \tag{E.42}$$

where

$$\mathrm{d}\Omega \equiv \mathrm{d}(\cos\theta)\,\mathrm{d}\phi = \sin\theta\,\mathrm{d}\theta\,\mathrm{d}\phi\,. \tag{E.43}$$

This is the *orthonormality condition* for spherical harmonics. Note that for $m = 0$, spherical harmonics reduce to Legendre polynomials, up to a normalization factor:

$$Y_{l0} = \sqrt{\frac{2l+1}{4\pi}}\, P_l(\cos\theta)\,. \tag{E.44}$$

**Exercise E.11** Show that

$$Y_{l,-m}(\theta,\phi) = (-1)^m Y^*_{lm}(\theta,\phi)\,, \tag{E.45}$$

and

$$Y_{lm}(\pi-\theta,\phi+\pi) = (-1)^l Y_{lm}(\theta,\phi)\,. \tag{E.46}$$

The first equation tells you how to get $Y_{l,-m}$ from $Y_{lm}$; the second equation relates the values of the spherical harmonic $Y_{lm}$ at *antipodal* (i.e., opposite) points on the 2-sphere.

The first few spherical harmonics are given by:
$\underline{l=0}$:

$$Y_{00}(\theta,\phi) = \sqrt{\frac{1}{4\pi}}\,, \tag{E.47}$$

$\underline{l=1}$:

$$\begin{aligned} Y_{11}(\theta,\phi) &= -\sqrt{\frac{3}{8\pi}}\sin\theta\,\mathrm{e}^{\mathrm{i}\phi}\,,\\ Y_{10}(\theta,\phi) &= \sqrt{\frac{3}{4\pi}}\cos\theta\,,\\ Y_{1,-1}(\theta,\phi) &= \sqrt{\frac{3}{8\pi}}\sin\theta\,\mathrm{e}^{-\mathrm{i}\phi}\,, \end{aligned} \tag{E.48}$$

$\underline{l = 2}$:

$$
\begin{aligned}
Y_{22}(\theta, \phi) &= \frac{1}{4}\sqrt{\frac{15}{2\pi}} \sin^2\theta \, e^{2i\phi} , \\
Y_{21}(\theta, \phi) &= -\sqrt{\frac{15}{8\pi}} \sin\theta \cos\theta \, e^{i\phi} , \\
Y_{20}(\theta, \phi) &= \sqrt{\frac{5}{4\pi}} \left(\frac{3}{2}\cos^2\theta - \frac{1}{2}\right) , \\
Y_{2,-1}(\theta, \phi) &= \sqrt{\frac{15}{8\pi}} \sin\theta \cos\theta \, e^{-i\phi} , \\
Y_{2,-2}(\theta, \phi) &= \frac{1}{4}\sqrt{\frac{15}{2\pi}} \sin^2\theta \, e^{-2i\phi} .
\end{aligned}
\tag{E.49}
$$

Since $Y_{lm}(\theta, \phi)$ differs from $P_l^m(\theta)$ by only a constant multiplicative factor and phase $e^{im\phi}$, the magnitude $|Y_{lm}(\theta, \phi)|$ has the same shape as $|P_l^m(\theta)|$ (See Fig. E.5).

## E.4.1 *Some Properties of Spherical Harmonics*

### E.4.1.1 Completeness

Spherical harmonics are complete in the sense that any square-integrable function $f(\theta, \phi)$ on the unit 2-sphere can be expanded in terms of spherical harmonics:

$$
\begin{aligned}
f(\theta, \phi) &= \sum_{l=0}^{\infty} \sum_{m=-l}^{l} A_{lm} \, Y_{lm}(\theta, \phi) , \quad \text{where} \\
A_{lm} &= \int_{S^2} d\Omega \, f(\theta, \phi) \, Y^*_{lm}(\theta, \phi) .
\end{aligned}
\tag{E.50}
$$

Equivalently, the completeness property can be written as

$$
\sum_{l=0}^{\infty} \sum_{m=-l}^{l} Y^*_{lm}(\theta', \phi') Y_{lm}(\theta, \phi) = \delta(\hat{\mathbf{n}}, \hat{\mathbf{n}}') , \tag{E.51}
$$

where $\hat{\mathbf{n}}$ and $\hat{\mathbf{n}}'$ are the unit (radial) vectors

$$
\begin{aligned}
\hat{\mathbf{n}} &= \sin\theta \cos\phi \, \hat{\mathbf{x}} + \sin\theta \sin\phi \, \hat{\mathbf{y}} + \cos\theta \, \hat{\mathbf{z}} , \\
\hat{\mathbf{n}}' &= \sin\theta' \cos\phi' \, \hat{\mathbf{x}} + \sin\theta' \sin\phi' \, \hat{\mathbf{y}} + \cos\theta' \, \hat{\mathbf{z}} ,
\end{aligned}
\tag{E.52}
$$

and $\delta(\hat{\mathbf{n}}, \hat{\mathbf{n}}')$ is the 2-dimensional Dirac delta function on the 2-sphere:

$$\delta(\hat{\mathbf{n}}, \hat{\mathbf{n}}') = \delta(\cos\theta - \cos\theta')\delta(\phi - \phi') = \frac{1}{\sin\theta}\delta(\theta - \theta')\delta(\phi - \phi') \,. \tag{E.53}$$

In terms of spherical harmonics, the general solution to Laplace's equation $\nabla^2\Phi = 0$ in spherical coordinates is

$$\Phi(r, \theta, \phi) = \sum_{l=0}^{\infty}\sum_{m=-l}^{l}\left[A_{lm}\, r^l + B_{lm}\, r^{-(l+1)}\right] Y_{lm}(\theta, \phi) \,, \tag{E.54}$$

where the terms in square brackets is the solution to the radial part of Laplace's equation.

#### E.4.1.2 Addition Theorem

If one sums only over $m$ in (E.51), one obtains the so-called **addition theorem** of spherical harmonics,

$$\sum_{m=-l}^{l} Y^*_{lm}(\theta', \phi')Y_{lm}(\theta, \phi) = \frac{2l+1}{4\pi} P_l(\cos\gamma) \,, \tag{E.55}$$

where

$$\cos\gamma \equiv \hat{\mathbf{n}} \cdot \hat{\mathbf{n}}' = \cos\theta\cos\theta' + \sin\theta\sin\theta'\cos(\phi - \phi') \,. \tag{E.56}$$

Completeness of the spherical harmonics and the addition theorem imply

$$\delta(\hat{\mathbf{n}}, \hat{\mathbf{n}}') = \sum_{l=0}^{\infty}\frac{2l+1}{4\pi}P_l(\hat{\mathbf{n}} \cdot \hat{\mathbf{n}}') \,, \tag{E.57}$$

which is an expansion of the Dirac delta function on the 2-sphere in terms of the Legendre polynomials.

**Exercise E.12** Using the addition theorem, show that the $1/r$ potential for a point source can be written as

$$\frac{1}{|\mathbf{r} - \mathbf{r}'|} = \sum_{l=0}^{\infty}\sum_{m=-l}^{l}\frac{4\pi}{2l+1}\frac{r_<^l}{r_>^{l+1}}\,Y^*_{lm}(\theta', \phi')Y_{lm}(\theta, \phi) \,, \tag{E.58}$$

where $r_<$ ($r_>$) is the smaller (larger) of $r$ and $r'$. This expression is fully-factorized into a product of functions of the unprimed and primed coordinates.

### E.4.1.3 Transformation Under a Rotation

We can imagine rotating our coordinates through the Euler angles $\alpha, \beta, \gamma$ using the $zyz$ form of the rotation matrix $\mathsf{R}(\alpha, \beta, \gamma)$ (See Sect. 6.2.3.1). This is equivalent to a rotation of the 2-sphere. In this case, the spherical harmonics transform according to

$$Y_{lm}(\theta', \phi') = \sum_{m'=-l}^{l} D_{lm,m'}(\alpha, \beta, \gamma) Y_{lm'}(\theta, \phi), \qquad \text{(E.59)}$$

where $(\theta', \phi')$ are the coordinates of a point $P = (\theta, \phi)$ after the rotation of the sphere. The fact that $Y_{lm}(\theta', \phi')$ can be written as a linear combination of the $Y_{lm'}(\theta, \phi)$ with the *same* $l$ is a consequence of the spherical harmonics being eigenfunctions of the (rotationally-invariant) Laplacian on the unit 2-sphere with eigenvalues depending only on $l$,

$${}^{(2)}\nabla^2 Y_{lm}(\theta, \phi) = -l(l+1) Y_{lm}(\theta, \phi), \qquad \text{(E.60)}$$

where

$${}^{(2)}\nabla^2 f(\theta, \phi) \equiv \frac{1}{\sin\theta} \frac{\partial}{\partial\theta}\left(\sin\theta \frac{\partial f}{\partial\theta}\right) + \frac{1}{\sin^2\theta} \frac{\partial^2 f}{\partial\phi^2}. \qquad \text{(E.61)}$$

The coefficients $D_{lm,m'}(\alpha, \beta, \gamma)$ in (E.59) are called **Wigner rotation matrices**. They arise in applications of group theory to quantum mechanics (Wigner 1931). In terms of the Euler angles, the components of the Wigner rotation matrices can be written as

$$D_{lm,m'}(\alpha, \beta, \gamma) = \mathrm{e}^{-im\alpha} d_{lm,m'}(\beta) \mathrm{e}^{-im'\gamma}, \qquad \text{(E.62)}$$

where

$$\begin{aligned} d_{lm,m'}(\beta) \equiv & \sqrt{(l+m)!(l-m)!(l+m')!(l-m')!} \\ & \times \sum_s \left[ \frac{(-1)^{m-m'+s}}{(l+m'-s)!s!(m-m'+s)!(l-m-s)!} \right. \\ & \left. \times \left(\cos\frac{\beta}{2}\right)^{2l+m'-m-2s} \left(\sin\frac{\beta}{2}\right)^{m-m'+2s} \right], \end{aligned} \qquad \text{(E.63)}$$

and where the sum over $s$ is chosen such that the factorials inside the summation always remain non-negative. The Wigner matrices also satisfy

$$\sum_{m''=-l}^{l} D_{lm,m''}(\alpha, \beta, \gamma) D^*_{lm',m''}(\alpha, \beta, \gamma) = \delta_{mm'} \qquad \text{(E.64)}$$

as a consequence of

$$\int_{S^2} d\Omega \, Y^*_{lm}(\theta', \phi') Y_{l'm'}(\theta', \phi') = \delta_{ll'} \delta_{mm'} \,. \tag{E.65}$$

## E.5 Bessel Functions and Spherical Bessel Functions

Separation of variables of Laplaces's equation in cylindrical coordinates $(\rho, \phi, z)$ leads to either of the following two differential equations for the radial function $R(\rho)$:

$$\begin{aligned} R''(\rho) + \frac{1}{\rho} R'(\rho) + \left(k^2 - \frac{\nu^2}{\rho^2}\right) R(\rho) = 0 \,, \\ R''(\rho) + \frac{1}{\rho} R'(\rho) - \left(k^2 + \frac{\nu^2}{\rho^2}\right) R(\rho) = 0 \,. \end{aligned} \tag{E.66}$$

The two equations correspond to different choices for the sign of the separation constant, $\pm k^2$. These equations can be put into more standard form by making a change of variables $x \equiv k\rho$, with $y(x)|_{x=k\rho} \equiv R(\rho)$:

$$\begin{aligned} y''(x) + \frac{1}{x} y'(x) + \left(1 - \frac{\nu^2}{x^2}\right) y(x) = 0 \,, \\ y''(x) + \frac{1}{x} y'(x) - \left(1 + \frac{\nu^2}{x^2}\right) y(x) = 0 \,. \end{aligned} \tag{E.67}$$

The first equation is called **Bessel's equation** of order $\nu$; the second is called the **modified Bessel's equation** of order $\nu$.

**Exercise E.13** Show that if $y(x)$ is a solution of Bessel's equation, then $\bar{y}(x) \equiv y(ix)$ is a solution of the modified Bessel's equation.

### *E.5.1 Bessel Functions of the 1st Kind*

Since $x = 0$ is a *regular singular point* of Bessel's equation, the method of Frobenius (Appendix E.1) requires that we consider a power series expansion of the form

$$y(x) = x^\sigma \sum_{n=0}^{\infty} a_n x^n \,. \tag{E.68}$$

Substituting this expansion into Bessel's equation and equating coefficients multiplying like powers of $x$ leads to a quadratic equation for $\sigma$ (called the **indicial** equation) and a recursion relation relating $a_{n+2}$ to $a_n$ (and $\sigma$) for $n = 0, 1, \cdots$. Setting $a_1 = 0$ (which forces all of the higher-order odd coefficients to vanish) and choosing the normalization coefficient $a_0$ appropriately, we obtain the solution

$$J_\nu(x) = \sum_{n=0}^{\infty} \frac{(-1)^n}{n!\Gamma(n+1+\nu)} \left(\frac{x}{2}\right)^{2n+\nu} . \tag{E.69}$$

$J_\nu(x)$ is called a **Bessel function of the 1st kind** of order $\nu$. The function $\Gamma(n + 1 + \nu)$ which appears in the denominator of the expansion coefficients is the **gamma function** defined by

$$\Gamma(z) \equiv \int_0^\infty \mathrm{d}x\, x^{z-1} \mathrm{e}^{-x} , \qquad \mathrm{Re}(z) > 0 . \tag{E.70}$$

The gamma function generalizes the ordinary factorial function $n! = n(n-1)\cdots 1$ to non-integer arguments in the sense that

$$\begin{aligned} \Gamma(n+1) &= n! \quad \text{for} \quad n = 0, 1, \cdots \\ \Gamma(z+1) &= z\,\Gamma(z) \quad \text{for} \quad \mathrm{Re}(z) > 0 . \end{aligned} \tag{E.71}$$

**Exercise E.14** (a) Prove $\Gamma(z + 1) = z\Gamma(z)$ for $\mathrm{Re}(z) > 0$. (*Hint*: Integrate $\Gamma(z + 1)$ by parts taking $u = x^z$ and $\mathrm{d}v = \mathrm{e}^{-x}\,\mathrm{d}x$.) (b) Show by explicit calculation that $\Gamma(1) = 1$ and $\Gamma(1/2) = \sqrt{\pi}$.

### E.5.1.1 Asymptotic Form

To gain a better intuitive understanding of the Bessel function $J_\nu(x)$, it is useful to look at its *asymptotic* form—i.e., its behavior for both small and large values of $x$. Using the general definition (E.69), one can show that

$$\begin{aligned} x \ll 1: &\quad J_\nu(x) \to \frac{1}{\Gamma(\nu+1)} \left(\frac{x}{2}\right)^\nu , \\ x \gg 1, \nu: &\quad J_\nu(x) \to \sqrt{\frac{2}{\pi x}} \cos\left(x - \frac{\nu\pi}{2} - \frac{\pi}{4}\right) . \end{aligned} \tag{E.72}$$

Thus, $J_0(0) = 1$ and $J_\nu(0) = 0$ for all $\nu \neq 0$; while for large $x$, $J_\nu(x)$ behaves like a *damped sinusoid*, and has infinitely many zeros $x_{\nu n}$:

$$J_\nu(x_{\nu n}) = 0 , \qquad n = 1, 2, \cdots . \tag{E.73}$$

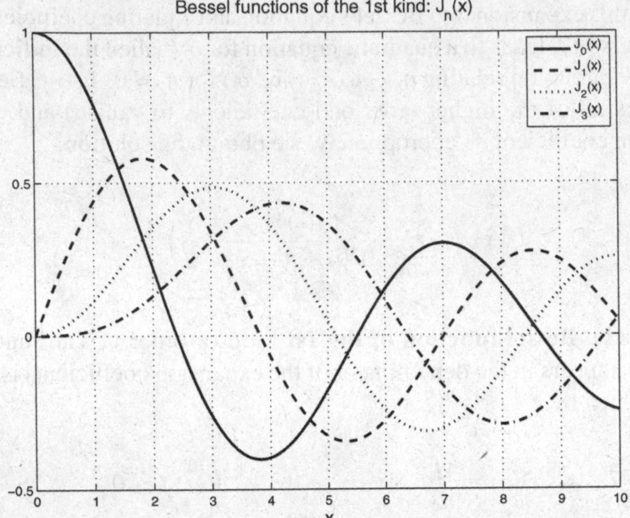


**Fig. E.6** First few Bessel functions of the 1st kind for integer $\nu$

Plots of the first few Bessel functions of the 1st kind for integer values of $\nu$ are given in Fig. E.6.

> **Exercise E.15** Using (E.72), show that the zeros of $J_\nu(x)$ are given by
>
> $$x_{\nu n} \simeq n\pi + \left(\nu - \frac{1}{2}\right)\frac{\pi}{2}\,. \tag{E.74}$$

### E.5.1.2 Integral Representation

It is also possible to write $J_n(x)$ for integer $n$ as an integral involving trig functions,

$$J_n(x) = \frac{1}{\pi}\int_0^\pi \cos(n\theta - x\sin\theta)\,\mathrm{d}\theta\,. \tag{E.75}$$

This result is useful for finding a Fourier series solution to Kepler's equation as discussed in Sect. 4.3.4.

## E.5.2 Bessel Functions of the 2nd Kind

If $\nu$ is not an integer, then $J_{-\nu}(x)$ is the second independent solution to Bessel's equation. But if $\nu = m$ is an integer, then

$$J_{-m}(x) = (-1)^m J_m(x)\,, \tag{E.76}$$

so $J_{-m}(x)$ is not an independent solution for this case. A second solution, which *is* independent of $J_\nu(x)$ for all values of $\nu$ (integer or not) is[2]

$$N_\nu(x) \equiv \frac{J_\nu(x)\,\cos(\nu\pi) - J_{-\nu}(x)}{\sin(\nu\pi)}\,. \tag{E.77}$$

$N_\nu(x)$ is called a **Neumann function** (or a **Bessel function of the 2nd kind**). In some references, $N_\nu(x)$ is denoted by $Y_\nu(x)$.

### E.5.2.1 Asymptotic Form

The asymptotic form of $N_\nu(x)$ is given by

$$\begin{aligned} x \ll 1: \quad & N_\nu(x) \to \begin{cases} \frac{2}{\pi}\left[\ln\left(\frac{x}{2}\right) + 0.5772\cdots\right], & \nu = 0 \\ -\frac{\Gamma(\nu)}{\pi}\left(\frac{2}{x}\right)^\nu, & \nu \neq 0 \end{cases} \\ x \gg 1, \nu: \quad & N_\nu(x) \to \sqrt{\frac{2}{\pi x}}\,\sin\left(x - \frac{\nu\pi}{2} - \frac{\pi}{4}\right). \end{aligned} \tag{E.78}$$

Note that for all $\nu$, $N_\nu(x) \to -\infty$ as $x \to 0$. In addition, just as we saw for $J_\nu(x)$, $N_\nu(x)$ behaves for large $x$ like a damped sinusoid, but is 90° out of phase with $J_\nu(x)$. Plots of the first few Bessel functions of the 2nd kind for integer values of $\nu$ are given in Fig. E.7.

With $J_\nu(x)$ and $N_\nu(x)$ as the two independent solutions to Bessel's equation, it follows that the most general solution to the radial part of Laplace's equation in cylindrical coordinates is

$$R(\rho) = A\,J_\nu(k\rho) + B\,N_\nu(k\rho)\,. \tag{E.79}$$

But since $N_\nu(x)$ blows up at $x = 0$, if $\rho = 0$ is in the region of interest, then all of the $B$ coefficients *must vanish* to yield a finite solution to Laplace's equation on the axis. Since both $J_\nu(x)$ and $N_\nu(x)$ go to zero as $x \to \infty$, there is no constraint on either $A$ or $B$ as $\rho \to \infty$.

---

[2]For $\nu = m$ an integer, one needs to use L'Hôpital's rule to show that the right-hand side of the expression defining $N_m(x)$ is well-defined.

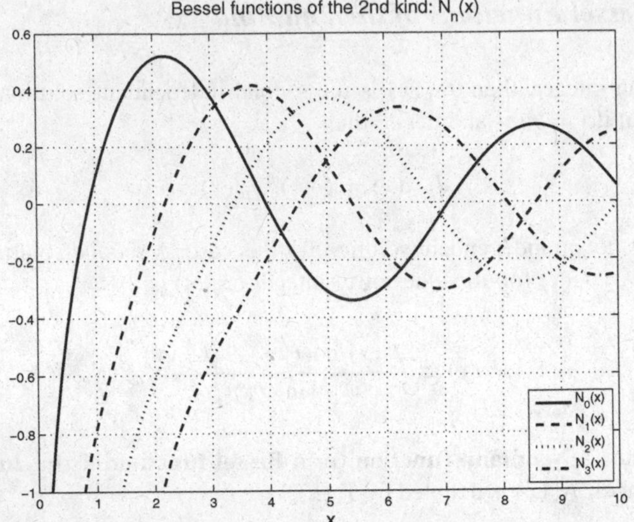


**Fig. E.7** First few Bessel functions of the 2nd kind for integer $\nu$

## E.5.3 *Some Properties of Bessel Functions*

### E.5.3.1 Recurrence Relations

The following relations hold for either $J_\nu(x)$, $N_\nu(x)$, or any linear combination of these functions with constant coefficients:

$$
\begin{aligned}
(x^\nu J_\nu(x))' &= x^\nu J_{\nu-1}(x)\,, \\
\left(x^{-\nu} J_\nu(x)\right)' &= -x^{-\nu} J_{\nu+1}(x)\,, \\
J'_\nu(x) &= -\frac{\nu}{x} J_\nu(x) + J_{\nu-1}(x)\,, \\
J'_\nu(x) &= \frac{\nu}{x} J_\nu(x) - J_{\nu+1}(x)\,, \\
2J'_\nu(x) &= J_{\nu-1}(x) - J_{\nu+1}(x)\,, \\
\frac{2\nu}{x} J_\nu(x) &= J_{\nu-1}(x) + J_{\nu+1}(x)\,.
\end{aligned}
\tag{E.80}
$$

### E.5.3.2 Orthogonality and Normalization

Bessel functions $J_\nu(x)$ satisfy the following orthogonality and normalization conditions

$$\int_0^a d\rho\, \rho J_\nu(x_{\nu n}\rho/a) J_\nu(x_{\nu n'}\rho/a) = \frac{1}{2} a^2 J_{\nu+1}^2(x_{\nu n})\, \delta_{nn'}\,, \tag{E.81}$$

where $x_{\nu n}$ and $x_{\nu n'}$ are the $n$th and $n'$th zeroes of $J_\nu(x)$. Note that the orthogonality of Bessel functions is with respect to different arguments of a *single* function $J_\nu(x)$, and not with respect to *different* functions $J_\nu(x)$ and $J_{\nu'}(x)$ of the same argument. (This latter case held for the Legendre polynomials $P_l(x)$ and $P_{l'}(x)$.) Thus, the orthogonality of Bessel functions is similar to the orthogonality of the sine functions $\sin(n2\pi x/a)$ on the interval $[0, a]$ for different values of $n$.

If the interval $[0, a]$ becomes infinite $[0, \infty)$, then the orthogonality and normalization conditions actually become simpler,

$$\int_0^\infty d\rho\, \rho J_\nu(k\rho) J_\nu(k'\rho) = \frac{1}{k}\delta(k-k')\,, \tag{E.82}$$

where $k$ now takes on a continuous range of values. This is similar to the transition from Fourier series (basis functions $e^{ik_n x}$ with $k_n = n2\pi/a$) to Fourier transforms (basis functions $e^{ikx}$ with $k$ a real variable):

$$\int_{-a/2}^{a/2} dx\, e^{i2\pi(n-n')x/a} = a\,\delta_{nn'} \quad \rightarrow \quad \int_{-\infty}^{\infty} dx\, e^{i(k-k')x} = 2\pi\,\delta(k-k')\,. \tag{E.83}$$

**Exercise E.16** Prove the orthogonality part of (E.81). (*Hint*: Let $f(\rho) = J_\nu(x_{\nu n}\rho/a)$ and $g(\rho) = J_\nu(x_{\nu n'}\rho/a)$ with $n \neq n'$. Then write down Bessel's equation for both $f$ and $g$; multiply these equations by $g$ and $f$, respectively; then subtract and integrate.)

**Exercise E.17** Prove the normalization part of (E.81). (*Hint*: You will need to integrate by parts and then use Bessel's equation to substitute for $x^2 J_\nu(x)$ in one of the integrals.)

### E.5.4 *Modified Bessel Functions of the 1st and 2nd Kind*

As mentioned previously, the *modified* Bessel's equation of order $\nu$ is given by:

$$y''(x) + \frac{1}{x}\, y'(x) - \left(1 + \frac{\nu^2}{x^2}\right) y(x) = 0\,. \tag{E.84}$$

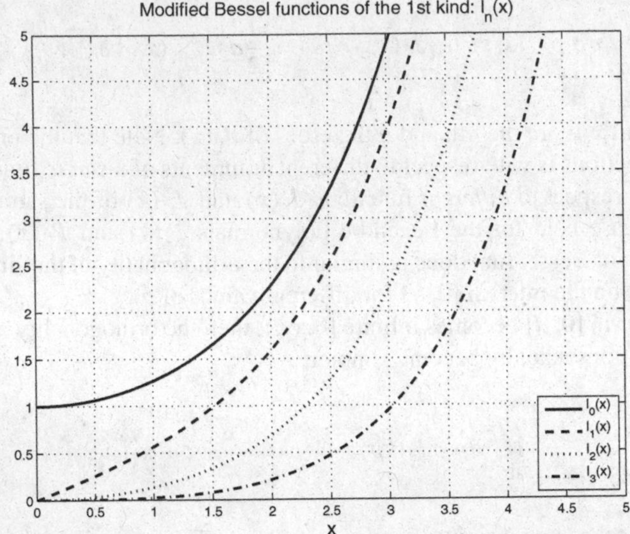


**Fig. E.8** First few modified Bessel functions of the 1st kind for integer $\nu$

It differs from the ordinary Bessel's equation only in the sign of one of the terms multiplying $y(x)$. **Modified** (or **hyperbolic**) **Bessel functions** (of the 1st and 2nd kind) are solutions to the above equation. They are defined by

$$I_\nu(x) \equiv \mathrm{i}^{-\nu} J_\nu(\mathrm{i}x)\,, \qquad K_\nu(x) \equiv \frac{\pi}{2}\mathrm{i}^{\nu+1} H_\nu^{(1)}(\mathrm{i}x)\,. \tag{E.85}$$

Note the pure imaginary arguments on the right-hand side of the above definitions, consistent with our earlier statement that if $y(x)$ is a solution of Bessel's equation then $y(\mathrm{i}x)$ is a solution of the modified Bessel's equation. Plots of the first few modified Bessel functions of the first and second kind, $I_\nu(x)$ and $K_\nu(x)$, for integer values of $\nu$ are given in Figs. E.8 and E.9.

### E.5.4.1 Asymptotic Form

The asymptotic behavior of the modified Bessel functions $I_\nu(x)$ and $K_\nu(x)$ are given by

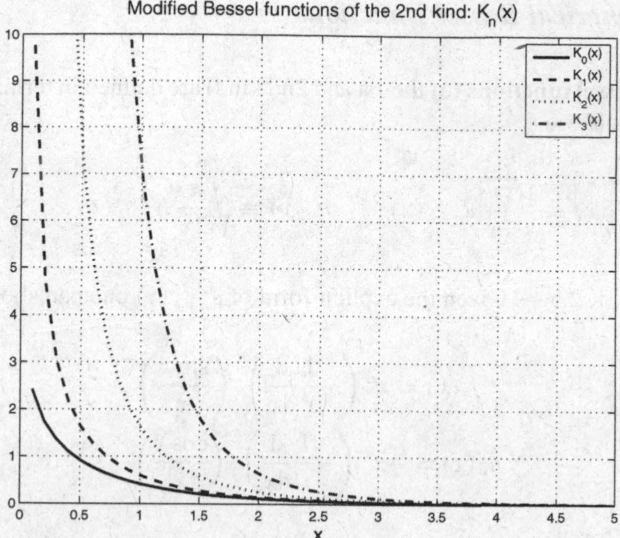


**Fig. E.9** First few modified Bessel functions of the 2nd kind for integer $\nu$

$$
\begin{aligned}
x \ll 1: \quad & I_\nu(x) \to \frac{1}{\Gamma(\nu+1)}\left(\frac{x}{2}\right)^\nu , \\
& K_\nu(x) \to \begin{cases} -\left[\ln\left(\frac{x}{2}\right) + 0.5772\cdots\right], & \nu = 0 \\ \frac{\Gamma(\nu)}{2}\left(\frac{2}{x}\right)^\nu , & \nu \neq 0 \end{cases} \\
x \gg 1, \nu: \quad & I_\nu(x) \to \frac{1}{\sqrt{2\pi x}}\,\mathrm{e}^x \left[1 + O\left(\frac{1}{x}\right)\right], \\
& K_\nu(x) \to \sqrt{\frac{\pi}{2x}}\,\mathrm{e}^{-x}\left[1 + O\left(\frac{1}{x}\right)\right].
\end{aligned}
\tag{E.86}
$$

Thus, $I_0(0) = 1$ and $I_\nu(0) = 0$ for all $\nu \neq 0$, while $K_\nu(x) \to \infty$ as $x \to 0$ for all $\nu$. For large $x$, $I_\nu(x) \to \infty$ while $K_\nu(x) \to 0$ for all $\nu$.

Given $I_\nu(x)$ and $K_\nu(x)$, the most general solution to the radial part of Laplace's equation for the choice of negative separation constant $-k^2$ is

$$
R(\rho) = A\, I_\nu(k\rho) + B\, K_\nu(k\rho)\,. \tag{E.87}
$$

Since $K_\nu(x)$ blows up at $x = 0$, if $\rho = 0$ is in the region of interest, then all of the $B$ coefficients must vanish to yield a finite solution to Laplace's equation on the axis. Similarly, since $I_\nu(x)$ blows up as $x \to \infty$, if the solution to Laplace's equation is to vanish as $\rho \to \infty$, then all of the $A$ coefficients must vanish.

### E.5.5 Spherical Bessel Functions

**Spherical Bessel functions** (of the 1st and 2nd kind) are defined in terms of ordinary Bessel functions via

$$j_n(x) \equiv \sqrt{\frac{\pi}{2x}}\, J_{n+\frac{1}{2}}(x)\,, \qquad n_n(x) \equiv \sqrt{\frac{\pi}{2x}}\, N_{n+\frac{1}{2}}(x)\,, \tag{E.88}$$

where $n = 0, 1, 2, \cdots$. Given the explicit form of $J_{n+\frac{1}{2}}(x)$ one can show that

$$\begin{aligned} j_n(x) &= x^n \left(-\frac{1}{x}\frac{\mathrm{d}}{\mathrm{d}x}\right)^n \left(\frac{\sin x}{x}\right), \\ n_n(x) &= -x^n \left(-\frac{1}{x}\frac{\mathrm{d}}{\mathrm{d}x}\right)^n \left(\frac{\cos x}{x}\right). \end{aligned} \tag{E.89}$$

In particular, it follows that

$$j_0(x) = \frac{\sin x}{x}\,, \qquad n_0(x) = -\frac{\cos x}{x}\,. \tag{E.90}$$

Plots of the first few spherical Bessel functions are given in Figs. E.10 and E.11.

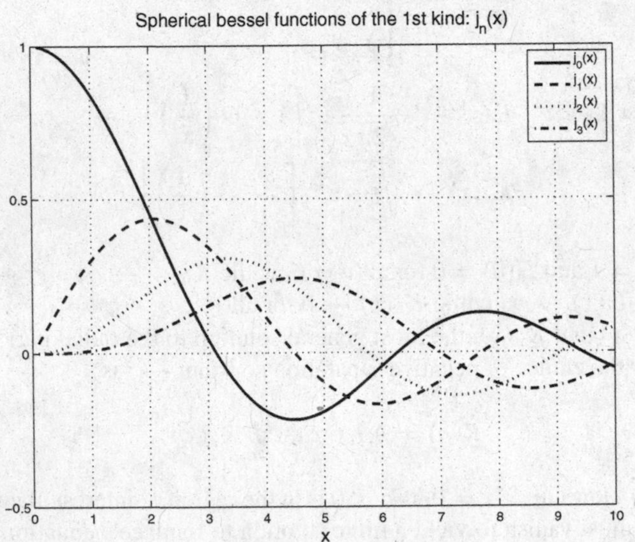


**Fig. E.10** First few spherical Bessel functions of the 1st kind

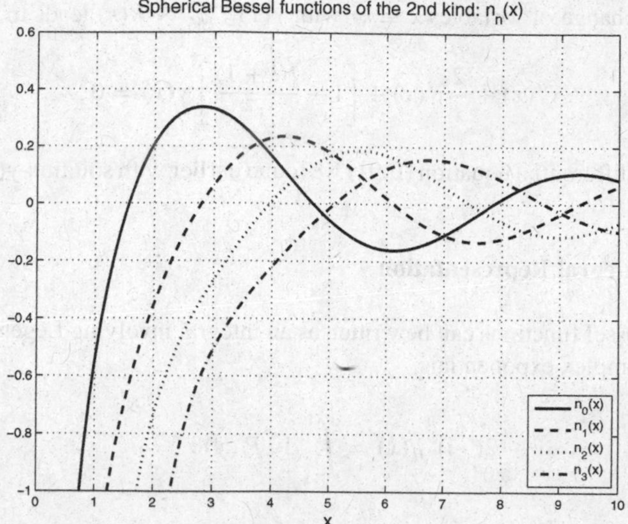


**Fig. E.11** First few spherical Bessel functions of the 2nd kind

> **Exercise E.18** Verify (E.90) for $j_0(x)$ directly from its definition in terms of the ordinary Bessel function $J_{1/2}(x)$.

### E.5.5.1 Spherical Bessel Differential Equation

Given the relationship between $j_n(x)$ and $J_{n+\frac{1}{2}}(x)$, one can show that the spherical Bessel functions satisfy the differential equation

$$j_n''(x) + \frac{2}{x} j_n'(x) + \left[ 1 - \frac{n(n+1)}{x^2} \right] j_n(x) = 0 \, . \tag{E.91}$$

Alternatively, one arrives at the same differential equation by using separation of variables in *spherical coordinates* to solve the **Helmholtz equation**

$$\nabla^2 \Phi(r, \theta, \phi) + k^2 \Phi(r, \theta, \phi) = 0 \, . \tag{E.92}$$

The $\phi$ equation is the standard harmonic oscillator equation with separation constant $-m^2$; the $\theta$ equation is the associated Legendre's equation with separation constants $l$ and $m$; and the radial equation is

$$R''(r) + \frac{2}{r} R'(r) + \left[ k^2 - \frac{l(l+1)}{r^2} \right] R(r) = 0 \, . \tag{E.93}$$

Making the change of variables $x \equiv kr$ with $y(x)|_{x=kr} \equiv R(r)$ leads to

$$y''(x) + \frac{2}{x}y'(x) + \left[1 - \frac{l(l+1)}{x^2}\right] y(x) = 0\,, \tag{E.94}$$

which is the differential equation (E.91) we found earlier with solution $y(x) = j_l(x)$.

#### E.5.5.2 Integral Representation

Spherical Bessel functions can be written as an integral involving Legendre polynomials and complex exponentials,

$$2(-\mathrm{i})^l j_l(x) = \int_{-1}^{1} \mathrm{d}y\; P_l(y)\mathrm{e}^{-\mathrm{i}xy}\,. \tag{E.95}$$

## E.6 Elliptic Integrals and Elliptic Functions

Elliptic integrals and elliptic functions arise in some simple applications, such as finding the length of a conic section (e.g., an ellipse) and solving for the motion of a simple pendulum when one goes beyond the small-angle approximation. In the following two subsections, we briefly define elliptic integrals and elliptic functions using the notation given in Chap. 12 of Boas 2006. Other references may use slightly different notation.

### *E.6.1 Elliptic Integrals*

**Elliptic integrals** of the 1st and 2nd kind are often written in two different forms; the **Legendre forms**:

$$\begin{aligned} F(\phi, k) &\equiv \int_0^{\phi} \frac{\mathrm{d}\theta}{\sqrt{1 - k^2 \sin^2\theta}}\,, \qquad 0 \le k \le 1\,,\\ E(\phi, k) &\equiv \int_0^{\phi} \sqrt{1 - k^2 \sin^2\theta}\,\mathrm{d}\theta\,, \qquad 0 \le k \le 1\,, \end{aligned} \tag{E.96}$$

and the **Jacobi forms**:

$$F(\phi, k) \equiv \int_0^x \frac{dt}{\sqrt{1-k^2t^2}\sqrt{1-t^2}}, \qquad 0 \le k \le 1,$$
$$E(\phi, k) \equiv \int_0^x \frac{\sqrt{1-k^2t^2}}{\sqrt{1-t^2}}\, dt, \qquad 0 \le k \le 1, \tag{E.97}$$

with $x \equiv \sin\phi$. The two arguments of these functions are called the *amplitude* $\phi$ and the *modulus* $k$. Note that the Jacobi and Legendre forms of elliptic integrals are related by the change of variables $t = \sin\theta$.

**Complete elliptic integrals** of the 1st and 2nd kind, $K(k)$ and $E(k)$, are defined by setting the amplitude $\phi = \pi/2$ (or $x = 1$) in the above expressions:

$$K(k) \equiv F(\pi/2, k), \qquad E(k) \equiv E(\pi/2, k). \tag{E.98}$$

**Exercise E.19** (a) Show that the arc length of an ellipse $(x/a)^2 + (y/b)^2 = 1$ from $\theta = \phi_1$ to $\theta = \phi_2$ can be written as

$$s(\phi_1, \phi_2) = a\,[E(\phi_2, e) - E(\phi_1, e)], \tag{E.99}$$

where $e \equiv \sqrt{1-(b/a)^2}$ is the eccentricity of the ellipse. (Here $\theta$ is defined by $x = a\sin\theta$, $y = b\cos\theta$, and we are assuming that $a \ge b$.) (b) Using the result of part (a), show that the total arc length $s = 4a\,E(e)$. (c) Show that for nearly circular ellipses (i.e., for $e \ll 1$), $s \approx 2\pi a(1 - e^2/4)$.

**Exercise E.20** (a) Show that the period of a simple pendulum of mass $m$, length $\ell$, released from rest at $\theta = \theta_0$ is given by

$$P(\theta_0) = 4\sqrt{\frac{\ell}{g}}\,K\left(\sin(\theta_0/2)\right). \tag{E.100}$$

Do not assume that the small-angle approximation is valid for this part of the problem. (*Hint*: Use conservation of total mechanical energy to find an equation for $\dot\theta$ in terms of $\theta$ and $\theta_0$.) (b) Show that for $\theta_0 \ll 1$, the answer from part (a) reduces to

$$P(\theta_0) \approx 2\pi\sqrt{\frac{\ell}{g}}\left(1 + \frac{1}{16}\theta_0^2\right), \tag{E.101}$$

which in the limit of very small $\theta_0$ is the small-angle approximation for the period of a simple pendulum, $P \approx 2\pi\sqrt{\ell/g}$.

### E.6.2 Elliptic Functions

The **elliptic function** sn $y$ is defined as the inverse of the elliptic integral $y = F(\phi, k)$ for a fixed value of $k$,

$$y = \int_0^x \frac{\mathrm{d}t}{\sqrt{1-k^2t^2}\sqrt{1-t^2}} \equiv \mathrm{sn}^{-1}x \quad \Leftrightarrow \quad x = \mathrm{sn}\, y\,. \tag{E.102}$$

Since $x = \sin\phi$, we can also write $\mathrm{sn}\, y = \sin\phi$ in terms of the amplitude $\phi$. Note that the above definition of sn $y$ is very similar to the integral representation of the inverse sine function

$$y = \int_0^x \frac{\mathrm{d}t}{\sqrt{1-t^2}} = \sin^{-1} x \quad \Leftrightarrow \quad x = \sin y\,. \tag{E.103}$$

In fact, when $k = 0$, $\mathrm{sn}\, y = \sin y$. In addition, sn $y$ is periodic with period

$$P = 4\int_0^1 \frac{\mathrm{d}t}{\sqrt{1-k^2t^2}\sqrt{1-t^2}} = 4F(\pi/2, k) = 4K(k)\,, \tag{E.104}$$

similar to the sine function. Plots of $x = \mathrm{sn}\, y$ for $k^2 = 0, 0.25, 0.5$, and 0.75 are shown in Fig. E.12. These have periods $P = 6.28, 6.74, 7.42$, and 8.63 to three significant digits.

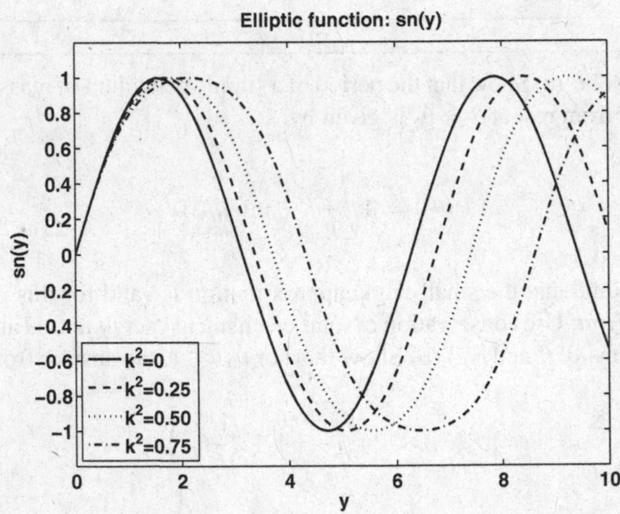


**Fig. E.12** Plots to the elliptic function sn $y$ for $k^2 = 0, 0.25. 0.5$ and 0.75. Recall that for $k^2 = 0$, $\mathrm{sn}\, y = \sin y$

Given sn $y$, one can define other elliptic functions using relations similar to those between trig functions,

$$\mathrm{cn}\, y \equiv \sqrt{1 - \mathrm{sn}^2 y}\,, \qquad \mathrm{dn}\, y \equiv \sqrt{1 - k^2 \mathrm{sn}^2 y}\,. \tag{E.105}$$

Using the above definitions, it is easy to show that $\mathrm{cn}\, y = \cos\phi$. In addition, using the Legendre form of the elliptic integral $F(\phi, k)$, it follows that $\mathrm{dn}\, y = \mathrm{d}\phi/\mathrm{d}y$. The proof is simply

$$\frac{\mathrm{d}\phi}{\mathrm{d}y} = \frac{1}{\mathrm{d}y/\mathrm{d}\phi} = \sqrt{1 - k^2 \sin^2\phi} = \sqrt{1 - k^2 \mathrm{sn}^2 y} = \mathrm{dn}\, y\,. \tag{E.106}$$

**Exercise E.21** Show that

$$\frac{\mathrm{d}}{\mathrm{d}y}(\mathrm{sn}\, y) = \mathrm{cn}\, y\, \mathrm{dn}\, y\,. \tag{E.107}$$

## Suggested References

*Full references are given in the bibliography at the end of the book.*

Abramowitz and Stegun (1972): A must-have reference for all things related to special functions.

Boas (2006): Chapters 11, 12, and 13 discuss special functions, series solutions of differential equations, and partial differential equations, respectively, filling in most of the details omitted in this appendix. An excellent introduction to these topics, especially suited for undergraduates. There are many examples and problems to choose from.

Mathews and Walker (1970): Chapters 1, 7, and 8 discuss ordinary differential equations, special functions, and partial differential equations, respectively. The level of this text is more appropriate for graduate students or mathematically-minded undergraduates.

# References


B.P. Abbott, R. Abbott, T.D. Abbott, M.R. Abernathy, F. Acernese, K. Ackley, C. Adams, T. Adams, P. Addesso, R.X. Adhikari et al., Observation of gravitational waves from a binary black hole merger. Phys. Rev. Lett. **116**(6), 061102 (2016). https://doi.org/10.1103/PhysRevLett.116.061102

B.P. Abbott, R. Abbott, T.D. Abbott, M.R. Abernathy, F. Acernese, K. Ackley, C. Adams, T. Adams, P. Addesso, R.X. Adhikari et al., The basic physics of the binary black hole merger GW150914. Annalen der Physik **529**, 1600209 (2017). https://doi.org/10.1002/andp.201600209

M. Abramowitz, I.A. Stegun, *Handbook of Mathematical Functions* (Dover Publications Inc, New York, 1972). ISBN 0-486-61272-4

G. Arfken, *Mathematical Methods for Physicists* (Academic Press Inc, New York, 1970)

V.I. Arnold, *Mathematical Methods of Classical Mechanics*, vol. 60, Graduate Texts in Mathematics (Springer, New York, 1978). ISBN 0-387-90314-3

M. Benacquista, *An Introduction to the Evolution of Single and Binary Stars* (Springer, New York, Heidelberg, Dordrecht, London, 2013)

R.E. Berg, D.G. Stork, *The Physics of Sound*, 3rd edn. (Pearson Prentice Hall, Englewood Cliffs, New Jersey, 2005). ISBN 978-0131457898

J. Bertrand, C.R. Acad. Sci. **77**, 849–853 (1873)

M.L. Boas, *Mathematical Methods in the Physical Sciences*, 3rd edn. (John Wiley & Sons Inc, United States of America, 2006). ISBN 0-471-19826-9

H. Bondi, *Relativity and Common Sense: A New Approach to Einstein* (Dover Publications Inc, New York, 1962). ISBN 0-486-24021-5

P. Dennery, A. Kryzwicki, *Mathematics for Physcists* (Dover Publications Inc, Mineola, New York, 1967)

S. Dutta, S. Ray, Bead on a rotating circular hoop: a simple yet feature-rich dynamical system. *ArXiv e-prints*, (December 2011)

A. Einstein, Zur Elektrodynamik bewegter Körper. Annalen der Physik **322**, 891–921 (1905). https://doi.org/10.1002/andp.19053221004

L.A. Fetter, J.D. Walecka, *Theoretical Mechanics of Particles and Continua* (McGraw-Hill Book Company, United States of America, 1980)

R.P. Feynman, *Surely You're Joking Mr. Feynman! Adventures of a Curious Character* (W.W. Norton & Company, New York, London, 1985). ISBN 0-393-31604-1

R.P. Feynman, R.B. Leighton, Matthew Sands, in *The Feyman Lectures on Physics*, vol. II (Addison-Wesley Publishing Company, Reading, Massachusetts, 1964). ISBN 0-201-02117-X-P




M.J. Benacquista and J.D. Romano, *Classical Mechanics*, Undergraduate Lecture Notes in Physics, https://doi.org/10.1007/978-3-319-68780-3

H. Flanders, *Differential Forms with Applications to the Physical Sciences* (Dover Publications Inc, New York, 1963). ISBN 0-486-66169-5

M.R. Flannery, The enigma of nonholonomic constraints. Am. J. Phys. **73**, 265–272 (2005). https://doi.org/10.1119/1.1830501

I.M. Gelfand, S.V. Fomin, *Calculus of Variations* (Dover Publications Inc, Mineola, New York, 1963). ISBN 0-486-41448-5. (Translated and Edited by Richard A. Silverman)

H. Goldstein, C. Poole, J. Safko, *Classical Mechanics*, 3rd edn. (Addison Wesley, San Francisco, CA, 2002). ISBN 0-201-65702-3

D.J. Griffiths, *Introduction to Electrodynamics*, 3rd edn. (Pearson Prentice Hall, United States of America, 1999). ISBN 0-13-805326-X

D.J. Griffiths, *Introduction to Quantum Mechanics*, 2nd edn. (Pearson Prentice Hall, United States of America, 2005). ISBN 0-13-111892-7

P.R. Halmos, *Finite-Dimensional Vector Spaces*, 2nd edn. (D. Van Nostrand Company Inc, Princeton, New Jersey, 1958)

J.B. Hartle, *Gravity: An Introduction to Einstein's General Relativity* (Benjamin Cummings, illustrate edition, January 2003). ISBN 0805386629

H. Hertz, *The Principles of Mechanics Presented in a New Form* (Dover Publications Inc, New York, 2004). ISBN 978-0486495576 (The original german edition Die Prinzipien der Mechanik in neuem zusammenhange dargestellt was published in 1894)

R.W. Hilditch, *An Introduction to Close Binary Stars* (Cambridge University Press, Cambridge, 2001)

K.V. Kuchǎr Theoretical mechanics. Unpublished lecture notes (1995)

J.B. Kuipers. *Quarternions and Rotation Sequences: A Primer with Applications to Orbits, Aerospace, and Virtual Reality* (Princeton University Press, 1999)

C. Lanczos, *The Variational Principles of Mechanics*, 4th edn. (Dover Publications Inc, New York, 1949). ISBN 0-486-65067-7

L.D. Landau, E.M Lifshitz, *Classical Theory of Fields, Course of Theoretical Physics*, 4th edn., vol. 2 (Pergamon Press, Oxford, 1975). ISBN 0-08-025072-6

L.D. Landau, E.M Lifshitz, *Mechanics, Course of Theoretical Physics*, 3rd edn., vol. 1 (Elsevier Ltd, Oxford, 1976). ISBN 978-0-7506-2896-9

J.B. Marion, S.T. Thornton, *Classical Dynamics of Particles and Systems*, 4th edn. (Saunders College Publishing, United States of America, 1995). ISBN 0-03-097302-3

J. Mathews, R.L. Walker, *Mathematical Methods of Physics* (Benjamin/Cummings, United States of America, 1970). ISBN 0-8053-7002-1

N.D. Mermin, *It's About Time: Understanding Einstein's Relativity* (Princeton University Press, Princeton, New Jersey, 2005). ISBN 0-691-12201-6

E. Noether, Invariante Variationsprobleme. *Nachr. D. König. Gesellsch. Wiss. Zu Göttingen*, 1918:235–257, 1918

E. Noether, Invariant variation problems. Trans. Theory Stat. Phys. **1**, 186–207 (1971). https://doi.org/10.1080/00411457108231446

J.D. Romano, R.H. Price, Why no shear in "Div, grad, curl, and all that"? Am. J. Phys. **80**(6), 519–524 (2012). https://doi.org/10.1119/1.3688678

T.D. Rossing, P.A. Wheeler, R.M. Taylor, *The Science of Sound*, 3rd edn. (Addison Wesley, San Francisco, 2002). ISBN 978-0805385656

F.C. Santos, V. Soares, A.C. Tort. An English translation o Bertrand's theorem. *ArXiv e-prints*, (April 2007)

H.M. Schey, *div, grad, curl and all that: An informal text on vector calculus*, 3rd edn. (W.W. Norton & Co., New York, 1996)

B. Schutz, *A First Course in General Relativity* (Cambridge University Press, May 2009). ISBN 9780521887052

B. Schutz, *Geometrical Methods of Mathematical Physics* (Cambridge University Press, Cambridge, 1980)

E.F. Taylor, J.A. Wheeler, *Spacetime Physics: Introduction to special relativity* (W.H. Freeman and Company, New York, 1992)

J. Terrell, Invisibility of the Lorentz contraction. Phys. Rev. **116**, 1041–1045 (1959). https://doi.org/10.1103/PhysRev.116.1041

C.G. Torre, *Introduction to Classical Field Theory*. All Complete Monographs (2016). http://digitalcommons.usu.edu/lib_mono/3/

E.P. Wigner, *Gruppentheorie und ihre Anwendungen auf die Quantenmechanik der Atomspektren* (Vieweg Verlag, Braunschweig, Germany, 1931)

# Index




M.J. Benacquista and J.D. Romano, *Classical Mechanics*, Undergraduate Lecture Notes in Physics, https://doi.org/10.1007/978-3-319-68780-3

Printed in the United States
By Bookmasters